The Great War's Finest

An Operational History of the German Air Service

Volume 1: Western Front 1914

Matt Bowden

The Great War's Finest

An Operational History of the German Air Service

Volume 1: Western Front 1914

Matt Bowden

Dedicated to Scott & Jackie Bowden

Interested in WWI aviation? Join The League of WWI Aviation Historians (www.overthefront.com), Cross & Cockade International (www.crossandcockade.com), and Das Propellerblatt (www.propellerblatt.de).

ISBN: 978-1-935881-58-2
© 2017 Aeronaut Books, all rights reserved
Text © 2017 Matt Bowden
Design and layout: Jack Herris
Cover design: Aaron Weaver
Cover art: Ivan Berryman
Color profiles: Bob Pearson
Digital photo editing: Aaron Weaver & Jack Herris

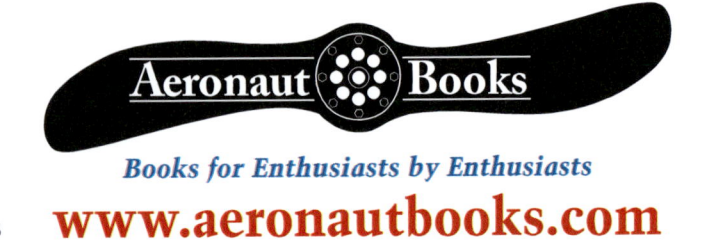

Aeronaut Books
Books for Enthusiasts by Enthusiasts
www.aeronautbooks.com

Contents

Acknowledgments		4
Preface		5
Glossary		6
Rank Structure		6
Map Key		7

Chapters

1	The Birth of German Military Aviation	8
2	Battle of the Frontiers	46
3	Battle of the Marne	154
4	The 'Race to the Sea'	260
5	The First Battle of Ypres	281
6	General Situation at the End of the Year	340

Appendices

A	Mobilization of Peacetime German Air Forces	354
B	Order of Battle: German Western Armies – August 1914	355
C	Aircraft Types by FFA, August 1914	359
D	Order of Battle: Western Allied Armies – August 15, 1914	360
E	Order of Battle During the Battle of the Marne	361
F	Order of Battle: *Fliegertruppe* – October 10, 1914	363
G	Order of Battle: *Fliegertruppe* – December 10, 1914	364
H	Aircraft Data	365
I	Roster of *Breiftauben Abteilung Ostend (BAO)* – December 1914	366

Bibliography	368
Index	373

Acknowledgments

I first want to thank my wife, Blair, who has faithfully supported me throughout this project. She has always understood how much this means to me and has selflessly taken on great burdens to allow me to complete it. This book would have never been possible if it wasn't for her. My children and I are lucky to have such a great woman in our lives.

Secondly, I'd like to thank my parents, Scott and Jackie, to whom this volume is dedicated. I would not be the man I am today if it wasn't for the love, compassion and support they have given me throughout my entire life. My fanatical interest in military history and World War 1 began at a very early age as a direct result of the classical education they provided, for which I am deeply grateful.

This study would have never been feasible without my access to A.E. Ferko Collection, held at the University of Texas at Dallas' Eugene McDermott Library Special Collections Department. I am eternally grateful to the department staff for their assistance and understanding during each of my trips to the library over the past several years. I would especially like to thank the Special Collections Department's Patrizia Nava and Thomas J. Allen who have both gone above and beyond in helping me with this project.

Next, I would like to express my gratitude to Marlea Leljedal of the United States Army Heritage Education Center in Carlisle, Pennsylvania for working tirelessly to make sure I quickly had copies of any documents I required from the United States Army War College Library. Likewise, from across the Atlantic, I'd like to acknowledge the services of Benjamin Haas, who gave me access to the critically important materials from inside the various German archives.

The beautiful maps found inside this book are the work of Florin Safner, Bogdan Cazacioc, and Elena Hallen. Each of these individuals deserve praise for turning my crude ideas and sketches into gorgeous fully colored masterpieces. Their patience in working with me to perfect each map is greatly appreciated.

Once again, I thank my father, Scott Bowden, for taking the time away from his own writing to carefully review and improve the book's text. This volume is infinitely better thanks to his expertise.

Finally, I must thank my publisher, Jack Herris, for having enough faith in me and my project to publish this series. Aeronaut Books is a leader in the World War I Aviation community; it is an honor to have them publish my work.

Matt Bowden
Arlington, Texas

Below: The Rumpler *Taube* shown here was the only aircraft available to the German garrison in China. The inherently stable *Tauben* were too stable for air combat and their high drag limited performance. The position of the wing also limited the observer's field of view. *Tauben* soon disappeared from the front, replaced by higher-performance biplanes. (Aeronaut)

Preface

World War 1 aviation has always attracted considerable academic attention with countless numbers of books, magazines, and journal articles dedicated to the subject. With few exceptions, the main focus of these texts has almost exclusively been the study of popular aircraft, aces, or successful fighter squadrons. While these works have been of an inestimable value, there remains a lack of larger, broader scoped histories which examine the various air services' operations and explain how their efforts impacted the overall course of the war. Indeed, a comprehensive operational level study of the German Air Service throughout the entirety of the war has never been written.

This series intends to address what in my opinion is a major void in the historiography of the air war by finally providing readers an opportunity to learn how the *Fliegertruppe* (later the *Luftstreitkräfte*) operated as a combined unit, with a focus on how they cooperated with the army to fight the battles on the ground. To that end, each volume will concentrate on four separate, yet interconnected topics. First, the reader is presented with an overview of the strategic situation with each of the opposing leadership's plans and objectives for the coming campaign. Secondly, the narrative seeks, at the operational level, to describe and explain the various battles on the ground in order to put the *Fliegertruppe's* mission into proper context. From there, the text provides a highly-detailed description of aerial operations using squadron histories, first-hand accounts, aviation staff officer reports and more. Furthermore, in order to better understand both how and why the German airmen acted as they did, each volume will explore the German Air Service's composition, organization and doctrine at the time of the battle. Finally, my study will define and analyze the impact of German aviators' actions on the outcome of each battle. Thus, it is my intention for this series to be the first published work to properly incorporate each of these factors to provide the reading public with a better understanding of how the larger air war was conducted.

Writing the first volume in this series, covering air operations on the Western Front in 1914, presented many unique challenges. Without any aerial dogfights, fighter planes, or aces, there has been very little written on the air war's first year. Indeed, the only English language study which has attempted to adequately cover the subject from the German perspective was John Cuneo's *The Air Weapon 1914–1916*, written in 1947. In the book, Cuneo devotes roughly 30,000 words to Western Front aerial operations in 1914—describing a select number of aerial reconnaissance flights without providing adequate context or fully explaining why they were important. Thus, with very little else to go on, I set out to gather as much primary source/contemporary writings on the subject as I could. Whether it was squadron war diaries, a *Luftwaffe* Historical Section study of German Air Operations during the month of August, unpublished manuscripts covering aerial operations during the Marne Battle, or other rare texts, my search uncovered a wealth of previously uncited sources which allowed me to slowly build upon Cuneo's ideas and create the core of the book. Likewise, in order to fully give the reader an understanding of the battles on the ground, I consulted dozens of deeply informative sources, often written by the participants themselves.

Drawing upon these rarely cited sources, I have constructed a balanced narrative that shifts back and forth between a description of events on the ground and in the air with occasional breaks to fully explain the impact of the *Fliegertruppe's* actions on the course of the battle. For added background, an entire chapter on the establishment and growth of military aviation in Germany during the pre-war years is included. This section contains a reasonably detailed description of how the process of aircraft design and production worked, including details regarding the organization and training of German army airmen prior to the war. Finally, the issues of aircraft production as well as the organization of the *Fliegertruppe* are later revisited several times throughout the book.

By integrating the history of German aerial operations with a detailed narrative of the battles on the ground, it is my intention for this series to provide readers unique insight into how the First World War's war in the air was waged and to highlight the impact of aerial operations on the outcome of each major battle. I sincerely hope that each volume will challenge existing beliefs and will drive historians and history buffs alike to reevaluate the importance of aerial operations throughout the war.

Matt Bowden
Arlington, Texas

Glossary

AFP	Armee Flugpark	Army Air Park
AK	Armeekorps	Army Corps (German)
BEF		British Expeditionary Force
CA	Corps d'Armée	Army Corps (French)
	Cuirassier	French Heavy Cavalry
DC	Division de cavalerie	Cavalry Division (French)
DI	Division d'Infanterie	Infantry Division (French)
Fest.FA	Festungsflieger Abteilung	Fortress Flying Sections
FFA	Feldflieger Abteilung	Field Flying Sections
FLA	Feldluftschiffer Abteilung	Field Airship Section
GHQ	Große Hauptquartier	Great Headquarters (of HM the Kaiser)
HKK	Höherer Kavallerie-Kommandeur	Senior Cavalry Command (Cavalry Corps)
ID	Infanterie-Division	Infantry Division (German)
Idflieg	Inspektion der Fliegertruppen	Inspectorate of Flying Troops
IB	Infanterie-Brigade	Infantry Brigade (German)
Ilust	Inspektion der Luftschiffertruppen	Airship Troops Inspectorate
	Jäger	German light infantry battalion
KD	Kavallerie Division	Cavalry Division (German)
NAF	Nationalflugspende	National Aviation Fund
NCO		Non-Commissioned Officer
OHL	Oberste Heeresleitung	Supreme Army Command
RD	Reserve Division	Reserve Division (German)
RK	Reserve Korps	Reserve Corps (German)

Rank Structure

	German Rank	U.S. or British Equivalent
	Generalfeldmarschall	Field Marshal
	Generaloberst	Colonel General
	General der Infanterie	General of Infantry
	General der Kavallerie	General of Cavalry
	General der Artillerie	General of Artillery
	Generalleutnant	Lieutenant General
	Generalmajor	Major General/Brigadier General
	Oberst	Colonel
	Oberstleutnant	Lieutenant Colonel
	Major	Major
Hptm.	Hauptmann	Captain
Rittm.	Rittmeister	Captain from a mounted unit
Oblt.	Oberleutnant	Lieutenant
Ltn.	Leutnant	Second Lieutenant
OfStv.	Offizierstellverterer	Officer Deputy
Fw.	Feldwebel	First Sergeant
Vzfw.	Vizefeldwebel	Sergeant First Class/ Staff Sergeant (U.K.)
	Sergeant	Sergeant
Uffz.	Unteroffizier	Corporal
Gefr.	Gefreiter	Private First Class/ Lance Corporal (U.K.)
	Musketier	Private
	Pionier	Private in technical troops

Map Key

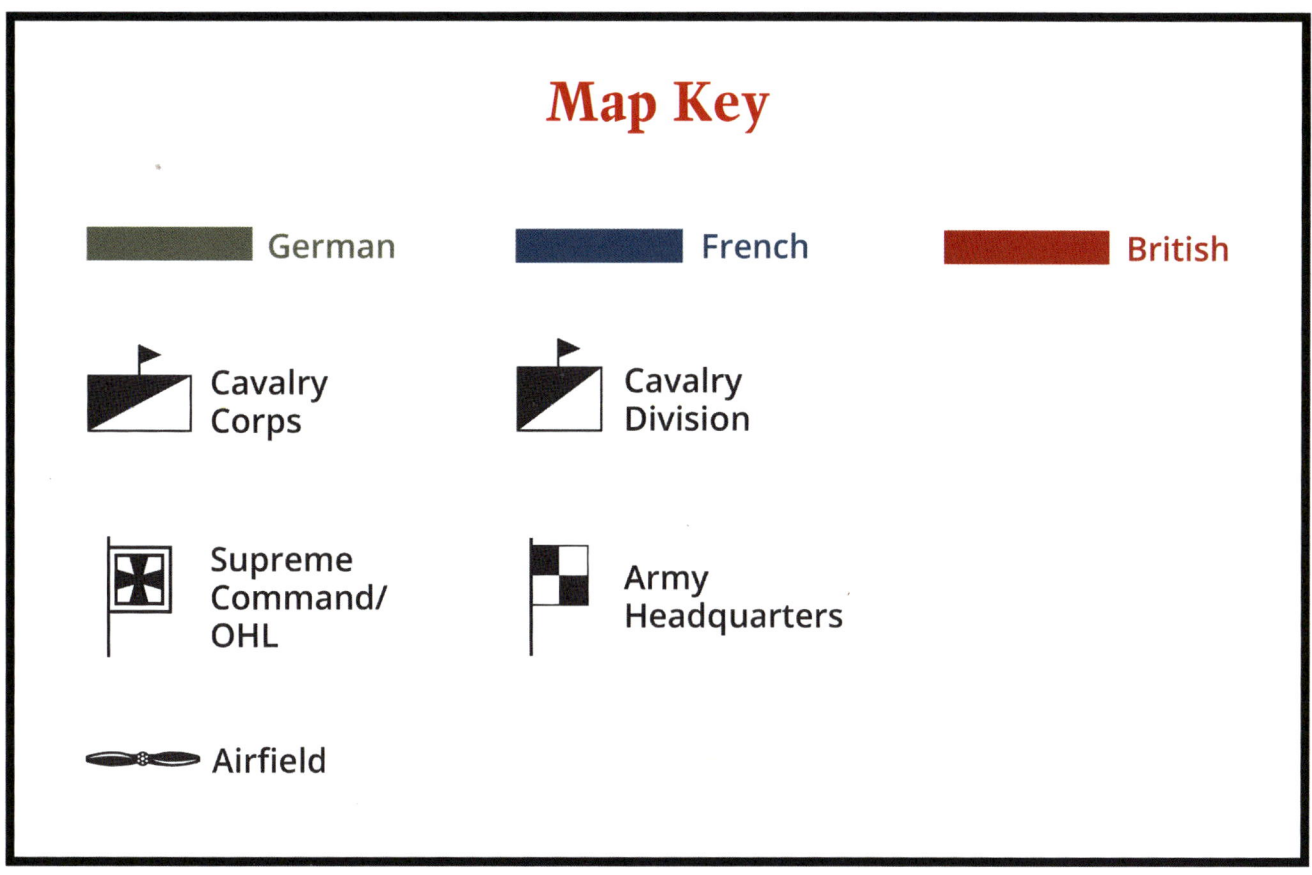

▬ German ▬ French ▬ British

⚑ Cavalry Corps

⚑ Cavalry Division

⚑ Supreme Command/ OHL

⚑ Army Headquarters

✈ Airfield

Above: The beginning of the transition; a German cavalryman watches an airplane. (Author's Collection)

1
The Birth of German Military Aviation
1908–1914

"Friends of German aviation breathe a sigh of relief. At last! At last we again have a day when our feeling of inferiority to our neighbors in the West has disappeared."

Berlin Newspaper: B.Z. am Mittag[1]

July 23, 1913

The history of German military aviation can be traced back to 1870 with the formation of the Prussian Army's first balloon section during the Franco-Prussian War. Over the next 38 years the army's fledgling balloon force steadily grew as technological advances increased their potential impact on the battlefield.[2] Beginning in 1893, these improved balloon sections were successfully used under simulated wartime conditions during the yearly Kaiser maneuvers, thus proving their worth under those circumstances. By 1901 the army's balloon force had been expanded to allow a balloon section to be attached to each of Germany's eight field armies.[3]

Unfortunately, the army's early successes with lighter-than-air aviation hindered the overall growth of German military aviation in the years immediately before the Great War. Building upon the balloon troops' latest achievements, the Prussian War Ministry started placing orders in 1908 for the army's first dirigibles, believing airships to be far superior to heavier-than-air airplanes.[4] This erroneous belief allowed the French, who concentrated their efforts on aircraft, to quickly gain a near overwhelming superiority in military aviation. Consequently, once the military feasibility of the dirigible was exposed to be untrue, Germany's aviation troops were compelled to spend the remainder of the pre-war years catching up with their French rivals. The period 1908–1914 was therefore a race against time for the Germans, with their sole objective being to reach technological and quantitative parity with their western neighbors.

Trial & Error: The First Steps in Building a German Air Service (1908–1910)

Internal politics within *Wilhelmine* Germany's peacetime army can only be described as incendiary. Multiple bureaucratic departments within the army were constantly fighting with one another for added influence and the right to set policy according to each department's interests. Thus, with no long-term strategy or direction from the Kaiser, these departments, including the Prussian War Ministry, General Staff, and Military Cabinet, unwittingly worked to undermine the army's preparedness in the years leading up to the war.[5]

The largest and most significant inter-departmental feud during this time was between the Prussian General Staff and the War Ministry.[6] While the two sides most famously quarreled over the size of the army, they repeatedly clashed over a multitude of other issues. One of these disagreements was in the growing field of new military technologies. Throughout the 1890s and early 1900s, the opposing sides had repeatedly battled over raising the number of the army's machine guns, modernizing the artillery, increasing artillery ammunition supply, and allocating funds to subsidize the construction of dirigibles—each time with the General Staff in favor of technological change/improvements and the War Ministry in opposition. Thus, by the time the airplane had entered the picture, the opposing battle lines were already drawn for a lengthy fight that would solve nothing but put Germany's army at a disadvantage in the event of war.

Despite their belief in the dirigible's supremacy over the airplane, the Prussian War Ministry made the decision in 1907 to assign a leading airplane expert, *Hptm.* Wolfram de le Roi, to the Inspectorate of Transport Troops' Research Unit, which was tasked with looking into the airplane's military potential.[7] The Transportation Inspectorate was in theory independent from the General Staff and War Ministry. However, it was controlled indirectly by the War Ministry's budget decisions, which the Ministry used to its advantage.[8] Suspicious of the often technophobic War Ministry, the General Staff

Above: German balloon section directs artillery fire during pre-war training exercise. (Author's Collection)

quickly created their own technical section under the command of *Hptm.* Hermann von der Lieth-Thomsen which was charged with "following foreign and domestic progress in aviation." In October 1908, Thomsen's unit was placed under *Hptm.* Erich Ludendorff's mobilization department. As advocates for "the simultaneous and wholehearted promotion of airships and airplanes," Thomsen and Ludendorff immediately began calling for a larger aviation force within the army, establishing a personal relationship that would later have a profound impact on the German Air Force during the war.[9]

Airship and dirigible development continued to be the army's priority throughout 1908 after Count Ferdinand Zeppelin's famous dirigible achieved some notable successes during trials. The War Ministry therefore decided to not get financially involved with the seemingly risky airplane in favor of concentrating on balloons and airships. In response to the decision, *Hptm.* de le Roi submitted a report entitled, "The Attitude of The Military Authorities of Various States toward the Flying Machine Question and Proposals for the Introduction of Tests in The Field of Aviation Technology." In his paper, de le Roi described French and American progress with the airplane, detailing how military leaders in each country favorably viewed the airplane's military prospects. Next, de le Roi addressed the lack of progress within the field in Germany. It was his opinion that Germany's failure to produce successful airplanes was attributed to German industrialists' unwillingness to devote their own funds and resources to such a project. Thus, de le Roi suggested that the army begin its own project to produce an aircraft as a way to solve the problem.[10]

The Transport Inspectorate's Airship Battalion (which was charged with the practical development of military aviation for the field formations) quickly rebutted de la Roi's conclusion. In a written response, the Battalion's commander argued that the French and American's progress was "frivolous" and not worth emulating.[11] Instead, it was the Airship Battalion's opinion that the task of developing and producing airplanes should be left entirely to private industry. After reading both side's reports *Generalleutnant* Alfred von Lyncker, the Inspector of Transport Troops, settled the dispute in the battalion's favor, stating:

The German Army administration is presently of the opinion that its own work in the field of aviation technology is not yet absolutely necessary, since no type of flying machine has yet achieved the success that would demonstrate its suitability for military

Above: As Commander of the General Staff's Mobilization Department, Erich Ludendorff tirelessly worked for the expansion of the army's aviation forces.. (Author's Collection)

purposes. The solution to the problem should therefore be left to private enterprise, respectively factories, with which constant contact should be maintained.[12]

The War Ministry agreed with Lyncker's conclusion, rejecting de le Roi's proposal as well as a similar suggestion put forward by the General Staff. As early as the fall of 1908, the General Staff had begun affirming the airplane's military value for reconnaissance and communication purposes. The War Ministry on the other hand, believed that the Zeppelin's greater range and payload would give Germany the upper hand against an opponent with airplanes. If airplanes were indeed necessary, the War Ministry fully believed that German private industry should develop and produce them.

It is easy to understand the cause of the War Ministry's unwillingness to support heavier than air aviation in 1908. As the office responsible for the army's budget (which was strictly limited by the large number of social democrats in the *Reichstag* as well as increases in funding to the navy), the War Ministry was simply unwilling to risk committing funds to the airplane, whose military value was still questionable. Thus, with the Zeppelin appearing to be superior, the Ministry felt it could afford to rely on private enterprise to develop the airplane if it ever came to pass that they would be needed.

The War Ministry's faith in German private industry was not entirely unfounded. News of great successes in France and the United States had sparked great interest in airplanes amongst the German Empire's people. The 'German Air Fleet League,' a new civilian group formed to encourage and promote military aviation, had quickly gained 3,000 members by the end of 1908. Together with other groups, the Air Fleet League began joining and influencing the German Airship Association, which controlled all civil aviation matters within Germany. More importantly for the army, the increased interest in heavier-than-air flight resulted in the creation of ten private airplane construction projects around Germany by November 1908.[13] After receiving input from experts, the War Ministry decided to fund several of these projects. However, none of the projects bore fruit in producing a militarily viable airplane—thus confirming the War Ministry's suspicions regarding the airplane's viability.

Nonetheless, upon reviewing the recent accomplishments of French and British aviation as well as the comparatively substandard state of affairs in Germany, the War Ministry decided in January 1909 to adopt de le Roi's proposal of having the army directly subsidize the development of an airplane. According to the army's specifications, the new flying machine was required to carry two men and fuel for at least 125 miles, reach a speed of 37 miles per hour and obtain a ceiling of 1,640 feet (500 meters).[14] After some internal debate, the War Ministry ultimately chose a triplane design submitted by civilian builder W.S. Hoffman as the basis for the new project.[15] However, with no vertical stabilizer or steering device, Hoffman's machine was outdated and deeply flawed. Unsurprisingly, despite 13 months of cooperation and labor between Hoffman and the Transportation Inspectorate's Research Unit, the aircraft never flew. On both its test flights performed in March 1910, the machine barely got off the ground before crashing— permanently ending the army's experiment in developing its own airplane.[16]

Interest in heavier-than-air aviation received a considerable boost in September and October 1909 after Orville Wright came to Germany to perform a

Above & Right: Photographs from Louis Blériot's famous Channel crossing flight. (Courtesy of the Library of Congress)

series of flying demonstrations for the Imperial Family and the German people. Wright completed 19 flights during his stay, drawing crowds as large as 200,000 that saw him break world records for endurance and altitude.[17] Members of the military, however, were unimpressed with Wright's performances. In his diary, future Inspector of Aviation Troops Walter von Eberhardt revealed how military observers in attendance viewed the demonstrations and the airplane's military potential:

Having access to the flying apparatus I had the opportunity of observing the airplane and the type of takeoff. The plane with its motor turning at its maximum revolutions was propelled along a rail by a weight, and ascended easily into the air, circling over the field–it was a huge success; Wright flew for almost an hour, often over the heads of the hundred thousand spectators who acclaimed him with frenzied shouts. The altitude of his flight was slight, scarcely 350 feet; so I consider the thing at present to be merely a clever circus stunt without military value. It is however a great sport and after all, the principle is known and a great step has been made.[18]

Nevertheless, Wright's trip had a positive impact on public opinion, and on future policy regarding the development of airplanes in Germany. For example, membership in civilian aviation associations sharply rose after Wright's visit. These associations, such as the Air Fleet League and German Aviators Association, steadily gained power and lobbied for the adoption of airplanes into the Prussian Army throughout 1910 and 1911.[19] Putting Orville Wright's tour into perspective, future German Air Force officer Adolf-Victor von Koerber described the demonstrations as an "incredibly important event that united the people, army, press, industry and finance around the idea [of expanding the nation's aviation program]."[20]

Arguably the most important event in the early progress of German aviation was the creation of Germany's first aircraft manufacturing firms. The early failures of individually led airplane projects had taught aviation enthusiasts throughout Germany a valuable lesson. Regardless of how wealthy they were, an individual lacked the talent, knowledge and resources required to successfully produce something as technologically complex as an airplane. Indeed, it was discovered that only well-organized businesses could deploy the resources and talent necessary to design, develop and eventually mass-produce large numbers of aircraft. Thus, a number of Germany's first aviation companies were formed between the fall

Organization of Prussian Army Aviation Agencies to April 1, 1911[1]

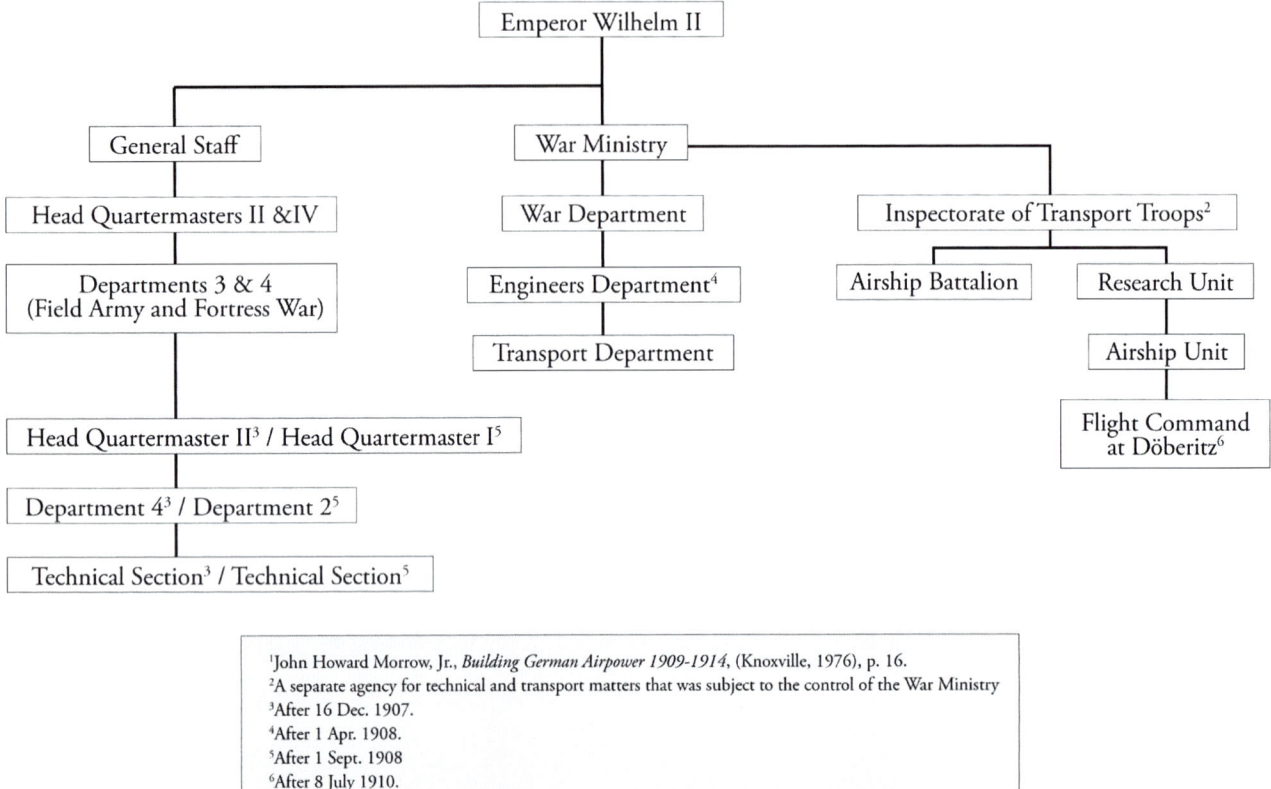

```
                          Emperor Wilhelm II
                          /              \
                  General Staff         War Ministry
                       |                 /         \
         Head Quartermasters II & IV   War Department   Inspectorate of Transport Troops[2]
                       |                 |                 /              \
              Departments 3 & 4     Engineers Department[4]  Airship Battalion   Research Unit
            (Field Army and Fortress War)   |                                        |
                       |              Transport Department                       Airship Unit
                       |                                                              |
         Head Quartermaster II[3] / Head Quartermaster I[5]                   Flight Command
                       |                                                      at Döberitz[6]
              Department 4[3] / Department 2[5]
                       |
         Technical Section[3] / Technical Section[5]
```

[1] John Howard Morrow, Jr., *Building German Airpower 1909-1914*, (Knoxville, 1976), p. 16.
[2] A separate agency for technical and transport matters that was subject to the control of the War Ministry
[3] After 16 Dec. 1907.
[4] After 1 Apr. 1908.
[5] After 1 Sept. 1908
[6] After 8 July 1910.

of 1908 and the winter of 1909. *Euler Works* was the first, established in October 1908 in Darmstadt. In the twelve months that followed, the Rumpler, Wright, Aviatik and Albatros firms were also established.[21]

By early 1910 the airplane's military usefulness was no longer disputed. News of Louis Blériot's English Channel crossing during the summer of 1909 had demonstrated that the airplane's performance capabilities had dramatically improved since the first days of flight. Furthermore, it proved France's superiority over the Germans in heavier-than-air aviation to be nearly insurmountable. Thus, with a rapidly deteriorating situation, the critical question facing the German leadership was: how could the army subsidize aircraft development and production with its meager military aviation budget? Fortunately for the Germans, the newly created aviation firms presented the answer.

On March 15th, *Hptm.* de la Roi published an important memorandum dealing with "the development of aviation and its uses for the army." De la Roi's conclusions expressed in the paper were extremely far-sighted, making it one of the founding documents of the German Air Service. According to the memo, airplanes would soon surpass airships in all military uses, including reconnaissance, communication, and even attack.[22] The development and production of airplanes were therefore considered absolutely vital to the army's interests. Considering the army's past failures in attempting to build a machine of their own, de la Roi recommended that the army create an organization that would coordinate with airplane and engine manufacturing firms to produce aircraft suited for military operations. This recommendation was based on the army's success with the motor vehicle industry, which failed to build military grade vehicles until the War Ministry established a similar organization. In both cases, the army would work with manufactures, test prototypes, and give compensation/contracts to the firms with the best machines.[23]

To achieve this objective, de la Roi recommended that the army create its first combined flying school/aviation unit. This unit would be charged with the tasks of inspecting and evaluating the various aircraft manufacturers' incoming prototypes and submitting an official report on each machine. Moreover, it would be used to train officers in using aircraft for military purposes. In an effort to keep costs down, all incoming army pilots were originally expected to obtain their initial flight training and receive a civilian pilot's license at their own expense.[24] This

Organization of Prussian Army Aviation Agencies, April 1, 1911 to July 31, 1914[1]

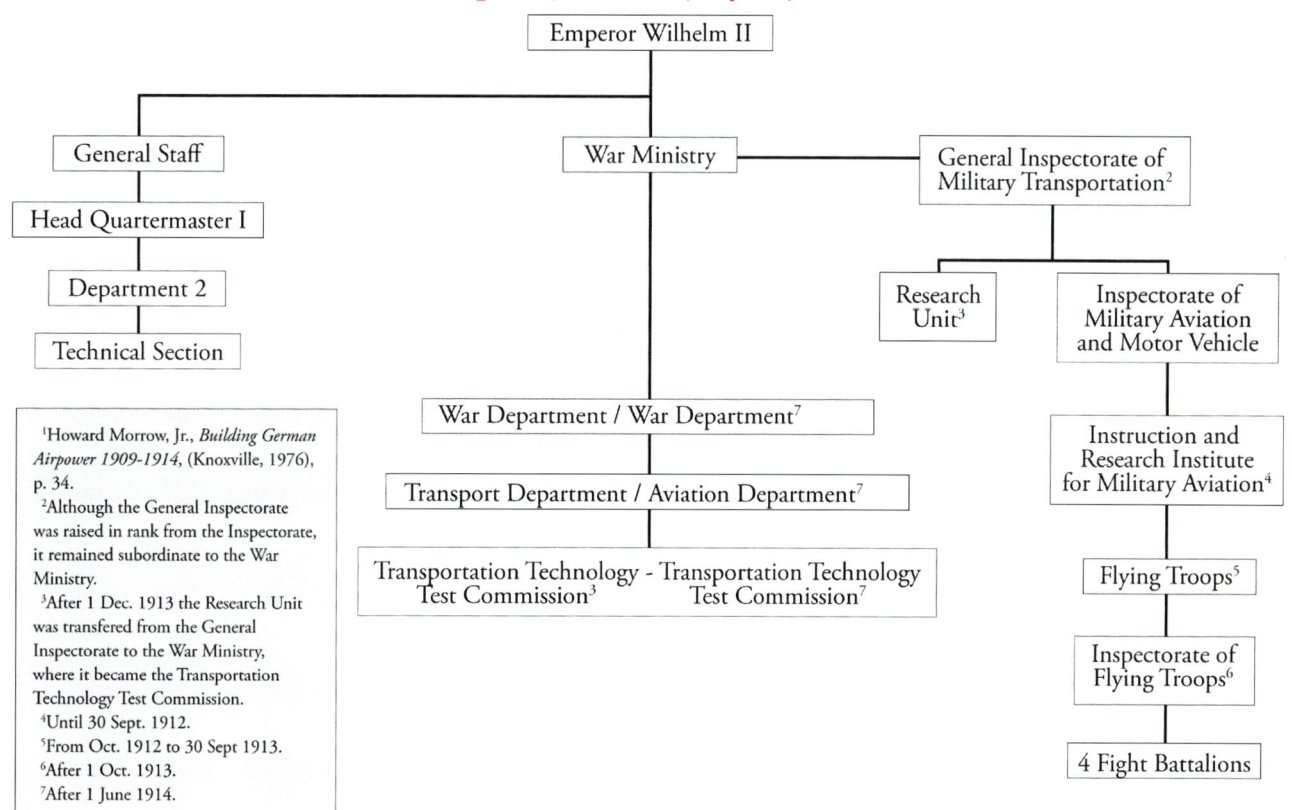

```
                          Emperor Wilhelm II
                                 |
        ┌────────────────────────┼────────────────────────┐
   General Staff              War Ministry         General Inspectorate of
        |                          |                Military Transportation[2]
   Head Quartermaster I            |                          |
        |                          |                ┌─────────┴──────────┐
   Department 2                    |          Research              Inspectorate of
        |                          |          Unit[3]           Military Aviation
   Technical Section               |                            and Motor Vehicle
                                   |                                     |
                         War Department / War Department[7]        Instruction and
                                   |                             Research Institute
                         Transport Department / Aviation Dept.[7]  for Military Aviation[4]
                                   |                                     |
                    Transportation Technology - Transportation       Flying Troops[5]
                    Test Commission[3]    Technology Test Commission[7]  |
                                                                   Inspectorate of
                                                                   Flying Troops[6]
                                                                         |
                                                                   4 Fight Battalions
```

[1] Howard Morrow, Jr., *Building German Airpower 1909-1914*, (Knoxville, 1976), p. 34.
[2] Although the General Inspectorate was raised in rank from the Inspectorate, it remained subordinate to the War Ministry.
[3] After 1 Dec. 1913 the Research Unit was transferred from the General Inspectorate to the War Ministry, where it became the Transportation Technology Test Commission.
[4] Until 30 Sept. 1912.
[5] From Oct. 1912 to 30 Sept 1913.
[6] After 1 Oct. 1913.
[7] After 1 June 1914.

arrangement quickly proved impractical however, due to a lack of pre-licensed officers to fill the unit's roster. Therefore, in order to solve the problem, the army reached an agreement with several aviation firms to pay for prospective aviation officers to be trained at the various factories by the firms' instructors. This mutually beneficial arrangement allowed officers to complete their training while the manufacturers earned additional income. Among all the firms that participated, Dr. Walther Huth's *Albatros Works* was the most successful and important. Huth had seen the training program as an opportunity to increase his firm's standing with the army's leadership and was among the first to offer his aircraft and instructors for the army's use. Thus, by the end of 1910 Albatros had already trained ten officers and had established itself as the army's biggest ally within the industry.[25]

Hptm. de la Roi's proposal for the creation of an aviation unit was quickly implemented. On July 8th, the Research Unit appointed de la Roi himself the commander of the newly created *Flieger Kommando Döberitz* (Flight Command Döberitz). This important new aviation department initially had four pilot officers, who had all obtained their pilot training from a newly licensed instructor from the Albatros firm. By the end of the year, Albatros had trained five additional airmen for de la Roi's command. The final exam for these pilots consisted of circling over a pre-designated point near the airfield three times in the unit's first aircraft—an Albatros-Farman biplane.[26]

Flight Command Döberitz garnered very little support from the army it was charged with supporting. Flying was forbidden if any ground troops were using the nearby training grounds on account of the airplane's motor possibly scaring any horses. Even the act of hiring an instructor and acquiring their first aircraft proved problematic due to lack of available funds. Fortunately, Dr. Huth loaned the unit one of his Farman biplanes and allowed his chauffer, Simon Brunnhuber, to provide training under the fictitious name "Dr. Brück." Later that fall, the unit obtained some additional machines and instructors to train more students.[27] One of those pupils later described his experiences with the unit:

At dawn the officers inquired of the instructors whether or not there was to be flying. If the leaves even shook lightly in the breeze, we immediately closed the shutters and crept under the bed covers. You can see what a wretched state flying was still in at that time. But flying operations had to be done with great care. Each crash did a great deal of damage to the young aviation [program] and all who opposed it—and these were not few in number—rejoiced, professing to see in each accident proof for their claim that it was impossible to perfect aviation and to get the slightest bit of use out of it. If the weather was suitable for flight training, officers, non-commissioned officers and men got into the truck waiting in the camp at Döberitz or in the personnel bus and with sleepy faces we went through the dawn and cold morning mist to the airfield. As a rule after two or three hours, flying was discontinued because of increasing heat and we again returned to camp to finish our sleep. About five in the afternoon we again went to the field. Here time not spent aloft, was consumed by working on motors or in the workshop, by tactical or motor instruction or else we made up a friendly game of cards or practiced pistol shooting or some other sport.[28]

Despite its problems, the creation of Flight Command Döberitz demonstrated that the army had learned from its previous mistakes and was now capable of working with private industry to build up its aviation forces. As Germany's military leadership had already discovered with its dealings with the motor vehicle industry, airplane manufacturers required direct funding from the army in order to rapidly expand and improve their product to better fit the military's needs. To that end, the army's various departments all began working to assist domestic German aircraft firms grow so that Flight Command Döberitz could obtain better aircraft and serve its true purpose in the event of war.

Traditionally being the army's biggest supporter of airplane development, the General Staff began publically advocating for a major reorganization of the army's aviation forces with an added emphasis on heavier-than-air aircraft. The Chief of the General Staff, Helmuth von Moltke (known as Moltke the Younger), was a believer in the airplane from the beginning and had tried repeatedly to use his position and influence to build up the army's aviation forces.[29] In a memorandum written in January 1910, Moltke lobbied for additional funding and (among other things) the creation of an independent aviation agency with control over all aviation matters including influence over German airplane and motor firms.[30] The War Ministry predictably demurred, believing the airplane's importance to be too small to warrant these changes. However, news from German agents in France indicated that the French aviation industry had begun to produce large numbers of militarily viable machines capable of daily operations with their field armies. Thus, upon receiving word that the French were to have a fleet of 100 of the world's best aircraft by March 1911, the War Ministry was forced to concede that a reexamination of the airplane's role in the German Army was sorely needed.[31]

To properly study the matter, representatives of the General Staff, War Ministry and Inspectorate of Transport Troops were brought together in October 1910 to form a commission. After two months of research and reviewing aviation experts' testimony, 'the Cologne Commission' (as it became known) released its conclusions and submitted their official report. First, the airplane's potential military value as a reconnaissance platform was admitted by all parties to be indisputable. Secondly, with all their aircraft holding world records and with their domestic industry producing large numbers, the French were declared the world's preeminent aviation power, with Germany far behind. Thus, agreeing that French superiority in any field posed a major risk, the commission took action to improve domestic German aviation by raising the budget and reorganizing the aviation troops under the 'Inspectorate of Military Aviation and Motor Vehicles'—a newly created department which would have the power to oversee all military aviation matters.[32] Unfortunately, the War Ministry blocked all additional proposals out of their continued bias in favor of airships and dirigibles. Airships therefore continued to receive almost half (48.6%) of the army's upcoming 1911 aviation budget despite having experienced little technological growth or success the previous year.

1911: A Period of Indecision

The year 1911 marked the beginning of a systematic shift in thinking throughout the army toward seriously supporting Germany's aviation industry. With very little funding from the military and a small civilian market, German aircraft firms simply did not possess the funds necessary to use the proper research and development, testing and production techniques. Consequently, the earliest successful German firms were compelled to reproduce French and American designed aircraft rather than their own machines.[33] It was obvious, however, that this arrangement could not continue if Germany's domestic aviation industry was to truly improve and thrive.[34] Thus, in an attempt to increase the German aviation industry's access to capital, the army's leadership decided that only <u>German</u> designed aircraft could compete in domestic airplane competitions and be eligible to win prize money. Previously, French produced aircraft had come to Germany and won every competition—taking with them badly needed prize money that could have been used by German manufacturers.

Nonetheless, despite the increase in funds from these competitions, the German firms' dependency on foreign designs continued throughout 1911. Lessons learned from the War Ministry's dealings with other new technological industries such as automobile producers strongly suggested that German airplane manufacturers required a much larger budget in order to compete with the French. The War Ministry however, continued to cling to the lost cause of German airship superiority. As a result of the War Ministry's bias, a staggering 92% of the army's aviation budget was dedicated to airships in 1910. The remaining 8%, totaling just 300,000 Marks, had been evenly distributed amongst the various aircraft manufacturing firms by the War Ministry and General Inspectorate of Military Transportation with the positive intention of fostering competition and discouraging the formation of any future monopolies. Thanks to the rising popularity of the airplane, the War Ministry was compelled to allocate roughly half 1911's aviation budget to heavier-than-air aviation. Unfortunately, that budget of 2.25 million Marks was simply not enough to stimulate adequate growth within an industry that was already far behind its main competitor.[35] For example, only 29 aircraft were ordered in 1911. By the end of the year the army possessed just 30 airworthy machines—all of which had to be repaired at the various firms' factories because the army lacked the facilities and personnel to do the work themselves.[36] Meanwhile, French army aviation was continuing to increase their lead over the Germans in terms of both funding and quantity and quality of aircraft.[37] Recognizing the seriousness of the situation, Major Thomsen and the General Staff once again called upon the Cologne Commission to take action. Representing the General Staff at the meeting was *Oberst* Ludendorff, who pleaded for a portion of the dirigible's part of the budget to be reassigned to airplane development. Although some concessions were made, such as a grant for an additional credit of 1.3 million Marks to create two air stations at Metz and Strasburg, the War Ministry refused the General Staff's requests to prioritize airplanes and increase funding to the empire's airplane manufacturing firms.[38]

What was the source of the War Ministry's misguided position? As Minister of War, General Josias von Heeringen's opinion was that airships were a safer and all around better reconnaissance platform than airplanes. As such, Heeringen mandated that his subordinates continue the department's commitment to airships regardless of the latest technological advances or information presented by any of the army's other departments. Even the most reasonable requests were to be refused or contested. For example, in November 1911, Moltke proposed the army increase its inventory to 116 aircraft by the following

Above: During his tenure as Minister of War, Josias von Heeringen vigorously opposed the growth of the army's heavier-than-air forces. (Author's Collection)

October. Heeringen's War Ministry initially rejected Moltke's sound proposal; then, after significant political pressure, they approved raising the number to just 64 by the same date.[39]

1911 had seen the military begin to establish a positive relationship with manufacturers by creating policies that nurtured the overall growth of German domestic industry as a whole. Competition was encouraged and all forms of cronyism were disallowed. Firms of all sizes were assigned a proportionally fair number of new aviation officers for pilot training, which served as each firm's second largest source of income. By winter 1911–12 however, the lack of orders for aircraft had caused many firms to send requests to the army asking for additional officers to train. Without the income from pilot training, many of the firms would have been compelled to

Above: Kaiser Wilhelm II. (Author's Collection)

lay off workers and shut their doors. The army ultimately acceded to these requests and sent more students despite having no intention of ordering more aircraft. The subsequent income from these new trainees nonetheless proved insufficient, as German manufacturing firms continued to operate on the brink of insolvency throughout the winter of 1911–12. Indeed, two companies, Dorner and Wright, were forced out of business during this period due to a lack of contracts.[40]

It was a dreadful time for the German aviation industry. Lack of income and support from the War Ministry stifled development and production, which kept the army's aviation troops underequipped and ill-prepared to support the ground troops in the event of war. Worse still, the airplane's military legitimacy continued to be mocked by many high ranking members throughout the army—resulting in little internal pressure to change the status-quo. Elsewhere, France, England and even Russia were each ahead of the Germans in heavier-than-air aviation with plans to further expand their domestic programs in the near future. Thus, with their potential enemies on the verge of obtaining an unassailable superiority within an important technological field, Germany's military aviation program unknowingly found itself on the precipice of disaster in late 1911.

Kaiser Manöver 1911

Only a dramatic shift in the War Ministry's attitude toward the airplane would produce the necessary budget increases required to grow domestic manufacturing firms. Fortunately, the Ministry's opinion started to change in the fall of 1911 after army aviators were given the opportunity to demonstrate the airplane's military value under simulated war-time conditions at the 1911 *Kaiser Manöver* field exercises. Performed each year in September, these maneuvers were multi-day exercises where large formations (at least one corps per side) would simulate an active campaign and battle against one another in order to give the officers and men valuable experience as well as to test the army's latest doctrine and technology under realistic circumstances.

At the request of the Kaiser, de la Roi's Flight Command Döberitz was ordered to participate in the 1911 maneuvers, which were held in Mecklenburg from September 11–13. Participating in that year's exercise was the Guard Korps, II *Armee Korps* [AK] and IX AK. As always, the opposing sides were divided into "teams": the Blue Team under the command of *Feldmarschall* Freiherr von der Goltz; and the Red Team, commanded by *Generaloberst Prinz* Friedrich Leopold—the Emperor's brother in law.[41] The decision to incorporate aircraft into the maneuvers was based upon several factors including the latest advances in aircraft design as well as a number of notable domestic achievements that had been achieved earlier in the year. The most important of these early successes occurred when a crew from the research unit at Döberitz safely completed the army's first long-range aerial reconnaissance flight during a smaller field exercise near Darmstadt. The crew consisted of Otto Reichardt, a NCO pilot at the Euler Firm's pilot school, and Reserve *Leutnant* Heyne serving as observer. During the course of the exercise, the pair were ordered airborne where they promptly located and reported the opposing team's position in time to impact the outcome of the mock battle in their side's favor.[42] This small, seemingly insignificant episode demonstrated the airplane's unique value on the modern battlefield by performing a mission that an airship or cavalry squadron would have been unable to do. Consequently, there were many calls from within the military after reading reports of the

Darmstadt exercise to test airplanes on a larger scale during the Kaiser maneuvers later that fall.

De la Roi understood that the future of German domestic aviation hinged upon a good showing during the fall maneuvers. Thus, all of his men were made to understand the importance of the event and its possible implications on the future. Hans Jürgen von Giebenhain, a student serving in the unit at the time, later recalled the changes in training once it had been confirmed that the airmen would participate:

Again it appeared that an extension was inevitable because the preparations for the Kaisermanöver from 11–13 September in Mecklenberg and Pommern took a large part of the training period at the school. The Kaiser had explicitly called for the attendance of the aviators with their aeroplanes in these maneuvers, and the officers from Döberitz realized the opportunity to demonstrate their skills before personalities of the highest grade positions and importance. For this reason preparations were very carefully and scrupulously carried out with the best students being trained for this important event.[43]

According to orders from the General Staff, eight aircraft were to be disassembled and shipped by rail to Mecklenburg where they would be used each day in operations over 'enemy' lines. The Blue team was given four Albatros–Farman biplanes and the airship "M.II" while the Red team was allocated four *Taube* monoplanes and the airship "M.III." In addition to its regular compliment of pilots, observers, and enlisted men, each team's flying section was given one General Staff officer, who was ordered to observe the flying sections' actions and make a formal report at the maneuvers' conclusion.

Excellent weather conditions allowed flight operations to be performed on each of the exercises' three days. De la Roi's airmen were directed throughout the event to fly no lower than 1600 feet (500 meters) while over 'enemy' columns. This rule was enacted to determine if it was indeed possible to identify large masses of troops from the "seemingly enormous height" of 1600 feet, which many theorists had believed was too high for any observer to accurately discern what was occurring below with the naked eye. As it happened, the flying sections quickly proved that they could deliver valuable intelligence for the army leadership while flying at the prescribed altitude. Indeed, observers from both sides delivered accurate reports each day showing the positions of the opposing force's troops. Flying with the Red Team's flying section, pilot *Ltn.* Braun described his experiences at the event:

Kaiser Manöver 1911	
September 11–13, 1911 Mecklenburg	
Blue (*Oblt.* Geerdtz)	**Red (*Hptm.* Koppen)**
4 various B-Type Biplanes	4 *Taube* Monoplanes

Excluding test flights, I made two reconnaissance flights during the Kaiser Maneuvers. The first flight with Leutnant *Fink as observer was in the area Strasburg—Mecklenburg. An enemy cavalry division was located and reported to be at Strasburg. During the flight a nut broke on the carburetor intake pipe, so that the engine's revolutions slackened considerably. Ltn. Fink climbed out of his seat onto the wing and managed to repair the damage during the flight.*

My second flight of the maneuvers was with Hptm. *Stülpnagel [of the General Staff]. As we reached a height of 200 meters, the front cylinder block of the Mercedes-Motor began to jump, so I had to make an emergency landing which was successful. Since the motor could not be repaired in the field, my otherwise undamaged aircraft was unavailable for use on the Maneuvers' third and final day.*[44]

Meanwhile another member of Red Team's flying section, observer *Ltn.* Fink, had on his own initiative established a small ad-hoc photo lab inside a tent on the section's airfield. Using a small wooden box with a 3.5f=25cm lens and a focal plane shutter as his camera, Fink took several photographs of the Blue Team's troops during the exercise that were never used by the Red Team's leadership. Nevertheless, Fink's actions were important in proving the possibility of aerial photography as a legitimate means to gather intelligence.

During the second day of the maneuvers, *Ltn.* Hofer's Taube was made ready for a flight with General Staff Officer Rundstedt serving as the observer. While Fink was Hofer's usual observer, a critically important development near the frontlines had occurred, causing Rundstedt to demand to go aloft to personally make the observations himself. However, shortly after Hofer and Rundstedt took off, their Taube was seen to violently shutter before going into a nose dive and crashing. Hofer was immediately taken to a hospital where he died of his wounds while Rundstedt managed to escape with only a leg fracture.

Despite the Hofer/Rundstedt crash, the aviation troops' participation in the maneuvers was an overwhelming success. Both flying sections combined to log about 24 hours of flight time while travelling 1180 miles (1900 kilometers)—making it the largest military aviation exercise in history at that time. More importantly, heavier-than-air aircraft had significantly outperformed airships throughout the event. De la Roi's airmen proved that a flying section consisting of just four aircraft could promptly reconnoiter multiple assigned sectors and return with valuable intelligence in less time than an airship could do the same job for a single location. Furthermore, the maneuvers proved that the army's cumbersome airships were much more difficult to move forward with advancing ground troops than their airplane counterparts. For example, during the last day of the maneuvers the Red Team's airship: "M. III" was hastily moved and left unsecured, allowing a gust of wind to blow the ship into the rural village of Groß-Below, where it burst into flames.[45] This was a major problem for army leaders, whose war plan required highly mobile armies aided by airships to decisively defeat the French in a war of maneuver.

A newspaper article written shortly after the maneuvers' conclusion stated, "As we were told, the Kaiser openly praised the excellent work of the aviation officers during his critique at the end of the maneuvers and in particular pointed out that Blue Team commander, *FeldMarschall* von der Goltz, said that he was able to deploy his forces more effectively than what he otherwise would have done because of the messages delivered by his aviators."[46] The Kaiser's aforementioned example occurred on the first day of the maneuvers when observer *Oblt*. Geerdtz, flying with pilot *Oblt*. Mackenthun in Biplane B.15, reported the strength and route of march of the Red Team's entire western wing. Armed with that information, Goltz successfully redeployed his troops to check the red team's advancing forces.

The Kaiser was in fact so pleased with his aviators' performance that he personally thanked each of the participating officers in a special awards ceremony put on to recognize the exercise's most noteworthy participants.[47] After speaking to each of the airmen, the Emperor awarded the *Kronenorden* 4th Class to pilots: *Oblt*. Bahrends and *Leutnants* Braun, Canter, Mahncke, Justi, Carganico, and Vogt. *Hptm*. Koppen, the new commander of the Technical Section for Military Aviation, also received the award in recognition for his leadership and contributions to training the airmen in preparation for the exercise.[48]

German military historians commonly name the 1911 Kaiser Maneuvers as a turning point in the army's aviation program.[49] No longer could critics claim that the airplane's value on a battlefield was only "a baseless theory." De la Roi's flying sections had proven under simulated war-time conditions that airplanes were excellent reconnaissance platforms capable of quickly moving with the troop's forward supply columns and getting airborne to deliver valuable intelligence. The army's airships on the other hand were exposed to have some serious flaws that restricted their use with the army during mobile operations. Thus, upon the maneuvers conclusion, it was no surprise that a shift in opinion within the army's leadership began to make itself felt. However, as with so many bureaucratic organizations, the army took time to implement the sorely needed changes.

1912–1913: The Establishment of Germany's Aircraft Industry & Formation of the Air Service

Despite the successes achieved during the fall maneuvers, aircraft development and production in Germany remained stifled by the War Ministry's 1911 decision to continue heavily funding airships in favor of airplanes. Thus, after the positive showing during the Kaiser Maneuvers, Helmuth von Moltke decided to once again formally request that airplanes be given a larger share of the aviation budget. The General Staff Chief's request was made in April 1912 while representatives from the War Ministry, General Transportation Inspectorate, and the Inspectorate of Military Aviation and Motor Vehicles were meeting to discuss potential changes to the aviation budget for the upcoming "military calendar year."[50] In addition to the budget changes, Moltke called for an increase in contracts for German aircraft manufacturers and the creation of an independent organizational agency for military aviation. Regrettably, the War Ministry rejected Moltke's proposals and announced that it would stand by the Cologne Commission's decision to continue allowing the General Inspectorate to oversee aviation development.[51] A dejected Moltke wrote upon hearing the decision that: "We will not overtake the French in this manner."[52]

The only positive change made during the meeting came with the decision to abolish the "Instruction and Research Institute for Military Aviation" and replace it with the 'Flying Troops Command.' The new 'Fliegertruppe Kommando' was formed at Döberitz (near Berlin) with subordinate units on the eastern and western frontiers in order to concentrate the aviation troops in peacetime and expedite their mobilization in the event of war.[53]

With the War Ministry stubbornly refusing to increase funding to the airplane firms, many frustrated patriotic Germans decided to take matters into their own hands. One Berlin magazine editor exemplified the typical feelings of nationalist German civilians throughout the empire when he wrote:

"Are we deaf and dumb? Are we blind to what is going on before our eyes? —In France there are great stirrings; France is increasing its lead in aviation. The national sacrifice, the zeal for sacrifice is unbelievable.

Above: National *Flugspende* (Aviation Fund) fund-raising postcard. (Author's Collection)

From all sides without exception, whether rich or poor, money flows to the War Ministry... Where are Germany's men?"[54]

Similar pleas for change were made all over Germany. Thus, out of this climate the "National Aviation Fund" (*Nationalflugspende*) was established in April of 1912 as an organized attempt to supplement the meager budget allocated to the aviation industry and increase domestic design and production. This tactic had already been attempted successfully in 1908 with the "Zeppelin Fund," which raised money for Count Zeppelin to establish his company and place pressure on the War Ministry to accept more airships. As the fund's own literature stated, the fund existed "because the active support of the entire nation is due the men, who, as pioneers in a great new cultural task, risk their lives in the patriotic endeavor to secure for Germany in this area an equal place in the universal struggle of nations."[55]

The National Aviation Fund (NAF) was the brainchild of August Euler, an airplane manufacturing firm owner who had personally trained the Kaiser's brother, Prince Heinrich, how to fly. Prince Heinrich's fascination and enthusiasm for aviation had gained him the reputation as the airplane industry's biggest advocate within the Hohenzollern Court and a leading figure in the NAF. In a speech at the General Aircraft Show in Berlin on April 3rd 1912, Heinrich said: "Aviation is necessary! A strong German corps of fliers is necessary!"[56] Using his influence, Prince Heinrich established an imperial committee to help sponsor the fund and get it started. The committee was made up of bankers and former government bureaucrats who all desired to see a flourishing domestic aviation industry and an end of French supremacy within the field. The committee issued a call for the NAF's formation stating in part:

With pride we Germans can call our own the man who first achieved the longing of centuries: The Zeppelin...But the turbulent development which overtook aeronautics with the introduction of the flying machine, compels us to make the utmost effort in order not to be left behind by the sacrifices and activity of other nations. If at all, here too it must be said: 'Germans to the front!'[57]

Almost immediately, the NAF was up and running with members of the Imperial Government lobbying

Above: A younger brother to the Kaiser, Prince Heinrich was fascinated by aviation. He was taught how to fly by August Euler and ultimately became known as the "Patron of German aviation." (Author's Collection)

on its behalf—first targeting lower level government agencies to act as intermediaries between the fund and the populous. As a result, many cities and German states quickly established their own sub-committees for fund raising within their local areas. In order to encourage Germany's royal families to contribute, the Kaiser personally sponsored a national aircraft motor competition using 50,000 Marks from his private treasury. Furthermore, the Emperor saw to it that Prince Regent Luitpold of Bavaria and Duke Ernst Guenther of Schleswig-Holstein were added to the fund's national committee.[58]

Donations were received in varying amounts from individuals representing all of Germany's economic classes. Patriotic Germans, rich and poor alike, eagerly contributed what they could afford in the name of improving German aviation and national defense. By December, the fund had secured an impressive 7.2 million Marks.[59] To best determine how the funds were dispersed, the national committee established a board of trustees consisting of 50 men. In an effort to ensure decisions were being made in the country's best interests, the German Chancellor, Bethmann-Hollweg, was allowed to directly appoint 10 members actively serving in the military or government to the board. Non-voting military/government personnel were also attached as advisors to help further influence internal discussions.

The board first addressed the issue of flight training. Due to the small number of pilots then serving in the army, it was decided that all licensed pilots in Germany be compelled to serve in the army's flying sections. Any licensed pilot or flight school pupil who wasn't in the active service would be required to enter the flying troop reserves and participate in four separate three week maneuvers that were scheduled over the following two years. All incoming pilots were also required to have a secondary education and be physically fit for military service. The selection and training of the students was still to be done by the various firms, who would provide modern military aircraft for training. In exchange the firms earned 8,000 Marks for each student who completed the course, but were to be held financially responsible for any pupils who failed to finish the program. This arrangement ultimately worked well. A total of 71 pilots were trained by October 1913 using the fund's donations, with the total rising to 186 by the outbreak of war in August 1914.[60]

The Fund's 7.2 million Marks were of inestimable value. The total sum of donations was more than *double* the military's own budget for airplane related expenses in 1912 (3,391,250 Marks). In addition to the aforementioned pilot training expenses, 62 airplanes were purchased in 1912 using the Fund's donations. Considering this number represented nearly half of the total number of aircraft (129) delivered to the army in 1912, it is quite obvious why the creation of the National Aviation Fund has been named as "the turning point" in German aeronautics.[61] Without the NAF's income, pilot training would have remained low and contracts for new aircraft would have been cut in half—keeping the *Fliegertruppe* dangerously understrength while denying many domestic firms the necessary income to remain profitable.

Meanwhile, 1912 saw the battle between the General Staff and the War Ministry reach its climax. After the War Ministry made the decision in April to continue prioritizing airships over airplanes, Moltke decided to go on the offensive and shatter the Ministry's unfounded distrust of the latter. Flurries of reports and memorandums were sent to the War Ministry, General Inspectorate of Military Transportation and amongst senior military leadership that soundly demonstrated the airplane's capabilities and disproved the War Ministry's fears. The theme of these documents were mostly the same: 1) airplanes were unquestionably the future of military aviation, 2) France's growing superiority in aviation was extremely dangerous, and 3) a strong independent German Air Service supported by a well-funded domestic German aircraft industry was a necessity in order to modernize the army and improve its chances of success in the coming war. One of the first of these memorandums, sent in April, was a scathing rebuttal to the War Ministry's written decision to continue backing the airship. The memo stated in part:

In face of the assertions in the memo under consideration I must point to the systematic and wholesale organization of the aviation service in France... Against the [War Ministry's] view, the existence of which has occasionally come to my knowledge, that the aircraft organization in France really exists only on paper, or will remain there, I must protest that this is a dangerous piece of self-deception. Even if there are still lacunae in the 1912 scheme of organization, they still have so great a head start in comparison with our measures that the French have a perfect right to look upon their extraordinary superiority in this department with proud satisfaction. It needs no argument to show that in a war that superiority will be associated with all kinds of disadvantages for us. It is therefore in the highest degree regrettable that with us the same attention is not being devoted (and if I am to accept the memo in question, will not be devoted) to the air service.

In view of that statement I will refrain from going into details. I will, however, insist that in my view we should proceed as systematically with the development of our air service as with the organization of all other formations of the army. As long as we are working without definite objectives we shall find that in a crisis we have not the resources at our disposal on which we are relying on paper in peace. The difficulties in the mobilization of our flying sections last autumn [during the 1911 Kaiser Maneuvers] is a serious lesson in that respect. Moreover, the scheme of concentration and the associated preparations for the employment of all our forces against the enemy have reached such proportions that they can only be mastered if we have definite aims in view...

It is peculiarly regrettable that decisions have not yet been taken with regard to the development of military aviation for the years 1913 to 1916. The new Army Bill establishes the army strength for that period. It is quite impossible for us to be satisfied for the five year period with the number of people available on October 1st 1912. We can count with certainty on fresh detachments and reduction of the establishment, even though the Army estimates be not discussed and we have no men to spare. I cannot regard as valid the reasons which the War Ministry puts forward for its attitude.[62]

Four months later, Moltke sent another memo, this time with a revolutionary proposal detailing what was necessary for the army to catch up with the French and establish an effective air service. This memorandum would ultimately serve as the general outline for the air service's expansion.

...I enclose in the appendix a summary of my views and wishes on the basis of our further organization of the flying service. The principles which are outlined in the attached summary will be vital for the establishment and further evolution of this new and important weapon, even though progress is made in the science of aviation.

I fully realize that my demands, which exceed the views of the Inspector-General, will make extraordinary inroads on money and personnel. We must proceed to intensify our efforts if we wish to overtake the obvious and material inferiority under which we suffer in this department of our defenses in comparison with our western neighbor. I have never ceased by frequent and continuous reports to draw attention to the War Ministry to the progressive achievements of France in this department during the last few years. It will only be by extraordinary efforts on our part that we can get the lead again, a lead which is of vital importance for our operations. I do not doubt that the resources required will be granted by the Reichstag, *even though they exceed previous demands, so that the question of money will form no obstacle to the establishment of the proposed organization. I referred to the importance of the question of personnel at the conferences on the last Five Years' Bill.*

(Signed)
von Moltke

Organization of Military Aviation
I. General Fundamental Proposals
The first goal which must be reached as soon as possible in the organization of our military flying service is the provision:
1. Of several (two or three) mobile field reconnaissance sections and an aeroplane park for the army headquarters.
2. One mobile field reconnaissance section for each corps, including reserve corps
3. One mobile field reconnaissance section for the cavalry corps

4. Fortress reconnaissance sections for the important fortresses.

If the present fluctuating experiments with aeroplanes for the purpose of directing artillery fire show that special flying squadrons will become necessary, to this list must be added:

-Artillery sections for the army corps.

Our minimum demand in aviation formation up to April 1st 1914 is thus:

– 8 mobile reconnaissance sections for Army HQs
– 26 mobile reconnaissance sections for corps HQs
Total: 34 sections[63]

– 8 Aeroplane Parks

– Cadres for 13 Fortress reconnaissance sections at Cologne, Mainz, Diedenhofen, Metz, Strasburg, Germersheim, Neu-Breisach, Breslau, Posen, Thorn, Graudenz, Lötzen, Königsberg

– 8 Depot Detachments

II. Strength of the Aviation Service

Mobile Field Flying Sections— …strength will be fixed at: One commander, 8 aeroplanes, 8 pilots and 8 observers, plus the required subordinate personnel. The squadrons must be mobile and formed in such a way that rapid entraining and detraining, as well as quick advance by road in connection with operations of the army, are possible. The motor transport attached to each section for the purpose of carrying tents, accessories, spare parts, workshops, personnel, as well as the machines themselves, must be organized in such a way that there will be not the slightest difficulty about dividing the sections into half sections of four aeroplanes.

It will thus be possible that at the outset of half sections can be attached to the army cavalry.

Fortress Sections—The squadrons to be formed can be of varying strength. Königsberg, Lötzen, Thorn, Strasburg, Metz, Diedenhofen and Neu-Breisach must have aeroplanes available from the first day of mobilization. I suggest that at first the cadres be formed of half squadrons, i.e. four planes with personnel, as well as workshops and supply depots, so that in case of a siege the cadres can be strengthened.

Aeroplane Parks— For the Armies these will be mobile, and will be used to supply, supplement and effect repairs to all aviation units at the disposal of Army Headquarters. Moreover, certain flying formations attached to the army cavalry are to be referred to the aeroplane parks of appointed Army Headquarters for their requirements in those respects. With that end in view the aeroplane parks are to be supplied with reserve equipment, spare parts, workshops and other usual necessaries, which the field aeroplane sections are unable—or only in limited quantities—to take with them for fear of hampering their mobility. The aeroplane parks are therefore to be regarded primarily as a great reserve of matériel for their field sections and must be made so mobile that they can always follow the armies in case of need (by the use of motor transport) be brought right up to the aerodromes.

III. Establishment of aeroplane depots in peace time

In accordance with the review in the appendix, aeroplane depots must be established in peace in such a way that by April 1st 1914, if possible, every corps must have one available. For peace purposes (tactical training, employment with troops) this will be under the control of the Corps Headquarters, while in case of mobilization, in addition to the aeroplane sections of the corps in question, it will have to supply the additional units for the Army Headquarters…

IV. Material, Personnel, Training and Instruction

The full number of aeroplanes required for all the fighting formations must always be assembled at the aviation depots. But beyond that it is desirable to have certain percentage of fighting planes for the purpose of at once making good damaged and lost machines and thereby assuring the possibility of mobilization.

For all practice flights in peace only war machines [military approved/deployed with the flying sections] shall be used on principle, while instructional and learners' machines can be employed for first instruction. Future experience must decide what "the life" of an aeroplane is. In view of natural depreciation and the rapid progress in motor construction, a continual renewal of our establishment of aeroplanes will be inevitable if we are always to have our active formations equipped with the most efficient material.

For this purpose, and that of replacing the high rate of wastage we must expect in the field, the extension and higher efficiency of the aircraft and aircraft motor factories which we have in mind for army supplies is an extremely important question. We must make all possible efforts to secure those objects, if necessary by state subsidies.

Besides material, the personnel must be in existence in peace time at full war establishment, especially for those aviation formations which it is intended to employ on the first days of mobilization. For the formations which come into service subsequently, the existence of a cadre of about half of the existing establishment of pilots will suffice, while the other pilots and observers who will come out in the case *of mobilization are to be sent on four or six weeks' courses at different times of the year...*

With a view to covering our requirements at an early stage, the training of officer pilots must be more decentralized than it has been hitherto. It is the business of the competent authorities to say whether it will be preferable to carry out the training at private factories or at the larger military aerodromes. It is probable that for the present both courses will have to be adopted.

Training—Thanks to the subordination of aircraft stations to the Corps Headquarters for tactical purposes, that cooperation of the aeroplane force with the other arms and the command which is absolutely vital will be greatly facilitated...Civil aviators who are suitable and appointed to employment with the field formations will be called upon in peace for certain tactical courses with them...

V. Concentration and Command of the Whole Air Force
The very great development which is bound to take place in the flying service when the measures already discussed have been put into operation, and the inevitable increase in the numbers, duties and importance of that force, even now raise the question whether the complete independence of the Air Force (and its separation from the Department of the Inspector-General of Military transport) is not a step which will become necessary in the future and, therefore, should be taken now.

An independent "Inspectorate of Military Aviation" at the head, under its "G.O.C. Air Force" to have the control of a definite number of aviation stations in technical matters and be responsible for training, would be a simple central authority such as is required by the needs and developments of the new arm.[64]

Moltke's proposals were remarkably far sighted. His memorandum offered detailed solutions to each of the problems facing the air service at the time. First, Moltke recognized the necessity of supporting the aircraft firms' growth and prosperity. Under the proposed system Germany's domestic manufacturers would be continually offered large numbers of contracts for new aircraft in order to fully equip the 34 proposed flying sections and keep them continually stocked with the latest aircraft. This would boost revenue flowing into the country's best aircraft firms, allowing the companies to grow and invest in research and development to produce better aircraft in the future. Furthermore, it would allow these same companies to increase production in wartime to keep up with the inevitable losses that would be experienced by the flying sections at the front.

The memorandum also concluded that the army would greatly benefit by having a flying section attached to each army, corps, and cavalry corps headquarters upon mobilization. During peacetime, Moltke proposed that the air service train directly with the ground troops to establish an effective combined arms doctrine. In the meantime, basic training for all incoming pilots, observers, and ground personnel was to be improved and expanded to ensure the army's high quality training standards were consistently followed. Finally, in order to properly and efficiently control his proposed air service, Moltke recommended that the army's aviation forces be reorganized into an independent department with its own Inspectorate to oversee internal matters in peacetime and a Commanding General to prepare and coordinate all aviation forces in wartime.

Unwilling to make a single meaningful concession or abandon any power, the War Ministry outright refused each of Moltke's proposals. The Ministry's written response was full of accusations impugning the General Staff's statistics/reasoning as well as a series of half-truths and excuses to justify their need to maintain control over the air service and aviation industry. Thus, the Ministry countered the General Staff's proposal with a modest increase in the number of yearly aircraft contracts that totaled about 50% of what Moltke had suggested. Unsurprisingly, the General Staff Chief disapproved of the decision saying, "I do not share the hope that we shall catch up with France within a calculable period... As regards the program of the War Minister, as I have said it falls very far short behind my proposals and therefore leaves us a long way behind Russia and France."[65] Fortunately, the income from National Aviation Fund was able to assist the manufacturers design and produce more aircraft in order to maintain the General Staff's hopes of catching up with the French.

The War Ministry was also forced to concede that a reorganization of aviation forces was finally necessary. Thus, "The Royal Prussian Air Service" was created on October 1st 1912 with a compliment of 21 officers and 306 men on its original roster.[66] However, despite its impressive sounding name, the newly formed Air Service failed to resemble the force in Moltke's proposal, nor did it have the necessary authority to change any of the organizational deficiencies that had previously existed. The aircraft section remained under the control of the General Inspectorate of Military Transportation and there continued to be no connection between the aviation forces and the army's ground troops. Instead, the front line airplane units were to continue being grouped with the General Inspectorate's rear-area organizations such as the balloon/dirigible units, mechanical transport

> **Kaiser Manöver 1912**
>
> September 8–13, 1912
> Central Germany
>
> | Blue | Red |
> |------|-----|
> | B-Type Biplanes | A-Type Monoplanes |
> | (Albatros & Breguet) | (*Taubes* and Bristol) |
> | FFA 1 (*Hptm.* Wagenführ) | FFA 3 (*Rittm.* von Hantelmann) |
> | FFA 2 (*Oblt.* Geerdtz) | FFA 4 (*Oblt.* von Dewall) |

service, and the telegraph and railroad troops.[67]

The folly of the War Ministry's decision was promptly confirmed by reports of the *Fliegertruppe's* actions at the yearly Kaiser Maneuvers—held in central Germany from September 8–13, 1912. The exercise saw both sides given two flying sections, each consisting of six aircraft and 14 officers. After having earned the General Staff's trust during the previous year's maneuvers, these flying sections were rewarded with conducting the bulk of long-range reconnaissance for both sides throughout the exercise.[68] This was done as an experiment by the General Staff to determine the limits of what the flying sections could do. Fortunately, the airmen performed beyond their leaders' expectations—delivering valuable intelligence during the maneuver's most important moments which influenced the outcome.[69]

The 1912 Kaiser Maneuvers were also noteworthy in that they were the first Prussian military exercises where German-designed and produced aircraft were used exclusively. The fact that German manufacturers had been making French, British and American designed aircraft had been unsettling for many members of the army leadership. However, with increased funding brought in through the National Aviation Fund, German produced aircraft had replaced foreign designed machines in all front line units. By far the most famous and important of these early designs was the Etrich Taube monoplane, which would ultimately become the most popular German monoplane of the prewar era.[70] Although the Taube's creator—Igo Etrich—was an Austrian, the rights for licensed production had been sold in 1910 to the Rumpler firm where it was exclusively produced for the Prussian Army until 1912 when the Albatros and Euler firms began constructing their own versions of the craft. In the spring of 1913, the army implemented a policy of standardization and named the Taube as one of only two types of aircraft the army would accept (the other being "tractor design" biplanes). In accordance with this decision, the Jeannin, LFG and Gotha firms quickly began Taube production later that summer.[71] The subsequent competition between these firms for Taube contracts lowered costs while simultaneously increased the quality of the airplanes—thus creating the conditions for a healthy and diverse German aviation industry.

Reacting to the technological advances made by domestic airplane manufacturers, the aviation troops' outstanding performances during the 1911 and 1912 Kaiser Maneuvers, and the outbreak of the First Balkan War, the War Ministry finally decided to substantially increase the aviation budget from 7 million Marks in 1912 to 31.2 million Marks in 1913 (15.6 million Marks appropriated for airplanes—still less than half of the budget). However, despite this increase the Ministry remained committed to continuing the existing organizational structure which isolated the aviation troops from the rest of the army's front line formations. In response, the General Staff immediately released a series of memorandums calling for an immediate reorganization of aviation forces along the lines previously proposed by Moltke as well as an expansion of the airplane sections' existing duties. For example, in a letter sent to the War Ministry, Moltke predicted the need for specialized artillery cooperation squadrons—requesting they be established as part of his greater organizational scheme as soon as possible. Thus, in light of this request and others like it, Erich von Ludendorff sent the following memorandum detailing the General Staff's proposal for an increase in the size and role of the future air service:

To: Section 7 of the War Ministry

As soon as my program of 26/9/17, No. 12751 F. (secret) i.e., the establishment of:
7 Reconnaissance Flights for Army Headquarters (one for each Army HQ)
23 Reconnaissance Flights for Corps Headquarters (one for each Corps HQ)
each of 8 aeroplanes with personnel
Cadres for 12 Fortress Reconnaissance Flights
each of 4 aeroplanes with personnel
7 Aircraft Parks
7 Reserve Flights
(excluding Bavaria) has been completed... we must aim at the following establishment:

The provision of a second reconnaissance flight of 8 machines for each army
7 flights = 56 aeroplanes
The provision of a reconnaissance flight of 8 machines for each reserve corps
12 flights (on present establishment) = 96 machines
The provision of two half flights, each of four machines, for the Army Mechanical Transport Staffs
8 half flights, each with four machines = 32 aeroplanes
The provision of special artillery flights, each of 6

General Staff Proposals — Total Requirements		
	By 1/4/14	By 1/4/16* (If Possible)
Armies	56	112
Corps	184	184
Reserve Corps	—	96
Army Mechanial Transport	—	32
Artillery Flights (Including Reserve Divisions)	—	426
Aircraft Parks (60% of whole establishment)	144	500 (in theory)
Reserve flights (40% of whole establishment)	96	350 (in theory)
Fortress Reconnaissance Flights	48	96
Total	528	1,796

* We must decide later on our requirements up to 1/4/15 or 1/4/16. If the results of the 1913 experiments with artillery flights are good the artillery flights must have first consideration.

machines with personnel (the troops themselves to supply observers) for the divisions and reserve divisions
46 flights of 6 = 276 machines

For Reserve Divisions: 25 flights of 6 = 150 machines

For aircraft parks in each army, comprising 60 percent of the machines:
about 500 machines
For Reserve flights comprising 40 percent of the machines:
about 350 machines

For Fortress Flights for 12 fortresses: (Cologne, Metz, Mainz, Diedenhofen, Strasburg, Neu-Breisach, Königsberg, Lötzen, Graudenz, Thorn, Posen and Breslau.
8 machines a piece: deducting the 48 machines with the existing cadre = 48 machines

Total Requirements: (See table above.)

Of course, the figures are only a quite general basis. They take no account of any further increases in the peace establishment of our army, and moreover Bavaria is excluded, though she is to have a corresponding aviation establishment.

Whether more will be desirable later on must be left to the future and technical developments.
(Signed)
Ludendorff [72]

While the numbers of aircraft Ludendorff requested were unrealistically high, it nonetheless conveyed the General Staff's desire for additional aircraft. Just as importantly, Moltke and the General Staff recognized the need to reorganize all aviation forces into units that could properly support the ground troops in wartime. Moltke personally attempted to address this matter in a November 1912 memo that was written as an attempt to convince the War Minister to adopt his aforementioned reorganization proposal. The memo read in part:

To the War Ministry
...The figures given by me or my representatives referring to the French Aviation Services are based on the reports of our military attachés in Paris and the careful work of Section III of the Great General Staff. I should be glad if all the material on French military aviation, and any other material unknown to me which our attachés have in Paris, could be submitted to me personally so that I could give it to Section III to work on and harmonize the conflicting views.

But first of all I will send the Ministry of War the memorandum on the condition of French aviation to which I have already referred. It also reveals the future developments which are intended. On the 1st of April, 1914, the assumption of 450 military aeroplanes and somewhere about 350 aviators will not be too high. For that time I consider we should require 324 aeroplanes, including Bavarian formations, while the War Minister, if I counted in three sections of six machines for Bavaria, intended to have 156 machines with personnel.

In view of these figures I do not share the hope that we shall catch up with the France within a calculable period...

Above: As Minister of War, Erich von Falkenhayn was a major proponent of military aviation. (Author's Collection)

In my view the obstacles to the realization of my program are... simply the question of money and personnel.

My memorandum of 8/11/1911 [November 8 1911] was written before the conference on the Five Year Bill and dealt with a restricted program covering the period up to 1/10/1912.

My proposals for the five-year period are contained in the secret memorandum of 22/12/1911 I.N. 960. In spite of the last sentence of the memorandum of 19/12112, No.122, and my numerous requests I have only just been given definite figures for the organization of our air service after 1/10/1912 or 1/12/1912.

I realize now to my extreme regret that what has been done does not correspond to my proposals for the five-year period. The solution of the personnel and money question now makes it much more difficult of course. Yet, as before, I have no doubt that, in spite of all the difficulties, in these serious times the Reichstag will give us everything we ask and that we shall be spared from solving the problem of personnel by reducing the establishment of other arms. In any case the Five Year Bill must not be allowed to become an obstacle to this side of our military development also.

As regards the program of the War Minister, as I have said it falls very far short (about 50 percent in the matter of manned aeroplanes) behind my proposals, and therefore leaves us way behind France and Russia.

The absence of reconnaissance flights at Army Headquarters will be severely felt. The Headquarters staff will be without the machinery required for tactical reconnaissance from the air. Further, we shall be unable to assign aircraft to the army cavalry. Both omissions will place our commanders at a serious disadvantage in comparison with enemy commanders.

The construction of Zeppelin and rigid airships of equal value is no substitute. Like the army cavalry, Zeppelins will supply the Supreme Command with data for their operations and be employed in long-range bombing. The duties of Zeppelins are therefore quite different to those of aeroplanes. Our maneuvers do not illustrate those differences accurately.

... I do not like the reduction of the establishment of the corps sections to two machines with personnel. The equipment of our corps with machines is already inferior to that of the French in proportion of one to eight. The artillery will not be assigned anything like sufficient machines for its purposes, while we ourselves will have to reckon with excellent artillery reconnaissance on the part of the enemy.

The fortress maneuvers at Thorn have clearly revealed the importance of air reconnaissance in conjunction with photography. It will be difficult for airships to defend themselves. They or their sheds will soon fall victims to hostile artillery or bombs. Captive balloons also are in great danger from artillery fire. Thus the aeroplane remains the only method of aerial reconnaissance. The fortress flights must be drawn upon for sieges and in this case put under orders of the Reserve Corps.

It is patent that replacement also will left behind.

Fliegertruppe—May 1913[*]

Commander:	*Major* Roethe		*Hptm.* Bartsch
Adjutant:	*Ltn.* Canter		*Oblt.* Georg Keller
			Ltn. Carganico
Flieger Station Döberitz,			*Ltn.* Weyer
Commander:	*Major* Roethe		*Ltn.* Schulz
	Hptm. Wagenführ		*Ltn.* Knofe
	Hptm. Grade		
	Oblt. Vogel von Falkenstein	Flieger Station Darmstadt,	
	Oblt. Solmitz	Commander:	*Hptm.* Dewall
	Ltn. von Scheele		*Ltn.* Bartsch
	Ltn. Förster		*Ltn.* von Mirbach
	Ltn. Canter		*Ltn.* Reinhardt
	Ltn. Mahncke	Flieger Station Strasburg,	
	Ltn. Engwer	Commander:	*Oblt.* Barends
			Ltn. Schmikaly
Reserve:	*Oblt.* von Detten		*Oblt.* Höpker
	Ltn. Kastner		*Ltn.* Laun
	Ltn. Frhr. von Thüna		*Ltn.* von Beguelin
	Ltn. Coerper		*Ltn.* Wulff
	Ltn. von Buttlar		
	Ltn. Fink	Flieger Station Cologne,	
	Ltn. von Hiddessen	Commander:	*Ltn.* Hantelmann
	Ltn. Blüthgen		*Ltn.* Joly
Command:	*Hptm.* Gundel	Luftschiffer and Airship Troop:	
	Oblt. von Oertzen		
	Ltn. Frhr. von Freyberg	Luftschiffer-Battalion #1	
		(Berlin-Tegel)	*Major* Neumann
Surgeon:	Dr. Sergeois	Luftschiffer-Battalion #2	*Oblt.* Gross
		Company 1 (Berlin-Tegel)	
Paymaster:	Kaiser	Company 2 (Königsberg)	
Flieger Station Metz,			
Commander:	*Major* Siegert	[*] H.J. Nowarra, *50 Jahre Deutsche Luftwaffe*, I, p. 40	

The measures proposed by the War Minister seem to me inadequate. I have serious objection to leaving the aeroplane parks behind. The extra seventh machine of each flight will not make any difference. The question of replacements must be taken in hand on the broadest principles in order to cope with the heavy wastage of material. In exceptional cases the most important workmen in the airship establishments which deliver to the army may be temporarily exempted from military service. We shall have to have efficient private factories.

In the foregoing I have clearly set forth the reasons which show that my program both can and should be carried out. I regard it as the minimum which we can put at the disposal of the armies, corps and cavalry divisions, the artillery, fortresses and siege troops, for the purposes of tactical reconnaissance from the air, if they are not to be at a disadvantage compared to their enemies. Nor must we fall behind the requirements for my program if we are to feel secure in the matter of replacement.

I therefore adhere to my former standpoint, that my program must be carried into effect by April 1, 1914. Please see that this is done.

*(Signed)
von Moltke*[73]

Predictably, the War Ministry ignored the General Staff Chief's plea to reorganize, declaring it to be premature and unnecessary. Thus, once again it

> ### Fliegertruppe—October 1913*
>
> | | | | |
> |---|---|---|---|
> | Head Office: | Prussian Inspectorate of Flying Troops [*Idflieg*] | Nr. 4 Strasburg, | |
> | | | Commander: | *Major* Siegert |
> | | | Acting Commander: | *Hptm.* Haehnelt |
> | Inspector: | *Oberst* von Eberhardt | Company #1 Strasbourg: | *Hptm.* Genée |
> | Adjutant: | *Oblt.* Förster | Company #2 Metz: | *Hptm.* Hildebrandt |
> | | | Company #3 Freiburg: | *Hptm.* Barends |
> | Flieger-Battalion: | | | |
> | Nr. 1 Döberitz, | | Bavarian Flieger Battalion: | |
> | Commander: | *Major* Gundel | Munich, | |
> | Acting Commander: | *Hptm.* Wagenführ | Commander: | *Hptm.* Biller |
> | Company #1 Döberitz: | *Hptm.* Grade | Acting Commander: | *Hptm.* Stempel |
> | Company #2 Döberitz: | *Hptm.* von Oertzen | Company #1 Schleißheim: | *Hptm.* Petri |
> | Company #3 Großenhain: | *Hptm.* Vogel von Falkenstein | Company #2 Schleißheim: | *Hptm.* Pohl |
> | Detachment Jüterbog: | *Oblt.* von Ascheberg | Balloon and Airship Troops | |
> | | | Head Office: | Prussian Inspectorate of Airship Troops |
> | Nr. 2 Posen, | | | |
> | Commander: | *Major* Roethe | Luftschiffer-Battalion: | |
> | Acting Commander: | *Hptm.* Kuckein | Flieger Station: | |
> | Company #1 Posen: | *Hptm.* Bartsch | Nr. 1 Berlin | Commander: Krenzlin |
> | Company #2 Graudenz: | *Hptm.* von Poser | Nr. 2 Berlin | Commander: *Major* Schulze |
> | Company #3 Königsberg: | *Hptm.* von Lölhöffel | Company #1 Berlin: | *Hptm.* Zettelmeyer |
> | | | Company #2 Hannover: | *Hptm.* von Quast |
> | Nr. 3 Cologne, | | Company #3 Dresden: | *Hptm.* Gaissert |
> | Commander: | *Major* Friedel | Bat. 1: Kite Balloons | |
> | Acting Commander: | *Hptm.* Kirch | Bat. 2: Airships | |
> | Company #1 Cologne: | *Hptm.* Göbel | | |
> | Company #2 Hannover: | *Hptm.* Georg Keller | | |
> | Company #3 Darmstadt: | *Hptm.* von Dewall | * H.J. Nowarra, *50 Jahre Deutsche Luftwaffe*, I, p. 50. | |

appeared as if the General Staff and War Ministry were about to have yet another lengthy battle over the future of the German Air Service. Thankfully, it didn't happen. In what would prove to be one of the most significant events in the growth of pre-war German military aviation, Erich von Falkenhayn replaced the airplane-skeptic Josias von Heeringen as Prussian War Minister in July 1913. Falkenhayn was unlike any other high ranking officer within the army leadership. Instead of following the "usual path" of remaining in Germany to advance his career, he had adventurously chosen to go to China where he spent six years serving as an instructor for the Chinese Army and a staff officer on the German Expeditionary Force during the 1900 Boxer Rebellion.[74] During his time in Asia, Falkenhayn was free to develop his own ideas and form his own opinions outside the Prussian Army's prescribed dogma. Consequently, upon taking over as War Minister, Falkenhayn was better suited to see past the petty bureaucratic quarrel between the War Ministry and General Staff and act in what he believed was the best interest of the army and the country. While certainly not perfect, Falkenhayn's War Ministry became a much greater ally to the aviation industry and flying service than Heeringen's administration ever was.

Falkenhayn's arrival at the War Ministry was quickly felt throughout the army. First, the badly needed "Army Law of 1913" was passed. In addition to expanding the military budget, the law increased the size of the active army by 136,000 officers and men—raising the number to 866,000.[75] The law also sensibly reorganized the army's aviation forces. Under the new system, five aviation battalions (four Prussian and one Bavarian) were created and placed under the control of a new department named the "Inspectorate of Flying Troops" (*Idflieg*).[76] Each Prussian battalion was composed of three companies and a command staff, while the Bavarian Battalion was made up of a single enlarged company with a command staff.

By sub-dividing the airplane battalions into companies, each comprised of two or three flying sections, Falkenhayn's reorganization was undoubtedly a step in the right direction. For the first time in the young air service's history, aviation units were organized in peacetime into the same

> **Prussian Army Aircraft Procurement, 1911–1913**[*]
>
> **1911:** **11 Monoplanes:**
>   10 Rumpler
>   1 Harlan
> **17 Biplanes:**
>   12 Albatros
>   2 Aviatik
>   1 Euler
>   1 LVG
>   1 Wright
>
> **1912:** **60 Monoplanes:**
>   48 Rumpler
>   7 Harlan
>   2 Aviatik
>   2 Bristol
>   1 Dorner
> **79 Biplanes:**
>   46 Albatros
>   13 Euler
>   9 Aviatik
>   8 LVG
>   3 DFW
>
> **1913:** **183 Monoplanes:**
>   66 Rumpler
>   37 Albatros
>   36 Gotha
>   26 Jeannin
>   12 Fokker
>   3 Euler
>   2 Mars
>   1 LFG
> **278 Biplanes:**
>   98 Aviatik
>   88 LVG
>   48 Albatros
>   24 Euler
>   18 Mars (LFG)
>   2 AEG
>
> [*] Kriegswissenschaftliche Abteilung der Luftwaffe, *Die Militarluftfahrt bis zum Beginn des Weltkrieges 1914*, Band III, p. 319; Morrow, *Building German Airpower*, p. 85.

units that would be deployed with the army upon mobilization. This enabled the officers and men of each flying section to start training and building a rapport with one another. Moreover, with the new battalions attached to various active army corps rather than fortresses along the frontier, the airmen were finally in a position to act on the General Staff's recommendation to begin combined arms training with the ground troops.

There were, however, some major flaws with the new system. *Idflieg*, which controlled all airplane units in addition to its duties to negotiate with the manufactures, was still under the Inspectorate of Military Aviation and Motor Vehicles—which was subordinated to the General Inspectorate of Military Transportation. By remaining under the control of a 'support troop command' such as the General Military Transportation Inspectorate, the German Army's flying troops (*Fliegertruppe*) were organizationally isolated from all other front line units. Thus, combined arms training and other liaison opportunities between the two groups were practically non-existent during the final 10 months of peace. Furthermore, by being under the control of the General Inspectorate, *Idflieg* was unable to have any control over its aviation battalions/companies in wartime. This system, established over the protests of the General Staff, *Idflieg*, and army aviators themselves, would ultimately become a major factor in limiting the *Fliegertruppe's* impact on the campaigns of 1914 and reducing the German Army's chances of quickly winning the war in the west.

1913–14: German Aircraft Become Among the Best in the World

Thanks to the increased funds brought in through the new military budget, the National Aviation Fund and various domestic aviation competitions, German airplane manufacturers had by 1913 acquired the necessary capital to start independently designing and producing their own aircraft capable of successfully competing on the world stage. Finally, after years of trial and error, the army leadership had successfully created the optimal method to grow the domestic aviation industry to the benefit of the army. The formula was simple: the army's aviation experts would announce required specifications for a desired aircraft and then the manufacturers—subsidized and regulated by the military—would compete with one another to create the best design. The firms with the best airplanes were then rewarded with contracts and were enabled to expand and improve their product, thus encouraging results and rewarding success.

To their credit, the War Ministry and the General Inspectorate of Transportation had acted numerous times to avoid any one firm from becoming a monopoly. From the very beginning the army actively encouraged competition amongst the firms, which resulted in the production of better aircraft at cheaper prices. For example, pilot training (the firms' second largest source of income) was fairly and proportionally assigned to all manufacturers. Moreover, foreign aircraft continued to be banned from domestic flight competitions where German firms had the potential to win prize money while showcasing their latest designs to military brass. As a result, many of the smaller manufacturers were able to bide their time and remain in business while researching and developing their own designs which the army would later go on to purchase. By early 1914 the War Ministry was purchasing aircraft from eleven different aircraft firms—most of which had existed prior to 1911 and had slowly grown into becoming formidable manufacturers.[77] With such a large number of viable companies competing for contracts, the army guaranteed innovation and growth across the industry and with it the production of better aircraft.

Profits for each of the Empire's most successful firms steeply rose in the years leading up to war. The War Ministry calculated in 1912 that a firm needed 30–35 contracts each year to remain profitable. In 1913, Rumpler, Albatros, Aviatik, LVG and Gotha were awarded 65, 85, 98, 88 and 36 contracts respectively—thus demonstrating how successful manufacturers were able to flourish under Falkenhayn's reign as War Minister.[78] Using their increased profits, the firms continued to expand and improve their manufacturing capabilities in order to create higher quality machines than their French counterparts.

The army's goal from the creation of the initial airplane technical section had always been to catch up with the French. However, as late as 1912 it was believed that the German aviation industry might not ever be able to achieve parity with its western neighbor. Fortunately, the substantial increase in funding and the military's belated prioritization of the airplane combined to unleash the awesome power of German engineering—allowing for record breaking aircraft to be produced by the summer of 1913. The first record-breaking flight occurred in May when a Rumpler Taube shattered the record for "duration and cross country-flying in a two-seater." Two months later the same machine captured the world altitude record.[79] Another record was captured later that October when an Aviatik biplane piloted by Viktor Stoeffler flew 1,400 miles in a single day. On July 11th 1914, Richard Boehm set the world record in his Albatros biplane for duration during a non-stop flight with a remarkable time of 24 hours and 12 minutes—a record that wouldn't be broken for another 13 years! Finally, on July 14th 1914, a DFW biplane broke the world altitude record when it reached 25,000 feet.[80] These records, among others, indicate just how far the German aircraft industry had come since 1911 when firms were still producing badly copied French aircraft for the army's use. Writing on the front page of a prominent Berlin tabloid newspaper, the editor of *Berliner Zeitung am Mittag* explained to German civilians the significance of the records:

Friends of German aviation breathe a sigh of relief. At last! At last we again have a day when our feeling of inferiority to our neighbors in the west disappears. From this point of view the flight yesterday... is a restorative designed to revive national pride and the eagerness of our pilots to accomplish great feats.[81]

As expected, the *Fliegertruppe* were well represented at the 1913 Kaiser Maneuvers, held from September 7–10 in Silesia. Given the airplane's increasingly important role within the army, the General Staff had decided to employ three flying sections on each side. Two of the flying sections on both teams were attached to corps HQs while the third remained under the control of the army commander. Each section was originally equipped with six airplanes and assigned the same number of enlisted men and support vehicles that a flying section would be assigned in wartime. Under these simulated wartime conditions, the airmen on both sides completed their reconnaissance assignments with positive results. Most importantly, the maneuvers proved the General Staff's theory that aviation units could be successfully used in a tactical role at the corps level. These corps HQ flying sections operated from airfields located close to the frontlines in an effort to quickly deliver tactical intelligence regarding the nearby "battle" to the corps commander and his staff. These messages assisted corps commanders on both teams to smartly deploy their troops and commit reserves at the opportune time.[82]

Meanwhile, the army continued to pursue its policy of improving the quality and safety of its aircraft. Following a fatal accident on September 4th involving a prominent young officer, the army formed a safety commission to investigate the cause of the crash and prescribe changes to prevent future

---

Kaiser Manöver 1913

September 7–10, 1913
Middle Silesia

Blue	Red
FFA 1 (*Hptm.* Goebel)	FFA 4 (*Hptm.* Wagenführ)
6 Albatros Taubes	8 LVG B-Types
FFA 2 (*Hptm.* Grade)	FFA 5 (*Hptm.* von Poser)
6 Taubes (various types)	6 Rumpler Taubes
FFA 3 (*Hptm.* Geerdtz)	FFA 6 (*Hptm.* Oertzen)
6 Albatros B-Types	4 Taubes (3 Rumpler & 1 Albatros)

* Ferko–UTD, Box 13, Folder 4, "Kaiser Manöver notes."

similar occurrences. After 10 days of listening to expert testimony, the commission ruled that the crash was the result of "a lack of expert designers" within the aircraft industry.[83] In the commission's opinion, many firms were beginning to cut corners to increase profits rather than placing the safety of the army's officers as their foremost concern. Thus, a memo was published entitled "Rules and Suggestions for the Aircraft Factories" to better standardize and improve the materials and methods used in future aircraft production. In order to guarantee that these new rules were thoroughly implemented, the War Ministry established the "Transportation Technology Test Commission" as a replacement for the Transportation Inspectorate's old research unit. Established on December 1st 1913, the newly formed test commission immediately took a leading role in improving the quality of aircraft being produced in Germany. The commission was given the power to regulate many aspects of design and production, which gave them unprecedented influence over the firms. For example, new regulations mandating that German firms use higher quality materials and better construction techniques were created and enforced by the commission which ultimately helped with the production of safer aircraft for the army's aviators. These regulations also steered the manufacturers into building the newly standardized two seat reconnaissance biplane rather than wasteful experimental designs that had the potential to fail. Although there were some concerns that this would stifle creativity and future competitiveness with the French, the commission's work successfully assisted the industry build the type of aircraft that the army strongly desired and needed.

By 1914 the army had ordered and received over 600 aircraft. 461 airplanes were delivered in 1913 alone—278 of which were the new, state of the art biplane designs produced by Aviatik, LVG and Albatros. While the famous Taube monoplane continued to be featured prominently throughout the *Flieger* Battalions, the machine was on its decline by 1914. The biplanes had better flight characteristics and were easier for ground crews to maintenance in the field while on campaign. Nonetheless, orders for the monoplanes continued to be placed through the outbreak of war.

The Prussian military's close involvement with domestic manufacturing firms was the key to rapidly creating a successful aircraft manufacturing industry capable of producing large numbers of record breaking aircraft. By early 1914 the army estimated that Germany's top eight firms could combine to deliver 103–112 aircraft per month in the event of war—a far cry from the meager 28 aircraft produced domestically throughout the whole of 1911.[84] This unprecedented growth was primarily the result of increased funds from the NAF as well as the 1913–14 military budgets, which enabled airplane manufacturers to quickly modernize themselves and expand their facilities and workforce. Thus, with the War Ministry encouraging competition by keeping as many firms open as possible, the quality of the aircraft rapidly improved—ultimately resulting in four separate German firm's aircraft achieving world aviation records in 1913 and 1914. Germany's engine manufacturers similarly took advantage of military involvement within their industry to produce aviation motors equal to the latest French models at the time of mobilization. As a result, the *Fliegertruppe* were able to quickly go from flying French designed and powered aircraft to operating large numbers of high quality domestically produced machines in just two short years.

Bavarian Aviation 1911–1914

The German Empire was created in 1871 after the successful conclusion of the Franco-Prussian War. Although history generally remembers the conflict as a Prussian victory, success would have never been possible without the contributions of the other German states. The largest of these was the Kingdom of Bavaria in southern Germany. The Bavarians performed admirably throughout the Franco-Prussian conflict, contributing two corps into the Prussian Third Army. Indeed, it was the Bavarian II Corps under the command of Jakob von Hartmann that famously stormed Wissembourg during the war's opening battle.[85] The Bavarians' success at Wissembourg allowed the German armies to advance across the frontier into France, creating the conditions for the strategically significant victory at the Battle of Wörth two days later.[86]

When the war finally ended, the various German states merged to form a unified German Empire. Although he initially resisted, Wilhelm I (King of Prussia) was persuaded to accept the Imperial Crown on behalf of the German Princes by the Bavarian King—Ludwig II. Thus, on January 18th 1871, Wilhelm I was proclaimed German Emperor in the Hall of Mirrors at the Palace of Versailles. That April the German Constitution was adopted by the *Reichstag* and proclaimed into law by the new Kaiser. Under the new constitution, the *Reich* was to be a federal empire composed of a number of separate constituent states; namely four kingdoms, five grand duchies, 13 duchies and principalities, as well as the free cities of Hamburg, Lübeck and Bremen.[87]

During the unification process it was agreed that the Kingdom of Bavaria and it's ruling Wittelsbach dynasty be given the right to retain limited autonomy

Above: The primitive Otto B pusher was used as a trainer in Bavaria before and during the war. (Aeronaut)

over their own affairs. Among other things Bavaria was granted limited independent diplomatic privileges, control over its own independent railway and postal systems, and military autonomy in peacetime. By remaining independent from the Prussian military caste the Bavarian Army was free to make its own policy decisions, including issues related to military aviation.[88]

As an independent force, the Bavarian Army had its own General Staff and War Ministry. All aviation matters were placed under the control Bavarian War Ministry's 'Inspectorate of the Engineer Corps.' In 1911, after a series of disastrous attempts to build their own machine, the Bavarians finally decided to follow the Prussian model and rely upon private companies for airplane production. However, the fiercely independent Bavarian officer corps only wanted domestic (i.e. Bavarian) firms to build their aircraft and train their pilots—often at their own expense. For example, in November the Engineer Inspectorate rejected the Albatros firm's offer to train Bavarian airmen and deliver aircraft to the Bavarian Army. Necessity, however, ultimately forced the Bavarians hand. The kingdom simply did not have any viable domestic firms in 1911, which compelled the Engineer Inspectorate to purchase seven aircraft from Euler and one from Albatros that December.[89]

Bavaria's first successful aircraft manufacturer, the Otto Works of Munich, produced their first aircraft in January 1912. With nowhere else to turn, the Bavarian War Ministry promptly rewarded the firm with the overwhelming majority of the army's contracts. The Bavarian army's aviation program therefore remained deeply tied to the Otto Firm throughout the prewar years. Indeed, Otto trained most of the Bavarian army's pilots and even loaned one of its mechanics to train all ground crews in a specialized course.[90]

Throughout 1912 and 1913, several north German firms began inquiring about establishing subsidiary or branch factories of their own inside Bavaria. The Bavarian leadership initially rejected these offers due to their desire to maintain Otto Works' profitability. However, it quickly became evident that increased competition was necessary to maintain ingenuity and growth and keep Bavarian flying units equipped with modern aircraft. As the only firm in Bavaria, Otto had become increasingly more complacent with its pusher-design airplane while the highly competitive North German firms were busy producing far superior tractor-design machines. Because the number of contracts awarded by the Bavarian War Ministry was initially so small, Otto needed to sell a limited number of additional aircraft to the Prussians. Unfortunately, the quality of the Bavarian firm's aircraft was too poor to ever be seriously considered by the Prussians. Consequently, despite maintaining a monopoly over the contracts awarded by the Bavarian War Ministry, the Otto firm was nearly bankrupt by 1914 due to its inability to produce a new design capable of competing with the North German firms' designs. Thus, in an effort to save itself, Otto entered into license negotiations with LVG, who was producing arguably the finest biplane on the planet at the time—the LVG B.I.[91] Shortly thereafter Otto began exclusively producing LVG biplanes under license and continued to do so until the outbreak of war. It was ultimately too little too late however, as the firm economically collapsed just months after the war began.[92]

Meanwhile a second Bavarian factory, Pfalz Aircraft Works, was seeking to replace Otto as the Kingdom's greatest homegrown aviation firm. Pfalz had informed the Bavarian War Ministry in early July 1914 that it

> **Bavarian Army Aircraft Procurement, 1911–1913***
>
> **1911:** 1 Wildt Monoplane
> 8 Biplanes:
>   7 Euler
>   1 Albatros
> **1912:** 23 Biplanes:
>   17 Otto
>   6 Euler
> **1913:** 9 Monoplanes:
>   5 Albatros
>   2 Etrich
>   1 Otto
>   1 Rumpler
> 58 Biplanes:
>   46 Otto
>   12 LVG
>
> *Kriegswissenschaftliche Abteilung der Luftwaffe, *Die Militarluftfahrt bis zum Beginn des Weltkrieges 1914*, Band III, p. 104; Morrow, *Building German Airpower*, p. 94.

was producing parasol monoplanes exclusively for military use and was seeking contracts.[93] The aircraft was a copy of the highly capable Morane–Saulnier L, built under license. Intrigued, the Bavarian Engineer Inspectorate requested a completed machine be presented for trials in late July, just days before the outbreak of war. The Inspectorate ultimately liked what they saw and placed an order for 20 (later 60) aircraft, with the first two being delivered to units at the front in December 1914.[94]

Organizationally, the Bavarian Flying Service mirrored the Prussian system. First, an aviation company was established in April 1912 after a successful performance during the Bavarian Army's 1911 autumn maneuvers. Originally consisting of two officers, two NCOs and ten men, the newly formed company was placed under the command of Count Wolfskeel von Reichenberg.[95] The company was later reorganized in October 1913 into an independent battalion as part of the major Prussian reorganization instituted by Falkenhayn's War Ministry. Under the new system, the Bavarians' Battalion was to be comprised of one oversized company and a command staff. Furthermore, just as *Idflieg* had been created for Prussian units, the reorganization had mandated that all Bavarian aviation units be placed under a similar department known as the Bavarian "Aviation and Motor Vehicle Inspectorate."[96]

The Bavarians' desire for independence in aircraft design and procurement had blinded the army into supporting a single firm, Otto Works, whose substandard designs proved to be unworthy of an invitation to Prussian aircraft competitions. Despite their efforts, the firm failed to produce a suitable modern design in late 1913 and 1914—resulting in the Bavarian Army reluctantly inviting north German manufacturing firms to step in and provide assistance. Put simply, the lack of competition within Bavaria had placed its army qualitatively behind the Prussians by 1913. Fortunately, this failed to affect the Bavarian airmen's performance at the front once the war began. Many of the aircraft ultimately used by the Bavarian flying sections at the start of the war were Otto-built LVG B.Is. The others were the latest Otto pusher aircraft, which despite their flaws, were successfully used by Bavarian airmen during the war's opening three months.

Airships & Balloons

While the airplane had undoubtedly risen to become the *Fliegertruppe*'s most important platform, airships and balloons were still considered very important. Due to their low speed and vulnerability to ground fire, the army's Zeppelins were considered unfit for close tactical support. The army's airships were therefore charged instead with long-range strategic reconnaissance and bombing attacks. With a greater range than airplanes, the Zeppelin airships held the potential to be a useful means of gathering intelligence regarding the strength and location of enemy armies located deep behind the front lines. However, it was the opinion of most officers within the General Staff that the army's airships were unable to carry out long-range reconnaissance sorties in wartime. This was largely based on observations regarding the dirigibles' vulnerability as well as their poor performance record throughout the pre-war period between 1910–1913. For example, despite having two airships participate in the 1913 Kaiser Maneuvers, their merits failed to garner a single mention in the highly detailed post-exercise report.[97] Thus, the airship's potential as a producer of strategic intelligence was widely regarded by the summer of 1914 to be a fantasy.

On the other hand, the airships' value as a bomber was considered far more practicable and realistic. Using the cover of darkness to mitigate their vulnerability to anti-aircraft fire, the army's Zeppelins were expected to take advantage of their payload capacity and attack important enemy targets close to the front lines. That way, by not flying too far hinterland into enemy territory, an airship could make a successful landing on friendly soil in the event of an emergency—which the General Staff believed could be a distinct possibility. It should be noted, however, that the practicality of using dirigibles offensively was also widely questioned. Members of the General Staff and army officers alike were publically skeptical of the airship's offensive potential, pointing to the

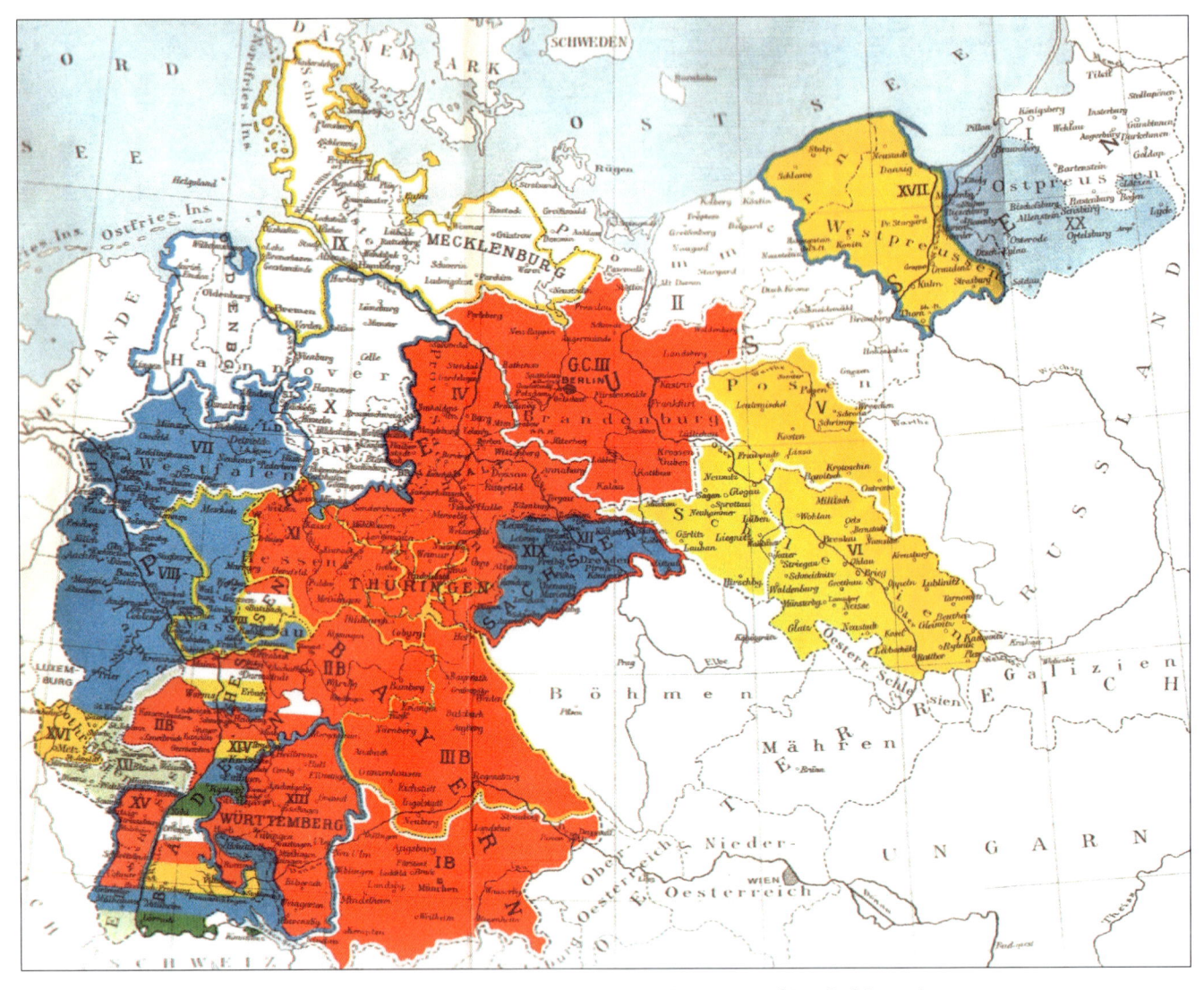

Above: Map of Germany Showing the Various Army Korps Districts (Courtesy of Joe Robinson)

existing crafts' vulnerability and unreliability as well as insufficient payload capacity.[98] Indeed, the issue was considered so unimportant that bombing experiments prior to the war were only conducted by one Zeppelin.[99]

Falkenhayn's major reorganization of aviation forces in October 1913 also shook-up the army's airship units. Just as the reorganization had created an aviation inspectorate for airplanes (*Idflieg*), it had also established a new airship inspectorate: The *Inspektion der Luftschiffertruppen* (*Ilust*). Moreover, an additional two airship battalions were created to facilitate the expansion of the airship troops.[100] With this objective in mind, the army acquired nine Zeppelins, two Parsevals and one Schütte-Lanz dirigible from 1913 to the outbreak of war. However, only six Zeppelins and the Schütte-Lanz were available for active service in August 1914.[101] Due to their vulnerability and lack of mobility, these seven airships were to be placed under the direct control of the Chief of the General Staff at Supreme Headquarters in the event of mobilization. It was widely hoped that in spite of the General Staff's skepticism, the airships would be able to positively contribute to the war effort. The truth was, however, that the dirigible units mobilized for war woefully unprepared and their military value was considered to be trivial by the majority of the army leadership.[102]

The army's captive balloon units were largely ignored in the years leading up to war. In fact, the balloon sections mobilized in 1914 using the same service manual that had been written for the army's balloon detachments in 1903. According to the manual the balloon sections were charged with alerting local ground troops of the position and deployment of the enemy during an engagement.[103] The balloon's lack of mobility and low ceiling, however, greatly diminished their potential to participate under the conditions of 'open warfare.' Thus, the balloon sections were destined to be largely forgotten and ignored until the onset of trench warfare during the winter of 1914–15.

Organization & Training of the Prewar *Fliegertruppe*

The *Fliegertruppe's* proposed role in support of the army continually evolved in the years leading up the outbreak of war. While long-range reconnaissance had originally been the airplane's sole purpose, the Prussian General Staff had begun actively campaigning in 1913 for a major expansion of the flying sections' duties. Significant advances in airplane design and performance, combined with the substantial growth of the German aircraft industry's production capabilities, had prompted the General Staff to plan for the use of the army's flying sections both tactically and operationally. Indeed, tactical missions such as artillery cooperation flights, short-range reconnaissance sorties and bombing of enemy ground forces were all possible by late 1913 and were proposed to be implemented as doctrine as quickly as possible by the forward thinking General Staff.

The October 1913 expansion and reorganization of the flying service was the first step in turning the General Staff's plans into a reality. For the first time in its history, the *Fliegertruppe* were organized in peacetime into organizational units that would be used in the field in the event of war. The organizational model was simple. Four Prussian battalions, each consisting of three companies, were stationed throughout the empire while the Bavarian battalion exclusively supported its own army in Bavaria. Each aviation company (*Flieger Kompanie*) was given the necessary personnel and equipment to form and maintain two or three independent flying sections.[104] With one exception, each company had its own aerodrome (*Fliegerstation*).[105] These *Fliegerstations*, equipped with the necessary number of mechanics and workshops to keep the company's aircraft airworthy, were then placed within different army corps districts across Germany—usually in the same town as the corps' headquarters. Thanks to their location near corps headquarters, the aviation companies and their subordinate flying sections were able to train and establish a working relationship with the army's ground forces. Thus, by gaining some familiarity with the local ground units, the men of the *Fliegertruppe* could begin building the foundation for a successful tactical support system.

Moltke's famous March 1913 organizational proposal serves as further evidence as to how the General Staff envisioned airplanes to be used in wartime.[106] In addition to the aforementioned call for the creation of 30 Field Flying Sections to be attached to each active corps and army headquarters before April 1914, Moltke's amended proposal recommended an additional 23 be established by April 1916. Furthermore, the General Staff recommended that there be a sufficient number of aircraft and trained personnel available by 1916 to create 46 artillery aviation sections to be attached to all active division commands immediately upon the outbreak of war. This revolutionary proposal exhibited the General Staff's desire (well before any of the other European army commands) to have specialized aviation units created to provide tactical support to the army as far down as division level.

Additional insight into the General Staff's intentions related to the employment of the *Fliegertruppe* can be found in the groundbreaking memorandum: "General Staff's Directive for "Use Of Aircraft With Ground Forces."[107] Although Germany's airmen deployed for war without an official doctrine, the guidelines and recommendations presented in the memo were widely followed by the majority of flying sections in 1914—making it arguably the most important document related to the German Air Service's actions in the first year of the war. Among other issues the document discussed: flying section organization, deployment of airfields, planning and executing reconnaissance flights, and communicating with ground troops during battle. The paper also defined the *Fliegertruppe's* role by naming the various missions that they could be assigned. These missions were: reconnaissance, messaging/communication, artillery cooperation, and aerial bombing.

Unfortunately, the General Staff's vision for close cooperation between aviation units and ground forces was an impossibility under the command structure established by the October 1913 reorganization. While Falkenhayn's War Ministry had smartly allowed for the creation of *Flieger* battalions and an expansion of the aviation budget, they had foolishly refused to organize them into an independent air force as recommended by the General Staff. The argument for an independent air force was rooted in the belief that, as a combat arm, the flying service needed to be separated from the supply and transport troops and given their own command structure. The War Ministry, in yet another effort to maintain power and control, decided to disregard the General Staff's proposals and keep the airplane battalions under the command of *Idflieg*, which due to its subordination to the General Inspectorate of Military Transportation had no powers to control units at the front in wartime. In other words, in the event of war there would be no officers at the front responsible for coordinating the employment and supply of the air service. Instead, each flying section would be responsible for themselves.

Beobachtungsstand.

Above: The natural limitations of the observation stand prompted many forward-thinking officers to begin planning the use of aircraft in a tactical role. (Author's Collection)

In a memorandum desperately written to persuade the War Ministry to change their philosophy, *Idflieg* argued the following:

Within the Aviation Inspectorate there is complete agreement that the previous organization of the aviation forces is insufficient for both peacetime training, as well as for mobilization.

The deficiencies that currently exist are as follows:

1. The close link of aviation forces to the transport troops, as mandated by the General Inspectorate for Military Transportation must be separated.

Justification: The **Fliegertruppe** *is a purely operational fighting force, while the object of the transport troops is to transport supplies and materials*

The subordination of the aviation forces under the transport troops is the wrong decision. The officers [of the **Fliegertruppe***] would benefit from being a part of the combat forces: such as Infantry, Cavalry, Artillery while the transport troops should be with the engineers, railway, telegraph, vehicle troops.*

2. The lack of independence given to the Aviation Inspectorate.

Justification: Due to intermediaries such as the Ministry's Military Aviation and automobile Inspectorate, there are many delays in the course of the Inspectorate's regular business. Especially during operations, the Fliegertruppe's needs the ability to act rapidly.

3. The Erroneous design and composition of the **Fliegertruppe***.*

Justification: At the head of the **Fliegertruppe***, there must not be only an "Inspectorate" but a command. This command should have the authority and the influence of what a current brigade commander enjoys and given the coming growth of the* **Fliegertruppe***—a Division commander. This commander and his staff must in peacetime as well as wartime be in charge of the technical and industrial growth of aviation matters. The expression—Flieger Battalion, Flieger company cannot just be coincidence.*[108]

Idflieg's arguments were simple. By keeping the growing aviation service under the thumb of the General Transportation Inspectorate, the War Ministry was seriously undermining the *Fliegertruppe's* ability to support the army in wartime. Sadly, the War Ministry refused to make any changes to the flawed organizational model—a decision which would later have major repercussions on the outcome of the war's first campaign.

Yet another part of the General Staff's plan for the buildup of the Air Service was to expand the number of advanced training centers for recently licensed airmen. Initially, specialized training courses were only offered to pilots. Observers were expected to simply rely on their previous military training and apply it as best they could while conducting aerial reconnaissance. However, by the fall of 1911, the army had realized this was a mistake and had started training observers alongside the pilots in a combined military aviation course. Shortly thereafter it was determined that the unique responsibilities assigned to pilots and observers required the two to be trained separately.[109] Thus, specialized observer schools were established in several locations throughout Germany. The subsequent training of aerial observers was administered according to a well-defined curriculum created by members of the General Staff and *Idflieg*. The initial phase was done in a classroom, teaching essential subjects such as: the basics of flight, aerial navigation, estimating the size and composition of ground forces, and the writing of succinct and detailed messages for army leadership during battle. Next, the student would apply what they had learned in the classroom during test flights performed over the local army corps' training area. Advanced subjects such as artillery cooperation, fire direction, aerial bombing and photography were also discussed and simulated.[110] Sadly, while the course curriculum was strictly codified and closely followed by members of the *Fliegertruppe*, it was not incorporated or even mentioned in current army regulations. Consequently, the opportunity for ground troops to practice tactical cooperation with their flying section or at a minimum gain an understanding of how the *Fliegertruppe* operated was never properly taken advantage of.

Nearly all the courses for observers were based on the *Fliegertruppe* supporting the ground forces during a war of maneuver. This coincided with the General Staff's view that victory in the coming war would need to be achieved quickly in open warfare before the front became static. The young observers therefore spent the majority of their time learning to identify enemy troops in column as well as practicing writing tactical reports for large battles in the open. Reconnaissance of field fortifications and trench systems were rarely discussed due to the army's overall strategy of preparing for a war of maneuver.

A number of experimental programs for observers were established in the spring and summer of 1914 as part of the General Staff's plan to expand the *Fliegertruppe's* tactical role. The most important of these were the artillery cooperation courses held at artillery firing schools such as Jüterbog, Wahn, and Thorn (East Prussia). These courses were established in order to test and determine the best method of communicating with artillery commanders from the air. *Ltn.* Andre Hug, who served as a pilot at Wahn and Thorn throughout 1913 and the spring of 1914, explained his experiences at the school.

Flight activity was poor during the wintertime of 1913–14. The aircraft on all bases had to be conserved and only flights around the base were allowed. The supply of new planes increased so that, in the spring of 1914, an end was put to the restrictions. I was then transferred to a half-squadron assigned to the gunnery range in Thorn, East Prussia, near the Russian border of that time. The half-squadron was composed of three Albatros with 100 h.p. Argus engines, and with three military pilots and three observers to assist the heavy artillery at the Thorn artillery range.

We were used as observation planes for the firing of the heavy artillery. Before the artillery would fire we had to fly over the target and report what observations we had made about the target. We also had to observe how the shells came down—whether over or under the targets—and to give the commander a general sketch of the effect of his artillery fire.

Information from our artillery-spotting aircraft, concerning the position of the enemy and the impact of the shells, was dropped by a little sandbag attached to a streamer of about ten yards long. The shooting exercise normally lasted for two hours during the early morning. For the rest of the day we were free and flew to the 120 mile distant beaches of the Baltic Sea. Or, we did a simulated emergency landing to pay a visit to one or another of the very promising big country estates.[111]

Although the experiment was shut down due to the outbreak of war before finding a definitive answer, the participating observer officers gained valuable experience working with the artillery, testing different theories and establishing ad-hoc communication methods for fire direction missions.[112] These officers would later teach what they learned to members of their unit during the war's opening weeks, thereby

Above: Crash of DFW military *Taube* A.183/13. Early national insignia are carried. This aircraft was assigned to *Flieger Bataillon* 1 in March 1913. (Peter M. Grosz Collection/SDTB via Aeronaut)

enabling many flying sections to regularly conduct successful artillery direction missions throughout the war's first year.

Meanwhile, another highly specialized course was established at Döberitz for testing the efficiency and practicality of aerial bombing. For weeks, observers tested different methods by attacking mock targets placed in fields around the aerodrome. Experimental bombsights and bomb racks were also tested during this time with mixed results. Even the creation of a specialized bomber, designed specifically to carry heavy payloads, was considered and discussed with several manufacturing firms.[113] Unfortunately, just like the artillery school in Jüterbog, the outbreak of war shut down the bombing course before anything significant could be accomplished.

Despite the growth of the air service and improvements in training, the *Fliegertruppe's* personnel faced serious problems during the final months before mobilization. Perhaps the most important was the army leadership's lingering prejudice toward the air service. Outside of a few far-sighted officers, the majority of senior commanders had little faith in the *Fliegertruppe's* ability to impact events during a real conflict. In almost every case, these officers' bias was due to a total unfamiliarity with airplanes and modern aviation. Very few of them understood the recent technological advances in aircraft production and the improved performance capabilities of modern aircraft. Furthermore, they were completely unaware of the tactics and modern doctrine being taught inside the various army aviation schools.[114] Noted German aviation historian John Cuneo later wrote that the leaderships' lack of interest regarding important documents like the *General Staff's Directive for "Use of Aircraft with Ground Forces"* was "notorious."[115] In other words, when these older officers thought of military aviation, one can reasonably assume that they envisioned something closer to what Orville Wright showcased in 1909 rather than what the *Flieger Battalions* were actually using during the summer of 1914.

The unfamiliarly and mystery associated with the air service was a product of the lack of combined arms training that occurred prior to the war. Outside of the flights performed during their initial training, aircrews rarely got the opportunity to practice combined operations with local ground troops. Whenever a *Flieger Kompanie* managed to participate with ground forces, it was most often during a small-scale exercise where the opposing strengths were no greater than a single brigade and often as small as a regiment.[116] While it is true that the Kaiser Maneuvers offered an excellent opportunity to see multiple corps on the field at once, the reality was that very few airmen were ever afforded the luxury of participating in the event. Most aircrews were instead compelled to take part in individual corps exercises held by each of the corps that *were not participating* in that year's maneuvers. These events were most often the largest and most realistic simulation that German airmen were to experience prior to war.

Unfortunately, the corps exercises were often set up by the corps commander to minimize or even exclude the airmen from participating altogether. When this wasn't the case, the maneuvers were established with a lack of oversight or understanding of how aviation units were to operate. This resulted in various flying sections often failing to receive orders or instructions regarding what the ground troops' commanders required of them. As a result, many flying sections were compelled to conduct flights throughout the exercise without any cooperation with the ground troops' leaders. As Cuneo argued: "These tricks did not contribute towards producing better observers."[117]

In fact, by wasting numerous valuable opportunities for the *Fliegertruppe* to gain experience and form a better relationship with ground troops, these exercises did the army serious harm. Common sense dictated that the advanced training of pilots and observers could only be obtained through experience. Thus, with the army and the flying service training to wage a quick and decisive war under the conditions of open warfare, it was imperative that the army's aviators get an opportunity to realistically simulate what they would encounter in wartime. This specifically meant monitoring and reporting enemy troop movements prior to battle as well as alerting army/corps leaders to any important details that could be seen once an engagement had started. To that end, observers would be required to report the location, route of march, strength, and composition of all enemy columns that were observed during a patrol. Furthermore, they would be expected to accurately document the enemy's deployment once the ground troops became engaged in battle. However, thanks to the lack of combined arms training opportunities, most German airmen never had the chance to experience what a corps sized force in column or in battle really looked like from the air. Instead, the majority of observers' pre-war experiences were limited to unrealistic sorties in support of an isolated brigade or regiment that was conducting the exercise at half strength with little to no artillery or support vehicles present.[118]

In what became an unfortunate self-fulfilling prophesy, the army's substandard combined arms training, which was partly caused by the senior leaderships' anti-aviation bias, directly perpetuated a systematic distrust in the *Fliegertruppe*. It is important to note that with the exception of the Kaiser Maneuvers, none of the large exercises used their aviation assets realistically. Corps maneuvers frequently saw troop leaders making impossible requests of their flying section and then questioned the *Fliegertruppe's* value after the desired intelligence failed to materialize. In other cases, officers neglected to even issue orders to their airmen, leaving the flying section to perform fruitless, self-directed ad-hoc reconnaissance flights without any prior knowledge of the opposing team's positions. As a result, many officers went to war in 1914 extremely prejudiced against their airmen, believing that their capabilities were limited to how they performed during the flawed corps exercises.

The young flying service's poor reputation had not gone unnoticed by its leaders. In an effort to quell any concerns, 50 of some of the army's brightest officers were sent from the General Staff College to the army aviation school at Döberitz for a shortened aerial observation course.[119] It was hoped that this move, along with the publication of several memorandum describing the flying section's duties and capabilities, would boost the *Fliegertruppe's* reputation and educate the army's brightest officers of the flying service's true abilities.[120] Events during the war's first weeks quickly proved, however, that these measures had little impact on changing these officers' unqualified opinion.

Overview: The General Situation on the Eve of War

The German Army's aviation program had made great strides since the creation of the first airplane unit at Döberitz in 1910. As late as 1911, German manufacturers had been compelled to produce badly copied French airplanes due to the inferiority of their own designs. However, thanks largely to the 7.2 million Marks in donations brought in through the National Aviation Fund, Germany's domestic firms began to close the gap in late 1912. Soon thereafter, the army aviation budget was raised after the airplane's military value was once again proven during the 1912 Kaiser Maneuvers. The number of aircraft purchased by the Prussian Army rose accordingly from 139 aircraft in 1912 to 461 higher quality machines such as the Albatros and Aviatik biplanes the following year.[121] By the outbreak of war the goal of catching the French had been achieved, with the German aircraft producing some of the world's finest flying machines.

Credit for the modernization of Germany's aviation forces should undoubtedly be given to the Prussian General Staff. From the formation of the army's first airplane research unit in 1908, the General Staff had publicly called for the "simultaneous and wholehearted promotion of airships and airplanes."[122] Over the next five years the General Staff waged a relentless campaign to fund and expand the flying service. With a majority of the army holding a differing opinion, Moltke's General Staff kept the important debate over the airplane's military value alive at a time when their opponents attempted to minimize or ignore the issue. From 1909–1912, these

Above: DFW *Mars* monoplane powered by a six-cylinder engine. The four-wheel undercarriage was a characteristic of the *Mars* monoplanes and biplanes. The *Mars* monoplane may have been used at the front in the war's early weeks. (Aeronaut)

actions continually forced a reluctant War Ministry into creating new organizational departments and allocating additional funds to Germany's fledgling aviation industry. Without these increased funds, many domestic manufacturers would have been forced to close, thus allowing the French superiority to grow. Indeed, without the creation of new organizations such as the Flight Command Döberitz, the training of personnel and overall growth of the army's young flying service would have remained stunted in its infancy.

Fortunately, the War Ministry's position finally softened in the spring of 1912 when, in response to the outbreak of the Balkan War, it decided to increase the aviation budget and seek better relations with domestic manufacturers. With added pressure from the General Staff, the War Ministry substantially raised the annual appropriation for aviation in 1913 and again in 1914. Using these funds, aviation firms across Germany expanded their facilities, hired new workers and initiated mass production of aircraft, engines, and other important products. By the outbreak of war, 118 million Marks had been spent on training and equipment in successfully making Germany's Air Service comparable in size and quality to its enemies.[123] The army was therefore able to mobilize for war in August 1914 with 450 aircraft, of which 295-320 were available for active service at the front.[124]

As the Chief of the General Staff, Helmuth von Moltke deserves the lion's share of the recognition for the advanced state of German aviation at the start of the war. Despite strong opposition, Moltke remained personally committed to championing the airplane and the growth of the army's flying service throughout his tenure during the pre-war period. Along with Ludendorff, Moltke penned a number of important memorandums that forced the War Ministry into reevaluating their position on all matters related to aviation. The most important of these were Moltke's farsighted proposals regarding the organization of the air service and the training of its officers. Based largely upon Moltke's recommendations, the *Fliegertruppe* were organized in 1913 into aviation battalions and companies that would allow for flying sections to be promptly attached to each of the various army and active corps commands upon mobilization. Moltke's later memos influenced doctrine and how the army's aviation units would be used in wartime. Indeed, whether advocating tactical reconnaissance flights to assist corps and division commanders during battle or artillery cooperation missions, Moltke's recommendations consistently forced the army and its aviators into accepting new ideas that would later pay dividends on the battlefield. These efforts, often overlooked by historians, greatly improved the quality of the *Fliegertruppe* personnel, which was the prime difference between Germany's fully functioning air service and the inept Russian Air Force that mobilized for war with a similar number of available aircraft.[125]

There nevertheless remained some serious flaws in the *Fliegertruppe*'s wartime organization. Despite repeated pleas from high ranking members the General Staff, the War Ministry had neglected to create an independent organization for the air service. This force would have had a command staff present at General Staff Headquarters with the authority, among other matters, to establish doctrine for all flying sections as well as improve liaison between its airmen and the army's leadership. Instead, under the existing system, *Idflieg* was to remain in control of the army's military aviation program from its offices attached to the General Inspectorate of Transportation in Berlin. As a subordinate organization to the General Inspectorate, *Idflieg* had no ability to direct or influence the actions of the flying sections at the front. This left the individual flying sections completely under the control of their respective army or corps commanders—many of whom mobilized for war believing that airplanes would prove to be useless under the stresses and realities of modern warfare.

And so, with the quality of its officers and airplanes as their greatest strength and with organizational deficiencies as their biggest weakness, the German Air Service mobilized for war eager to support the army and validate their worth on the battlefield while many of the army's senior commanders continued to have their doubts. In less than one short month the question of the *Fliegertruppe's* true military value would be settled once and for all.

Chapter 1 Endnotes

1 Peter Supf, *Das Buch der deutschen Fluggeschichte*, Vol. 2 (Berlin, 1935), p.102.
2 Richard von Kehler, "Frei- und Fesselballon im Dienste der Kriegführung vor dem Weltkrieg," *Unsere Luftstreitkräfte*, Edited by Walter von Eberhardt (Berlin, 1930), pp.27–38.
3 John R. Cuneo, *Winged Mars: The German Air Weapon 1870–1914* (Harrisburg, 1942), pp.9–18.
4 Two years earlier, the American Wright brothers offered to sell one of their airplanes to the War Ministry but were rejected in favor of fully concentrating on airship construction and production.
5 Eric Dorn Brose, *The Kaiser's Army: The Politics of Military Technology in Germany during the Machine Age 1870–1914* (New York, 2001), p.6.
6 According to the German Army Handbook, the Prussian General Staff was responsible for the "preparation of plans of campaign and mobilization and for the movements and methods of employing the army in war." Meanwhile the War Ministry was charged with controlling military expenditures and acting as the head of the military's administrative branches. In this purely administrative role, the War Ministry was to carry out "a continuous financial and military policy." Although the two departments were considered to be equally as powerful as one another, the War Ministry wielded more influential in peacetime because they were responsible for the budget. John H. Morrow Jr., *Building German Air Power 1909–1914* (Knoxville, 1976), p. 6; *Handbook of the German Army 1914* (Nashville, 2002), p.74, p.79.
7 De le Roi's assignment was made on the recommendation of Hermann von der Lieth-Thomsen, an aviation enthusiast who was serving at the time in the War Ministry's Engineering & Fortress Department. This was just the first of many positive actions taken by Thomsen, who would later play a major role in the growth of the Air Service during the war. Note: De le Roi's research unit was organized under the War Ministry's Engineering Department. Morrow, *Building German Air Power*, pp.14–16.
8 This is an enormously important fact that should be remembered while studying German aviation during the pre-war period. The Transportation Inspectorate (later the General Inspectorate of Military Transportation) remained organizationally above the flying troops through the outbreak of war. Morrow, *Building German Air Power*, pp.15–16.
9 Morrow, *Building German Air Power*, p.15.
10 Kriegswissenschaftliche Abteilung der Luftwaffe, *Die Militarluftfahrt bis zum Beginn des Weltkrieges 1914*, 3 Bände (Freiburg im Breisgau, 1965), Band II, pp.109–113.
11 Morrow, *Building German Air Power*, p.17.
12 Morrow, *Building German Air Power*, p.19.
13 Morrow, *Building German Air Power*, p.19.
14 Cuneo, *Winged Mars*, p.89.
15 Alfred Mahnke, "25 Jahre Deutsche Luftwaffe: Zum Ersten Einsatz deutscher Heeresflugzuege im Kaisermanöver am 11. September 1911," *Der Deutsche Sport Flieger*, Okt. 1936, Heft. 10, p.12.
16 Morrow, *Building German Air Power*, p.23.
17 Dr. Richard Stimson, *Orville Flies in Germany*, Retrieved from http://wrightstories.com/orville-flies-in-germany/
18 Cuneo, *Winged Mars*, pp.86–87.
19 Morrow, *Building German Air Power*, p.23.
20 Adolf-Victor von Koerber, "Heute vor 25 Jahren!" *Luftwelt*, 2 Jahrig. 1935, #4, p.144.
21 The acronym GmbH stands for *Gesellschaft mit beschraenkter Haftung*, which translates to 'Limited Liability Company'. Morrow, *Building German Air Power*, pp.25–27.
22 Kriegswissenschaftliche Abteilung der Luftwaffe, *Die Militarluftfahrt bis zum Beginn des Weltkrieges 1914*, Band II, pp.118–125.
23 Morrow, *Building German Air Power*, p.27.
24 Morrow, *Building German Air Power*, p.27.
25 Kriegswissenschaftliche Abteilung der Luftwaffe, *Die Militarluftfahrt bis zum Beginn des Weltkrieges 1914*, Band I, p.121.
26 The four pilots who originally made up 'Flight Command Döberitz' were: *Hptm.* Wolfram de la Roi, *Oblt.* Geerdtz, and *Ltns.* Walter Mackenthun

and Eugen von Tarnnoczy. Hans Jürgen von Giebenhain, "The Aviator School in Döberitz and its First Semester and Courses," *Cross and Cockade US*, Volume 16(3) p.265; Cuneo, *Winged Mars*, p.90.

27 Supf, *Das Buch der deutschen Fluggeschichte*, Vol. 1, pp.320–321.
28 Cuneo, *Winged Mars*, p.90.
29 Cuneo, *Winged Mars*, p.96.
30 Kriegswissenschaftliche Abteilung der Luftwaffe, *Die Militarluftfahrt bis zum Beginn des Weltkrieges 1914*, Band II, pp.116–118.
31 Morrow, *Building German Air Power*, pp 30–31.
32 Kriegswissenschaftliche Abteilung der Luftwaffe, *Die Militarluftfahrt bis zum Beginn des Weltkrieges 1914*, Band II, pp.125–135.
33 Since the French did not have patents in Germany, this was legal. However, without blueprints the German firms were forced to build the machines by purchasing a machine from France, flying it to the factory and copying the machine via reverse-engineering. Morrow, *Building German Air Power*, p.29.
34 The strategy of simply copying French designs would continuously keep the German Aviation Industry at least a full year behind. More importantly, with France being Germany's principle rival, continuing this strategy would have guaranteed French aerial dominance throughout the coming conflict—a serious problem for the army. Therefore, it was considered imperative that German industry gain independence from French designs and produce German engineered aircraft to the army as quickly as possible.
35 As a frame of reference, the entire German military budget was over 900 million Marks at this time.
36 Morrow, *Building German Air Power*, pp.41–43.
37 The French military had allocated roughly 10 million Francs for aviation in 1911 and stood poised to have 100 airplanes available for the army's use in the spring of 1912.
38 Morrow, *Building German Air Power*, p.44.
39 Morrow, *Building German Air Power*, p.44.
40 Morrow, *Building German Air Power*, pp.44–45.
41 "Kaisermanöver 1911: Erster Einsatz der deutschen Fliegertruppe," *Luftwelt* (Berlin, 1936), Bd.3, Nr.11, p.430.
42 Mahnke, "25 Jahre Deutsche Luftwaffe", *Der Deutsche Sport Flieger*, Okt. 1936, Heft. 10, p.13.
43 Giebenhain, "The Aviator School in Döberitz and its First Semester and Courses," *Cross and Cockade US*, Volume 16(3) pp.273–274.
44 "Kaisermanöver 1911: Erster Einsatz der deutschen Fliegertruppe," *Luftwelt*, Bd.3, Nr.11, p.432.
45 After an investigation, the ground crew was not held responsible. Nonetheless, the accident highlighted the greater difficulties involved with moving airships as opposed to airplanes. "Kaisermanöver 1911: Erster Einsatz der deutschen Fliegertruppe," *Luftwelt*, Band.3 N.11 p.433.
46 "25 Jahre Deutsche Luftwaffe", *Der Deutsche Sport Flieger*, Okt. 1936, Heft. 10, p.14.
47 "25 Jahre Deutsche Luftwaffe", *Der Deutsche Sport Flieger*, Okt. 1936, Heft. 10, p.14.
48 "Kaisermanöver 1911: Erster Einsatz der deutschen Fliegertruppe," *Luftwelt*, Bd.3, Nr.11, p.434.
49 Hilmer Frhr. von Bülow, *Geschichte Der Luftwaffe* (Frankfurt am Main, 1934), pp.11–12.
50 The German military calendar ran from April 1–March 31. All new budgets, deployment plans, command assignments, etc. started on April 1.
51 Naturally, the War Ministry feared losing power and ignored the Chief of the General Staffs' recommendations.
52 Morrow, *Building German Air Power*, p.48.
53 The creation of the new *Fliegertruppe Kommando* was reluctantly enacted by the War Ministry–first as a provisional unit in April and later as a permanent agency in October—in response to growing tensions in the Balkans during the 1912 Balkan Crisis/1st Balkan War. Morrow, *Building German Air Power*, pp.48–49.
54 Cuneo, *Winged Mars*, pp.99–101.
55 Morrow, *Building German Air Power*, p.57.
56 Cuneo, *Winged Mars*, p.99.
57 Cuneo, *Winged Mars*, p.99.
58 Morrow, *Building German Air Power*, pp.58–59.
59 Morrow, *Building German Air Power*, p 59.
60 Morrow, *Building German Air Power*, p.62.
61 Morrow, *Building German Air Power*, pp.70–71.
62 Erich von Ludendorff, *The General Staff and Its Problems*, 2 volumes (New York, 1920), vol. 1, pp.35–36.
63 It should be noted that Moltke's minimum request for 34 Flying Sections to be created by spring 1914 was essentially what the Germans went to war with in August 1914. The German Army mobilized for war with 33 Field Flying Sections (30 Prussian and 3 Bavarian) and 8 Airplane Parks. In addition to these were 10 Fortress Flying Sections.
64 Ludendorff, *The General Staff and Its Problems*, vol. 1, pp.37–43.
65 Ludendorff, *The General Staff and Its Problems*, vol. 1, pp.44–45.
66 The Bavarians maintained an autonomous air service, which were not included in these numbers yet would be available in the event of war. Hilmer Frhr. von Bülow, *Geschichte Der Luftwaffe*, p.15.
67 Cuneo, *Winged Mars*, p.105.
68 Hans Ritter, *Der Luftkrieg* (Berlin, 1926), pp.25–26.

Above: Two DFW *Mars* biplanes are shown at the DFW flying school. The one in front is named *BULLI*. Both are carrying an early version of the German national insignia, indicating the photo was taken during the war. (Peter M. Bowers Collection/ Museum of Flight via Aeronaut)

69 *Ed Ferko Collection*, History of Aviation Collection, Special Collections and Archives Division, Eugene McDermott Library, The University of Texas at Dallas [hereafter: Ferko–UTD], Box 20, Folder 5, "Notes on pre-war aviation."
70 Morrow, *Building German Air Power*, p.75.
71 Morrow, *Building German Air Power*, p.76.
72 Ludendorff, *The General Staff and Its Problems*, vol. 1, pp.49–51.
73 Ludendorff, *The General Staff and Its Problems*, vol. 1, pp.43–47.
74 Robert T. Foley, *German Strategy and the Path to Verdun* (New York, 2007), pp.88–90.
75 Wallace Notestein & Elmer E. Stoll, *Conquest & Kultur: Aims of The Germans In Their Own Words* (Washington D.C., 1918), p.125.
76 Remaining fully independent, the Bavarian battalion reported to their own 'Bavarian Inspectorate of Flying Troops'. Cuneo, *Winged Mars*, pp.106–107.
77 Morrow, *Building German Air Power*, pp.83–84.
78 Morrow, *Building German Air Power*, p.84.
79 These records were later re-broken by other German aircraft in the months leading up to war.
80 Morrow, *Building German Air Power*, p.86; Cuneo, *Winged Mars*, pp.107–108.
81 Cuneo, *Winged Mars*, p.108.
82 'T'. "Unsere Fliegertruppen im Kaisermanöver," *Deutsche Zeitschrift für Luftschiffahrt* (Berlin, 1913) XVII, p.483; Supf, *Das Buch der deutschen Fluggeschichte*, Vol. 1, pp.112–113.
83 Morrow, *Building German Air Power*, p.79.
84 Morrow, *Building German Air Power*, p.85.
85 Hartmann was a *Legion d'honneur* recipient from his service in Napoleon's Army in 1814. He is therefore one of only a few officers to have been awarded the French Empire's *Legion d'honneur* and the Kingdom of Prussia's *Pour le Merite*.
86 Quintin Barry, *The Franco-Prussian War 1870–1871*, 2 volumes (West Midlands, 2009), vol. 1, pp.76–90. George Hooper, *The Campaign of Sedan* (London, 1914), pp.80–84.
87 Joe and Janet Robinson, *Handbook of Imperial Germany* (Bloomington, 2009), pp.68–71.
88 See: Harald Potempa *Die Königlich–Bayerische Fliegertruppe 1914–18*, (Frankfurt am Main, 1997).
89 Morrow, *Building German Air Power*, pp.93–94.
90 Morrow, *Building German Air Power*, p.95.
91 P M Grosz, *The LVG B.I* (Hertfordshire, 2003), pp.4–5.
92 Otto's successor, the Bavarian Aircraft Works, continued to support the army during the war as an aviation engine manufacturer. In the spring of 1916 the Bavarian Aircraft Works merged with another company and was renamed the *'Bayerische Motoren Werke'*, commonly known today as BMW. BMW's engines, especially in the final year of the war, would go on to become arguably the finest produced by either side throughout the conflict.
93 Morrow, *Building German Air Power*, p.101.
94 Jack Herris, *Pfalz Aircraft of WWI* (Reno, 2012), p.16.
95 Cuneo, *Winged Mars*, p.107.
96 Morrow, *Building German Air Power*, p.95.
97 Cuneo, *Winged Mars*, p.107.
98 Charles Bertrand, État Actuel de l'Aéronautique

Above: Early Rumpler *Taube* with very prominent national insignia. Soldiers were likely to fire on any aircraft and aircrew of both sides frequently experienced "friendly fire'" on that account. Airmen know that "friendly fire" isn't friendly. (Aeronaut)

    Militaire & Navale en France *et à l'Étranger* (Paris, 1913), pp.26–27.
99 Georg P. Neumann, *Die Deutschen Luftstreitkrafte im Weltkriege* (Berlin, 1920), p.341; Cuneo, *Winged Mars*, p.120.
100 Alex Imrie, *Pictorial History of The German Army Air Service* (Chicago, 1971), p.18.
101 Cuneo, *Winged Mars*, p.120.
102 The army dirigibles' best contributions occurred on the Eastern Front where the ships' vulnerabilities were not so easily exploited. Neumann, *Die Deutschen Luftstreitkrafte im Weltkriege*, p.341.
103 Friedrich Stahl, "Entwicklung und Verwendung der Luftschiffe und Fesselballone im Dienste des Feldheeres," *Unsere Luftstreitkräfte*, Edited by Walter von Eberhardt (Berlin, 1930), p.73.
104 Alex Imrie, *Pictorial History of The German Army Air Service*, p.19.
105 The lone exception was *Flieger Battalion #1's* 1st and 2nd Companies, who shared the same *Fliegerstation* at Döberitz. Ferko–UTD, Box 20, Folder 5, "Pre-war *Fliegertruppe* Organizational Papers."
106 A translation of this critically important memo can be found in the Appendices section as Appendix #78 at: Kriegswissenschaftliche Abteilung der Luftwaffe, *Die Militarluftfahrt bis zum Beginn des Weltkrieges 1914*, Band III, pp.180–182.
107 The critical document, "Denkschrift der Inspektion der Fliegertruppen," was written and circulated in February 1914. It can be found later in this book in Chapter 2. Kriegswissenschaftliche Abteilung der Luftwaffe, *Die Militarluftfahrt bis zum Beginn des Weltkrieges 1914*, Band III, pp.188–189.
108 Kriegswissenschaftliche Abteilung der Luftwaffe, *Die Militarluftfahrt bis zum Beginn des Weltkrieges 1914*, Band III, pp.184–185.
109 Hilmer Frhr. von Bülow, *Geschichte Der Luftwaffe*, p.20.
110 Hilmer Frhr. von Bülow, *Geschichte Der Luftwaffe*, p.20.
111 Andre Hug, "Carrier Pigeon Flieger," *Cross and Cockade Journal*, Volume 13(4), p.296.
112 Imrie, *Pictorial History of The German Army Air Service*, pp.20–21.
113 After observing several of the experiments, *Idflieg* and the airplane manufacturers decided to set aside the idea of creating specialized aircraft in favor of concentrating on producing larger numbers of the "standardized" B-Type biplanes and Taube monoplanes. Hilmer Frhr. von Bülow, *Geschichte Der Luftwaffe*, p.18.
114 An excellent account of an aviator's experience at a pre-war aviation school can be found in: Giebenhain, "The Aviator School in Döberitz and its First Semester and Courses," *Cross and Cockade US*, Volume 16(3) pp.263–276.
115 Cuneo, *Winged Mars*, p.118.
116 A concise description of these training exercises can be found in: Terence Zuber, *Ardennes 1914*, The History Press. (Gloucestershire, 2009) pp.68–73; According to Zuber, the best in-depth description of these events can be found in: Johannes Liebach, *Bataillons- und Brigade-Uebungen und Besichtigungten der Infanterie in praktischen Beispielen* (Berlin, 1914).
117 Cuneo, *Winged Mars*, p.129.
118 Elard von Loewenstern, *Die Fliegersichterkundung im Weltkriege* (Berlin, 1937), pp.7–8; Cuneo, *Winged Mars*, p.129.
119 The first aerial observation course at the War

Above: Gotha LE3 *Taube*, the best and longest-lived of the *Taube* types. (Aeronaut)

Right: A pre-war Aviatik P13 B-type biplane, the first of Aviatik's successful reconnaissance two-seaters. By 1912 Austria-Hungary had recognized the *Taube's* limitations and before the war Germany did also. German companies started to build biplanes before the war and these flew the majority of reconnaissance missions as the few *Taubes* soon disappeared from the front. (Aeronaut)

Academy began on February 1st 1914. Imrie, *Pictorial History of The German Army Air Service*, p.20.
120 Cuneo, *Winged Mars*, p.129.
121 Morrow, *Building German Air Power*, p. 85; Kriegswissenschaftliche Abteilung der Luftwaffe, *Die Militarluftfahrt bis zum Beginn des Weltkrieges 1914* Band I, p.186.
122 Morrow, *Building German Air Power*, p.85.
123 Imrie, *Pictorial History of The German Army Air Service*, p.21.
124 Morrow, *Building German Air Power*, p.87.
125 The Russians had a front line strength of 244 airplanes and 14 dirigibles at the start of the war. Unlike the Germans however, the Russian Army lacked a "champion of military aviation" prior to the war to push for increased funding and higher training standards. In fact, the Russian War Department refused to adopt any policy toward aircraft production and the growth of the army's air service. Consequently, many of their aircraft were outdated and spare parts were almost non-existent. Moreover, without any forward thinking leader to call for an institution of modern training practices, the quality of their personnel was also entirely inadequate. John H. Morrow, Jr. *The Great War in The Air* (Washington D.C., 1993), pp.47–48 and pp.81–83.

2
Battle of the Frontiers
August 1914

"The brilliant results of the reconnaissance missions performed by our aviators during the first weeks of the war caused a complete change in the valuation of the new arm.... Thanks to its excellent performance it enjoyed an esteem which raised it above the subordinate role to which it had been assigned in peace and made it the peer of the principal combat arms"

General Ernest von Hoeppner[1]
Commanding General of the *Luftstreitkräfte* 1916–18

Germany's war officially began on August 1st when Kaiser Wilhelm II signed the order for general mobilization. Over the course of the following two weeks, the Empire's armed forces deployed according to the deployment plan: *Aufmarsch 1914–15* (hereafter referred to as the German War Plan). The plan was the product of over 20 years of work conducted by the Prussian General Staff to solve the Reich's predicament of facing a prolonged two front war against both France and Russia.[2] According to the well-developed plan, seven field armies were to initially deploy for an offensive in the West while a single army remained on the defensive in East Prussia. Together, the seven western armies were to wage an aggressive campaign that would quickly and decisively defeat the French and British armies in the field, thus knocking them out of the war. Then, once Germany's western enemies had capitulated, the bulk of the army would be transported to the east to face Russia.

Once fully concentrated, the seven western armies were to be deployed along the empire's frontier on the line Krefeld—Aachen—Luxembourg—Metz—Swiss Border. Altogether, 1.6 million men were organized into 23 active corps, 11 reserve corps, 10 cavalry divisions, and 17½ *Landwehr* brigades.[3] In addition to establishing each army's initial area of concentration, the deployment plan contained special instructions (*Besondere Weisungen*), which detailed each army's strategic mission in the coming campaign by describing the General Staff's concept of operations and prescribing each army's initial movements accordingly. In consonance with these instructions, the seven western armies were configured into three unofficial groups—each with a uniquely important role to play.[4]

With a combined strength of 10 active, six reserve and two cavalry corps, the German right wing (First, Second & Third Armies) was entrusted with delivering the opening campaign's main attack. According to their instructions, the right wing's three armies were ordered to 'wheel' through Belgium in order to envelop and defeat the allied left in a meeting engagement near the Franco-Belgian border. Upon this engagement's conclusion, the right wing was to roll up and destroy the remaining allied forces in concert with the efforts of the other western front armies. Success in the opening campaign, therefore, ultimately depended upon the successful prosecution of the right wing's initial attack. Thus, the two other groups' plans of operations were designed to facilitate the success of the right wing.

The center group (Fourth and Fifth Armies) was ordered to concentrate on the line Trier—Luxembourg—Thionville—Saarbrucken and prepare for two eventualities based on the actions of the French. In the event of a major French attack in Lorraine, the Fifth Army was to launch a counterattack through Metz against the enemy's exposed left flank while Fourth Army protected its rear.[5] However, if the French main attack did not occur in Lorraine, the center armies were to move west toward the Meuse River in support of the right wing's attack while maintaining contact with the strategically important fortifications at Metz.[6]

To the south, the German left wing (Sixth and Seventh Armies) was given the initial task of defending Lorraine and the left flank of the German line. Like the center group, the left wing's armies were to be prepared for two scenarios based upon the actions of the French. If the French launched their main offensive into Lorraine, the German left was to strategically withdraw to strong defensive positions established on the Nied River where the advancing

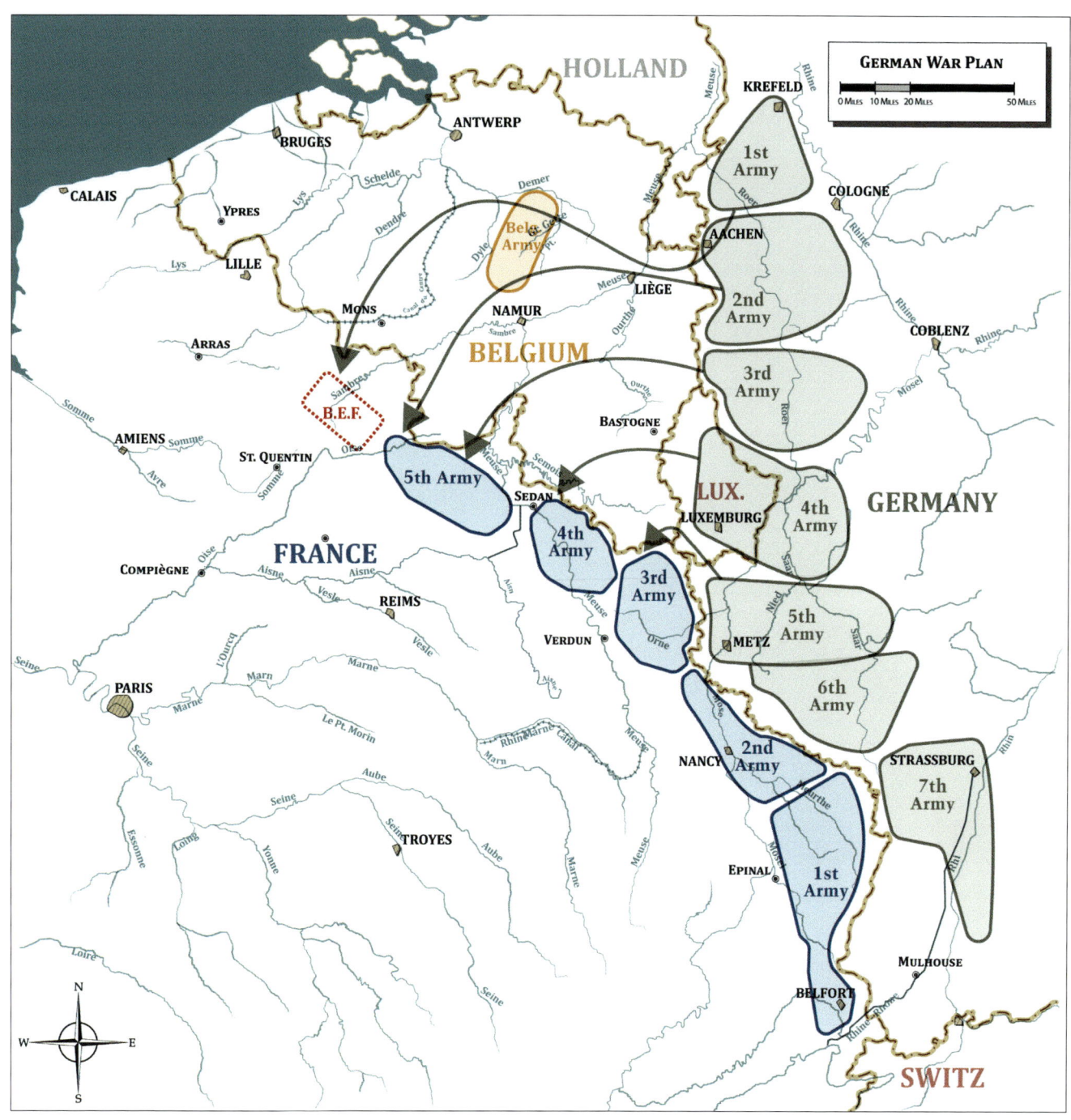

French would become vulnerable to counter-attacks from three directions. If, on the other hand, the French troops opposite the German left remained on the defensive, Sixth and Seventh Armies were to launch an offensive against the French fortress line. In either case, the left wing's primary objective was to "fix" or hold as many French divisions to the southern end of the front as possible and not allow them to be transferred to the battles taking place on the German right wing—where the campaign would ultimately be decided.[7]

Organization of the Army

Under the provision of Article 63 of the Constitution of the German Empire, the Kaiser was the supreme commander of all Germany's armed forces on land.[8] Command of the army, however, was assigned to Helmuth von Moltke who, as Chief of the General Staff of the Field Army[9] was authorized to issue orders in the Kaiser's name. Moltke's headquarters—known as Supreme Army Command (*Oberste Heeresleitung* or OHL)—was therefore responsible for conducting the land war and determining the nation's overall military strategy.[10]

Above: Helmuth von Moltke (Author's Collection)

Above: Cavalry played a critical role in the opening months of the war in the West. Here is an excellent depiction of a German Uhlan in campaign dress. (Author's Collection)

The largest operational command within an army was a corps. Each active corps (*Armee Korps* or *AK*) consisted of approximately 41,500 men of all ranks organized into two infantry divisions supported by a contingent of corps support troops (*Korpstruppen*). Reserve corps (*Reserve Korps* or *RK*) were similarly organized into two divisions, but functioned at a reduced strength with less than half the artillery of an active corps. Corps support troops consisted of pontoon trains, communications detachments, munitions columns, field hospitals, field kitchens, a searchlight section and transportation vehicle columns. Each active corps was also given a heavy field howitzer battalion comprised of 16 15cm howitzers as well as a Field Aviation Section. Fully mobilized and on the march, the corps' 1,500 officers, 40,000 men, 14,000 horses and 2,400 ammunition and supply wagons would cover up to 30 miles of road.[11]

The German infantry division (*Infanterie Division* or *ID*) was a combined arms force of two infantry brigades, an artillery brigade, a cavalry regiment, and various support units. Each division was organized on the "square pattern" of two brigades with two regiments each. An infantry regiment consisted of 3,000 men organized into three battalions—each with four companies. For fire support, a machine gun company of six guns was attached to each active infantry regiment. Divisional artillery possessed two weapons: the 7.7cm flat trajectory gun (*Feldkanone 96 n/A*) and the 10.5cm high trajectory light howitzer (*leichte Feldhaubitze 98/09*). An artillery brigade was organized into two regiments of six batteries. Each battery possessed six guns and caissons, an observation wagon, and four various supply wagons. In each active corps there were three field gun and one howitzer artillery regiments; one division had an artillery brigade of two field gun regiments while the other division had a 'mixed brigade' consisting of one field gun and one howitzer regiment. Unfortunately, a reserve corps' division (*Reserve Division* or *RD*) only possessed a single 7.7cm gun regiment. A division's cavalry regiment consisted of four squadrons to be used for short-range reconnaissance and security duties. For logistical support, each division was given a number of field hospitals and munitions columns as well as a pioneer company, pontoon train, telephone detachment and veterinary section.[12]

Finally, the western armies' ten cavalry divisions were divided into four cavalry corps (*Höherer Kavallerie-Kommandeur* or *HKK*) to be used mainly for reconnaissance. A cavalry division (*Kavallerie Division* or *KD*) had three brigades, each comprised of two regiments. Support troops assigned to each cavalry division consisted of a horse artillery battalion, one machine gun company, a *Jäger* battalion, six communications wagons, a pioneer company, and a motorized vehicle column.[13] Although a cavalry corps' primarily mission was to conduct reconnaissance, it was also expected, when called upon, to be used as a mobile fighting force capable of delaying a larger enemy force—either offensively or defensively—as circumstances dictated.

Wartime Organization of the *Fliegertruppe*

Imperial Germany's War Plan was based on the sobering reality of having to initially fight a two front war against numerically superior enemies. Overall success, according to the plan, was fully dependent upon the rapid conclusion of the initial campaign in the West before the situation in the East had deteriorated. Thus, with the German leadership uncertain of Franco-British operational intentions, strategic and operational reconnaissance was absolutely vital to helping Moltke's western armies conduct a swift and successful campaign. Germany's aviation troops (*Fliegertruppe*) were therefore expected to play a vital role in the army's operations from the war's first engagement.

Upon mobilization the army's five peacetime aviation battalions were divided into 33 flying sections (*Feldflieger Abteilungen* or *FFA*). These flying sections were attached to each of the eight Army HQs as well as each of the twenty-five active corps HQs. Unfortunately, there were insufficient quantities of aircraft and airmen at the time of mobilization to equip the Reserve Corps and Cavalry Corps HQs with their own flying sections. Each flying section was equipped with six aircraft, seven pilots and six observers. There were 116 enlisted men also attached to help prepare the airfield, maintain each aircraft and perform all the necessary labor to keep the section operational. In terms of *matériel*, each FFA initially deployed with a limited quantity of 3.5, 5, and 10 kilogram bombs as well as at least one camera.[14] Finally, each section maintained a fleet of five touring cars, six aircraft towing vehicles, one tender, and a number of lorries for transporting all the stores, munitions, fuel, and baggage the FFA required while on campaign.[15]

At the outbreak of war, the flying sections were

Frontline Inventory, August 31 1914	
Type	Number
Class A Unarmed Monoplanes	
Albatros A/13	5
Albatros A.I	2
Gotha A/13	13
Gotha A/14	3
Hirth A.	1
Jeannin A/13	6
Jeannin A/14	3
LVG A.	1
Rumpler A/13	2
Rumpler A/14	8
Subtotal Class A	**44**
Class B Unarmed Biplanes	
AEG B.I	3
Albatros B/12	1
Albatros B/13	6
Albatros B.I	18
Albatros B.II	5
Aviatik B/12	1
Aviatik B/13	34
Aviatik B.I	12
DFW B.I	2
Euler B/13	2
Euler B.I	3
Fokker B/13	1
LVG B.	22
LVG B/13	30
LVG B.I	32
Rumpler B.I	1
Subtotal Class B	**173**
Miscellaneous	
DFW Mars*	1
Total Inventory	**218**

* Perhaps a Mars monoplane since some Mars biplanes were given B-type serial numbers.

Above: LVG B.I two-seat reconnaissance biplanes were the most numerous German type at the front early in the war. Here several of these robust, reliable aircraft are shown at the front. Early LVG designs featured a kink in the aileron trailing edges that immediately distinguishes them from other manufacturers' aircraft. (Aeronaut)

equipped with a wide variety of aircraft including both biplanes (B-Types) and monoplanes (A-Types). The most numerous of which was the LVG B-series. With excellent flight characteristics and a highly reliable 100hp Mercedes D.I inline piston engine, the LVG was easy to fly for the pilot, a stable reconnaissance platform for the observer, and a favorite among ground crews. Similar B-Types were also produced by the Aviatik and Albatros firms. Both of these aircraft were equipped with same 100hp Mercedes engine as the LVG and were able to operate at great altitudes in relative impunity from enemy ground fire. Although there were other B-Types employed by the army's aviators, the triumvirate of LVG, Aviatik, and Albatros were by far the best and the most numerous. Together, they formed the backbone of the *Fliegertruppe* in 1914, combining to make up 74% of the German Air Service's inventory at the time of mobilization.[16] Future Royal Flying Corps ace James McCudden, who began the war as an observer, saw the *Fliegertruppe* in action during August 1914 and was greatly impressed with his German counterparts' aircraft:

As a rule they flew a deal higher than our machines which was not surprising to those who knew anything of what the new German machines had done in the way of height and long distance flying just before the war.[17]

The vast majority of the A-type monoplane aircraft used by the *Fliegertruppe* were the famous Taube design produced by a number of different firms. However, despite performing well during the Kaiser Maneuvers in the years prior to the war, it was widely accepted that the design was outdated at the time the war began. Nevertheless, the *Fliegertruppe* needed every aircraft that was available and the Taubes were still extremely capable reconnaissance platforms. Thus, they were pressed into service along with every other serviceable military aircraft in Germany.

To support the flying sections at the front, each field army possessed an army airpark (*Armee Flugpark* or *AFP*) located deep in its own rear-area. The airpark's role was to keep each of the army's flying sections at maximum strength by performing major repairs and providing replacements. In his treatise: *The Supply and Movement of German First Army During August & September 1914*, General Walter von Bergmann described the airparks' responsibilities as: "to hold in readiness, personnel, airplanes, and all manner of spare parts; to make necessary repairs; and to collect all of our own or enemy planes that had come down or been abandoned, either in the combat zone or in the communications zone." Bergmann elaborated further that: "such a park would be established at some suitable place within the communications zone, preferably on the premises of factories or power plants that could be utilized for the purpose."[18] In order to adequately perform its assigned tasks, each AFP was assigned 69 enlisted men and a generous number of large workshops, tools, and other equipment related to aircraft maintenance. They each also possessed

Above: Typical Aviatik P14 B-type. This photo was likely taken prewar due to the lack of insignia. Like all B-types, the pilot sat in the back seat to give the observer the best view forward and downward. The engine was a 100 hp Argus As.I cooled by side radiators. The wings have a slight sweepback, there are struts to the wingtips, and no fixed fin was fitted. (Aeronaut)

three aircraft, two pilots, and an observer for use as replacements at the front in the event one of the army's flying sections sustained heavy losses.[19] Finally, in addition to performing repairs, the airparks were to also act as a conduit between the aircraft firms in Germany and the flying sections at the front. As the war progressed, newly produced aircraft were to be delivered from the manufacturers to the airparks where they would be assembled, tested, and dispersed amongst the flying sections.

According to the deployment plan, eight field balloon sections (*Feldluftschiffer Abteilung* or *FLA*) were attached to each Army HQ. Whereas the FFAs were to fly over the enemy cavalry screen and conduct long-range reconnaissance, the FLAs were expected to solely provide their HQ with tactical intelligence once the ground forces had become fully engaged. Specifically, the FLAs were to inform local commanders of the enemy's deployment and movements on the battlefield. Furthermore, they were expected to take advantage of their telephone connection to work with nearby artillery batteries in coordinating indirect fire on enemy positions that were outside the ground troops' line of sight. Each balloon section consisted of one 600cc Parseval-Sigsfeld balloon, the section's commanding officer, a

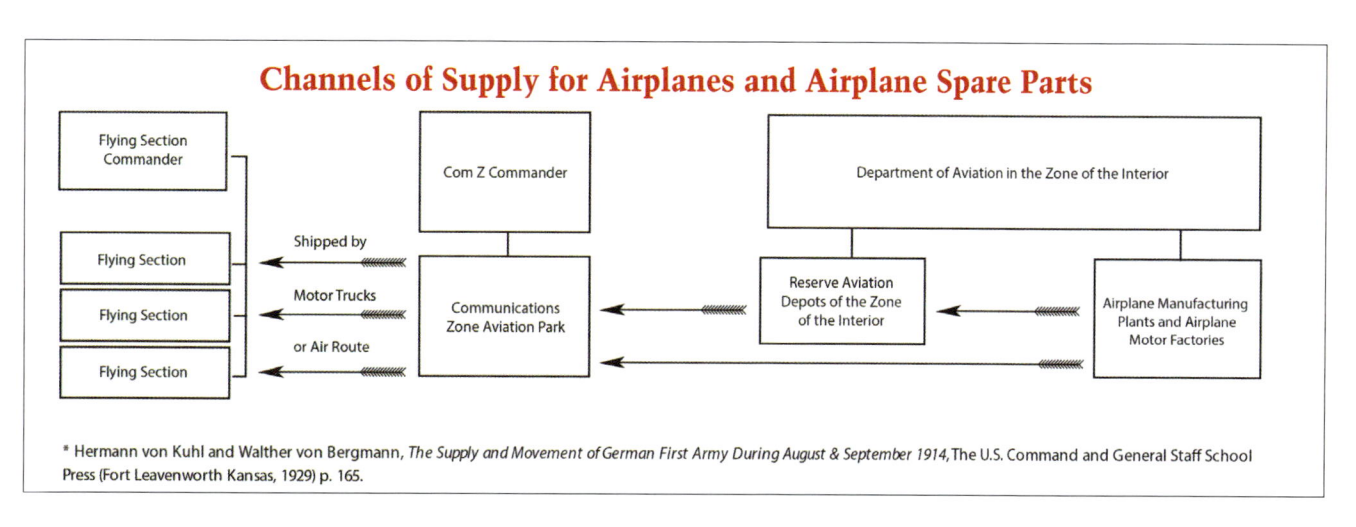

Channels of Supply for Airplanes and Airplane Spare Parts

* Hermann von Kuhl and Walther von Bergmann, *The Supply and Movement of German First Army During August & September 1914*, The U.S. Command and General Staff School Press (Fort Leavenworth Kansas, 1929) p. 165.

Above: Despite their mobility issues, balloons were often a useful observation platform during the period of open-warfare. Pictured here is the Parseval-Siegsfeld *Drachen* Kite Balloon belonging to *Feldluftschiffer Abteilung* 1, which was attached to German First Army throughout 1914. (Author's Collection)

veterinary officer, four observer officers, 177 enlisted men, 123 horses, one balloon wagon, one winch wagon, 12 gas wagons, two equipment wagons, and a telephone wagon. Each FLA also maintained an additional 13 gas wagons held far in the rear in the army's communication zone—usually at the airpark.[20] The large number of vehicles and equipment attached to a FLA severely restricted the section's mobility, making it less suitable for open warfare than their FFA brethren.

Attached to Germany's various fortresses were eight Fortress Flying Sections (*Festungsflieger Abteilungen* or *Fest.FA*) and 16 Fortress Balloon Units (*Festungsluftschiffer Trupps*). The fortress flying sections were each comprised of four aircraft, two pilots, two observers, and 37 men while the balloon units maintained a complement of five officers, 109 men, 53 horses, and eight wagons to keep its kite balloon and free balloons fully functioning.[21] Although these units were given the same roles and usually just as capable as their mobile brethren, they were rarely used during the opening campaign. As it happened, these formations were strictly confined to operations within their fortresses' sector and were not released for duties in the field until October.

The army's dirigible units (*Luftschiff Kommandos*) mobilized for war with only eight serviceable airships. Under the direct control of Supreme Headquarters (OHL), four ships were initially deployed for service on the Western Front. Indeed, three of the army's newest ships were stationed on the western frontier—the *Z. VI* at Cologne; *Z. VII* at Baden Oos; and *Z. VIII* at Trier. The *Z. IX*, which was in the midst of her acceptance tests when war was declared, was quickly cleared for service and deployed near Düsseldorf on August 10. Finally, the requisitioned civilian airship *Sachsen* arrived ready for operations late in the concentration period, bringing the total number of dirigibles committed in the west to five.[22] German military strategists had realized the dirigible's vulnerabilities and limitations as early as spring 1914 and had planned accordingly to use them on long-range strategic reconnaissance flights during the period of concentration before large scale operations had begun. Once the opposing forces were heavily engaged however, it was believed that the dirigibles could only be used sparingly in a reconnaissance role under certain conditions. The ships' main purpose by

Above: Although it was a formidable weapon, the Germans only possessed six truck mounted 7.7cm anti-aircraft guns at the outbreak of the war. (Author's Collection)

that point would therefore be in an offensive role—flying primarily at night under the cover of darkness against large static targets such as fortresses, supply depots, or rail yards.

Experience quickly proved that— at least on the Western Front—the army's dirigibles were a failure. Strangely, not a single strategic sortie was ever flown during the period of concentration—an important mission that only a dirigible could perform given the distance. While the German official history claims the inactivity was caused by poor weather and delays preparing the ships, close examination of the facts show that there were a number of days during the period of concentration that the ships were ready for operations and could have been employed without any safety concerns. Furthermore, the early dirigible bombing missions were also unsuccessful.[23] These substandard circumstances ultimately led to the losses of several ships and the order to discontinue dirigible operations in the West.

Anti-aircraft artillery (AA) was still in its infancy in August 1914. Indeed, the army only had 18 special guns at the time of mobilization—six of which deployed with the army for forward operations in the field. Those six were modified truck mounted 7.7 cm guns which possessed an automatic breech mechanism and separate range finder that allowed it to fire at a rate of 25 shots per minute. In an effort to give the various armies some protection, the six guns were divided as follows: two for XV AK and one each for I, VII, XVI and XXI AKs.[24] The remaining 12 AA guns (four horse drawn, eight stationary) were deployed in the rear-areas for the first two months of the war to protect important bridges and Zeppelin hangars.[25]

The *Fliegertruppe's* Role in the German Army of 1914

All European armies went to war believing that the coming conflict would be decided by large-scale offensives in open warfare. The opening campaign's first battles were therefore anticipated to take the form of a meeting engagement/encounter battle. Meeting engagements occurred whenever two opposing armies were both in motion and subsequently made contact with one another's forward units. In the days prior to the battle, the *Fliegertruppe* were directed to locate and monitor the enemy's main body and then report its location to German commanders on the ground. These long-range reconnaissance flights assisted the army and corps commanders maneuver their units so that they were fully concentrated and opportunistically deployed when contact with the enemy was finally made.

Once the opposing armies became engaged in battle, the flying sections were ordered to conduct short-range tactical reconnaissance sorties directly over the battlefield in order to keep the commanders informed

Above: War-time propaganda postcard depicting German infantry closely supported by a *Taube*. (Author's collection)

of the battle's progress. By circling above the battle, an observer was in position to discover weak points in the enemy line, identify the movement of any enemy reinforcements, locate concealed batteries, or take note of any other important information that could be of use to the commanders on the ground. Once an observer identified something of importance, he would either land at HQ and deliver a verbal report or drop a handwritten message over the side of the aircraft at a designated drop point in order to promptly return to the battlefield to continue the mission.

Whenever the ground forces were engaged in battle, the FFAs retained the option to perform artillery direction or artillery cooperation sorties. Because no standardized system of artillery cooperation existed prior to the war, the method and effectiveness of these flights varied greatly between the different flying sections. During the war's first weeks, attempts at artillery cooperation were performed through the use of flares or colored lamps, tipping wings, or dropping notes in a canister near the artillery commander's command post. By the end of August however, German aviators throughout the Western Front had almost unanimously adopted the method of dropping notes by canister after it had been proven to be the best available method. Indeed, this was the technique that had been discovered as the most practical by the attendees of the *Fliegertruppe* artillery course at Jüterbog in the spring of 1914.

After the meeting engagement's conclusion, the defeated enemy force would be compelled to withdraw to safety under the cover of a rear-guard. With the German army in pursuit, the flying section's aircraft would be sent aloft to resume their long-range reconnaissance duties in order to establish the condition and route of march of the retiring enemy force. With the aviator's intelligence in hand, army and corps commanders could develop operational plans to smartly pursue and destroy the defeated enemy's force in detail, thus turning a tactical victory into a larger, more decisive one.[26] If the retreating enemy managed to escape, they would fall back to a defensive position where they would reorganize and force a second general engagement whereupon the process would repeat itself.

Typical endurance for a German aircraft at the time of mobilization was three hours. Taking into account an average cruising speed of 55–70 miles per hour, each aircraft was roughly limited to within a 100 mile radius of its own airfield. Cruising altitude varied greatly depending upon both the type of aircraft and the mission type. For long-range flights,

B-Types were expected to cruise between 6,000–7,000 feet while A-Types generally operated around 5,000 feet. Once an enemy column was sighted however, German aviators typically dropped to around 3,500 feet to better identify the details regarding the enemy force. Pre-war guidelines set 1,000 meters (3,200 feet) as the minimum altitude allowed over enemy occupied territory. Tactical reconnaissance and artillery direction flights were therefore performed between 3,000-4,000 feet, although circumstances often compelled many intrepid airmen to ignore the rule to provide better support.[27]

The *Fliegertruppe's* aforementioned role in support of the army was defined by the Prussian General Staff's prewar memo "General Staff's Directive for the Use of Aircraft with Ground Forces." Written in February 1914, the document established in detail how the army's flying sections were to conduct operations and cooperate with ground forces. Although mostly directed at airmen, the memo also created guidelines for how army and corps command staffs would communicate and employ their aviation assets, something the General Staff felt compelled to do considering how willfully uninformed many of these officers were regarding military aviation at the time. Sadly, the regulations laid out in the document were never uniformly enforced due to the German organizational model's lack of aviation officers at Supreme Headquarters to enforce them. However, it should be noted that the majority of flying sections adhered to the guidelines put forth in the directive, making it arguably the most important document in understanding how the flying sections operated during the first year of the war.

Above: An excellent photograph showing an early-war German observer preparing for a flight carrying the tools of his trade. (Author's collection)

General Staff's Directive for Aircraft in Military Service

1) Airplanes assembled in numbers of six shall constitute an aviation section.

The flying sections are to be assigned to the Army HQs and Corps HQs. The strategic cavalry corps (HKK) as well as for artillery units can be established based on the need of the situation. Fortresses are to be equipped with fortress flying sections. The complement of personnel and equipment for the flying sections are to be kept at the army aviation parks.

The flying sections are to move by vehicle, while the aircraft are usually redeployed by air. If this is impossible due to weather, the aircraft can be disassembled and towed by vehicle.

For each aircraft, the observer is in command.

2) Aircraft's performance is based upon the weather and materials as well as the experience and aptitude of the crew. Flight can generally be performed at all hours, including at night. Landings in the dark or in unfamiliar places are, however, considered dangerous. In the event of high temperatures, it is desirable to fly in the mornings and at dusk when temperatures are at their lowest.

"Good flight performance" under wartime conditions is considered about 3 hours. Hourly endurance is generally due to weather (as well as wind) conditions; which can impact range by up to 100km. winter weather conditions dramatically lower performance capabilities. Performance should be analyzed on a case-by-case basis.

3) For the Flying Sections, all preparations for

flights begin at the airfield where aircraft are made ready for upcoming flights. Workshops, carpenters, mechanics, fuel and other material storage, is all done here. The general deployment of an airfield should be as close as possible to the [army or corps] HQ. The selection of the airfield is a matter solely for the Flying Section's Commander. A secondary forward airfield might also need to be established in order to deliver messages faster. This is especially advisable for the placement of any airfield for the army HQ's flying section because of how far behind the line the army HQ section would be from the corps HQs, where the army command staff might be located.

Telephone connection between the airfield and the commando post is required. It is important the airfield is moved if the flying sections do not have a working telephone equipment. It is inexpedient for the flying sections to temporarily borrow a telephone from a corps HQ telephone unit.

The allocations of cars, cycles and horses can be advantageous and must be used to their fullest extent in the absence of telephone communications. Apart from the airfield, there should be a landing area in case of the need for a temporary landing. This is especially necessary when entering into combat. The flying section commander shall select an appropriate location for this field, near where the fighting is to occur. The security of this field and its connection by either telephone, vehicle, cycle or horse should be guaranteed by the local HQ.

For takeoffs and landings [pattern altitude], it is agreed that traffic should be at 250m height; and on a soft meadow. The direction for traffic should be clearly visible through large white cloths laid down on the field that can be seen from great height.

If a suitable landing field is not found, the General Staff recommends that a "drop point" for messages be established. The same factors that related to the landing field are critically important for determining the location of the drop point. It must be smartly placed so that messengers and cyclists can quickly transport the messages to the intended troop commander.

The position of the airfield has to also be in accord with the desires of the army commander. Often, the placing of a landing field or later an airfield will be the decision of the Flying Section commander. The aircraft at that point are usually in the air [and in route to the new location].

4) Use of Aircraft: Aircraft are to mainly be used for reconnaissance. They are also to be used for messaging, battle, and attack. Aerial reconnaissance was to determine the position of enemy troop concentrations deep in hostile territory; looking specifically to note their strength and direction of march as well as tactical details if the group is preparing for battle. Aerial Messaging often can be used when it is large forces to determine their position or the location of adjacent units. An immediate communication between higher commands can allow for coordinated efforts despite the various commands being separated by a large area. Battle and attack/destroyer is to be done through bombing. This is to be done even when no express orders have been given.

The higher ranking artillery commanders have also requested that artillery cooperation flights that accomplish both target reconnaissance and fire direction missions take place.

5) All orders usually go to the flying section's commander, who also make most other arrangements for his men. In special cases it may be advisable for aircraft that the flying section commander delegate some of his power to an individual at the forward "landing airfield" to better carry out specific missions.

If the Flying Section commander received an order from the troop commander, he should relay the message to his men immediately.

Every aircraft is to have one pilot and one observer during a sortie. In special cases it may be advisable that the observer on a flight be a General Staff Officer.

It is also necessary to identify the exact location of the enemy camp or movements by monitoring suitable roads and railway lines, whether narrow gauge or large in the enemy's rear-area. However, it will also be necessary under certain circumstances to reconnoiter larger areas of terrain [away from the roads and rails].

The position of the enemy was initially not considered to be the sole responsibility of aerial reconnaissance; this was to be done in concert with the cavalry. Aerial reconnaissance required the identification and details of troops found in column, as well as anything else observed behind enemy lines (columns, supply trains and support troops).

Regarding reconnaissance during battle, it is advantageous to focus on specific areas of the enemy's position. For example: location of flanks, position of their artillery, details of their leadership staff in the rear, and available reserves. Commands [to the flying section] must contain specific and narrowly defined tasks.

For especially importance reconnaissance flights, multiple aircraft must be used either staggered or simultaneously. It is advisable to make sure that use thereof is only done under the most urgent circumstances so that the strength of a flying section is not prematurely exhausted.

6) The orders for aerial reconnaissance should be included in the operating instructions for the troops. This is generally the reconnaissance boundaries, the position of the airfields and any forward landing areas. The reconnaissance orders in detail, the relocation of airfields, tracking of flying section and their connections to the army/corps leadership are to be listed in "special orders."

It is recommended, that army units entering into activities inform the flying sections of the location of each corps in relation to one another even before a meeting engagement had occurred. To distinguish the forces the *Fliegertruppe* should provide support and important intelligence to win in all important engagements.

7. Definitive and detailed instructions intended for officers of the flying section about the position of friendly and enemy units is essential. This is to be issued by the army/corps leadership or by a general staff officer [on their behalf]. The troop leadership– either from the army commander himself or his chief quartermaster– issues suggestions for the employment of the flying section. Specifically, for the moving of airfields and the establishment of forward landing fields. There should also be communications to the airfield and forward airfield from the troop command that contains all commands to the flying section.

The use of aerial reconnaissance may require the presence of an officer from army/corps HQ at the airfield or forward landing area. On behalf of the troop leadership, this officer would receive and inspect the airmen's message and act as liaison between the Army/Corps HQ and the flying section.

8) For best results after the aircraft returns, have the observer personally discuss his findings with the troop leadership. The command staff should be present at the landing site when the aircraft lands. If that is not available, then dropping the message at the previously determined drop point, or nearby where the corresponding troops are located. Either way, [a written form of] the message is to be taken to the troop commander after landing. If the message had not been transmitted to the leadership, it is incumbent upon the observer to personally visit the commander and inform him of the reconnaissance flight's results.

When the aerial messages are dropped, it is the responsibility of the local troops to forward the message as fast as possible to the troop leadership's command post.

9) Out of the airplane's safety from anti-aircraft fire, a minimum altitude of 1000 meters [3,280 feet] is to be flown over areas occupied by the enemy except when clouds, fog, or darkness during flight compel the pilot to drop to a lower height...

10) Troops on the march are usually easily detected on the roads. However, marching under heavy foliage often can conceal them from view. Columns marching off the streets in open areas are easy to locate. Dust and rainy weather makes it difficult to observe.

Dormant troops are often overlooked, particularly buildings, forests, tree rows and shaded areas that block light all provide perfect protection from observation. Stacked arms and/or baggage can easily be detected in an open field; as well as collections of vehicles or horses. It is hard to distinguish vehicles including munitions wagons, machine gun sections or field kitchen. [Taking note of the size/number of] Baggage trains and mounted riders makes it easier to determine the size of the force. The mounting of [camouflage] tarps protects the ground troops only when it matches the color of the surrounding terrain. The bombardment of aircraft can in some cases, when it is unsuccessful, betray the ground troops' position.

The high command staff remaining together in an open field can attract attention and be pelted with bombs. For this purpose, it is inadvisable that the entire staff be located near an airfield or forward landing site.

Trenches and field fortifications can easily be detected while they are being constructed. Freshly erected coverings can also easily be

> spotted by their color and shape (long dashes).
>
> The strength of the discovered troops can be determined by the number of approximate of parked vehicles in the area. This is made more difficult when considering large defenses and wagon parks could easily be part of broken smaller units.
>
> Bivouacs have been said as a good identifier of resting troops.
>
> Whether and to what extent a dormant force would give up their position based on the appearance of an aircraft will only be decided by wartime conditions.[28]

The directive plainly defined why the *Fliegertruppe* were a critically important component to the army's operations. While strategic flights[29] to "identify and detail all enemy movements by road or by rail" remained the aviators' primary task, the document directed all flying sections to establish airfields as close to the front as possible in order to provide tactical assistance to the army whenever possible. Moreover, a secondary airfield or "drop point" near the corps or army commander's headquarters was also recommended to quickly deliver aerial reports for the commander's use. These directives, along with a number of additional guidelines written to maximize the flying sections' impact on operations, served as a blueprint for the *Fliegertruppe's* successes throughout 1914.

Constant communication between the *Fliegertruppe* and the army leadership was imperative if the outnumbered Germans were to quickly defeat the Franco-British forces as their war plan required. Although a standardized method of liaison between the aviation sections and their respective HQs did not exist at the time of mobilization, the typical daily routine of a flying section can be taken from the following example:

In the predawn hours, the order of the day would arrive from HQ. The order would describe the previous day's combats, position of local friendly units, the coming day's objectives/routes of march and the suspected position/intentions of the enemy. The order would also establish boundaries in which the section was to reconnoiter that day. Using the information provided in the order, the section's commander would brief all pilots and observers on the overall situation.

Next, if possible, the commander would immediately formulate a plan of operations. A number of the section's aircraft and aircrews were scheduled to go aloft at different times throughout the day to keep watch over their prescribed search area. Although as many as three aircraft could be airborne at one time, one aircraft was always kept grounded "at the ready" in the event HQ needed immediate aerial observation at another area of the front. Aerial operations would then continue throughout the day, with observers continually bringing back updated information regarding the situation at the front.

The overwhelming majority of all flights during the period of open warfare were long-range strategic reconnaissance sorties conducted to ascertain the location of unengaged enemy units. Before each flight, the pilot and observer reviewed the mission's objective and the situation at the front and created a flight plan accordingly. During the flight the observer took detailed notes of the location, strength and route of march of any enemy columns he had discovered. Other information such as the number of guns present and the proximity of friendly units were documented as well. Once the aircraft returned to the airfield, the observer quickly prepared a written report containing the results of the flight. The report was then driven to HQ in one of the section's touring cars, often accompanied by the observer himself so the contents of the report could be properly discussed with an HQ staff officer.

As the army advanced, divisional cavalry with the forward troops were ordered to locate and secure a suitable new airfield closer to the front for the following day. After the conclusion of the day's final flight, the entire section prepared to move forward with the army's supply column to its new airfield. Each aircraft was made ready for road transport and towed forward while the rest of the section's personnel and supplies followed in the unit's various trucks, wagons and service vehicles. If there was enough daylight left, the section's aircraft could also be flown to the new airfield. Nevertheless, upon arrival, the aircraft were secured and the airfield was prepared for operations to begin the following morning.

German Military Doctrine in 1914

In 1914, all European armies favored the offensive—both operationally and tactically. Military theorists across Europe believed that wars in the 20th century could only be won through bold offensive action in large-scale battles of annihilation.[30] According to German military doctrine, a decisive victory could only be achieved if the main attack was successfully directed against the enemy's flank and rear. Alfred von Schlieffen, the Chief of Prussian General Staff from 1891–1906 and the man primarily responsible for inculcating this belief throughout the army, succinctly explained the German offensive theory

Above: The Albatros *Taube* Type FT had large side radiators and a further simplified undercarriage. It was powered by a 100 hp Mercedes engine. The FT was Albatros's last *Taube* design and served in small numbers during the early months of the war as an Albatros A-type (unarmed monoplane). (Aeronaut)

that was ultimately employed in 1914 when he wrote: "the enemy's front is not the objective. The essential thing is to crush the enemy's flanks... and complete the extermination by attack upon his rear." Indeed, a successful attack and breakthrough against the enemy's center was viewed as potentially dangerous, for it exposed the attacker to a counter-attack against his flank or even an encirclement. Thus, whether at the operational or tactical level, the German Army was trained to constantly seek to maneuver around their opponent to strike at their flank and rear and achieve a major victory.

The nature of warfare, however, had dramatically changed by 1914. Magazine-fed small caliber rifles, machine guns, quick firing rifled artillery, and smokeless powder all forced European military theorists to rethink how their armies conducted themselves in battle.[31] For the Germans this meant modernizing infantry, cavalry, and artillery doctrine as well as training junior officers and NCOs to become the army's primary decision makers on the battlefield. All of these changes were made to negate the lethality of modern weaponry and allow offensive operations to succeed with minimal delays and/or casualties despite Germany's numerical inferiority.

The most important doctrinal change made prior to the outbreak of war was the army leadership's systematic adherence to the *Auftragstaktik*, or mission-type orders, system. Under this command style, commanders gave their subordinates a mission/ task without specifying how it was to be accomplished. In other words, subordinates were given the freedom to determine how the mission would be completed— provided it was compatible with the overall concept of operations.[32] The time and method of the attack would therefore be the decision of the subordinate officers closer to the front, who better understood the tactical situation. Thus, in an era where senior leaders could no longer take personal control over heavily engaged units, the *Auftragstaktik* system enabled lower level German commands to react quickly to changing circumstances on the battlefield in order to properly exploit any opportunities that arose during the course of the engagement.

Within the *Auftragstaktik* system, all NCOs and junior officers were trained to think and operate in adherence with German Army doctrine which encouraged officers to take the initiative and boldly seek to envelop the enemy. To that end, the Kaiser's forces developed the world's most sophisticated and realistic training program the world had seen. Here, troops learned how to properly maneuver on the battlefield and execute an attack during simulated live-fire exercises which exposed them to many different scenarios that they would later face in wartime. To simulate losses, an umpire would announce how many casualties the unit had suffered during various stages of the exercise— forcing the troops to learn how to adapt and overcome adversity. Various issues such as learning the effectiveness of

Ich hatt' einen Kameraden - einen bessern findst du nit - Die Trommel schlug zum Streite - er ging an meiner Seite - in gleichem Schritt und Tritt.

Above: After gaining fire-superiority, German infantry were expected to quickly deliver their final assault and drive the enemy back from their position in close combat. (Author's Collection)

massed rifle fire, determining when to dig in and wait for reinforcements, or perfecting how best to communicate with neighboring units during an assault were studied and practiced repeatedly. Thus, under these circumstances the German Army of 1914 possessed the confidence and ability to successfully launch repeated offensives under the decentralized, initiative driven *Auftragstaktik* system.[33]

To better study how the *Fliegertruppe* tactically supported the army, it is imperative to first understand how the ground troops operated during an engagement. Because German strategy was based on a quick and decisive victory through offensive action in the West before the Russians were able to mount a large-scale offensive of their own in the East, the overwhelming majority of battles in the West during the opening campaign featured the German army on the attack. Therefore, it is only important for this volume's purposes to review German offensive doctrine.

German theorists believed that the war would be decided by their infantry's tactical fighting prowess. The changes in infantry doctrine made prior to the war therefore had the greatest impact on the army's operations during the battles of 1914. According to tactical doctrine, the artillery's primary role was to support the infantry's advance by silencing the enemy's artillery with effective counter-battery fire. With the opposing artillery neutralized, the attacking German infantry would then be able to advance, envelop, and engage the enemy in a firefight, ultimately achieving fire-superiority and taking the position by assault.[34] The 1906 doctrinal document, *Exerzier-Reglement für die Infanterie*, specifically detailed how an infantry attack should be delivered in the face of lethal modern weaponry and thus, heavily influenced how the German Army as a whole operated on the battlefield in 1914.

Although no two situations were alike, all German attacks were conducted using the following guiding principles. During an advance, commanders would send cavalry, engineers, and light infantry ahead of the main body to gain information regarding the enemy's strength and position. Based on this intelligence, a decision would be made as to how the attack would proceed. Once the main body completed its approach march and a course of action had been determined, all lower commands and formations would deploy for battle in accordance with the commander's concept of operations.

Above: Large-scale cavalry charges were a rarity by 1914. However, the opening month of the war featured several successful mounted attacks, such as at the Battle of Longwy, shown here. (Author's Collection)

The main objective of German offensive tactics was to maneuver in order to place enough fire on the enemy in order to gain fire superiority and then close with the enemy by means of fire and movement in order to destroy him in close combat.[35] To that end German artillery batteries were directed to begin every engagement with preparatory counter-battery fire to silence the enemy's guns and allow the infantry to begin their advance. Next, under the cover of friendly artillery fire, the infantry would commence the attack—usually starting from a distance of 3,000–5,000 meters from the enemy's position. According to the attack orders, each unit (down to company level) was assigned its own "attack sector." For example, an infantry brigade maintained a 1,500 meter front, with each front-line company assigned its own 150 meter sector. As historian Terence Zuber explains: "the attack sectors were intentionally kept narrow to allow a deployment in depth, in order to have supports and reserves readily available to replace casualties in the firing line."[36] Although units would remain in march column as long as possible, the only formation used in combat was the skirmisher line, or "swarm." The density of this formation varied based upon the effect of the enemy's fire. However, the average distance between comrades was five paces per soldier.[37]

Once the attacking troops began to receive fire from the enemy, the forward platoons would spread out into thinner skirmisher lines. The remaining platoons would form at least two more lines that would follow in waves behind the first. Troops in the forward line would advance 'by bounds' (irregular movements from one piece of cover to the next). At some point during the advance, a junior officer would find a good position which offered some cover and would establish a firing line where the first wave was to concentrate its strength for the coming firefight. As the firefight was raging, the firing line would be continually reinforced from platoons bounding forward in the attack's second and third lines. The attacking infantry in the firing line would therefore slowly accumulate enough firepower to achieve fire superiority. If for any reason the attack had stalled, the firing line would be provided additional firepower, either with artillery, machine guns, or the commitment of additional reserves to their sector. Nonetheless, once fire superiority was achieved, a portion of the attackers in the firing line would be able to move forward and take the enemy's position in close combat.[38]

The mission of the army's cavalry, artillery and machine gun units was to fully support the infantry.

Above: The Albatros B.I, shown here with a 100 hp Mercedes, was the third of the three most numerous B-types used early in the war. Its long-span, three-bay wings are distinctive. (Author's Collection)

As previously mentioned, the artillery's primary role was to support the infantry on the battlefield by engaging enemy batteries through counter-battery fire. However, once the infantry had entered into a firefight, a select number of batteries would be used directly against the enemy infantry to help the attackers achieve fire superiority.[39] Similarly, the machine-gun companies were used exclusively as a support weapon to assist German infantry units achieve fire superiority.[40] Meanwhile, the cavalry's assigned task was to perform tactical reconnaissance and provide the army's leaders with important details regarding the enemy's location prior to the engagement. The four strategic cavalry corps (HKK) were also regularly employed as mobile combat forces where the men would fight dismounted using similar tactics that the infantry used.[41] Thus, whether to be used offensively as a means to quickly strike the allied flank and rear or defensively as a rear-guard to delay the progress of numerically stronger enemy forces, the cavalry corps ultimately proved to be more effective as a mobile fighting force than as an intelligence asset. Indeed, within six weeks of the outbreak of war, the *Fliegertruppe* had replaced the cavalry in the minds of most German leaders as the army's top reconnaissance asset.

As Germany mobilized for war, the majority of their commanders believed their aviation assets would have little impact on a tactical level. Experiences at the front quickly proved this to be an inaccurate assumption. Within weeks of the outbreak of war, German army and corps HQs were regularly calling upon their flying sections to conduct tactical flights over the battlefield. In most cases the airmen were used to determine the strength and deployment of nearby enemy forces before a battle or to monitor the rear-areas during an engagement for any newly arriving reserves. Other examples saw aircraft ordered airborne to search for gaps or weak points in the allied line or to conduct artillery cooperation above an important sector of the battlefield. Most importantly, using their aviator's reports, German leaders throughout 1914 were able to deploy their own reserves to key areas of the battlefield, thus preventing a collapse or hastening a victory.

Meanwhile the balloon sections were used almost exclusively for artillery direction. Upon their arrival on the battlefield, the FLAs would be attached to several heavy batteries, which fired indirectly at ranges that required an observer's correction. The telephone wagons attached to each FLA enabled the balloon's observer to maintain instantaneous contact with battery commanders on the ground. The observer would therefore be able to quickly inform the artillerymen of any new target and direct the battery's fire until it was neutralized. In rarer instances, a FLA observer would be used to sketch the deployment of nearby enemy infantry and artillery. Unfortunately, the balloon sections' mobility restrictions greatly reduced their ability to participate during the period of open warfare where the front lines often changed. As a result, the FLAs attached to the defensive left wing were significantly more active during the war's initial campaign than the sections attached to the more mobile right wing.

Mobilization & the Period of Concentration

Each of Germany's seven Western armies were mobilized and deployed to their assigned sectors swiftly and efficiently under the competent direction of the General Staff's Operations Department. Within just 13 days, 11,000 trains had transported the army's 119,754 officers, 2.1 million men and 600,000 horses to the various marshalling areas along the German frontier.[42] This speedy mobilization and deployment of the empire's armed forces resulted in all active corps being fully concentrated by August 12th with all reserve corps in position just two days later.

Likewise, Germany's aviation units speedily completed their initial deployments in accordance with the Operations Department's mobilization plan. *Ltn.* Andre Hug, a pilot with Seventh Army HQ's FFA 26, described his unit's mobilization:

In Freiburg I joined my newly-formed unit, Feldfliegerabteilung 26, which was attached to General von Heeringen, commander of the German Seventh Army. Six brand new Aviatik biplanes, with 100 h.p. Mercedes engines, six cylinders, and light metal pistons, were already waiting on the air field for the Staffel *leader,* Hauptmann *Walther, who was a trained observer and not a pilot. The six pilots and the six observers, the maintenance ground crew, all accessories, the* staffel *commander's car, trucks, and the military war chest well fitted with one million Marks for the payment of the* staffel, *were all ready. The war chest was just that— a war chest. Because it could not be predicted where we would travel, or what supplies we would need, our mobilization plans called for the* staffel *to have sufficient money on hand to pay for all its needs until supplies could reach it. The* Staffelführer *and I went to the National Bank where we cashed our checks from the War Ministry and received one million Marks. The money was placed in a big case which we took back to our* staffel.

…. Every plane was provided with two Mauser automatic pistols; one for the pilot and one for the observer. According to the mobilization plan, Feldfliegerabteilung 26 had to be loaded on the first day of mobilization at 7:00pm. In other words, the entire staffel *was to be on a supply train on 2 August 1914 and was to depart to Strasbourg where the Seventh Army Commander, together with his staff, had just established his headquarters.*

The aircraft had to be disassembled for loading onto the train. During the night the train travelled to Strasbourg were it stopped at a siding nearest to the aerodrome. The entire supplies had to be transported to the hangars which were one and a half miles distant from the airfield. The wings and fuselages had to be carried by hand. Twenty-two hours after the departure from Freiburg, Feldfliegerabteilung 26, was ready to operate again. If the mobilization order had decreed that the aircraft were to fly and that the crew and all the materials were to be trucked to Strasbourg, the transfer of the entire staffel *would have been accomplished within two hours rather than twenty-two hours.*[43]

Meanwhile, *Ltn.* Rudolf Berthold explained how FFA 23's mobilization went much more smoothly:

Unloading proceeded with feverish haste. Each two-man aircrew competes with the other because each wants to achieve the first flight over the enemy… Some of us already knew one another from our peacetime service. The Abteilung *is made up of six aeroplanes, an* Abteilungsführer *and adjutants, seven pilots and as many observers. They are really nice fellows. Our leader, Vogel von Falckenstein, has a true Prussian soldier's disposition and is an accomplished old pilot. Everything went smoothly thanks to his aviation experience and his tireless efforts.*[44]

The first four FFAs were ready for action on August 4th, while all the others were activated over the next five days.[45] The airfields in the mobilization area were in close proximity to the front along the line Aachen—St. Vith—Trier—Metz—Saarburg—Neu Breisach. This had been done to allow the first flights to penetrate as deep into allied territory as possible. However, although many of the flying sections had arrived nearly a week before the army, aerial activity during the period of concentration was practically non-existent due mostly to the ongoing preparations each unit was making for the coming campaign.

Making a flying section 'campaign ready' was no simple task. Once the men and machines arrived in the mobilization area, the section's enlisted personnel had to assemble and inspect each of the section's six aircraft. The inspection included a thorough checking of each machine's engine, flight surfaces, and controls in order to certify they were all fully capable of safely performing aerial operations under wartime conditions. Before finally being certified, each aircraft was given a test requiring it to climb to 1,000 meters (3,280 feet) within 15 minutes. Carrying out this assignment was of great benefit to the ground

Above: As Commander-in-Chief of the French Army, Joseph Joffre was the author of Plan XVII and was responsible for army operations once the war began. (Author's Collection)

crews, who used the opportunity to gain additional experience working together in wartime.[46]

Meanwhile, as the aircraft were being prepared for war. Each FFA sent several officers to their army's airpark in order to obtain sufficient quantities of spare parts, fuel, and other supplies required to begin mobile operations. Similarly, the balloon sections carried out their preparations by procuring large quantities of gas, forage, and other materials from the army quartermaster. The FLAs were declared ready for operation once the unit's gas columns were equipped with enough gas cylinders for two inflations. The officers and men of the FFAs and FLAs continued to spend the first week of the war working feverishly to ensure their units were fully operational as quick as possible and would remain airworthy throughout the coming campaign. The hard work ultimately paid off as the officers and men of the *Fliegertruppe* had all their units fully operational before the army began offensive operations.

The Western Allied Strategy

The wartime strategy for the Entente was built upon the foundation of the Franco-Russian military convention of 1892–93. By 1913, after years of negotiations, the Franco-Russian leadership had agreed upon the final coordinated war plan that the two nations would employ in 1914.[47] In accordance with this strategy, the French and Russian armies would both ultimately seize the initiative and attack Germany simultaneously within 15 days of mobilization.[48] Thus, by immediately applying overwhelming pressure through aggressive offensives on two fronts, the Entente's leaders believed that the outnumbered Germans would be quickly defeated and brought to the peace table.

The French war plan, known as Plan XVII, called for the formation and deployment of five field armies on the line Belfort—Namur. Overall command of France's land forces was entrusted to General Joseph Joffre, the Chief of the French General Staff and the author of Plan XVII. At Joffre's disposal was the combined strength of 21 army corps, 25 reserve divisions, eight third-rate territorial divisions and 10 cavalry divisions for a total of nearly 1.3 million men.[49] For additional support, a British Expeditionary Force (BEF) with the initial strength of six divisions and a cavalry division was expected to cross the channel and arrive on the French left flank prepared for offensive action around the 18th day of mobilization. Joffre claimed that the plan had no predetermined scheme of maneuver.[50] It was obvious, however, that the plan was conceived to carry out a massive offensive. According to the plan's most likely contingency, the French right wing (First and Second Armies) was to attack between the Vosges Mountains and Toul while the left wing (Third, Fourth, & Fifth Armies) attacked north of the line Verdun—Metz. Joffre's strategic objective was to therefore surround Metz and force a decisive battle with the armies of the German center, thus splitting the German Western battle-line in two.[51]

Organizationally, the French Army closely resembled the German model. A fully mobilized French Army Corps (*Corps de Armée* or CA) consisted of 40,000 men organized into two divisions of four regiments each. Each corps also possessed a brigade of reserve infantry, two regiments of cavalry, a company of engineers and various ammunition and supply trains. Regarding artillery, each corps had a total of 120 flat trajectory 75mm field guns (nine four gun batteries for each of the two divisions as well as twelve four gun batteries designated as 'corps artillery' to be used as the corps commander saw fit). Rather than being formed into reserve corps, French reserve divisions were organized into 'Reserve Division Groups,' each comprised of three divisions.[52] Finally, the BEF was initially organized into two corps consisting of two divisions organized into three brigades of four battalions. Later, during late August/early September, the British were able to add a third corps—bringing the total number of infantry divisions on the continent to six.

Determined not to repeat the mistakes of the Franco-Prussian War, French military doctrine fanatically emphasized the offensive. By the time Joffre ascended to power as Chief of the General Staff in 1911, the dogmatic philosophy of the "all-out offensive" (*L'offensive à outrance*) had been established throughout the army.[53] An army regulations manual written in 1913 stated that: "The lessons of the past have borne fruit: the French Army, returning to its traditions, accepts no law in the conduct of operations other than the offensive… Only the offensive yields positive results."[54] At its core, the ideology was based upon the belief that any attack could succeed as long as it was conducted with sufficient speed and determination (élan). Joffre and the French Army leadership believed this doctrine could be successfully employed due to the superiority of their field gun, the famed "French 75." With great range and the ability to fire twice as fast as the German and British field guns, *mademoiselle soixante-quinze* was undoubtedly the best field gun in the world. French war planners believed the 75's superiority would silence the German artillery while their infantry—attacking with exceptional skill and determination—would ultimately secure victory. Joffre echoed these sentiments, believing his offensive doctrine would be used to quickly defeat the Germans and bring the war to speedy conclusion.[55] The French *Generalissimo's* confidence in this strategy was evident in his *General Order #1*, which directed each of his armies to take to the offensive as soon they were deployed and concentrated.[56]

With the stage set, and in accordance with their respective war plans, the opposing armies would both take to the offensive as quickly as possible in order to achieve a decisive victory that they envisioned would bring the war to a speedy conclusion.

★ ★ ★ ★ ★ ★ ★ ★

The Start of Hostilities: Liège

Located in the northeastern corner of Belgium some 18 miles from the German frontier to the east and 10 miles from the Dutch border to the north, the fortress city of Liège stood directly in the path of German First and Second Armies' assigned route of march.[57]

Above: An officer of engineers by training, General Gérard Leman served as a professor at the Belgian War College for 25 years before being appointed both commander of the 3rd Infantry Division and Governor of Liège just prior to the outbreak of war. (Author's Collection)

Liège's primary defenses consisted of 12 modern concrete forts positioned at regular intervals around the city. In addition to the forts themselves, Liège was defended by the 3rd Belgian Infantry Division, the 15th Infantry Brigade, and various reserve and militia units, a total of 35,000 combatants under the overall command of General Gérard Leman. These 35,000 men were mostly deployed in trenches located in the intervals between the forts. Together, this perimeter of 12 modern forts connected by field works presented a formidable obstacle.[58] Nonetheless, with some 400 Belgian guns of various calibers commanding all roads and rail lines leading in and out of the city, Liège's capitulation was essential to the German right wing's advance into Belgium and the success of their war plan.[59]

Moltke initially entrusted Liège's capture to General Otto Emmich, commander of X *Armee Korps*

Above: Wartime postcard romantically depicting the "storming of Liège." (Author's Collection)

(X AK). With the fortress' rapid demise of paramount importance, Emmich was assigned a combined force of six brigades of active infantry, the 2nd Cavalry Corps (HKK 2) and two batteries of 210mm mortars to "launch a surprise attack, penetrate the outer works, and seize the town and its communication facilities."[60] Thus, on the morning of August 5th, after Leman rejected a German summons for the fortress' surrender, Emmich's guns opened fire on the eastern bank forts. Later that afternoon five German infantry brigades launched an assault against the eastern bank trenches. However, the attack was supported with minimal artillery and the assault failed with substantial losses.[61] Only one brigade, the 14th *Infanterie Brigade* (14 IB) under the temporary command of General Ludendorff, managed to successfully pierce the Belgian defenses. By mid-morning the following day, 14 IB had pushed its way beyond the Meuse River and into the city itself. Unfortunately, the other brigades' failures had left Ludendorff's men entirely cut off and surrounded without means of resupply. Running low on ammunition and food, the 14 IB's 1,500 remaining men were therefore compelled to remain inside the city until they could be relieved.[62]

Meanwhile to the north, *Generalleutnant* Georg von der Marwitz' HKK 2 had successfully crossed the Meuse and moved into a position northwest of the city which threatened the Belgian's line of communication. Acknowledging that an encirclement was inevitable, Leman released his mobile troops (3rd Infantry Division and 15th Infantry Brigade) during the evening of August 5–6 in order to join the Belgian Field Army located along the Gette River to the west. From that position the Belgian leadership intended for their field army to delay the German advance until the Franco-British armies could come to their aid.[63] In the meantime, Liège's 12 forts, henceforth operating independently without any communication with one another, would continue their resistance for as long as possible. However, without the support of the 3rd Infantry Division and 15th Infantry Brigade, Leman knew that it would only be a matter of time until the forts capitulated.[64]

Having learned the lessons of their initial failure, the Germans decided to change commanders and tactics. On August 8th, Emmich's original force was strengthened by the arrival of a new siege army consisting of the VII and IX AKs under the command of General Karl von Einem (commander of VII AK). To expedite the forts' surrender, Einem's army was given OHL's heavy siege artillery. Indeed, the Germans decided to simply bring up their siege artillery and

Above: Troops from the 14th Infantry Brigade fight their way into Liège. (Author's Collection)

pummel the Belgian forts into submission one by one. Thus, after their arrival on the 12th, the siege guns began systematically bombarding and destroying each of the forts until the last garrison at Fort Flemalle surrendered at noon on August 16th.[65] With Liège's surrender, German First and Second Armies were finally able to begin their advance to the Franco-Belgian border.[66]

The *Fliegertruppe* units of First and Second Armies were still in the process of mobilizing and preparing for war at the start of the siege. The first right wing flying section to become operational was IV AK's FFA 9 on August 6th.[67] Equipped with Rumpler *Taube* monoplanes, FFA 9 inexplicably remained inactive on the 6th and 7th because its commander, *Hptm.* Musset, had chosen to stay with Emmich and his staff during the initial assault. Consequently, the first flights over the battle zone were performed by Fortress Flying Section Cologne on the 6th and 8th. However, because of the great distances involved, Fest.FA Cologne's aircraft were only able to operate over the Liège sector for a maximum of 45 minutes—dramatically reducing their effectiveness.[68]

Late in the evening of the 5th–6th, long after Emmich's infantry attack had failed, Zeppelin *Z VI* was sent from its base at Cologne to bomb Liège's outer defenses. The ship was badly affected however by its 440 lbs. payload of modified artillery shells.[69] As a result, the crew was forced to fly at an altitude of only 4,700 feet, making them an easy target for ground fire. At 3:00am, *Z VI* dropped its projectiles near one of the eastern bank forts while under constant artillery and machine gun fire. Having taken many hits, the Zeppelin quickly lost altitude and eventually crashed in a forest near Bonn before it could reach its hanger outside Cologne.[70]

Musset's FFA 9 belatedly began operations on the 8th, conducting two tactical reconnaissance flights to survey the damage that had been inflicted upon several of the east bank forts. During the second flight, which had taken off at 5:00pm, the observer confirmed that Fort Barchon had fallen—the first of the twelve forts to do so. Meanwhile, earlier that afternoon FFA 9 was ordered to send an aircraft over Liège to try and determine the location and condition of 14 IB. The local German leadership had been unable to establish any communication with the brigade after its breakthrough and had begun to fear that they had been forced to surrender.[71] Those concerns were quelled when, despite strong shrapnel fire from the surrounding forts, the pilot skillfully landed his *Taube* in a Liège city park. The flight's observer then spoke

> **Friendly Fire**
>
> German aviators reported many instances of friendly fire during the first weeks of the war. In his post war memoirs, a junior officer in Second Army's VII AK, described how German infantrymen present at Liège reacted when they saw their first aircraft overhead:
>
> *Our company crossed the border between Germany and Belgium on the 7th day of August 1914. A short march after crossing the border brought us to a little village. The sun was just about to go down when an airplane suddenly appeared above the column. We had never marched into a hostile country before and we had never met an enemy. Everyone was nervous and excited over the events which they felt were sure to come. For this reason, they were at once certain that an airplane in hostile territory could only belong to the enemy. Suddenly a shot was heard; then several; and in a few seconds the entire company was shooting. Next, machine guns could be heard somewhere in the distance taking up the firing; then artillery shells were seen bursting in the air. Now every doubt disappeared. If the artillery was firing, the aviator must be an enemy. Even the driver of the field kitchen shot at the poor aviator with his pistol. The aviator continued to fly for some time and then seemed to sink to earth. The psychological excitement was satisfied by the wild shooting. The men raised a loud cry of joy because they believed they had hit him, and nearly the entire company started off to take the aviator prisoner. One by one the men returned with foolish expressions on their faces, like a young hunting dog which has been vainly chasing a rabbit. Next day a division order was received which began as follows: "It is only due to the very bad shooting of the troops that today two German aviators are still alive ... "*[73]

Facing Page, Top: Aerial photograph of the city of Liège. Note: the destroyed bridge at the center of the photo.. (Bundesarchiv, Bild 146-2014-0033)

Facing Page, Bottom: Photograph of Fort Boncelles, located south of the city. (Bundesarchiv, Bild 146-1978-102-12A)

to several officers that had gathered near the aircraft. After obtaining all the necessary information, the airmen climbed back into the aircraft and flew back to their airfield from where the observer's written report was later dispatched via touring car to army command.

In the days that followed, as the attacks at Liège continued, FFA 9 was ordered to conduct long-range reconnaissance flights over the Namur sector. The flights' main purpose was to confirm or disprove the rumors of French troops being quickly rushed north into Belgium to assist the Liège garrison or the Belgian field army. Two excellent reports produced by the fliers of FFA 9 on the 9th found no French forces in the area, thus allowing the German leadership to confidently continue their attacks on Liège without needing to dilute their strength by detaching a covering force to deal with a potential French relief sortie.[72] Indeed, the only action FFA 9's observers could report was the strengthening of the field works around fortress Namur itself.

On August 11th the German VII, IX and X AKs all arrived in the area of Liège prepared for offensive action. Their flying sections—FFAs 18, 11, and 21 respectively—all immediately began conducting tactical reconnaissance flights over Liège, giving the army leadership a clearer picture of the overall situation. To further assist Einem and his staff plan the reduction of the fortress, these three flying sections successfully performed aerial photography sorties of the entire fortress zone. The airmen's photographs were promptly developed and handed over to Einem's staff who utilized them when the heavy siege artillery finally arrived several days later.[74] Meanwhile as these flying sections were ordered to concentrate on tactical missions, another section—FFA 9—was directed to perform long-range reconnaissance flights all over Belgium. On the morning of the 11th FFA 9 was ordered to locate the Belgian Field Army and determine its intentions. Later that morning one of the section's *Taubes* performed a sortie along the line Aachen—Antwerp—Brussels—Namur, covering a distance of 202 miles. The flight proved a success, as the observer located the Belgian Field Army near Tirlemont and the Gette River. After landing, the observer personally delivered his written report to Second Army HQ.[75] This impressive reconnaissance sortie would ultimately be the longest distance flight performed by a German aircraft during the Marne Campaign.

Unfortunately, August 11th also brought setbacks for the *Fliegertruppe*. In addition to the four aforementioned flying sections, the Guard Korps' FFA 1 had also been ordered to participate in the action at Liège. Throughout the morning, the section's overzealous commander, Hptm. Jasper von Oertzen, quickly dispatched four aircraft in quick succession from the section's airfield at Thirimont with orders to conduct tactical operations over the Liège battle-area. The fourth aircraft crashed on takeoff, instantly killing its crew. One of the others experienced engine failure and was forced to make an emergency landing in Belgian occupied territory from where the crew managed to avoid capture and rejoin the section the following day. Finally, due to the oppressive heat and

69

excessive turbulence, the other two aircraft were compelled to turn around and return to the airfield before ever reaching Liège.[76]

By any measure, the morning was a disaster for FFA 1. The loss of two aircraft in addition to its failure to get a single machine over the battle-area was cause for alarm. Moreover, at a time when mobile operations had yet to begin, FFA 1 was already reduced from six to just four airworthy machines, prompting Second Army's Army Air Park #2 (AFP 2) to release its only remaining machine for the section's use. That evening a deeply concerned Second Army HQ decided to take action to prevent any further weakening of its flying sections before mobile operations began. A new directive was issued to Einem requesting that each flying sections' activity be strictly limited for the remainder of the Liège operation. According to Einem's subsequent order, FFAs 11, 18 and 21 were to continue to provide tactical reconnaissance in a limited fashion over Liège while FFAs 1 and 9 were assigned long-range reconnaissance sorties deep into Belgium. Both FFAs 1 and 9 were ordered to concentrate their search efforts on the line Diest—Tirlemont—Namur—Dinant in addition to regular checks around the important cities of Antwerp and Brussels.[77]

Throughout the final three days of the siege, FFAs 11, 18 and 21 continued to perform useful tactical reconnaissance by providing detailed reports regarding the damage that had been inflicted on the various Belgian forts.[78] The value of these reports was enhanced due to the scarcity of the German heavy artillery batteries' ammunition. By confirming which fort's turrets were already out of action, the German airmen often saved the siege batteries from needlessly wasting their valuable armor-piercing heavy caliber shells on an already neutralized target.

Upon the final fort's capitulation on August 16th, all participating FFAs of the Liège operation were ordered to begin their journey past the captured fortress city into the Belgian countryside to begin the mobile campaign. Within 48 hours, each of these flying sections would be airborne in the search for the allied left wing in order to assist First and Second Armies complete their envelopment maneuver as prescribed by the German War Plan.

German Left Wing: First Battle of Mulhouse

As the Germans attacked Liège in preparation for their right wing's offensive, the French launched a preliminary offensive of their own some 250 miles away in Alsace. On August 7th, as the rest of Joffre's forces were completing their deployment, a small force comprised of the French VII Corps (VII CA) and the 8th Cavalry Division (8 DC) departed Belfort under

Above: The former Minister of War, Josias von Heeringen, served as commander of German Seventh Army. (Author's Collection)

the overall command of *Général* Louis Bonneau with orders to invade the formerly French controlled region of Alsace and capture the city of Mulhouse. The operation's objective was to seize important territory covering the French Army's right flank in preparation for Plan XVII's main assault.[79] Moreover, by successfully recapturing "the lost territory of Alsace," Joffre could announce France had won the war's first victory, thereby boosting civilian morale at home as well as instilling confidence amongst the troops in the field.

Bonneau's combined force began their advance at 5am in two columns—each with orders to converge upon Mulhouse by nightfall. Marching near the Swiss border, the 14th *Division d' Infanterie* (14 DI) was directed to attack the German forces at Altkirch while the 41 DI was to proceed further north through the town of Thann.[80] German border security in the area was charged to the 58th *Infanterie Brigade* (58 IB), who put up a spirited defense at Altkirch but was eventually forced to withdraw in the face of overwhelming French numerical and artillery superiority. The German delay action at Altkirch was successful in prohibiting Bonneau's men from reaching their objectives for the day while alerting German Seventh Army HQ of French intentions in the region.[81] Nonetheless, Bonneau's troops entered Mulhouse unopposed at 3pm the next day, parading through the streets with their regimental bands playing *"La Marseillaise"* to cheers of *"Vive la France!"* from the French Alsatian population.[82]

German Seventh Army commander *Generaloberst* Josias von Heeringen and his Chief of Staff *Generalleutnant* Karl Hanisch immediately began making preparations to expel Bonneau upon hearing of Mulhouse's capture. Heeringen and Hanisch were both fearful that Bonneau's force would be quickly reinforced and had decided accordingly to employ the army's two active corps (XIV and XV AKs) to immediately counter-attack and expel the French from Alsace.[83] Heeringen proposed his two corps would act in concert and move in such a way as to split the French forces north of Mulhouse. OHL consented to Heeringen's plan in a message sent to Seventh Army HQ at 10:30pm on the 8th stating: "With your army you have full freedom of action. Future large-scale operations planned by OHL are not affected by your proposed operation."[84] Thus, upon receiving approval to proceed, Heeringen's staff dispatched directives to XIV and XV AK HQs ordering the counter-attack to begin around noon the following day.[85]

Seventh Army's flying sections were the first on the Western Front to begin active flight operations.[86] From the 4th-6th, XIV AK's FFA 20 dispatched at least two flights per day to monitor the rail traffic at Belfort. On the morning of the 7th, one of FFA 20's crews spotted Bonneau's columns on the march and decided on their initiative to land near the front to inform 58 IB's commander of what they had seen. The verbal report given by the flight's observer confirmed for the brigade commander that he was facing a much larger force and not just a probe. This allowed the small German force to smartly conduct a delaying action that temporarily halted Bonneau and his men.[87] Meanwhile, FFA 3 (attached to XV AK) continued to patrol the area around Belfort to determine if additional French troops would be sent to reinforce Bonneau's invasion force. Finally, in order to assist with the planning of the counter-attack, Seventh Army HQ's own FFA 26 was ordered late in the afternoon of the 8th to reconnoiter the French positions around Mulhouse to "locate positions suitable for an attack."[88] Delivered during the evening hours in conjunction with a number of messages from his cavalry, FFA 26's reports furnished Heeringen with a general knowledge of the French deployment throughout the area including several gaps in Bonneau's line north of Mulhouse.[89] This intelligence prompted Seventh Army HQ to create an offensive plan to exploit the gaps and envelop the French left wing. According to Heeringen's design, XV AK would assault and envelop the French left flank while XIV AK split their line and recaptured Mulhouse. It was hoped that the two corps' attack, delivered in concert, would roll up Bonneau's forces and push them against the Swiss border where they would be annihilated.

Present during this operation, FFA 26 pilot Andre Hug described how dangerous these first flights of the war were:

On August 3rd 1914, the day of the declaration of war by Germany upon France, I was ordered to reconnoiter to determine whether or not the southern flank of the French Army had already started to march towards German Alsace. Curious, because there was not yet any experiences concerning the activity of aircraft in a war; so in anxious expectation, I started on my first war flight. I wondered whether one could be shot down while flying in the supposedly safe height of 3,000–4,000 feet over hostile country. When I approached the French border a French plane drew near. I considered what course of action to take. As weapons of attack we had only two Mauser automatic pistols, not capable of shooting accurately at a distance over thirty yards. I did not trust the French pilot and he did not trust me. We both flew a curve simultaneously and continued our flight. On our return we met again and made way for each other. After the landing I was embarrassed with my cowardly behavior.

The following day, on a reconnaissance flight to the same area, a French plane again arrived. Immediately I made a sharp turn and flew straight toward the Farman biplane, as if I intended to ram him. I kept my nerve, and in the last moment the French pilot did a nose dive and disappeared. The axiom that attack is the best defense again turned out to be true.

Making a reconnaissance flight at 3,300 feet over the Vosges Mountains, I had the opportunity to prove at which height bullets of rifles or machine guns are no longer a danger for aircraft. When I flew over a group of a few hundred French soldiers, a white cloud suddenly appeared over the troops. The French soldiers had fired volleys at my plane much in the manner in which German infantry had practiced before the war. Apparently my plane was not hit. After landing, I noticed that one rifle bullet had pierced the lower layer of linen of the left wing and

Above: Wartime sketch depicting the German attack at Mulhouse. (Author's Collection)

was laying between the lower and upper linen in the wing.[90]

The morning of the 9th passed with relative calm. As Seventh Army HQ requested, both XIV and XV AKs waited until midday until all their forces were in position to attack. At 12:30pm XV AK's commander, *General der Infanterie* Berthold von Deimling, gave the order for his 39 *Infanterie Division* (39 ID) to assault the French 41 DI at the village of Cernay (Senneheim) on the extreme left of the French line. The attack quickly bogged down, requiring the corps' other division (30 ID) to be committed into action around 3:00pm. Soon thereafter, despite a fierce and valiant defense, the French line collapsed, prompting the whole of 41 DI to flee. With the French line broken, XV AK quickly dispatched a reinforced column to march against the flank and rear of the French troops still fighting in Mulhouse.

To the south, XIV AK's attack was delivered against Mulhouse itself. The corps' flying section, FFA 20, was sent up in the early morning hours to reconnoiter the French line around the city. The information gathered by the airmen helped secure some early minor gains. However, XIV AK's assault rapidly slowed due to several groups of French artillery firing from covered positions on the reverse slope of a hill just outside Mulhouse. Indeed, the effect of the French 75s was so great that XIV AK's commander: *General der Infanterie* Ernst von Hoiningen felt compelled to prematurely deploy all his reserves and ask Seventh Army HQ for the assistance of XV AK. Nonetheless, despite committing the corps' reserve, XIV AK continued to make very little progress. Finally, thanks to the work of some forward observers, a battery of German howitzers was able to silence the French artillery and give the German infantry an opportunity to breakthrough into the French rear.

XIV AK's assault continued into the night with brutal hand to hand fighting taking place in the streets of Mulhouse. Repeated French counter-attacks were beaten back with heavy losses before eventually stopping due to the arrival of XV AK's troops, which opened up a second front on the north side of the city. Thus, as Heeringen envisioned in his attack plan, German forces were able to take advantage of weak points in the French line around Cernay to threaten Bonneau's main force at Mulhouse with encirclement.[91] Bonneau was therefore compelled to immediately order all his forces to withdraw from Mulhouse and take up defensive positions to the southwest to cover the group's retreat back to the

Above: German troops reclaim Mulhouse after brutal hand-to-hand fighting.. (Author's Collection)

border. Over the next two days, the French fought a rear-guard action to allow their troops to safely reach the safety of Fortress Belfort.[92]

During this time, aircraft from FFAs 20 and 26 maintained continuous visual contact with the retreating French columns, reporting to their HQs that "the enemy was headed south and west in full retreat" back towards the French frontier.[93] At the same time, XV AK's FFA 3 resumed long-distance reconnaissance around Belfort to ascertain the strength of French forces in the area and determine whether another invasion of Alsace was imminent. The subsequent reports delivered to Seventh Army HQ found no evidence of a pending offensive, which enabled Seventh Army's forces to reorganize and later return to their assigned positions further north near Sixth Army's left flank.[94]

The actions at Liège and Mulhouse proved for the first time that the *Fliegertruppe* were able to provide meaningful operational and tactical support during an engagement. The artillery spotting work performed by Second Army's flying sections assisted Einem's siege artillery quickly and efficiently reduce the Belgian fortresses, enabling the right wing armies to begin the wheeling maneuver imperative to the German War Plan. Similarly, the sorties flown by Heeringen's aviators helped Seventh Army plan and conduct a successful battle resulting in the war's first victory over the French.

The Frontier Battles on the German Left Wing

The French War Plan was based upon two large-scale offensives focused on surrounding the German fortress city of Metz and destroying the nearby German field armies. According to Joffre's General Order No. 1, the first of these offensives was to begin in Lorraine immediately upon completion of the armies' concentration while the second was to occur further north in the Ardennes several days later. Located in the south on the extreme right of the French line, *Général* Auguste Dubail's First Army was to form the spearhead of the opening offensive. Dubail's four corps were ordered to move north, capture Sarrebourg then push the Germans east into the region of Strasbourg. On Dubail's left, *Général* Édouard de Castelnau's Second Army was directed to march on Morhange *en échelon* to protect First Army from a possible counter-attack out of Fortress Metz. French First Army's right flank was to be protected by the newly formed 'Army of Alsace' composed of the VII CA, 44 DI, four reserve divisions, and 8 DC under the overall command of *Général* Paul-Marie Pau.

Above: Auguste Dubail, commander of French First Army. (Author's Collection)

Above: Although he was given command of German Sixth Army as a result of being Crown Prince of Bavaria, Rupprecht quickly proved himself to be a capable commander. . (Author's Collection)

Together, all five French armies, totaling nearly 1.3 million men, completed their deployments on August 13th. Thus, in accordance with Joffre's orders, the Lorraine offensive began the next day when his First and Second Armies crossed the German frontier.[95]

The German forces opposed to the French right wing consisted of the German Sixth and Seventh Armies. Under the command of the Bavarian Crown Prince Rupprecht, German Sixth Army was composed of the XXI AK, Bavarian I, II III AKs (b. AK), Bavarian I *Reserve Korps* (b. RK) and HKK 3. Heeringen's Seventh Army consisted of the XIV and XV AKs and the XIV RK. To ensure operational solidarity, Moltke subordinated Seventh Army under Rupprecht's command—thereby giving Sixth Army HQ control over all operations on the left wing.[96] In accordance with the German War Plan, the left wing's mission was to tie down as many enemy forces in the region as possible, either through offensive action against the enemy fortress line or by conducting a series of strategic withdrawals followed by counter-attacks. Rupprecht wisely chose the latter course of action when his fliers confirmed on the afternoon of the 14th that the French were sending a large force, at least seven corps in strength, through the Charmes Gap into the German Imperial Territory of Lorraine.[97]

In consonance with the German War Plan, Rupprecht ordered his forces to conduct a fighting withdrawal as the French forces pursued. Over the next four days the Germans continued their withdrawal further into Lorraine, baiting the French to overextend themselves and become vulnerable to a counter-attack. A series of skirmishes were fought throughout the withdrawal between the leading French columns and the German rear-guard without a major result.

German Sixth Army's flying sections were heavily active during this period, providing valuable operational reconnaissance reports for Rupprecht at Sixth Army HQ as well as tactical support for

Above: Bavarians on the march in Lorraine. (Author's Collection)

the rear-guard troops. Sixth Army HQ's FFA 5 had already been actively monitoring the frontier area since the 11th, producing reports that detailed significant rail traffic at Épinal and Toul as well as large troop concentrations in the environs of Nancy and Lunéville.[98] Sixth Army's various corps flying sections, specifically FFA 8 (XXI AK) and FFA 1b (I b. AK), had also kept busy during this period conducting successful tactical flights on the 11th and 12th.[99] On the 13th, FFA 5 was able to provide Sixth Army HQ with a general outline of the French deployment along the line: Nancy—Raon l'Étape—St. Dié—Toul—Neufchâteau—Lunéville—Épinal.[100]

The morning of the 14th saw aircraft belonging to each of Sixth Army's five flying sections airborne. FFAs 6, 8, 2b and 3b each submitted reports describing in detail the impressive scale of the French offensive. Together, these messages gave Sixth Army HQ a complete understanding of the general situation. Rupprecht and his staff were now certain that the French were going to launch a major offensive into Lorraine and began drawing up proposals for a counter-offensive. Moreover, in addition to these early long-range reconnaissance flights, Rupprecht's corps flying sections continued to provide excellent tactical support to the troops on the ground. For example, FFA 3b conducted artillery cooperation sorties for III b. AK's heavy batteries. Meanwhile, an aircraft belonging to FFA 2b attacked Ferdinand Foch's XX CA by dropping bombs and aerial darts[101] on one of its columns. Finally, FFA 1b spent the day sending multiple machines airborne to assist I b. AK's troops repulse the attacks of two French corps in the area around Blâmont and Cirey.[102] Each flight cooperated with the corps' leadership by dropping notes or sketches at forward command posts detailing the French strength and deployment in the area as well as the location of their reserves.

Seventh Army's fliers were equally as busy on the 14th. At 6:30am a flight belonging to Seventh Army HQ's FFA 26 had spotted troops belonging to French First Army's XIV and XXI CAs engaged with German *Landwehr* units deployed near the frontier. Later that morning, a second patrol reported that several large French columns had seized Thann and were moving to retake Cernay. However, before they returned to their airfield to report to Seventh Army HQ, the flight's crew landed near the front to warn the local *Landwehr* commander of the large numbers of hostile troops in the area that were marching in his direction. Likewise, XIV AK's FFA 20 completed two flights during the 14th in support of the various

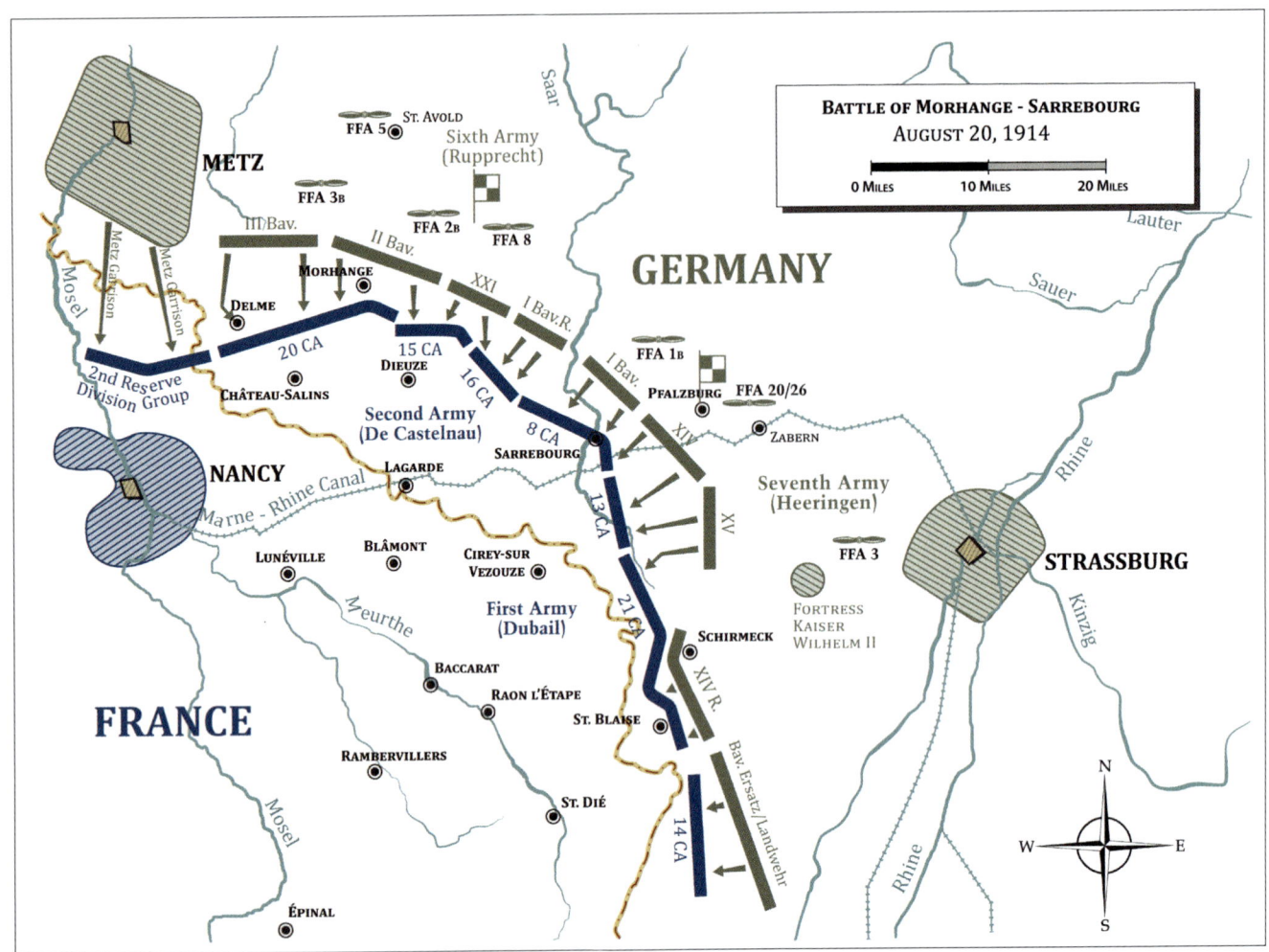

Landwehr brigades that were covering Seventh Army's withdrawal and redeployment. The section's first flight discovered two columns marching from Belfort in the direction of Altkirch and Mulhouse. After taking note of his observations, the observer directed his pilot to land near Mulhouse in order to warn the commander of the 55th *Landwehr* Brigade of the impending assault. Next, the crew flew back to their airfield and delivered a formal report to XIV AK HQ, which was forwarded to Seventh Army HQ later that evening. FFA 20's second flight, conducted shortly after noon, found three divisions marching in two columns—one in the direction of Altkirch and the other directly toward Mulhouse. The flight's observer dropped a handwritten message containing his observations over the side of the aircraft near the town of Mulhouse for the *Landwehr* commander's consideration.[103]

Early German intelligence reports concluded that as many as 15 French corps (more than 60% of their army) were being used for their Lorraine offensive.[104] The German leadership, eager to destroy or at least tie down these corps in a sector far away from the decisive right wing, ordered their left wing armies to withdraw and entice the pursuing French forces to recklessly expose themselves to a counter-attack from three directions—Fifth Army through Fortress Metz, Sixth Army from the east and Seventh Army from the south—thereby creating the conditions for two French armies' defeat in a large decisive battle under extremely favorable conditions.[105] This counter-attack plan was quickly cancelled however on August 16th when Sixth Army HQ learned that the original intelligence estimates of the French right wing's strength were grossly exaggerated. Instead of 15 corps, it was now believed that the French had committed seven into Lorraine.[106]

Consequently, once it was determined that the left wing was opposed by just seven corps, the counter-attack plan using Fifth Army was taken off the table.[107] Nonetheless, Rupprecht and his Chief of Staff, *Generalmajor* Konrad Krafft von Dellmensingen, still advocated an offensive against French First and Second Armies.[108] It was at this moment that the mentally weak Moltke started to lose control over his subordinates and the overall situation. Instead of firmly ordering Rupprecht to continue the withdrawal and take up a defensive position (the

Above: German infantry breakthrough the French line. (Author's Collection)

strategy he preferred), he dispatched a cryptic message that allowed Sixth Army HQ to take any course of action they saw fit. Thus, unwilling to concede more German territory without a fight, Sixth Army HQ was granted the ability to launch a counter-offensive whenever the opportunity presented itself.[109]

Aerial activity from the 15th—17th was extremely light due to the circumstances around the left wing's withdrawal combined with poor weather.[110] On August 18th however, a reconnaissance patrol flown by XIV AK's FFA 20 discovered that French First and Second Armies had lost contact with each other. The observer's message declared unequivocally that de Castelnau's Second Army had turned north up the Sarre River while Dubail's First Army had continued eastward.[111] FFA 20's report therefore confirmed for Rupprecht that the French were vulnerable and that the time had indeed come for his long awaited counter-attack.

Rupprecht's forces spent the 19th moving into position for the offensive which was scheduled to begin before dawn the following day. From the towns of Delme and Morhange in the north to Mont Donon and Saint-Blaise in the south, the two German left wing armies planned to attack in concert with eight corps along a 62 mile (100 km) front.[112] Because the attack plan largely featured frontal assaults, Rupprecht stressed to his subordinate commanders the importance of using their heavy artillery, which could take advantage of their superior range to silence the French 75s. Thus, as a means to further improve the artillery's effect, Sixth Army HQ attached Bavarian FLA 1b to I b. AK's heavy artillery regiment located north of Sarrebourg near the small village of Oberstinzel. In the early evening hours of the 19th, the unit's balloon and attending vehicles were moved into position where the final preparations were completed for its ascent the following morning. Likewise, all of Sixth and Seventh Armies' FFAs were notified of the impending counter-offensive and were ordered to support the attacks as soon as the morning mist dissipated.[113]

The German counter-stroke began as scheduled shortly before 4:00am when the XXI and Bavarian II AKs delivered the first attacks of the morning against the XX and XV CAs on the extreme left of the French line. By sunrise the entire front was alive with activity as the Germans began to pierce the French front in several locations. Having had plans to renew their advance for the sixth consecutive day, the French troops neglected to establish defensive positions during the evening of the 19th. The following morning's battles therefore saw the unprepared French forces quickly surprised and pushed back by

Above: A German division commander observes his troops deliver an attack at the Battle of Morhange-Sarrebourg. (Author's Collection)

determined German infantry attacks supported by massed artillery. Located on French First Army's left flank, VIII CA was compelled to withdraw at 9:00am due to losses sustained during its predawn battle with the I b. AK. Here, as Rupprecht's Sixth Army HQ had suspected, the German heavy artillery was able to take advantage of their superior range to silence the French batteries and support wave after wave of successful infantry attacks that were carried far into the French rear-area. By mid-afternoon VIII CA was once again pushed back, resulting in the loss of the city of Sarrebourg with extremely heavy losses.[114]

Although VIII CA had suffered terribly, the primary focus of Rupprecht's attack was delivered to the north against de Castelnau's Second Army. In total, eight German divisions supported by elements of the Fortress Metz Garrison, attacked around the cities of Delme, Morhange, and Dieuze with the intent of encircling and destroying de Castelnau's command. Shortly after dawn, the German artillery began shelling the French positions with great success, covering their own infantry as they successfully penetrated the French line in multiple locations all across the front. Firing from positions on the Morhange heights, Rupprecht's heavy artillery decimated the forward elements of Ferdinand Foch's XX CA, which were deployed in a valley on French Second Army's extreme left. Further east, infantry from the XXI and II b. AKs as well as the I b. RK routed the XV and XVI CAs—severing contact between French First and Second Armies. At 7:15am, less than two hours after the fighting in their sector had begun, de Castelnau ordered the badly mauled XVI CA to fall back in the face of the overwhelming strength of German Sixth Army's attack. Shortly thereafter XV CA was obliged to do the same, leaving Foch's XX CA temporarily isolated and exposed to counter-attack. By 10:00am it had become clear to De Castelnau that the situation was hopeless and that prompt action was required to save his army from total annihilation. Thus, at 10:10am he dispatched a general retreat order for his entire army—citing XV and XVI CAs' situation as "critical."[115] The remainder of the afternoon saw the battered French troops safely breakaway from Rupprecht's forces and begin their retreat back to the frontier.

Thick fog prevented aerial activity at daybreak as the German attacks began. By 7:00am however, the fog had burned off and all of Rupprecht's flying sections were airborne over the battle-area. With its first aircraft taking off at 6:35am, FFA 8 (XXI AK) was the first to go into action. The flight's report confirmed for XXI AK HQ that the corps' two infantry divisions were making good progress and were both heavily engaged near the town of Vergaville (1.5 miles northeast of Dieuze). More specifically, the observer's message accurately pinpointed the location and deployment of XV CA's two divisions—one of which was observed in road column being redeployed to shore up a failing spot in their line. The report also gave the location of a French artillery battery that had been seen by the airmen moving forward "at a trot" into a new firing position closer to the front line. This intelligence was promptly forwarded to and utilized by several local German artillery commanders to engage and drive off these French batteries before they could have a significant impact on the battle.[116] A second flight conducted by FFA 8 around 7:30am pinpointed the location of a French artillery group that was noted by the crew as holding up the German infantry's advance. Thinking quickly, the flight's observer quickly dropped a sketch of the French positions at the site of a nearby battery who ultimately engaged and silenced the troublesome enemy guns.[117]

A good example of the embryonic state of military aviation at this time presented itself during the battle when one of FFA 8's crews reported operating in close proximity to three French aircraft flying over the battlefield performing tactical reconnaissance of their own. Unarmed and more interested in assisting the ground troops, the opposing airmen left each other alone to do their work.[118] Thus, while the age of military aviation had clearly arrived, air combat was still months away.

Meanwhile, Sixth Army HQ's FFA 5 was ordered to "observe the army's front on the line: Nied—Château-Salins—Morhange—Avricourt and report on the enemy's artillery positions and deployment of reserves." However, the section's airfield was inconveniently located far away from the front at St. Avold where Sixth Army HQ had previously been located. Consequently, there was some delays in transmitting and receiving orders, resulting in the section's first report of the day being delivered to the Sixth Army HQ staff at 11:00am. This comprehensive report documented the general location of the XV, XVI and XX CAs—listing the positions of where the three corps' troops were heavily concentrated. The report particularly focused on the situation in XX CA's sector on the extreme left of the French line. In addition to defining where the corps' flank was located, it described in great detail the position of its reserves, the largest being an infantry brigade near Burlioncourt (3 miles northeast of Château Salins).[119] Finally, the airman's message reported the movement and concentration of men and artillery that had been

Above: Nicknamed 'the fighting friar', Edouard de Castelnau led French Second Army throughout 1914 (Author's Collection)

witnessed moving toward the front—actions that the observer opined were signs of a pending counter-attack.[120] Nonetheless, while FFA 5's 11:00am report was indeed full of excellent information, none of the enclosed intelligence was forwarded to the corps commands in time to make any impact on the course of the battle. It was successful, however, in keeping the Sixth Army HQ staff fully updated on the general situation at the front.

Elsewhere along the front, the Bavarian aviators were giving an excellent account of themselves in their first major engagement. As a unified Bavarian force under the command of the Bavarian Crown Prince, Sixth Army possessed four Bavarian corps as well as all the men and aircraft from the pre-war Bavarian Aviation Battalion—which upon mobilization had been divided into three flying sections. In true Bavarian fashion, all three Bavarian FFAs were equipped with Bavarian made aircraft, either Otto B.I biplanes (FFA 1b) or LVG B-types built by Otto under license (FFAs 2b & 3b). Supporting the spearhead of the offensive, FFA 2b performed extensive tactical reconnaissance for II b. AK's attack around Morhange. Throughout

Above: German artillery and supply columns meander through the Vosges mountains. The length of the columns during this period practically guaranteed detection by airplane. (Author's Collection)

the battle, the section's airmen kept close watch over XX CA's rear-area, reporting on the movement of French reserves and the location of troublesome artillery batteries. At a pivotal moment in the battle, one of the section's observers noted a French infantry brigade marching in road column to participate in a local counter-attack. After making the discovery, the airmen flew back to their airfield where a report was handed to a II b. AK HQ staff officer. Within minutes, the observer's information was forwarded to the artillery commanders at the front who used the intelligence to concentrate their fire and break up the French assault.[121]

In I b. AK's sector, FFA 1b and FLA 1b both worked with the corps' artillery to silence the French 75s in their area. Indeed, VIII CA's defeat outside the city of Sarrebourg was partly due to the excellent artillery spotting performed by FLA 1b, which had begun around 5:30am and continued throughout the day.[122] Aloft in the section's balloon near the town of Oberstinzel, the observer was able to speak directly to the local artillery commander on the ground by telephone, informing him of the location of French batteries and providing accuracy correction once the target was engaged. This ultimately drove off numerous French field batteries throughout the morning that had been brought forward to check the German advance. In many instances, the French guns were engaged before they ever got into action, either when they were in road column or while they were being unlimbered. Consequently, with the enemy's guns silenced, the I b. AK and I b. RK were able to freely maneuver, break through the French positions, and force VIII CA out of Sarrebourg.[123]

Meanwhile, airmen of FFA 1b circled over the Sarrebourg battlefield conducting tactical reconnaissance sorties to report I b. AK's progress. The section's observers delivered multiple reports between 9:00am and 1:00pm to Corps HQ, giving their commander, *General der Infanterie* Oskar von Xylander, an accurate account of his men's progress and the deployment of the French troops opposed to him.[124] These messages enabled Xylander and his staff to smartly deploy their reserves and ultimately force VIII CA out of Sarrebourg.

By any measure the German counter-offensive at Morhange–Sarrebourg was a total success. In a single day of action Rupprecht's men had inflicted 10,000 casualties on the French right wing armies, shattering three of their corps.[125] By late morning, two of Second Army's three corps had been routed with the third, Foch's XX CA, stubbornly holding the army's vulnerable left flank despite having taken incredible punishment from nearby German heavy batteries. Thus, in order to avoid being encircled and annihilated, de Castelnau was compelled to break off contact and order a general retreat for his army towards the frontier under the cover of a rear-guard. This in turn completely uncovered First Army's left, exposing their flank and rear to an attack. First Army was therefore obliged to disengage from Heeringen's Seventh Army and begin a withdrawal on the evening of the 20th with the intention of quickly regaining contact with Second Army and continuing the offensive.

The French retreat resumed late in the morning of the 21st after a series of attacks delivered by Rupprecht's emboldened troops convinced de Castelnau that he would be unable to resume offensive operations.[126] Retreating with great speed, Second Army quickly opened a 12 mile gap between themselves and their German pursuers, who were themselves overly cautious and exhausted from days of fighting. This ultimately allowed de Castelnau's badly demoralized troops to safely escape across the Meurthe River without suffering any further damage. By August 24th, Second Army was entrenched along the heavily fortified heights east of Nancy known as the *Grand Couronné* de Nancy. Meanwhile, First Army, with its left wing deployed *en échelon* to maintain contact with De Castelnau, remained in the Vosges Mountains. From these positions, the French right wing armies intended to rely upon their field fortifications and, if necessary, the nearby fortresses directly in their rear to accept battle and tie down as many German forces as possible while simultaneously withdrawing some of their own units for combat in the other sectors.[127]

Sixth Army HQ's FFA 5 monitored the French retreat from Morhange—Sarrebourg in the days immediately following the battle. Each day, the airmen's reports kept Rupprecht and the Sixth Army HQ staff fully updated regarding the French columns' strength and position. On the morning of the 21st, the section was ordered to reconnoiter along the line Nancy—Badonviller—Sarrebourg in order to ascertain the condition of the retreating French troops and confirm whether any reserves were coming out of Fortress Nancy. At 9:30am the morning fog dissipated and the first flight took off from the section's airfield near Hellimer. During the subsequent flight, the observer discovered several large columns of troops retreating in haste between Lagarde—Lunéville. To the south, the observer found that First Army had halted their withdrawal and were now facing Seventh Army. This report was corroborated by a message delivered to Sixth Army HQ by XIV AK's FFA 20. While on his morning patrol, the observer had discovered VIII CA on First Army's left flank withdrawing in a northerly direction to reestablish contact with Second Army.[128] Together, these two reports confirmed for Rupprecht that French Second Army was indeed retreating back to their original positions across the Meurthe River while First Army appeared committed to holding the territory they had won in upper Alsace. Using this intelligence as well as other reports provided by troops on the ground, Rupprecht informed his men at 2:14pm that the French were retreating back to the Meurthe. He also consulted his aviators' intelligence and ordered the army to pursue *en échelon* in order to guard against Dubail's First Army on their left flank.[129]

The important day of August 21st also saw Sixth and Seventh Armies' corps flying sections continue to provide excellent tactical reconnaissance. Due to early morning fog, the majority of these flights occurred after 12:00pm against the French rear-guards that were covering Second Army's withdrawal. At 12:45pm for example, one of FFA 2b's aircraft reported a column comprised mostly of artillery moving out onto a large plain located near Blâmont. This ultimately turned out to be a large field artillery group that was charged with covering XIV and XV CAs withdrawal. Thanks to FFA 2b's timely report, German heavy batteries in the area quickly engaged the group and drove the French off the field before they ever got into action. Later that afternoon, crews belonging to FFAs 3b and 6 conducted ad-hoc artillery direction sorties around Château-Salins.[130]

In the area west of Sarrebourg within Seventh Army's sector, the balloonists of FLAs 1b and 6 were both used throughout the late-morning and afternoon to direct artillery fire and help breakup several counter-attacks mounted by French First Army. In each case, the observers were able to promptly report the location of French troop concentrations as well any active artillery batteries in the critically important minutes before an attack. German commanders on the ground were therefore notified in advance when and where any French counter-attack would be attempted, allowing for reserves to be moved into advantageous positions to defeat the upcoming assault. During an attack, the FLA observers cooperated with battery commanders on the ground to primarily keep the French artillery silent and in some cases to directly engage the opposing infantry. Consequently, without any tangible support from their own artillery, each of the French counter-attacks were doomed to failure—thus compelling Dubail's First Army to ultimately withdraw the majority of their troops back across the frontier to stronger defensive positions.

Citing the great distance between Rupprecht's

Above: DFW B.I B.451/14 on display in Paris in September 1914 after being captured; a captured *Taube* is in the background. The early national insignia on the rudder and both sides of all wings are well shown from this angle. Despite being liberally marked with national insignia many aviators were subject to 'friendly fire' due to poor type recognition. (Aeronaut)

forces and the French defensive positions west of Nancy, OHL dispatched two Zeppelin airships, the Z VII & Z VIII, to gather additional intelligence for Rupprecht and his staff. Both ships were accordingly ordered airborne on the morning of the 21st to perform long-range patrols to the Toul—Épinal fortress line. Taking off from Baden–Oos towards the Vosges, the heavily weighted down Z VII was only able to reach an altitude of 5,000 feet. After dropping its bombload on a French camp, the ship unexpectedly found itself over a large number of French troops where it was repeatedly hit by shrapnel and rifle fire. Steadily losing gas, the Zeppelin rapidly lost altitude and ultimately crashed in Lorraine behind German lines. The Z VIII was similarly shot down by ground fire, crashing between the French and German lines near Badonviller. The ship had inexplicably been cruising at a height of only a few hundred feet when it was fired upon and hit over a thousand times by French rifle and artillery fire. With its controls shot off, the large dirigible simply hovered in place suffering even more damage until it slowly crashed to earth near a forest. After escaping from the wreckage and entering into a firefight with French cavalry, the Z VIII's crew managed to flee the scene and make their way back to German controlled territory.[131] The flights of the Z VI at Liège as well as the Z VII and Z VIII in Alsace-Lorraine proved that the Prussian General Staff's pre-war theory regarding the dirigible's inferiority was correct. Due to their large size, slow speed, and low ceiling, each of the ships had been brought down by ground fire on their first flight of the war. Consequently, OHL refused to ever again use the army's dirigibles during daylight hours on the Western Front.

Over the next several days, Sixth and Seventh Armies' flying sections continued to monitor the withdrawal of the French right wing. However, with their own troops unable to catch the French columns before they crossed the Meurthe and reached the relative safety of the *Grand Couronné*, Rupprecht's airmen dramatically scaled back the number of flights performed each day. Between the 22nd and 24th, the German left wing flying sections conducted no more than two flights per day, with the majority occurring in the early morning. Shortly after dawn on the 22nd, Sixth Army HQ's FFA 5 reported observing "enemy columns retreating in full flight at Blainville" (1 mile west of Lunéville).[132] On the 23rd, Seventh Army HQ's FFA 26 confirmed that large masses of troops of the French First Army were withdrawing through Baccarat and St. Dié towards Épinal while the troops comprising the army's right wing appeared to be entrenching in the Vosges. Later that afternoon, FFA 1b proudly returned to the same airfield that it had been forced to abandon on August 15th at the start of the French advance. When they arrived the section's men unearthed 2,500 liters of aviation fuel that they had hidden from the French at the beginning of the withdrawal.[133]

By the morning of the 24th Rupprecht's men had reached the French positions along the *Grand Couronné* and had reentered into a general engagement. Sixth Army had, against Rupprecht's wishes, been ordered by OHL to "relentlessly pursue" the French troops in the direction of Épinal. Rupprecht was therefore compelled to order his army's left wing, comprised of XXI and II b. AKs, to push well beyond the Meurthe River to the Charmes—Rambervillers road. Early that morning, FFA 2b delivered an excellent report to II b. AK HQ that outlined the French deployment in the corps' sector. The observer's message stated that the French were indeed positioned *behind* the corps' lead columns along the heights on both sides of river. Alerted to the serious threat on their right flank, II b. AK HQ decided to push the looming French units back before resuming the advance as ordered.[134] The corps' two divisions were therefore split as follows: 3 ID attacked from Einvaux—Moriviller while 4 ID covered the right flank between Blainville—Méhoncourt. Later that morning III b. AK's FFA 3b reported that the corps' own 4 ID had become heavily engaged southwest of Lunéville. Upon receiving the report, the corps' commander, General Ludwig von Gebsattel, requested and received permission from Sixth Army HQ to move his corps south to support 4 ID. Due to heavy road traffic, the corps was unable to reach the battlefield until midnight whereupon it became heavily engaged the following morning.[135] Together, these two aerial reports helped prevent Sixth Army's lead corps from seriously endangering themselves, thus creating the conditions for the ensuing battle to be waged under more favorable conditions. In the days and weeks that followed the fighting on the German left wing devolved into static "position warfare" that saw both sides consigned to merely holding the opposing forces in the theater and away from the decisive battles occurring on the German right wing.[136]

The Ardennes: Plan XVII's Main Assault

Undeterred by his defeat in Lorraine, Joffre quickly moved forward with the second phase of Plan XVII. The French Commander-in-Chief strongly believed the upcoming offensive could deliver a campaign

Above: Fernand de Langle de Cary, commander of French Fourth Army. (Author's Collection)

winning victory, which in concert with the ongoing Russian offensive in East Prussia would win the war. Joffre's confidence was based upon inaccurate intelligence estimates regarding the overall deployment of German forces. Reports from Belgium after the fall of Liège stated that the German right wing had a strength of at least seven active corps and four cavalry divisions. Meanwhile, French First and Second Armies reported similar numbers of troops in Lorraine on the German left wing. Thus, drawing upon his latest intelligence summaries, Joffre believed that the German center was weak and susceptible to attack.

The French main assault north of Metz was to therefore be directed towards the supposed "weak German center" with the intent of breaking the German line into two pieces. Specifically, French Third and Fourth Armies were ordered to attack and defeat the German center, *then turn west* against the flank and rear of the advancing German right wing. While the Third and Fourth Armies were executing this attack, the French Fifth Army, with the assistance of the BEF, was to engage and hold the German right

Above: Albrecht, Duke of Württemberg and commander of German Fourth Army. (Author's Collection)

Above: Crown Prince Wilhelm, eldest son of Kaiser Wilhelm II and commander of German Fifth Army. (Author's Collection)

wing until the German center could be defeated. The French commander-in-chief therefore believed that the German right wing would become easily isolated and encircled after the German center armies were destroyed.

The main thrust of Joffre's offensive against the German center was entrusted to French Fourth Army under the command of *Général* Fernand de Langle de Cary. Langle's army consisted of six army corps, two reserve divisions and two cavalry divisions—the largest Allied army in the Western Theater at the time. Under Joffre's concept of operations, Fourth Army would spearhead the advance in the direction of Neufchâteau. Meanwhile, *Général* Pierre Ruffey's Third Army, comprised of three army corps and one cavalry division, was directed to initially screen fortress Metz before joining Fourth Army's offensive.[137] Thus, after carefully passing through the Ardennes forest, the two armies were to move into Luxembourg where they expected to engage and defeat the forces of the German center (that were not expected to be advancing), which Joffre estimated to have a combined total strength of just six corps.[138]

Unfortunately for Joffre, the German center did not consist of a weak group of six corps, but was instead comprised of two large armies with a total strength of 10 infantry corps, three cavalry divisions and the Metz garrison. Consisting of three active and two reserve corps, German Fourth Army was under the command of Duke Albrecht of Württemberg. Deployed on Albrecht's left was the German Fifth Army under the command of Crown Prince Wilhelm—heir to the German Imperial throne. The Crown Prince's command was composed of three active and two reserve corps, the garrison of Metz and three cavalry divisions organized into the 'IV Cavalry

Corps' (HKK 4). Together, the two German armies were a formidable force with a combined strength much greater than Joffre and his subordinates had anticipated.[139]

On August 21st the French Third and Fourth Armies began their advance into the Ardennes as the first step in what Joffre believed would be the campaign's decisive action. The French commander-in-chief had established the towns of Arlon and Neufchâteau in Belgian Luxembourg as Third and Fourth Armies' initial objectives. These armies were ordered to execute their marches through the heavily forested region *en échelon* in order to better facilitate a change in facing from north to east if circumstances required them to do so. Consequently, the group's nine corps entered the Ardennes on the morning of August 22nd in a stair-step configuration descending to the right. While this deployment allowed the armies to more easily maneuver, it left the entire battle-line vulnerable to a large-scale attack. Indeed, the security of each corps' right flank would be fully dependent upon the successful advance of its right hand neighbor. In other words, a single corps' failure to advance would not only threaten its neighbor to the left, but could also potentially put the entire army at risk of being encircled. Joffre nonetheless confidently believed the *en échelon* deployment was sound based upon his latest intelligence estimates that concluded there were no significant German troops in the Ardennes. Thus, the two French armies were therefore expected to easily move through the forested region uncontested and then engage the weaker German forces on the other side. Indeed, Joffre confidently dispatched an official order to both Ruffey and de Langle on the evening of the 21st stating: "No serious opposition need be anticipated on the day of August 22nd." Within 24 hours, the French generalissimo learned that he couldn't have been more wrong.[140]

****

In accordance with the German War Plan, the armies of the German center were to refrain from any large-scale offensive action until the right wing had completed its wheeling maneuver and reached the Franco-Belgian border. As the "hub" of the right wing's giant wheel, Fourth and Fifth Armies had remained on the defensive until the 18th whereupon they slowly began to move northwest in order to maintain contact with Third Army on their right. The center group's mission was to advance together into the Ardennes and then proceed to the Meuse River with Fifth Army occupying the territory between Metz and Neufchâteau and Fourth Army between Neufchâteau and Givet. Most importantly, the group was to maintain contact between the attacking right wing armies and Fortress Metz—securing the attacking armies' inner flanks and lines of communication. The speed of Fourth and Fifth Armies' advance was subject to the right wing armies' progress. According to the German War Plan, once the right wing had completed its turn through Belgium and arrived on the French border, the center was to have arrived on the Meuse fully prepared to attack and cross the river.[141]

Reconnaissance sorties assigned to both the cavalry and *Fliegertruppe* between the 10th–17th had completely failed to provide an accurate picture of the French deployment opposite the Fourth and Fifth Armies. Indeed, the cavalry divisions of HKK 4 had been entirely unable to pierce the French cavalry screen while unnecessary delays and poor weather had kept the center armies' flying sections grounded throughout the seven day period. Consequently, OHL knew nothing of French intentions in the region prior to the start of the advance on the 18th. The German leadership did understand, however, given the numbers of troops that had already been engaged in Lorraine, that the French strategy in the Ardennes could only be one of two possibilities—either Joffre was planning his main attack there, or the French were to assume the defensive and dig in on the western bank of the Meuse.[142] Fortunately the center armies' flying sections quickly brought clarity to the issue beginning on the morning of the 18th.

Fourth Army's *Fliegertruppe* operated between Chiers—Othain—Montmédy throughout the 18th, resulting in a number of significant discoveries. First, the army HQ's own FFA 6 reported large numbers of French troops, estimated to be two or three corps, marching north in column from the towns of Montmédy, Stenay, and Dun.[143] Meanwhile, both FFAs 13 (VI AK) and 27 (XVIII AK) informed Fourth Army HQ of the heavy rail traffic they observed in the French rear around Sedan. To the south, Fifth Army's flying sections were also busy reconnoitering the French positions around the Meuse River. Here, the most promising message of the day came in from XIII AK's FFA 4, whose morning patrol had discovered both heavy rail traffic and a large column of French infantry moving north. Although certainly not definitive, these initial reports strongly suggested that the French were gearing up for an offensive in the direction of the Ardennes and Luxembourg.

August 19th was yet another extremely productive day for the German center group's *Fliegertruppe*. For the second day in a row Fourth Army's flying sections monitored the Meuse River area directly to the west of the army. While patrolling the Sedan—Montmédy line between 10 and 11am, one of FFA 6's crews once again discovered that large numbers of French

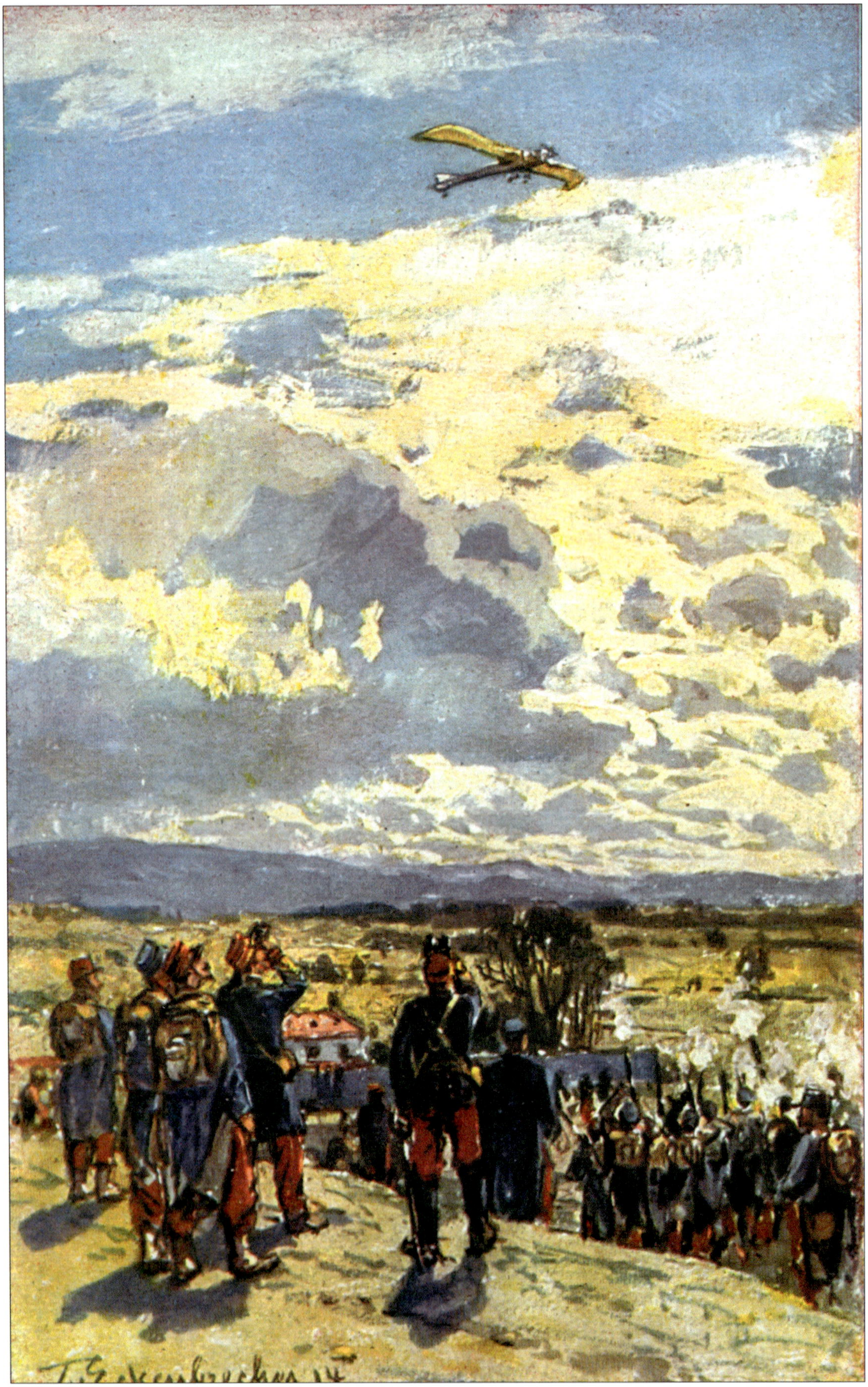

troops were moving northeast in the direction of Belgian Luxembourg. The contents of the observer's report, which was sent to Fourth Army HQ and later forwarded to Moltke at OHL stated:

Military bivouacs observed along all localities between the Chiers and Meuse on the road Stenay—Montmédy. Entrenchments are being dug northeast of Mouzon. Vanguard observed east of the Chiers. The observer estimates the strength of the force to be two or three corps. West of the Meuse lay isolated bivouacs along the road Stonne—Mouzon and Nouart—Stenay. The roads north of the line Mouzon—Jandun to the Meuse and the road Tannay—Le Chesne—Semuy was free [of traffic] between 10:20 and 11:10am.[144]

FFA 6's report disclosed the position of three corps belonging to French Fourth Army—the XVIII, XII and Colonial Corps. The airmen's message gave the leadership legitimate cause to assume that the French were going to cross the Meuse in large numbers and attempt an offensive in the Ardennes.

Later that afternoon, Fourth Army HQ learned from VIII AK's FFA 10 that two large columns of troops had been seen moving north around noon to bolster French Fifth Army in the area of Dinant. The report also confirmed the presence of at least one division west of Dinant in the area of Philippeville. Thus, it was clear from the airmen's observations that the French were strengthening their left wing. Therefore, due to their overwhelming importance, the contents of the observer's message were immediately forwarded to both Third Army HQ and OHL.

In Fifth Army's sector, FFA 25 patrolled the area Arlon—Rulles—Florenville—Mouzon—Brouennes—Marville—Thionville (Diedenhofen).[145] During their search, the airmen found a large column of troops south of Montmédy marching in the direction of the Belgian border. To the south, two French divisions were seen moving northeast out of Fortress Verdun. This further suggested that Joffre was launching a major offensive into the Ardennes and Luxembourg. However, there were still significant numbers of French troops unaccounted for and the German leadership remained hesitant to react to these initial aviation reports until Joffre's intentions could be confirmed.

Fourth and Fifth Army's *Fliegertruppe* were therefore very active on the 20th in the attempt to gather additional data regarding the French movements and intentions in the Luxembourg—Ardennes sector.

Facing Page: A *Taube* from FFA 4 discovers the lead elements of French Third Army marching toward the Ardennes. (Author's Collection)

Both of the German center armies were to continue their westerly movements towards the Meuse river in order to support the right wing armies, who were in the midst of their advance through Belgium to strike the allied left wing on the Franco–Belgian border. Consequently, with German troops moving closer each day towards the Meuse, it was imperative to determine whether the French were going to launch an offensive into the Ardennes or not. To ensure the group's security, Moltke ordered Fourth Army to maintain contact with Fifth Army's flank. The weather in Fourth Army's sector was ideal for aerial reconnaissance and each of the army's sections had aircraft airborne during the late morning hours. VIII AK's FFA 10 reported that one of its flights had once again found significant redeployments taking place in the French rear-area to strengthen their left wing. First, a large concentration of bivouacs—estimated strength of one corps—was located west of Dinant along the road to Phillipeville. In the Couvin—Rocroi area, a previously unobserved division was identified moving north to reinforce French Fifth Army's positions on the Meuse opposite German Third Army. Significant rail traffic between Florennes and Givet was also reported, further suggesting a massive strengthening of the allied left. Finally, FFA 10's afternoon flight discovered French cavalry east of the Meuse at Fumay.[146]

Additional intelligence, provided by XVIII AK's FFA 27, confirmed the existence of three French corps in the region directly west of Charleville near the Belgian border. This group was seen marching northeast at the time it was found between 6–7:00am. Thus, over the previous 48 hours German airmen had identified at least five French army corps between Charleville and Montmédy. Furthermore, Fourth Army HQ's own FFA 6 reported a mixed detachment of French troops north of Florenville at 10:15am marching northeast towards Neufchâteau in the heart of Belgian Luxembourg. Although they were decisively defeated later that evening at the village of Longliers, the French detachment's presence was serious cause for concern amongst the German leadership. There could no longer be any doubt whether the French were attempting an offensive into the Ardennes and Luxembourg. Fourth Army HQ's order for the 21st reflected this view:

The enemy has several corps between Hirson—Charleville marching in a northerly direction. Several additional French army corps also detected along the line Douzy-am-Chiers—Montmédy—Spincourt—Gondrecourt. Advanced positions are east of the Chiers. There are only columns and trains west of the Meuse on the line Sedan—Dun.[147]

Above: German and French cavalrymen fight over a French regimental standard. German cavalry greatly struggled to penetrate the French cavalry screen, enhancing the *Fliegertruppe's* importance throughout the campaign. (Author's Collection)

August 21st proved to be a critically important day for the German center. In light of the reports from the previous day, Fourth Army's airmen were assigned new sectors to patrol. On the army's right wing, FFA 10 was given the area around the Meuse between Dinant and Revin. Meanwhile, FFAs 27 and 13 were directed to reconnoiter the line Chiny—Rethel while Fourth Army HQ's own FFA 6 was to concentrate along the Meuse between Douzy and Sivry. Unfortunately, the region was plagued with thunderstorms, heavy rain, and dense fog which kept the airmen grounded for most of the day. As a result, the corps' flying sections were unable to provide reports on the latest movements of the large groups of French troops. Only a small break in the weather around 7:30am allowed a single aircraft belonging to FFA 27 to patrol the area south of Montmédy. The observer's report, which was delivered directly to Fourth Army HQ between 9:45 and 10:45am, stated that two columns of troops, estimated strength of a division, was seen marching north at Remoiville at 8:00am. Another column of similar size was also seen marching east on the Stenay—Montmédy road. Thick cloud cover prevented the observer from making any further observations regarding the French troops previously seen to the west of Charleville along the Belgian border. Consequently, Fourth Army HQ was completely unaware of the whereabouts and intentions of the larger French force that had been seen by their airmen during the 19th and 20th.[148]

Crown Prince Wilhelm's Fifth Army had halted on the 21st to properly cover their siege at Longwy. Although it was a small and outdated fortress, Longwy was a critically important component to the French defenses in the region and was a potential thorn in Fifth Army's side if bypassed. To hasten the siege and capture Longwy, the Crown Prince ordered FLA 4 to the scene to direct the siege artillery's fire. The section's balloon accordingly made several ascents throughout the day but was unable to be of any assistance due to "heavy haze."[149] Meanwhile, Fifth Army HQ's FFA 25 was ordered to patrol the roads into Belgium from Verdun and Montmédy. At 9:00am, one of the section's patrols discovered a French division marching north through Spincourt. The most important message of the day, however, came from XIII AK's FFA 4 which disclosed the position of French Third Army's leading columns along the line: Virton—Cosnes—Beuveille—Fléville in direct proximity to German Fifth Army's troops near Longwy.

The unfavorable weather had greatly impacted Fourth and Fifth Army leadership's ability to recognize the scale of the French offensive taking place to their front. Albrecht had been desperate to determine whether the large concentration of French troops that his airmen had seen on the 19th and 20th were marching north or east in order to adopt the necessary measures to check their advance. However, with no cavalry reconnaissance and with their flying sections largely grounded, Fourth Army HQ remained entirely ignorant of the dangerous situation building in their front.

Fifth Army's flying sections similarly suffered due to the weather. However, a single report delivered to army headquarters late in the afternoon by FFA 25 confirmed "a general enemy advance on the line Breux (north of Montmédy)—Landres."[150] The army's cavalry corps, HKK 4, also provided details regarding the nearby French formations. Positioned on the army's flanks, each of HKK 4's two divisions made contact with several regiments of French infantry marching ahead of their respective corps' main bodies. Additional cavalry patrols found strong French columns moving northeast across the front of the German center.[151] Thus, by the evening of the 21st,

Above: French POWs shortly after the fall of Fort Longwy. (Author's Collection)

Fifth Army HQ had become aware of French Third Army's exact location and axis of advance.

Fifth Army HQ's chief of staff, *Generalleutnant* Schmidt von Knobelsdorf, seized upon the opportunity presented by this intelligence and ordered a preemptive offensive to strike the French columns as they were advancing through the region's hazardous terrain. However, attacking into the heavily wooded Ardennes was incredibly risky. Geographically, the region was heavily wooded with sprawling layers of thick undergrowth, small streams and only a few sporadic areas of open ground. Few roads passed through the area, which was devoid of industrial development and totally unsuited for agriculture. Due to the nature of the terrain, important factors such as command and control, unit cohesion, and flank security were very difficult if not impossible to maintain. Consequently, the various units involved were vulnerable to ambushes and/or encirclement. Nevertheless, Fifth Army HQ ordered its corps commanders to proceed the following morning.

According to the attack order, V AK would defend the army's right flank at Virton while XIII AK and VI RK attacked around Longwy towards Longuyon. Meanwhile to the south, V RK and XVI AK would strike due west in the direction of Spincourt. Together, Fifth Army's forces were to use the advantages of surprise and terrain to annihilate a large portion of the French forces that were currently advancing into Belgian Luxembourg—thus ending their offensive before it could begin.[152]

Fourth Army HQ did not learn of Knobelsdorf's plan until 2:00am on the 22nd when a staff officer belonging to V AK formally requested Fourth Army's VI AK move south and close the gap which had opened in the line as a result of Fifth Army's latest movements. Although he was still ignorant of the situation in his own sector, Albrecht ultimately agreed to the move in order to maintain contact with Fifth Army's flank.[153] Thus, between 3:00 and 4:00am of the 22nd, VI AK's two divisions began their march towards the towns of Tintigny and Rossignol where they were to play a pivotal role in the coming battle.

Apart from the shifting of VI AK to the south to protect Fifth Army's flank, Albrecht's Fourth Army was expected to advance westwards through the Ardennes on the 22nd while simultaneously reducing its front in anticipation of making contact with the French main body—possibly as soon as the 23rd. Although Fourth Army's airmen had correctly alerted Albrecht and his staff to the presence of French troops marching into southern Belgium on the 19th and 20th,

Above: German Uhlans charge French infantry. (Author's Collection)

their lack of activity on the 21st due to the weather had prevented them from pinpointing their latest position and axis of advance. As a result, Fourth Army HQ believed the enemy's main force was further west on the evening of the 21st–22nd than it actually was. This left the still widely dispersed Fourth Army in an unenviable position on the eve of battle. Indeed, through no fault of their own, Fourth Army would only have three of its five corps available to repulse French Fourth Army's five corps. Fortunately for the Germans however, the French were even more uninformed of the overall situation in the Ardennes sector than they were. Whereas Albrecht's troops were at least aware of the possibility of making contact with the enemy and had been issued orders to conduct their advance accordingly, French Fourth Army confidently entered the Ardennes fully expecting to find it empty.

****

The morning of the 22nd saw the opposing armies begin their respective prescribed movements prior to dawn. By 5:00am the columns of French Third and Fourth Armies were underway marching *en échelon* to the northeast. Indeed, many of the French columns had started their approach marches prior to dawn in order to conceal themselves from German aerial observation. By sunrise, the French right and center had already crossed the Chiers River and was prepared to continue the day's advance to their respective objectives.[154] Meanwhile, the Crown Prince's Fifth Army initiated their approach march shortly after 5:00am. Because of the close proximity of the opposing forces, German Fifth Army and French Third Army made contact with one another almost immediately, beginning the bloodiest day of combat in French military history.

Although German Fifth Army HQ's decision to attack into the Ardennes would ultimately be proven correct, it nonetheless presented great challenges both on the ground and in the air. Unlike the battles fought two days before in Lorraine, the army leadership would be unable to maintain control of their units once the fighting began. Limited visibility due to the terrain would compel the various corps, division, and often even brigade commanders to lead attacks without being able to view their men's progress. Furthermore, the region's thick undergrowth made speedy movement of artillery and reserves nearly impossible. As a result, the battle would assuredly degenerate into countless clusters of small unit actions outside the control of the higher commands. These conditions would similarly restrict the *Fliegertruppe's* potential impact on the battle. Even if the airmen were to fly at dangerously low altitudes, the observers would be simply unable to distinguish the opposing battle-lines and would therefore be incapable of offering any tactical assistance once the armies had become engaged. The German center's various flying sections were therefore directed to reconnoiter the French rear-area west of the wooded terrain as well as the region's various roads in order to give the army and corps leadership a better understanding of the enemy's overall strength in each sector while there was still time to react.

The morning of the 22nd was humid and foggy due to moisture left from the previous afternoon's rainstorms. Fourth and Fifth Armies' flying sections were therefore compelled to remain inactive until the morning fog had burned off around 8:30am. On the ground, the heavy fog helped to conceal the opposing armies' movements from one another's advance guards. As a result, the first serious engagement of the day occurred when the main bodies of German Fifth Army and French Third Army blindly stumbled into one another. Over the following ninety minutes,

Above: Wartime postcard depicting the combat at the Battle of Neufchâteau. (Author's Collection)

the opposing armies struggled through the haze to determine the strength of their enemy. Between 8:30–9:00am, the fog finally dissipated to reveal the French army's positions. Located west of Longwy in the center of French Third Army's battle-line, it was discovered that a number of V CA's batteries had unlimbered in the open directly behind their infantry. Within minutes, Fifth Army's batteries opened fire on the exposed French artillery, followed by a large-scale infantry assault that pierced the unsupported French infantry's line. By 11:00am all of V CA was retreating in a panic—leaving a gaping hole in the French battle-line.[155]

The single gravest threat to Joffre's *échelon* deployment scheme quickly became a reality with V CA's collapse. The corps' retreat had uncovered the neighboring IV CA's right flank, forcing it to halt its advance. For the rest of the day, the IV CA was compelled to fight a defensive battle at Virton against the V AK with no chance of success.[156]

Present at this battle was future fighter ace Manfred von Richthofen. At the time Richthofen was a cavalry officer in command of a small patrol belonging to *Uhlan Regiment* Nr. 1 which was attached to V AK's 9 ID. Early that morning, while on a reconnaissance mission, the future ace led his men toward the French lines to determine the proximity of the enemy infantry's main body. During their advance Richthofen and his men were ambushed by large numbers of French *cuirassier*. Fighting dismounted, the French cavalrymen quickly opened fire on the small German patrol, leaving only Richthofen and four other men capable of escape. The episode was later retold in Richthofen's memoirs:

I had been ordered to find out the strength of the enemy occupying the large forest near Virton. I started with fifteen Uhlans and said to myself: 'Today I shall have the first fight with the enemy.' But my task was not easy. In so big a forest there may be lots of things hidden which one cannot see.

We approached the margin of the forest. As we could not discover anything suspicious with our field glasses we had to go near and find out whether we should be fired upon. The men in front were swallowed up by a forest lane. I followed and at my side was one of my best Uhlans. At the entrance to the forest was a lonely forester's cottage. We rode past it.

The soil indicated that a short time previously

considerable numbers of hostile cavalry must have passed. I stopped my men, encouraged them by addressing a few words to them, and felt sure that I could absolutely rely upon every one of my soldiers. Of course no one thought of anything except attacking the enemy. It lies in the instinct of every German to rush at the enemy wherever he meets him, particularly if he meets hostile cavalry... After a sharp ride of an hour through the most beautiful mountain dale, the wood became thinner. We approached the exit. I felt convinced that there we should meet the enemy. Therefore caution! To the right of our narrow path was a steep rocky wall many yards high. To the left was a narrow rivulet and at the further side a meadow fifty yards wide surrounded by barbed wire. Suddenly the trace of horses' hoofs disappeared over a bridge into the bushes. My leading men stopped because the exit from the forest was blocked by a barricade.

Immediately I recognized that I had fallen into a trap. I saw a movement among the bushes behind the meadow at my left and noticed dismounted hostile cavalry. I estimated that there were fully one hundred rifles. In that direction nothing could be done. My path right ahead was cut by the barricade. To the right were steep rocks. To the left the barbed wire surrounded the meadow and prevented me attacking as I had intended. Nothing was to be done except to go back. I knew that my dear Uhlans would be willing to do anything except to run away from the enemy... A second later we heard the first shot which was followed by very intense rifle fire from the wood. The distance was from fifty to one hundred yards. I had told my men that they were to join me immediately when they saw me lifting up my hand. I felt so sure we had to go back so I lifted up my arm and beckoned my men to follow. Possibly, they misunderstood my gesture. The cavalrymen behind me believed me in danger, and they came rushing along at great speed to help me get away... Then, a panic because the noise of every shot was increased tenfold by the narrowness (of the road). The horses of the two men ahead of me rushed away in hollow way. The last I saw of them was as they leaped the barricade. I never heard anything of them again. They were no doubt made prisoners.

I myself turned my horse around and gave him the spurs, probably for the first time during his life. I had the greatest difficulty to make the Uhlans who had rushed towards me understand that they should not advance any further, that they were to turn around and get away. My orderly rode at my side. Suddenly, his horse was hit and fell. I jumped over them as horses were rolling all around me. In short, it was a wild disorder. The last I saw of my servant, he was lying under his horse, apparently not wounded but pinned down by the weight of the animal. The enemy had beautifully surprised us. He had probably observed us from the very beginning and had intended to trap and catch us unawares, as is the character of the French.[157]

Despite Richthofen's setback, the German infantry in V AK's sector were successful in halting the IV CA's advance—causing serious difficulties further down the French battle-line. Indeed, IV CA's inability to advance had deeply affected the operations of its left hand neighbor—*Général* Gérard's II CA. Given only a single road for their advance, II CA had remained unable from the very beginning of day's combat to breakthrough the resistance of a single German infantry brigade at Bellefontaine. Moreover, the corps' leadership was already fearful of exposing its unprotected right flank and rear to the German troops engaged with IV CA. Thus, recognizing his corps' predicament, Gérard made the command decision to shift his focus east to give aid to IV CA on his right, whose situation was quickly deteriorating in the face of V AK's prolonged attacks. Although IV CA was indeed assisted by Gérard's decision, II CA's turn to the east completely uncovered the neighboring Colonial Corps' right flank, thus sealing its fate.[158]

Comprised of the best and most experienced soldiers in the French Army, the Colonial Corps was considered by some to be the finest corps command in Europe.[159] The unit was therefore placed in the middle of the French battle-line and expected to form the spearhead of the attack once the army had passed through the Ardennes. The French objective for the 22nd, however, was the critically important town of Neufchâteau located in the heart of the forest. Using the only two roads available to him, the corps' commander—*Général* Lefèvre—deployed his men into two columns. The right column, consisting of the 3rd Colonial Division, was to march through the town of Rossignol before later reaching the outskirts of Neufchâteau. The left column, comprised of a mixed brigade, was to reach the corps' objective after first passing through the small town of Suxy. Meanwhile, the corps' other division, the 2nd Colonial Division, was kept in the rear as Fourth Army's strategic reserve.[160]

Without waiting for news from II CA on its right, the Colonial Corps' two columns set out for their objective totally unaware that their right flank was uncovered. Less than an hour into the advance, the lead brigade of the corps' right column took fire from German troops located in the woods north of

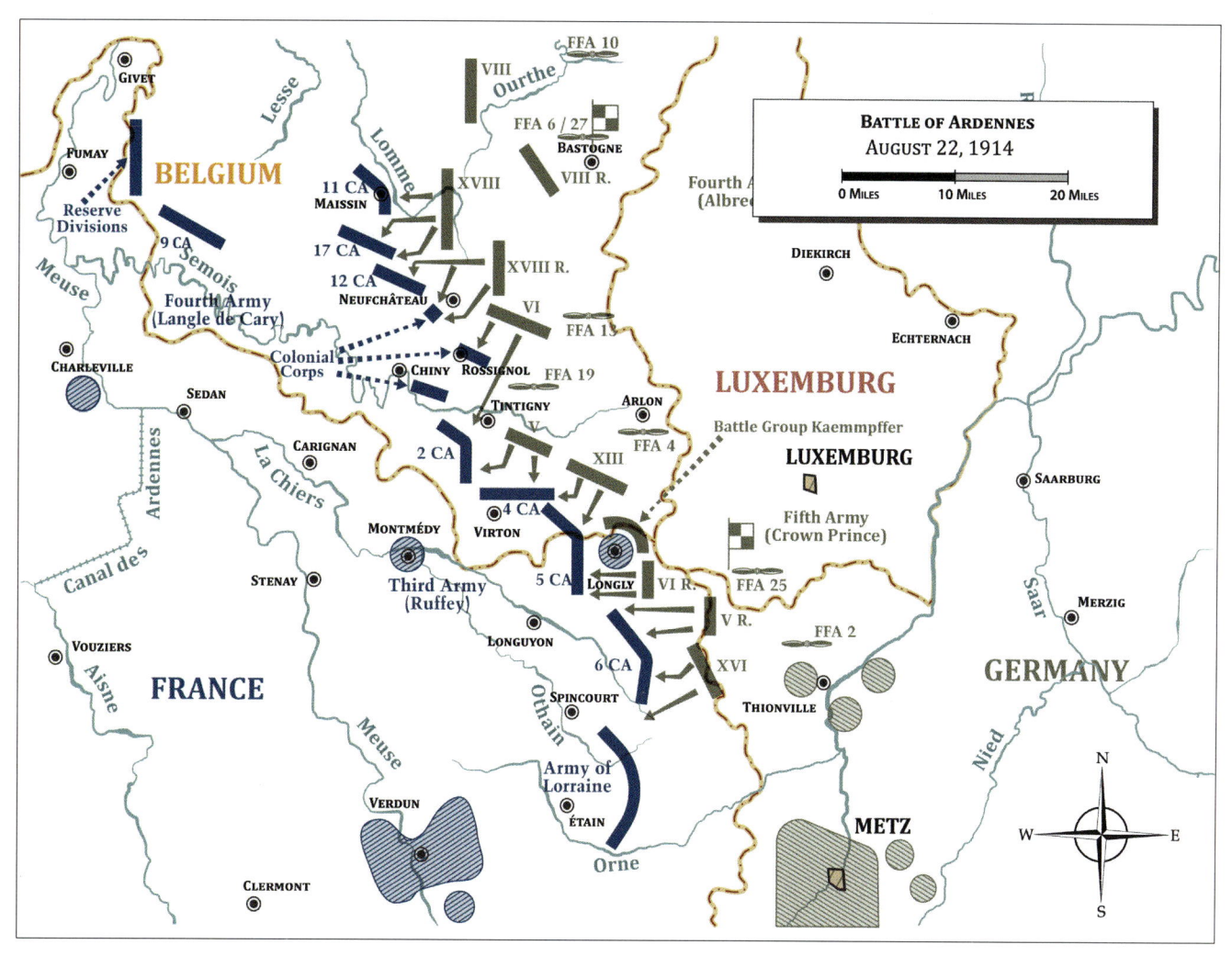

Rossignol. At that time the brigade was separated from the rest of the column, which was to the south across the Semoy River at the town of St. Vincent. Within minutes German artillery opened fire and destroyed the area's only bridge, thus permanently isolating the lead brigade from the rest of the column. Sensing that an opportunity had fallen into his grasp, the local German commander ordered his men to attack the French troops trapped in the town. Thus, with the support of several batteries, three German infantry regiments belonging to the VI AK attacked the French positions at the northern edge of the village. Advancing by bounds, the artillery and infantry worked together to quickly gain fire superiority over the French defenders. Once fire superiority was achieved, the German infantry were able to capture the town and its surviving French defenders. This scenario was repeated later that afternoon at St. Vincent. Using artillery support to achieve fire superiority, the German attackers captured the town after inflicting horrific casualties on the French troops who proved their valor by holding their ground almost to the last man. By nightfall the 3rd Colonial Division had practically ceased to exist. In a single day of fighting the division suffered 11,000 casualties— a staggering 73% casualty rate.[161]

The Colonial Corps' left column fared little better. After advancing through Suxy and reaching the outskirts of Neufchâteau without incident, the 5th Colonial Brigade was unexpectedly attacked by two brigades of the XVIII RK. The subsequent fighting lasted until dark as the superb French soldiers valiantly repulsed numerous German assaults. Outnumbered more than 2:1, the Colonial Corps' soldiers stubbornly held their positions throughout the day, believing the arrival of the corps' other column would turn the tide. That evening, with most of their men out of ammunition and with no word from the other column, the brigade's survivors had no choice but to conduct a withdrawal back to the positions they had eagerly set out from that morning.[162]

The *Fliegertruppe* played a minor role in the German victories over the Colonial Corps. Unaware of the close proximity of French Fourth Army's forces, VI AK HQ had ordered its own FFA 13 to monitor the area west of Neufchâteau around Florenville throughout the morning and afternoon of the 22nd. The situation changed however around 9:00am after the section's commander, *Hptm.* Alfred Streccius, heard the nearby sounds of artillery and rifle fire. With the morning fog now burned off, Streccius ordered two aircraft airborne to report on the strength of French forces that were currently engaged as well as moving up from the rear. The first aircraft successfully located the Colonial Corps' left column and followed its progress through the village of Suxy. Once it was determined that the column was headed for Neufchâteau, the observer returned to the section's airfield where he hastily handed over a written report to a VI AK HQ staff officer. The airman's message was taken to VI AK HQ who then immediately forwarded a copy to XVIII RK HQ. Although XVIII RK had already been notified by their cavalry of the enemy column's existence and had been drawn into battle by the time the airman's report arrived, the message was still important because it confirmed that only a mixed brigade and not a larger force opposed XVIII RK's troops at Neufchâteau.[163]

While patrolling the French rear-area, the observer of the second aircraft confirmed the location of the 2nd Colonial Division that was being held in reserve near Jamoigne. Upon learning the division's whereabouts, VI AK HQ ordered their flying section to continue patrolling the area and to immediately report any troop movements. As directed, FFA 13's aircraft took turns keeping watch throughout the day over the important French division, which never moved despite being just 3 miles from the bloody fighting at Rossignol where the rest of their corps was being annihilated. Nevertheless, if the French leadership had indeed chosen to move the division forward, FFA 13's airmen were in position to observe and report the action to corps headquarters in time to afford them the ability to adopt the necessary counter-measures.

The destruction of the Colonial Corps dramatically impacted the actions of the three corps echeloned to their left. The first to be affected was *Général* Roques' XII CA deployed on the Colonial Corps' immediate left. News of the Colonial Corps' problems had caused Roques to alter his men's route of march to the east out of concern for his right flank's safety. Sensing danger, the corps' two divisions advanced timidly and were ultimately stopped by only two underequipped reserve infantry regiments belonging to XVIII RK. Despite being heavily outgunned and outnumbered, the German reservists stood their ground and were able to successfully delay XII CA's advance until circumstances on both flanks forced Roques to order a withdrawal. It was ultimately XII CA's inability to achieve a breakthrough and turn the tide of the battle despite being opposed by just a brigade of reserve infantry that sealed the fate of French Fourth Army by creating the conditions for yet another corps-level disaster—this time for the XVII CA.[164]

Believing their right flank to be sufficiently protected, *Général* Poline's XVII CA had confidently set out for their objective—the town of Ochamps—around 9:00am. By 2:00pm, after easily passing through the Forest of Luchy, the corps' lead units became heavily engaged just west of their objective. Thirty minutes later XVII CA's exposed right flank, which had been uncovered by XII CA's failure to advance, was struck by a ferocious counter-attack delivered by elements of XVIII AK's 21 ID.[165] The skillfully executed German attack quickly moved against the French supply lines where a massive column of artillery and supply wagons was overtaken and captured as it was slowly meandering through the forest's only road. Panic quickly spread through the ranks as news of the German attack was passed from unit to unit along the front. By 4:30pm the entire corps was fleeing in disorder under destructive German rifle and artillery pursuit fire—sustaining heavy casualties in the process.[166] Poline's troops finally rallied and halted their retreat later that evening in an area to the west of their original starting positions.

While XVII CA was taking its beating, the XI CA on the extreme left of the French battle-line attempted to breakthrough the German positions which constituted the right flank of German Fourth Army's active front. The aggressive attacks against XVII CA to the south had left a single unengaged German division available to resist XI CA's advance, giving the French a great opportunity to turn German Fourth Army's flank and restore the overall situation.[167] However, despite being outnumbered 2:1 in infantry and outgunned 5:3 in artillery, the Hessian 25 ID (XVIII AK) performed heroically throughout the day, delaying the French advance around Maissin and refusing to allow the army's flank to become enveloped.[168] By 8:00pm news of XVII CA's retreat had finally compelled XI CA to abandon their gains and withdraw to the town of Paliseul about three miles to the southwest, thus ending the threat of envelopment amidst German fears that they might return and resume the attack the following day.[169]

****

Having been grounded by weather the day before, the center armies' *Fliegertruppe* awoke on the morning of the 22nd anxious to get airborne. Thick fog, however, kept the airmen grounded for several additional hours until 8:30-9:30am when it finally dissipated across the region. Nowhere was aerial intelligence more important at the time than at Fourth Army HQ. The army's senior officers had just learned of Fifth

Army's plans to attack nearby French units and had chosen to send one of their own corps to assist them. However, the absence of reconnaissance flights on the 21st had kept Albrecht and his staff ignorant of the close proximity of French forces in their own sector. This was highly problematic because Fourth Army was not yet fully concentrated and was not prepared to fight a large-scale engagement. Thus, recognizing their potentially dangerous situation, Fourth Army HQ ordered their flying sections on the morning of the 22nd to patrol the line Mézières—Montmédy and to directly report any observations to Fourth Army HQ at Bastogne as quickly as possible.

At 11:00am Fourth Army HQ's own FFA 6 delivered the first and what turned out to be the most important message of the day. The report stated that an estimated five French divisions had been observed advancing in the area of St. Medard—Suxy and would soon come in contact with the army's center. Minutes later, an observer from FFA 27 arrived at headquarters and delivered another report that confirmed "strong enemy forces" were in the area of Bouillon marching northeast in the direction of Bertrix. Operating from the same airfield, FFA 6 and 27's observers discussed their findings with each other and came to the conclusion that the total strength of the newly discovered French forces was between three and four corps. The two men were then called to give a verbal report to a Fourth Army HQ staff officer who condensed their respective observations into a single brief message that was forwarded to Albrecht, who was at the front visiting VI AK HQ.[170] The German Official History of the War described how Albrecht interpreted the report:

...he received an air-reconnaissance report at about midday which blindingly illuminated to him the extreme dangerous situation into which Fourth Army had fallen… The short, but significant report showed that Fourth Army was facing an immediate and serious crisis: strong enemy forces were slipping past them in a northerly direction and in the very near future superior forces would strike their center. The fighting might even have already begun. The priority now for Fourth Army was to protect the left flank of Third Army. Only when this was guaranteed could Third Army cross the Meuse in order to help the right wing deliver the decisive blow; otherwise a substantial part of Third Army might have to be diverted south. And for the time being Fourth Army only had its two central corps available to oppose the attack.[171]

Thanks to his aviators, Albrecht now fully understood how dire his army's predicament truly was. By sending VI AK to the south to assist Fifth Army and by keeping VIII AK in the north to cover Third Army's flank, Albrecht had only two corps available (XVIII AK and XVIII RK) to check the French advance.[172] Consequently, upon receiving the aviation report, the army commander and his staff quickly left VI AK HQ to take up a more central position at the Fourth Army HQ forward command post at Libramont. While enroute, Albrecht stopped at Neufchâteau to personally inform XVIII RK commander *General* Kuno von Steuben of the situation and to ensure that he and all of his subordinates understood that retreating would open up a massive gap in the line and was therefore not an option. Once at Libramont, the Fourth Army HQ staff worked as best they could to manage their reserves and to avoid any breakthroughs in their sector.

Due to the nature of the terrain, the Fourth Army *Fliegertruppe* were unable to offer any meaningful tactical assistance during the fighting of the 22nd. The only flights conducted during that afternoon were over the French rear-areas in order to give the army and corps leadership better estimates of the strength of enemy reserves. The amount of intelligence gathered using this method, however, was extremely limited. Acting on the orders of XVIII AK HQ, FFA 27 had dispatched an aircraft to Bertrix to try and determine the strength and location of the various French columns marching towards the corps' right hand division—21 ID. Unfortunately, the aircraft was shot down by ground fire while circling over XVII CA's 33 DI, thus denying critically important intelligence from reaching the German commanders on the ground.[173] Nevertheless, 21 ID was able to take advantage of the situation and deliver a decisive assault against the flank and rear of 33 DI, ultimately causing the entire XVII CA to retreat in panic. Elsewhere, FFAs 6 and 13 each carried out patrols over the French communication zone. In both cases, the observers reported that the French rear-area was entirely empty of reserves—indicating that Albrecht's troops had just repulsed the full brunt of the French advance and did not need to worry about a renewed attempt against the army's center the following day. Finally, VIII AK's FFA 10 reconnoitered the Meuse sector between Givet and Charleville. VIII AK had been left to the north on the 22nd to maintain contact with Third Army and cover their flank as they approached the Meuse. The airmen of FFA 10 had therefore been ordered to monitor the area directly west of the corps and to report any attempt by the French to envelop either Third or Fourth Army's flanks to the appropriate army commander. As it happened, the section's airmen discovered two large columns east of Fumay marching in the direction

Above: *Hptm.* Florian von Poser, commander of FFA 19 (Courtesy of the Library of Congress)

of Maissin against XVIII AK and Fourth Army's vulnerable right flank. Although the French troops were too far to the rear to impact the fighting on the 22nd, the report deeply worried Albrecht, who understood how dangerous the situation still was on his flank. Thus, later that evening, Fourth Army HQ took precautions to strengthen their positions around Maissin while drafting orders for their *Fliegertruppe* to concentrate their patrols the following morning over the right wing in order to determine the position and intentions of any French troops in the area.

Meanwhile circumstances were frustratingly similar for Fifth Army's *Fliegertruppe*. Despite the Crown Prince's decision to seize the initiative and launch an offensive, the army's airmen were powerless to intervene tactically. Once again, the close proximity of the opposing units as well the nature of the terrain prevented aerial cooperation. As a result, Fifth Army's corps flying sections remained mostly inactive throughout the day. One exception, however, was XVI AK's FFA 2. With XVI AK being deployed on the army's left flank, it was critically important that the corps' open flank did not become enveloped. Thus, the corps' commander, General Bruno von Mudra, was originally ordered to hold his corps back as a flank guard throughout the offensive. At 9:30am however, an observer from FFA 2 delivered a report to XVI AK HQ stating that "the area Landres—Briey—Conflans was free of enemy troops" and that Fifth Army's left flank was therefore not threatened.[174] Mudra immediately forwarded the report to Fifth Army HQ, who after receiving a similar message from their own FFA 25 decided to unleash XVI AK for an offensive role. For the remainder of the day Mudra's troops were used to threaten and eventually envelop French Third Army's right flank, causing their most successful corps—VI CA—to order a tactical withdrawal.[175]

The flight activity of FFA 25 was also greatly restricted throughout the 22nd. Under direct orders from Fifth Army HQ, the section's airmen reconnoitered French Third Army's rear-area as well as the situation on both of the army's flanks. The most important of these reports was delivered late in the day around 6:00pm. The message stated that the area between Stenay and Verdun was free of French forces and that the retreating units of French Third Army were taking up positions west of Arrancy at St. Laurent-sur-Othain. In other words, just as had happened in Fourth Army's sector, the Crown Prince's aviators confirmed that there were no French reserves available to renew the attack the following day.

Finally, although not in the air, it should be noted that the officers and men of FFA 19 saw their first combat of the war near Virton during the afternoon of the 22nd. Operating from an airfield outside the village of Étalle near V AK HQ, FFA 19's commander, *Hptm.* Florian von Poser und Großnadlitz, was informed at 4:00pm that a major French breakthrough had occurred at the front consisting of possibly two entire French divisions. Almost immediately, the section's enlisted personnel began packing up the equipment and preparing for a withdrawal, thus eliminating the possibility for aerial activity. However, within the hour news had reached the flying section that the situation at the front was favorable and that no breakthrough had occurred. Nevertheless, at 5:30pm Poser learned that a small force of French infantry (three companies in strength) was moving toward the airfield. Without hesitation, Poser gathered five officers along with 12 enlisted men and took up defensive positions at the end of the airfield.[176] After a brief firefight, the French soldiers abruptly surrendered. Among the POWs were a captain, two lieutenants, and 140 men of an infantry company belonging to the French 103 *Regiment d' infanterie* (IV CA). It was later determined that the unlucky company had become isolated and cut off from the rest of their regiment earlier in the morning and were attempting to make their way back to their own lines when they were discovered and fired upon by Poser and his men. With many of his men out of ammunition and water, the French captain quickly surrendered his unit to the German airmen. The episode proved not only Poser's bravery but also the *Fliegertruppe* ground crew's competency in defending the airfield from traditional attacks.[177]

* * * * * * *

As darkness fell over the Ardennes the two German center armies began the task of evaluating the extent of the day's victory and determining how to proceed. While the initial reports of POWs and captured war materiel suggested a victory had indeed been won, there was very little available information regarding the status of the German units at the front, many of whom were badly damaged and disorganized. Thus, with their troops scattered throughout the forests and a possible renewal of the French offensive looming, both Fourth and Fifth Army HQs recognized the need to reorganize before any further action could be taken. Specifically, Albrecht and the rest of Fourth Army HQ was extremely worried about the situation on their right flank, which remained extremely vulnerable after the 25 ID's heroic stand at Maissin had left the division severely weakened. Although the army's reserve (VIII RK) had arrived in the area shortly after dusk, there was no way to determine whether the French reinforcements seen late in the day by FFA 10 would also be sent into the area to resume the offensive the following morning. If indeed the French decided to resume the attack on Fourth Army's flank, the armies of the German center would collectively be unable to begin any pursuit. Consequently, the evening of the 22nd–23rd saw the Fourth and Fifth Army HQ staffs work frantically to reform and strengthen their lines, closing as many gaps as possible. By sunrise the two armies' front line troops were dug in and well prepared to repulse another French attack which ultimately never came.

Both Fourth and Fifth Army HQ's were anxious on the morning of the 23rd to determine the position and intentions of the French formations in their front. Orders were therefore drafted and given to all of the center armies' flying sections demanding reports of the enemy's front and communication zone to be forwarded to army headquarters as quickly as possible. What the German leadership needed to know was whether the French were going to attack, retreat or if they had entrenched during the evening. Only after a definitive answer could be obtained could the armies of Moltke's center take action and continue the advance in support of the right wing's attack.

Ideal weather conditions allowed the *Fliegertruppe* to begin flight operations as early as 5:00am. In Fifth Army's sector, the army command had become optimistic over the course of the night and had issued a directive at 6:25am ordering their corps to "energetically pursue the enemy, push him away from Verdun and transform yesterday's victory into a catastrophe." In other words, Fifth Army HQ wanted its left wing corps, XVI AK, to turn French Third Army's right and push them away from Verdun.[178] This meant the majority of Fifth Army *Fliegertruppe's* flights were concentrated opposite the army's left wing. Unfortunately, the Crown Prince's airmen struggled throughout the first half of the day to make any meaningful discoveries. Performing flights between 6–8am and 11:20am–12:40pm, XVI AK's FFA 19 found nothing outside of a single westbound train and some cavalry near Spincourt. Located on the army's right wing, V AK's FFA 2 was similarly unable to deliver any noteworthy reports during the morning. Finally, around noon, one of FFA 2's flights reported a strong French force—at least one division—along the line Houdelacourt—Senón—Amel—Éton near Spincourt. This report, delivered to Fifth Army HQ at 2:00pm, was promptly confirmed by HKK 4's 6 KD, who informed Fifth Army HQ that the French troops were not advancing. Additional flights performed soon thereafter by FFA 2 found another large group of French troops standing on the line Muzeray—Éton—Gondrecourt. Finally, late in the afternoon between 4 and 6pm, FFAs 2, 4 and 25 each carried out flights that revealed at least two corps between Stenay and Azannes with another corps deployed along the Othain River.[179] By nightfall Fifth Army HQ had compiled ample evidence from both their aviators and their cavalry that proved French Third Army was in the midst of a general withdrawal. Reports, however, of French forces advancing out of Verdun had caused XVI AK to halt their advance—permanently dashing the Crown Prince's hopes of pushing French Third Army away from Verdun.

Meanwhile, Fourth Army's airmen were employed with much greater success. Concerned about their right flank, Fourth Army HQ issued a directive late in the evening of the 22nd requesting all flying sections focus on reconnoitering the territory opposite the army's right wing. Although he had issued orders for his army to renew the attack on the 23rd, it was absolutely vital for Albrecht to determine whether the French troops in his front were going to resume the attack or retreat. If they were to retreat, Fourth Army could shift forces north, begin a pursuit to the Meuse and resume its duty as Third Army's flank guard. If,

Above: Max von Hausen, former Chief of the Saxon General Staff and the commander of German Third Army during the Marne Campaign. (Author's Collection)

on the other hand, the French were going to attack, Fourth Army needed to be in position to quickly throw them back. By 8:00am Albrecht and his staff received their answer. An observer's report from XVII AK's FFA 27 had arrived stating that the French troops along the line Bertrix—Petitvoir were withdrawing. At 9:15am however, word arrived from VIII RK on the right flank stating that contact had been made with "strong enemy forces" south of Maissin. Thus, with two conflicting reports in hand, Albrecht decided to wait for further developments before making a decision. Thirty minutes later a messenger from Fourth Army HQ's own FFA 6 delivered a hasty report which read in part: "Enemy between Neufchâteau and Florenville in a general retreat, but is still holding in the area of Bouillon to the northeast."[180] The observer's message had thus confirmed the veracity of both of the aforementioned messages. French Fourth Army was indeed retreating—its left wing was simply acting as a strong rear-guard.

Fourth Army HQ received additional reports at 10:30am from both FFAs 10 and 27. The messages described a "large 12 kilometer column, ending seven kilometers north of Bouillon, marching south on the Paliseul—Sedan road at 9:00am" and "strong formations retreating beyond the Semois River."

Together, the two reports removed all doubt as to whether the French were withdrawing out of the Ardennes. Indeed, another report from FFA 10 confirmed that all roads leading towards the German right wing were free of enemy troops.[181]

Thanks to their aviators, Albrecht and his staff were fully informed of the situation in their front by 11:00am. With Langle's Fourth Army confirmed to be retreating, orders were quickly issued to begin a pursuit. For example, XVIII AK's 25 ID was informed of the *Fliegertruppe's* findings at noon and was directed to advance through Maissin towards Bertrix.[182] Further south, VI AK and XVIII RK also began their pursuits of the retreating French at noon. Throughout the rest of the day the badly exhausted men of both corps struggled to gain ground against the French rear-guard's rapid artillery fire—allowing the opposing French army to gain some separation.

The French defeats in the Ardennes sounded the death knell of Joffre's Plan XVII. Instead of quickly marching through the Ardennes unopposed, the entire French battle-line was decisively defeated, resulting in massive casualties and the precipitous retreat of Joffre's two center armies. Two French corps (the Colonial and XVII CAs) were severely mauled and reduced to well below half strength. A third (IV CA) was heavily damaged with the near destruction of an entire division. Five additional French corps had suffered considerable casualties and were in no shape to conduct offensive action of any kind for the foreseeable future. Thus, with his two center armies defeated and in full retreat, Joffre could no longer realistically plan any offensive action into Germany. Instead, the French commander-in-chief would be forced to order a general retreat and look for an opportunity to launch a successful counter-attack when the situation presented itself. That task, however, became infinitely more difficult when he learned of the situation on his left wing, where the three armies of the powerful German right wing had begun to execute their attack.

Belgium: The German Right Wing Armies Begin Their Advance

The fall of Liège on August 16th enabled the three armies of the German right wing to begin their advance through Belgium in preparation for delivering the German War Plan's main assault. During the evening of the 17th, Moltke's OHL issued a directive to the right wing commanders which firmly established the German supreme commander's concept of operations for the coming offensive. According to the order, the armies of the right wing were to commence their advance at dawn on the 18th, initially pushing west then, in the case of First and Second Armies, wheeling

Above: Karl von Bülow, commander of German Second Army. (Author's Collection)

Above: Alexander von Kluck, commander of German First Army. (Author's Collection)

south in the direction of Namur and the Franco-Belgian border. Together, the First, Second, and Third Armies were to attack and ultimately defeat the combined forces of the numerically inferior allied left wing.

Each of the right wing's three armies had a unique role in the coming operation. Positioned on the group's left flank, *Generaloberst* Max von Hausen's Third Army was to initially march west and attack the French positions along the Meuse River near Dinant. Altogether, the army was comprised of just four corps (three active and one reserve)—the smallest of Germany's eight armies. In accordance with a pre-war agreement with the Prussian-led imperial government, three of Third Army's four corps, as well as Hausen and most of his staff, were Saxon, thus honoring the Kingdom's desire to field its own troops together as an independent army. On account of its size and location in the battle-line, the Saxons were to play the role of a support force. Accordingly, Third Army's initial attack and subsequent operations were often conducted to facilitate the success of its neighbors—primarily that of Second Army on it's right.[183]

Charged with conducting the right wing's main attack was the Second Army under the command of *Generaloberst* Karl von Bülow. An excellent soldier and a descendent of a well respected family within the Prussian nobility, Bülow had quickly moved up through the ranks to become one of the most highly regarded senior officers in the empire. Indeed, many suspected that he would serve as successor to Schlieffen as Chief of the General Staff upon the latter's retirement in 1906. Nonetheless, despite not getting the job, Bülow continued to serve in important posts, earning the confidence of the Kaiser and other superiors.[184] Thus, given his reputation, it seemed natural that command of the most important army on the Western Front be given to Bülow. According to OHL's concept of operations, Second Army was to

Above: Alexander von Kluck and the First Army HQ Staff. Chief of Staff Hermann von Kuhl is pictured third from the left. (Courtesy of the Library of Congress)

initially march west, wheel to the south and attack the Sambre line with its left flank brushing against Namur. According to German intelligence estimates, strong allied forces were expected to be encountered between the powerful fortresses of Namur and Maubeuge. It was here where the powerful German right wing—Second Army in particular—was expected to defeat the forces of the allied left, creating the conditions for a decisive envelopment and encirclement soon thereafter. In order to increase the likelihood of the attack's success, Bülow's army was afforded six corps (three active and three reserve) as well as HKK 1.

While Second and Third Armies' roles were seemingly straightforward, First Army's mission was incredibly complex. After Liège's capitulation, the army was ordered to move west and clear out all allied forces operating in its front. As the advance continued and Second Army began its turn to the south, First Army was to then assume the role as a flank guard. Finally, once Second Army was fully engaged and the location of the allied left was known, First Army was to quickly attack and envelop the allied left flank. In other words, Second Army was to assault and hold the allied left's main body while First Army maneuvered against its flank and rear.[185] Therefore, because the missions of First and Second Armies were so closely linked, Moltke decided to subordinate First Army under Second Army HQ's orders. Thus, beginning on the 17th, Bülow assumed authority over both First and Second Armies as well as HKKs 1 and 2.[186]

The actions of First Army would ultimately determine the campaign's success. Prior to the war Count von Schlieffen had taken great pains to teach to his subordinates about the differences between winning "an ordinary victory" versus seeking a "battle of annihilation." It was critically important, given Germany's predicament of facing a numerically stronger enemy on two fronts, that the German forces in the west not only defeat, but encircle and destroy a large portion of the western allied armies.[187] Thus, while Second and Third Armies' attacks were fully expected to succeed at defeating the numerically weaker allied left wing's line, it was still expected to be an 'ordinary victory.' Only First Army's envelopment attack could deliver the quick and decisive knockout blow that was required by the German War Plan. Accordingly, in order to properly carry out its various missions, the army was given six infantry corps (four active and two reserve) as well as the powerful HKK 2—making it the largest of Germany's eight field armies.[188]

Command of First Army was given to *Generaloberst* Alexander von Kluck, one of the army's most proven and aggressive senior officers. Born outside of the nobility into a middle-class Westphalian family, Kluck served in the wars against Austria and France and had rapidly rose through the ranks based solely upon merit. Kluck's chief of staff was *Generalmajor* Hermann von Kuhl, arguably the finest officer on the German General Staff at the outbreak of war and a man who Schlieffen predicted would be a "great captain of the future."[189] Together, the extremely capable Kluck and Kuhl were to draw upon their unique talents and experiences to lead First Army throughout the campaign—first in its role as flank guard and finally, once conditions were right, to deliver the anticipated fatal blow to the western allies' left flank.[190]

The allied forces opposed to the German right wing were comprised of French Fifth Army, the BEF and an independent group of French reserve and territorial divisions, deployed for defensive purposes on the extreme left of the allied line. Unfortunately for the Entente, the BEF and French Fifth Army had a combined numerical strength of 17½ infantry divisions—exactly seven divisions fewer than the German right wing armies opposed to them.[191] To make matters worse, the BEF was not expected to be fully concentrated and in position alongside Fifth Army until the 21st or 22nd.[192] As a result, the two armies would be forced to fight their opening battles separately. Deployed in the area: Charleroi—Namur—Dinant with the strength of four corps, French Fifth Army would ultimately be charged with fighting off the combined attacks of German Second (six corps) and Third (four corps) Armies with only the help of fortress Namur and its garrison. Meanwhile, the BEF's five infantry divisions and one cavalry division (with nominal assistance from a French cavalry corps) would be left to fight off the powerful German First Army.

Aerial reconnaissance was a critically important component to operations on the German right wing. Indeed, as of August 17th the German leadership was entirely unaware of where the allied forces opposite their right wing were located. Only the position of the Belgian Field Army, standing opposite First Army along the Gette River, was known at the time of the right wing's advance.[193] Therefore, in order for the German right wing's grand enveloping attack through Belgium to be successful, it was absolutely critical that the combined forces of the allied left wing were located as quickly as possible. With this intelligence in hand, the right wing commanders could properly plan their corps' approach marches so that the allied line was speedily enveloped and defeated in accordance with Moltke's concept of operations. For instance, while the position of the French left could be safely assumed to be in the vicinity of Namur, the location of the BEF was a complete mystery. Back at OHL, Moltke believed the British would initially concentrate around Lille, allowing them to threaten the German flank once the French were engaged. Other possibilities, however, included the area around Maubeuge in close proximity to French Fifth Army, or a landing on the Belgian coast to immediately threaten the German rear.[194] As flank guard, First Army had to be prepared for any of these eventualities, compelling its corps to be inconveniently deployed *en échelon* during its advance through Belgium. Only after the BEF was located could Kluck begin to concentrate his widely dispersed forces and take the offensive. Thus, with the success of the right wing's offensive at stake, it was imperative for the *Fliegertruppe* to locate the BEF as well as French Fifth Army as quickly as possible.

The Actions of the Right Wing Armies: August 17–20

On the morning of August 17th, German First and Second Armies moved beyond the still smoldering battlefields around Liège and continued west into the Belgian countryside. That afternoon First Army HQ learned that it was to be officially subordinated under General Bülow for the remainder of the operation. The order, issued by Moltke at OHL, stated:

His Majesty commands: The First and Second Armies and the II Cavalry Corps are hereby placed under the orders of the commander of Second Army for the advance north of the Meuse. The march will begin on the 18th.

The objective of the movement is to drive the enemy [Belgian] troops, reported to be in the position between: Diest—Tirlemont—Wavre, back from Antwerp, while covering on our left flank toward Namur. We have in view employing the two armies later from the line: Brussels—Namur, covering toward Antwerp.

Further orders will be issued for the taking of Namur by the left flank of Second Army and the right flank of Third Army. The artillery assigned to Second Army is to be brought forward for this attack.

The Third Army, with its flank joining the left of Second Army, will proceed through Durbuy toward the southeast front of Namur.[195]

General Kuhl immediately traveled to Second Army HQ at Liège upon receiving the news to review the situation in person with Bülow. The primary purpose of the meeting was to discuss the pending attack

Above: Albert I, King of the Belgians. (Author's Collection)

against the Belgian field army, which was deployed along the line Diest—Tirlemont—Jodoigne directly in the path of First Army's westward advance. Under the command of King Albert I, the Belgian army was a second-rate force of reservists and militia with the numerical strength of five divisions.[196] Although they were qualitatively inferior, the Belgians nonetheless posed a considerable threat to the German supply lines if they were not destroyed before First and Second Armies wheeled south to face the Franco-British forces along the Sambre. The German leadership was therefore anxious to dispatch the Belgian threat as quickly and with as little loss of life as possible.

To accomplish this aim, Kuhl proposed committing First Army in a sudden frontal attack using the shortest possible approach routes in order to better deny the Belgians a chance of escape. Bülow, with full command authority over the right wing, wished instead to wait another 24 hours for his Second Army to move into position so that *both armies* could launch a concerted enveloping attack. Kuhl strenuously objected to this strategy, stating that "the enemy would not wait for the execution of this plan, but would evade the enveloping attack in plenty of time."[197] In response to this valid concern, Bülow offered a compromise. First Army would be permitted to immediately carry out its attack the following day with the caveat that II AK on the army's right wing be sent north to envelop and encircle the Belgian left flank before the army's other front line corps (III, IV and IX AK) launched their frontal attack. Kuhl replied that the delays caused by II AK's approach march would still allow the Belgians to retreat before the other corps' attacks could be made. Indeed, he knew that if the Belgians escaped that they would retreat north into the fortifications at Antwerp where they would pose a constant threat to the German supply lines—forcing the German right wing to leave behind an entire corps to cover the fortress for the remainder of the campaign.

Nonetheless, over Kuhl's objections, Bülow ordered the enveloping attack to proceed on the morning of the 18th. As predicted, II AK was severely delayed in the marshy terrain near Beeringen where they were spotted by Belgian scouts. Belgian commanders were therefore quickly notified of the German movements and ordered a general retreat accordingly. This allowed the bulk of the army to withdraw before II AK could move into position to attack. By nightfall, the entire Belgian Army was in full retreat and moving north towards Antwerp as Kuhl predicted. Thus, the Belgians were able to withdraw under the cover of a rear-guard and avoid the decisive engagement the Germans had desired.[198]

The Belgians continued their retreat over the following two days, staying well ahead of their German pursuers. By the afternoon of the 20th, all five divisions of Albert's command were safely installed inside the defenses of Fortress Antwerp. Nevertheless, although they had been removed from the field, the Belgian presence inside Antwerp still posed a threat to the German line of communications.[199] As a result, First Army was compelled to detach General Hans von Beseler's III RK to cover the fortress and keep the Belgians pinned up—reducing the strength of the right wing's most important army by an entire corps just days into the operation.

First Army's aviators were airborne early on the 18th in hopes of playing a significant role in the anticipated battle with the Belgians. As soon as the morning fog had burned off, each of the army's flying sections sent aircraft to reconnoiter the Belgian positions in preparation for the attack. While II AK was making its long approach march, its flying section, FFA 30, scouted the positions on the Belgian left flank. FFA 30's airmen were ordered to take note of the Belgian deployment in order to give corps headquarters a better idea of where to focus their attack. The section's airfield was located directly next to the corps HQ near the town of Hasselt. The close proximity allowed each aerial report to be personally delivered and discussed by the observers with members of the II AK HQ staff. The section's first report of the day was delivered at 10:15am in an unmistakably negative tone. The observer reported already seeing significant movement indicating the entire Belgian left was preparing to withdraw. Within minutes a second report arrived bringing more bad news. The second observer had confirmed the first's speculations by testifying that he observed an entire brigade retreating northerly through Louvain in road column—an unmistakable sign that a general retreat had been ordered.

The airmen of FFA 30 continued to deliver reports throughout the rest of the afternoon keeping II AK HQ abreast of the latest Belgian movements. The first kept watch over the various retreating columns while the second attempted to ascertain the strength of the rear-guard. During their flight, the second patrol witnessed the only major engagement of the day at Tirlemont between "strong Belgian forces" and IX AK on First Army's left wing. Unfortunately, II AK HQ was unable to pass along any of their aviator's valuable information to First Army HQ because their radio communications were temporarily inoperable.[200] Consequently, all of FFA 30's information wasn't delivered to the army commander until the following morning.

The flying sections of First Army's other forward corps were similarly busy throughout the 18th. FFAs 7, 9 and 11 (III, IV and IX AK's respectively) each sent aircraft to reconnoiter the areas where their corps were expected to attack. Arriving between 10 and 11am, the first reports of the day outlined the Belgian strength and deployments. By midmorning, each of the sections had discovered signs of the Belgian retreat, a fact which many front line units had already discovered on their own.[201]

In concert with the cavalry, First Army's *Fliegertruppe* closely monitored the Belgian retreat over the following two days. Once it became clear that the Belgians would safely reach Antwerp, all flying sections were redirected to begin scouting the areas of northern and western Belgium in search of the BEF. The most notable of these flights were conducted by FFA 12. As First Army HQ's flying section, FFA 12 was charged with performing the bulk of the army's long-range reconnaissance sorties. On August 19th, FFA 12 sent one of their six Rumpler Taubes to patrol the line: Namur—Ostend—Bruges. During the flight the aircraft ran out of fuel and was forced to land in neutral Holland where the crew was interned for the duration of the war. Undeterred by this loss, the section sent another aircraft the following day on a similar 190 mile long journey over Ghent—Bruges—Ostend. Upon returning the observer reported the entire area "free of the enemy." Also during the 20th, a flying section from Second Army—X AK's FFA 21—conducted a 230 mile flight to search the Belgian coast line from Knokke—Ostend—Nieuport, finding no signs of allied troops. Altogether, First and Second Army's aviators were unable to detect any allied troop concentrations beyond the line Brussels—Maubeuge.[202] Nevertheless, despite being unable to locate any signs of the BEF, First Army continued moving west in accordance with the campaign's concept of operations. However, because the British Army's position remained unknown, First Army HQ was compelled to slow their advance on the 21st and "assume a strongly echeloned formation right and left so as to be prepared for all emergencies."[203]

Meanwhile to the south, Second Army was able to make its approach march without any noteworthy allied resistance. By the evening of the 20th Bülow's army had moved west past Namur to the line Ninove—Gembloux, indicating that the right wing's wheel around the Belgian fortress was underway. During their advance, Second Army's aviators had kept a close watch over the area directly in the army's front. Consequently, there were few patrols of the region south of the Sambre where French Fifth Army was actually located. Fortunately, Third Army's *Fliegertruppe* had spent the period actively searching the Meuse positions between Dinant and Givet. Flights on the 19th discovered trenches, artillery batteries, and bivouacs near Onhaye as well as along the river's west bank. Further south, a large "18 kilometer long column," appearing to be a supply train, was seen moving through Couvin from Rocroi at 11:20am.[204] By nightfall on the 19th, Third Army HQ was in possession of aviation and cavalry reconnaissance reports as well as documents taken from a captured French officer which concluded that at least two French corps were in the area southwest of Dinant. This information was forwarded to OHL who accordingly sent out the following intelligence estimates the following day to all army commanders:

As of 20 August, the French distribution of forces is assumed to be as follows: On the Meuse between Namur and Givet, French I, II, perhaps also X Corps; south of the Sambre between Namur and Maubeuge, enemy forces are approaching between Namur and Charleroi, and already today one or two corps at most have reached the vicinity of the Sambre, west

Above: The man Joffre called "a veritable lion," Charles Lanrezac skillfully led French Fifth Army through multiple crises.

of the line Charleroi—Fumay, about three corps are advancing north and these are probably reserve divisions (presumably these have yet to reach the line Philippeville—Avesnes). The landing of the British at Boulogne and their employment from the direction of Lille must be assumed. However, it is the OHL's opinion that large-scale landings have yet to be carried out...[205]

In lieu of all the available intelligence, Moltke ordered the right wing to begin the 'decisive offensive' prescribed by the War Plan. In order for the operation to be successful it was critically important that each of the three armies properly coordinate their attacks. To that end OHL transmitted the following directive to all three army commanders at 2:30pm on the 20th:

First and Second Armies must close up along the line reached on August 20, while still safeguarding against Antwerp. The attack upon Namur has to begin as soon as possible. Both army headquarters must reach agreements to synchronize the imminent attack against the enemy west of Namur with Third Army's attack on the Meuse between Namur and Givet. In the further operations of the right wing, the employment of strong cavalry west of the Meuse will be required. Therefore, HKK 1 must notify Third and Fourth Armies that he is vacating their fronts and begin to move north around Namur. When arriving on the northern bank of the Meuse, HKK 1 will be subordinated to Second Army's commanding general.[206]

In accordance with Moltke's order, the First and Second Armies used the 21st to draw closer together in preparation for the impending offensive. After a short march, Second Army's left wing reached the Sambre with its right wing following close behind. Acting as flank guard, Kluck's First Army advanced to the southwest in echelon in order to check any potential British offensive from the west, southwest or south. Meanwhile, Hausen's Third Army remained a full day's march away from the Meuse and was not expected to be in position to attack until the evening of the 22nd.

It had been OHL's intention that the three right wing armies deliver their attacks simultaneously.[207] Second Army was to seize the Sambre crossings between Namur and Charleroi and launch an offensive against the French forces deployed on the southern bank while Third Army attacked across the Meuse against their flank. Thus, because Third Army was still a day's march from the Meuse, Second Army HQ disallowed any movement beyond the Sambre on the 21st—issuing a direct order to their corps commanders stating:

X AK and Guard Korps [Second Army's left wing] will advance to the Sambre as advised to today, but there will be no attack. Instead, I intend to continue with the wheel in a southwesterly direction with First and Second Armies so that, in combination with Third Army, we can strike as uniform a blow as possible against the enemy forces which are reported to be south of the Sambre and west of the Meuse. Until then, the reports which come in regarding the enemy will be decisive in seeing the date for the attack; at any rate, I shall inform Third Army in time.[208]

Indeed, the only active operations anticipated on the 21st was to be the shelling of Namur's northeastern and eastern forts by heavy artillery and siege batteries attached to the Guard RK, who had been selected by OHL to reduce the formidable Belgian fortress. The rest of Second Army was to spend the 21st completing their approach marches around Namur and preparing for the coming offensive. However, as is often the case in war, the plan was overcome by events.

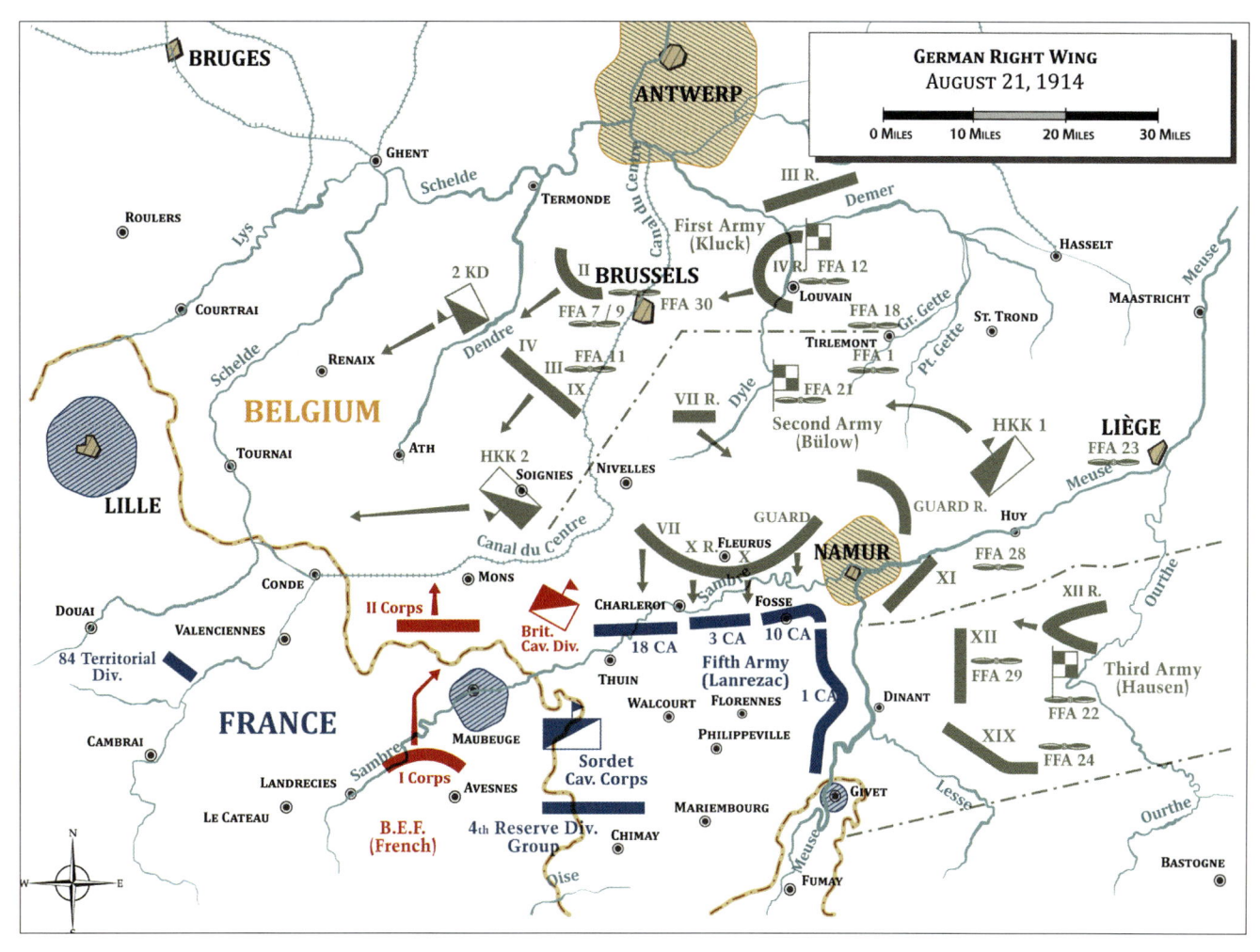

The Battle of Namur: The Right Wings' Initial Attack

The target of Second and Third Armies' planned assault was the French Fifth Army under the command of *Général* Charles Lanrezac. Known by his peers as "the lion of the French Army," Lanrezac was an exceptionally capable officer who had established a reputation as a brilliant teacher and theoretician while teaching at the Saint-Cyr Military Academy.[209] Lanrezac believed from the first day of mobilization that the Germans' main thrust would be directed from the north against Namur, Dinant and Givet. Joffre, however, contended that his adversary's offensive would be delivered further south through Sedan and would not commit any large numbers north of Namur on the west bank of the Meuse. Consequently, over Lanrezac's repeated objections, French Fifth Army was originally deployed opposite the Ardennes between Mézières and Mouzon. On August 15th, after reviewing intelligence reports from Belgium after the fall of Liège, Joffre finally acquiesced to Lanrezac's wishes and allowed Fifth Army to shift to the north toward Namur.[210] From that position, Fifth Army and the BEF were to attack and fix the German right wing in place until the armies participating in Joffre's Ardennes offensive could turn against its rear.[211]

Delayed by Joffre's wishful thinking, French Fifth Army did not reach the Sambre until the 20th.[212] Upon its arrival, the army was deployed facing two directions in order to check potential attacks from the north and the east. On the army's right, the I CA was positioned along the Meuse between Namur and Givet. Meanwhile, Lanrezac's remaining three corps— X, III and XVIII CAs—were deployed along the Sambre's southern bank on the line Namur—Charleroi—Thuin. Lanrezac spent the evening of the 20th contemplating whether to order his three corps on the Sambre to move across the river and seize the heights on the northern bank or to hold them back and wait until the BEF arrived on his left. In either case, Fifth Army would need to wait for the British to move into line on their left before beginning the offensive. Lanrezac therefore believed it was a huge risk to leave his forces completely isolated with a river at their back against an enemy force of unknown strength.[213] Thus, after careful consideration, the commander chose the latter course of action—sending weak detachments forward to defend the sector's many bridges while the rest of the army deployed along a series of heights seven miles south of the river.²¹⁴

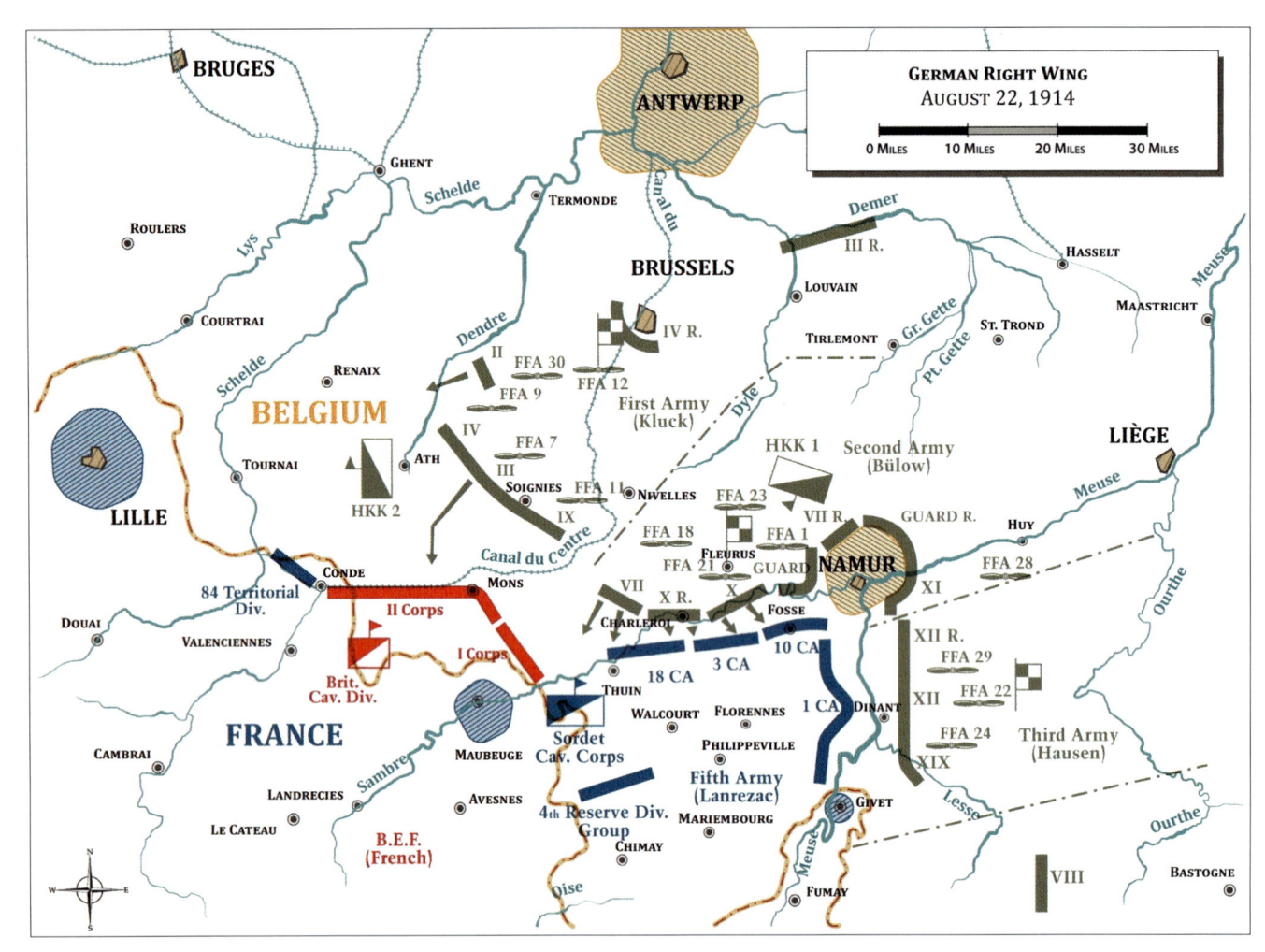

Bülow's Second Army began the final stage of their approach march around Namur early on the morning of the 21ˢᵗ. On the army's left wing, the Guard Korps and X AK quickly reached the Sambre's northern bank and halted as ordered. Arriving at noon, the Guard Korps' lead division—2ⁿᵈ Guards Infantry Division—was the first to reach the river. The division's commander, *Generalleutnant* Arnold von Winckler, personally observed the French defenses on the southern bank, considering them to be vulnerable to attack. Thus, recognizing the importance of seizing the river crossings for subsequent operations, Winckler immediately requested permission to cross the river and seize control of the opposite bank. After a lengthy debate at Guard Korps HQ, Winckler was finally given permission to proceed with the attack.²¹⁵

Immediately after gaining his Corps HQ's consent to advance, Winckler's artillery opened fire from their positions on the heights north of the river. Using the heavy bombardment to cover their movements, Regiment *Augusta* attacked across the Sambre at the town of Auvelais, driving the French back to Arsimont one mile to the south. That evening, German reinforcements consolidated Regiment *Augusta's* gains on the southern outskirts of Auvelais. Elsewhere, elements of X AK's 19 ID crossed an undefended bridge at Tergné and advanced to Roselies, which was taken in hand to hand fighting.

Second Army's aggressive action on the 21ˢᵗ had achieved a great operational advantage for the right wing armies. By establishing multiple bridgeheads on the Sambre's southern bank, Bülow's troops were now ideally situated to begin their offensive upon Third Army's arrival. However, late that evening Second Army HQ learned that Hausen's army would not be in position to attack until the morning of the 23ʳᵈ. Therefore, Bülow ordered Second Army not to advance on the 22ⁿᵈ. Instead, the day would be used to bring up the army's main body while the advance guards expanded and strengthened the bridgeheads on the southern bank. Second and Third Armies would then, according to Bülow's orders, deliver their attacks in concert at dawn on the 23ʳᵈ.

At his headquarters in Chimay, Lanrezac received the news of the day's fighting with indifference. Indeed, having ordered his troops to withdraw from the river crossings in the event they were attacked, French Fifth Army's commander believed the loss of the bridges to be of minimal importance. Outnumbered and facing two converging enemy armies on two fronts, Lanrezac wished to conserve his army's strength and fight a defensive battle on the heights south of the Sambre while waiting for the BEF to arrive on his left and French Fourth Army to be successful in its offensive on his right. Lanrezac's corps commanders believed differently, however, and had expressed their intentions in their nightly reports of the 21st–22nd to order an immediate counter-attack at dawn to reclaim the lost bridges. Despite vehemently disagreeing, Lanrezac felt it was too late in the evening to countermand his subordinates' orders before the attacks would begin. Consequently, Fifth Army launched a large-scale attack on the 22nd against the German positions between Charleroi and Namur instead of remaining on the defensive as Lanrezac had intended.[216]

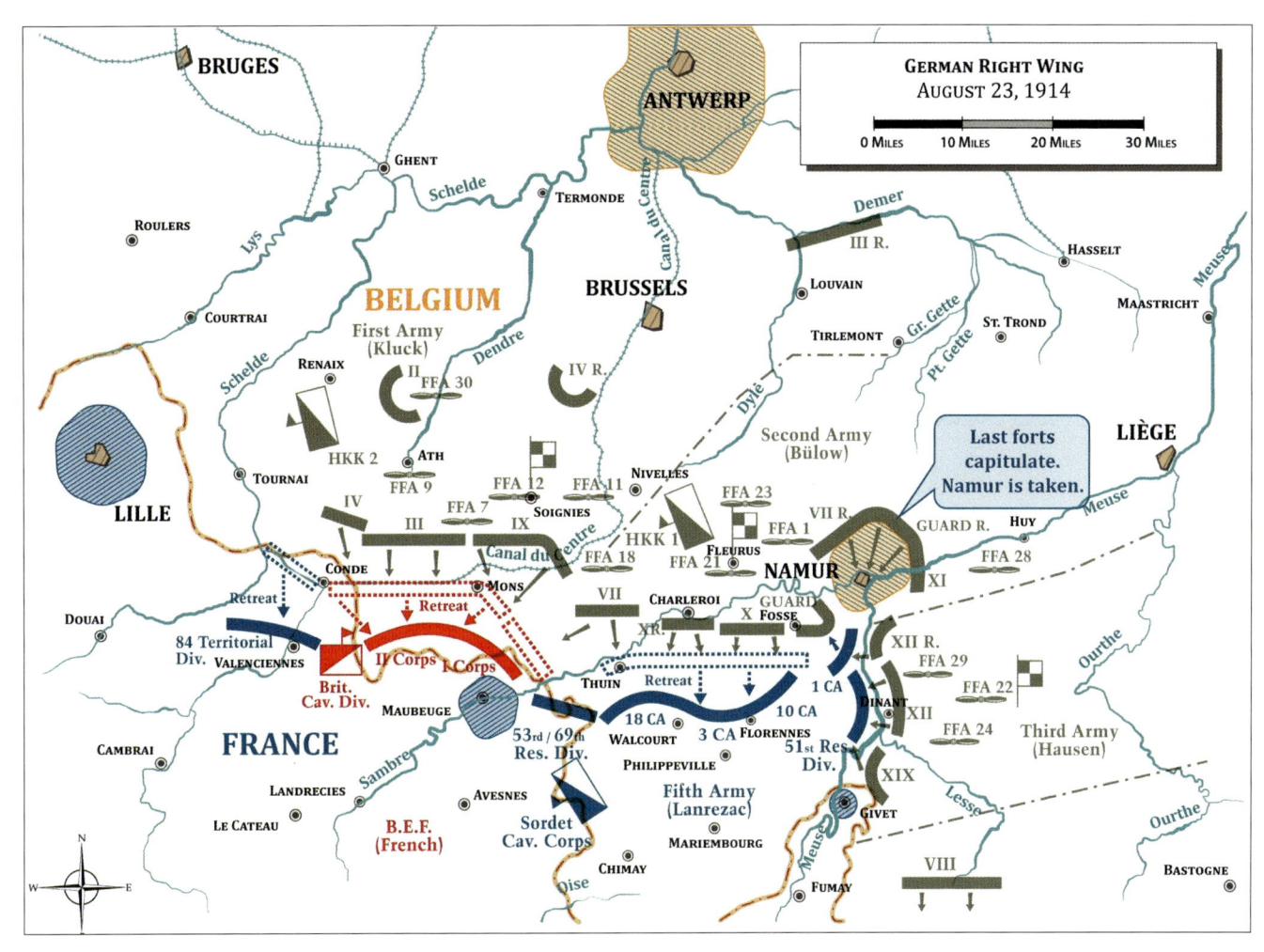

The French III and X CAs began the French counter-stroke at 6:00am against the Guard and X AKs on the left of Bülow's battle-line. Unbeknownst to the French, the Germans had strengthened their positions during the night, establishing a series of formidable defenses covered by artillery, machine guns and rifle pits. Thus, believing the German lines to be weak, the French corps commanders chose to try and surprise the Germans with an attack without any artillery preparation or support. With flags unfurled and the men rushing forward in tightly pressed ranks, the French assault closely resembled an attack of the Napoleonic era. Predictably, it failed with "staggering losses" as the tightly packed French infantry were quickly brought down by rapid artillery and machine-gun fire. Faced with the question of abandoning the attack and falling back on the heights to the south, the French corps commanders chose instead to redouble their efforts and bring up reinforcements and artillery to resume the attack.[217]

Watching from the heights on the Sambre's northern bank, X AK's commander—General Otto von Emmich—recognized that his men could only hold the river crossings if he committed his entire corps to the southern bank. Therefore, at 9:00am, Emmich ordered both of his divisions to cross the Sambre and throw back the attacking French forces trying to gain possession of the bridgeheads. With the Guard Korps'

Above: German and French infantry clash inside a village. (Author's Collection)

2nd Guards Division holding its ground on their left, Emmich's troops turned back the French divisions in their front, forcing their corps commanders to order a general retreat.[218]

Positioned on Second Army's right wing, the X RK and VII AK spent the morning completing their approach marches toward the Sambre. On the army's extreme right, the VII AK pushed back Sordet's Cavalry Corps and a British cavalry brigade defending towns north of the river. In order to cover the army's flank, VII AK halted their advance three miles north of the Sambre along the line Piéton—Haine-St.Paul. Meanwhile on VII AK's left, General Count von Kirchbach's X RK got bogged down in bitter street fighting in Charleroi's suburbs. Unable to advance any further, Kirchbach's two divisions split up and marched around the city, attacking elements of III CA that had gained control of the area's bridges. Although the French defenders at Leernes to the west of Charleroi stoutly held their ground and maintained control over their crossings, X RK's 19 RD attacked east of the city and successfully fought their way across the Sambre to establish a bridgehead at Couillet.[219]

By nightfall, French Fifth Army had been pushed back with 'heavy losses.' The defensive positions they eventually occupied were further south than the ones Lanrezac had intended the army to hold the previous evening. Indeed, III and X CAs—each three divisions strong—were collectively defeated by just three German divisions. Consequently, instead of remaining on the defensive and preserving their strength, Lanrezac's army was now badly damaged and incapable of the offensive action assigned by Joffre.

Lanrezac was in an unenviable situation on the evening of the 22nd. To the north, Bülow's now fully concentrated Second Army had crossed the Sambre and would attack in full force the following morning on a wide front. In the east, French scouts reported a large body of troops (German Third Army) closing against the Meuse. Thus, recognizing his army's dire situation, Lanrezac skillfully redeployed his remaining available forces to meet the new threat. First, he redirected the eastward facing I CA away from the Meuse to the area between Sart-St. Laurent and St. Gerard facing northwest. This move served to relieve pressure from the badly mauled X CA by threatening Bülow's left flank. Next, the 51st Reserve Division was installed along the Meuse sector opposite German Third Army with orders to energetically defend the west bank and delay any crossings for as long as possible.

The long awaited German offensive began at dawn on the 23rd across the entire sector. On Second Army's right, VII AK tried but failed to reach the Sambre's southern bank. On their left, X RK was ordered to launch an attack against French Fifth Army's left wing at Marbaix—Nalinnes. After struggling throughout the morning and afternoon to gain any ground, the German reservists finally seized their objective during the final hour of daylight. To the east, X AK and the Guard renewed their attacks against the French positions directly to the south. Deployed on the army's extreme left, the Guard Korps was compelled to shift the axis of its attack to the southeast in order to meet the newly redeployed I CA, which menaced their flank. After more than eight hours of heavy fighting, the Guard's elite soldiers finally forced a French withdrawal. Meanwhile, X AK advanced against the strong French defenses on the heights of Hanzinelle. However, without the assistance of X RK on its right, the corps' two divisions were unable to properly execute their commander's order to envelop the French line. Instead, they were only successful in pushing back Lanrezac's troops after multiple costly frontal assaults that did nothing to change the overall situation. Indeed, each of Second Army's four attacking corps were compelled to halt for the evening well short of their objectives with French Fifth Army's various units still intact in strong defensive positions.[220]

Meanwhile along the Meuse on the sector's newly opened eastern front, Hausen's Third Army attacked at dawn with the objective of taking control of the west bank and then pressing the attack against Lanrezac's flank and rear. However, the region's terrain was not at all suited for offensive operations. The river's western bank consisted of a series of 10-15 feet tall rocky cliffs which overlooked the sector's bridges as well as the approaching roads and staging areas on the eastern bank. Many parts of the western bank were also heavily wooded, enabling the French defenders to easily conceal machine guns and artillery. From these formidable positions, the badly outnumbered French 51st Reserve Division skillfully concentrated their forces at the major crossings and were able to delay Third Army for most of the day. Eventually Hausen's men gained control of a crossing and were able to send an infantry detachment across to establish a foothold on the west bank. By 6:00pm, two more bridgeheads had been established with large numbers of troops starting to flow across. Franco-Belgian[221] resistance was still strong enough, however, to keep Hausen's troops from participating in the battles to the west. Nevertheless, by nightfall Hausen's Third Army had seized full control of the west bank with the expectation to begin the offensive the following morning.[222]

Bülow and Hausen were extremely disappointed with the day's results. Instead of enveloping Lanrezac's army and creating the conditions for its destruction, their armies' attacks were sharply blunted, bringing the day to a close without a decision. Perhaps most troubling was the situation across Second Army's front. Both the X AK and the Guard had been engaged for three days and had suffered significant losses. Likewise, the X RK was reporting that it would be unable to hold its position if it were attacked again on the 24th. With his men tired, hungry and disorganized, Bülow was worried about his army's ability to continue the battle the following day. Nonetheless, despite the state of his men, he recognized the necessity of achieving a decisive victory in his sector and ordered the offensive to proceed on the 24th. However, Bülow felt it necessary to send a request to Hausen asking for Third Army to directly support the attack—a request that, as we shall see, would have major implications on the outcome of the campaign.

On the evening of the 23rd, as Bülow and Hausen planned the next day's attack, Lanrezac carefully analyzed his beleaguered army's situation. To the west, the BEF reported being driven back by superior forces at Mons. Lanrezac had little faith in the British and was now greatly worried about his own left flank which was defended by the exhausted and badly damaged XVIII CA as well as a group of lower quality reserve divisions. On his other flank, German Third Army was now free to pour across the Meuse and strike his flank and rear. Finally, it was learned during the early evening hours that Fortress Namur had capitulated earlier in the day. This would allow German Second and Third Armies' inner flanks to rapidly bring up reinforcements and deliver a concerted attack against the French right wing. Thus, with his line crumbling and with both flanks at risk of becoming enveloped by superior forces, Lanrezac saw no scenario that didn't result in the destruction of his army if he chose to hold his ground and fight on the 24th. Accordingly, the French army commander ordered a general retreat on the evening of the 23rd. The order was quickly and quietly distributed throughout the ranks, giving the French soldiers time to begin withdrawing from the sector before sunrise.[223]

Lanrezac's timely retreat order enabled French Fifth Army's main body to safely withdraw before the German attack was delivered. Consequently, when Bülow and Hausen's men moved to attack on the morning of the 24th, all they encountered was a rearguard. Although they were ultimately overpowered, Lanrezac's rear-guards successfully delayed the Germans and bought precious time for the army's main body to safely slip away to the line Beaumont—

Givet, thus bringing the four day battle to a close.

Second Army *Fliegertruppe's* Actions During the Battle of Namur

Both of the German right wing's cavalry corps were unavailable on the 21st as Second Army completed its wheel around Namur to the Sambre.[224] All incoming aerial reconnaissance reports were therefore regarded by Second Army HQ and its various subordinate corps commands as vitally important. The most noteworthy flights of the day occurred on the army's left wing where airmen from FFAs 1(Guard Korps) and 21 (X AK) were ordered to scout ahead of their corps' lead columns and report on the French deployments along the Sambre. Shortly after 8:00am, an observer from FFA 1 delivered a report to Guard Korps HQ stating the French strength in the area of the river to be weak while stronger forces, estimated two divisions, were deployed to the south. A similar report was also delivered to X AK HQ an hour later by its own FFA 21 illustrating in detail where the pockets of French resistance were observed along the Sambre's southern bank. Together, the two reports accurately described the French deployment opposite Second Army's left wing, confirming that an attack on the river's southern bank would succeed. However, by the time the two corps commanders received the reports, their ground troops had already witnessed with their own eyes the weak French defenses and begun their attack.

Aerial activity on Second Army's front increased significantly during the afternoon hours as the struggle for the river crossings intensified. While elements of the Guard and X AKs established themselves on the Sambre's southern bank, Bülow's aviators were dispatched south of the river to the area of Phillipeville in order to determine the location and strength of the French main body. Aircraft from FFAs 1, 21, and 23 collectively performed 10 flights during the afternoon with the objective to survey the French positions and provide the German leadership with a better picture of French Fifth Army's location. The most revealing report, delivered directly to Second Army HQ at 6:00pm by its own FFA 23, described the following:

Significant enemy bivouacs along the line Beaumont—Phillipeville and to the east. Three strong enemy columns, at least two divisions, also seen beyond this line moving north. Enemy entrenchments also observed north of the Beaumont—Phillipeville road.

The aforementioned flight was conducted more than 30 miles south of the Sambre along the line: Chimay—Fumay. The "three columns" described in the report were troops belonging to XVIII CA and Lanrezac's reserve division group. Two of the columns (XVIII CA) were reported to be marching in the direction of Thuin to extend the army's left wing but were still a great distance from that destination. Meanwhile, a report delivered by an FFA 1 observer confirmed that "the Sambre positions between Charleroi and Namur were still only weakly occupied."[225] Thus, with multiple reports stating that Lanrezac's main strength was well south of the Sambre, Bülow felt tempted to quickly push his army across to the southern bank the following morning with all available forces that were in position to do so. However, after considerable discussion with his staff it was decided to stick to the original plan and wait another day for both Second and Third Armies to be fully concentrated before beginning the attack. This would also give First Army time to complete its turn to the south and "execute the blow to the French forces which were reported [by FFA 23] south of the Sambre and west of the Meuse."[226]

August 21st was an extremely successful day for Bülow's *Fliegertruppe*. After completing a combined 14 sorties, the army's four flying sections had provided their commander with enough intelligence to make the correct decision regarding the army's employment for the following day. Bülow had initially considered pulling all his troops back to the Sambre's northern bank upon hearing that elements belonging to X AK and the Guard had successfully crossed over. Fortunately, multiple aerial reconnaissance reports arrived that afternoon which confirmed the area directly around the southern bank to be weakly defended.[227] These reports convinced Bülow to not only keep the forward units on the southern bank, but to reinforce them. By midnight all relevant aerial reports were in Second Army HQ's possession and had been thoroughly analyzed. Thus, upon reviewing the general situation and consulting all available intelligence, Bülow's orders to the army for the 22nd were for the troops on the southern bank to strengthen and expand their positions while the rest of the army moved up to begin the great offensive the following morning.

Second Army's *Fliegertruppe* spent the final hours of the 21st moving their airfields closer to the front in anticipation for the coming battle. On the army's left wing, FFAs 1 (Bothey) and 21 (Fleurus) were moved as close to the Sambre as possible in order to quickly provide tactical assistance to the ground troops. Because the army's right wing had not yet completed its wheel to the Sambre, FFA 18's airfield remained north of Charleroi near Quatre Bras. Likewise, Second Army HQ's FFA 23 continued its operations from an airfield next to army headquarters at Vieux-Sart (two miles south of Wavre). However, after learning that Bülow would have a forward command post at Gembloux, FFA 23's commander, *Oblt.* Vogel von Falckenstein, sent some men to the town where a small forward airstrip was established during the evening of the 21st–22nd.[228] With direct access to the army headquarters staff, this makeshift airfield was heavily used by FFAs 1, 21 and 23 over the course of the next three days.

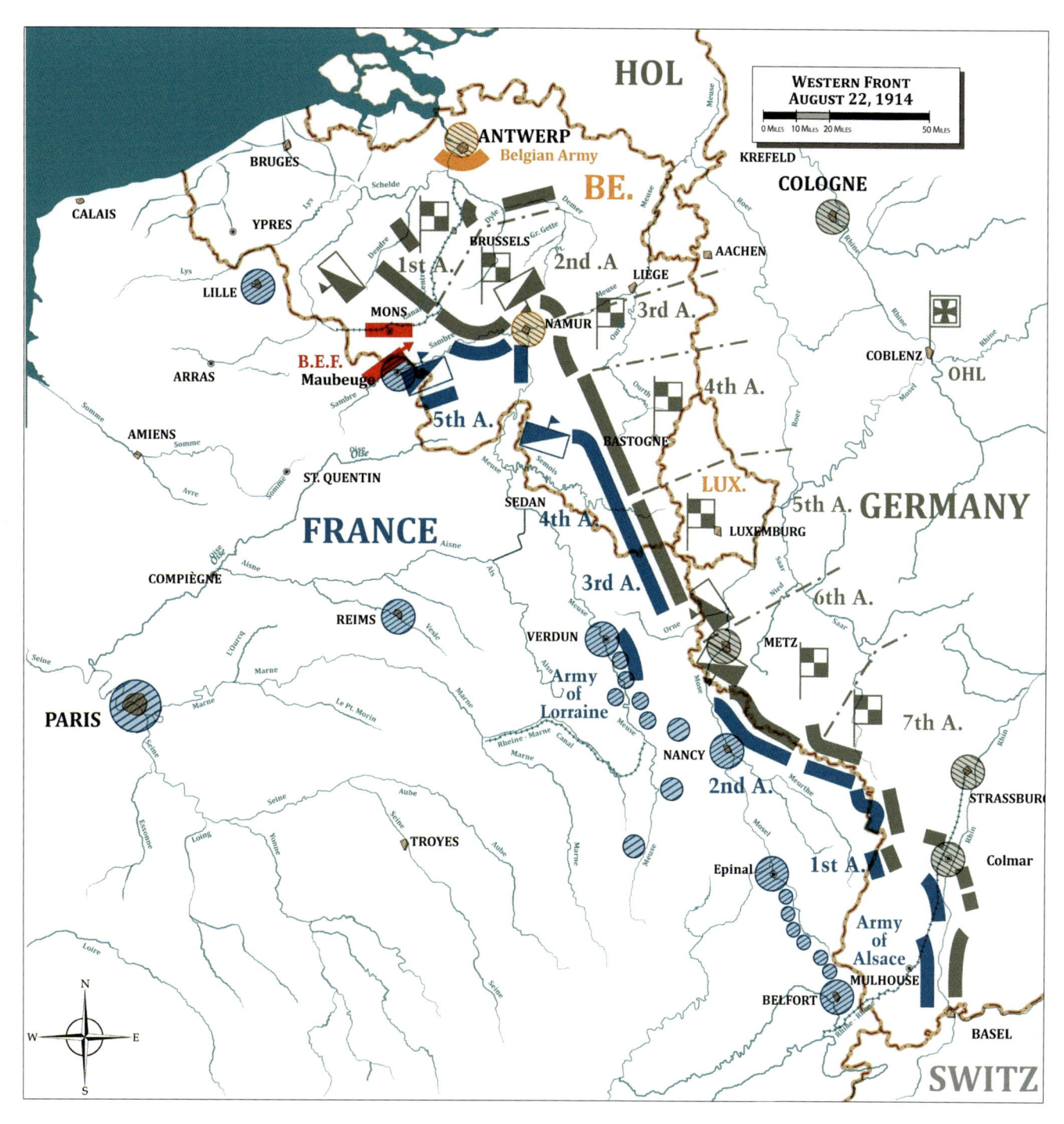

On the morning of the 22nd, Second Army HQ dispatched a communiqué to each of its four flying sections. The message's contents informed the airmen of Bülow's decision to not attack on the 22nd in favor of waiting for Third Army's arrival the following day. The army's various flying sections stopped planning to provide tactical reconnaissance and artillery support sorties and began planning instead for a day of longer range reconnaissance flights.

Even at this early stage of the conflict, the opposing sides were wise enough to use the cover of darkness to shield all major redeployments from the eyes of enemy airmen. Aware of this fact, Second Army HQ required updated aerial reports on the morning of the

22nd regarding the French troop's latest dispositions. Thus, the new plan of operation on the 22nd for FFAs 1, 21, and 23 was to maintain surveillance over Lanrezac's main body and rear. By doing so Bülow's aviators would be able to assist the army leadership better prepare for and plan the following day's attack. On the army's left wing, the Guard and X AK HQs also conveyed to their flying sections the need for immediate appraisals of the French strength in close proximity to the newly won bridgeheads. As the first German units to cross the Sambre, the two left wing corps commanders were worried that a French attempt to retake the bridges would succeed and therefore required confirmation whether any attempt would be made on the 22nd. Thus, FFAs 1 and 21 both planned to simultaneously have multiple aircraft airborne during the morning hours, each performing its own distinct task. In addition to the longer range sorties ordered by Second Army HQ, both sections would have flights reconnoiter the French forward positions in order to determine whether they had been reinforced during the night or if they showed any signs of pending offensive action.

Both of Bülow's left wing FFAs dispatched their first aircraft as the morning fog burned off between 7:00 and 7:30am. The sounds of gunfire had already been heard throughout the sector since 6:00am, causing both Corps HQs to send a staff officer to their section's airfield requesting tactical reconnaissance flights along the line Fosse—Sart-St.-Laurent be immediately ordered airborne. Shortly before 8:00am, FFA 21's flight discovered strong French columns moving south in the direction of Auvelais—Arsimont to deliver the main thrust of French Fifth Army's counter-attack. Recognizing the importance of what they had just witnessed, the airmen quickly returned to the north bank and landed near X AK HQ to personally inform the corps chief of staff, *Oberst* Gustav von Lambsdorff, of their discovery.[229] The observer's report was immediately forwarded to X AK's Commander, General Emmich, who was on the heights north of Tamines personally observing the combat on the southern bank. The aviator's information, combined with what he had just personally witnessed, prompted Emmich to order both of his divisions forward at 9:00am to check the impending French attack.[230] FFA 21's report was therefore greatly responsible for Second Army maintaining possession of the Sambre's southern bank at the end of day.

Around 9:00am, FFA 1 delivered its first report of the morning to members of the Guard HQ staff. The airmen had not seen the advancing enemy columns witnessed by FFA 21's crew (presumably because of clouds) and had taken the initiative to continue south to report on the strength of the French main body. The subsequent report described entrenchments north of the Beaumont—Phillipeville road where French troops had been observed the night before. The area northeast of Phillipeville near Mettet was also reported to be "strongly occupied by the enemy." When verbally pressed about the presence of French troops near the Guard's bridgehead at Jemeppe, the observer erroneously replied that the area was weakly held by French cavalry.

The inaccuracy of the observer's statement was quickly proven by a sudden influx of reports at Guard HQ describing strong French counter-attacks concentrating against Auvelais. With the majority of the 2nd Guard Division deployed south of the river, it became imperative to quickly determine the strength of the enemy's attack in order to sufficiently counter the threat and avoid a catastrophe. Thus, FFA 1 was again ordered at 9:20am to send an aircraft to the area to clarify the situation. Within 30 minutes the observer returned to verbally report that the troops on the southern bank were deployed irregularly and now heavily engaged. He also stated that the "majority of the enemy's forces, estimated combined strength of at least two divisions, were located northeast of Mettet—St. Gérard at Fosse advancing in three columns."[231] Listening to the observer's testimony, the Guard Korps' commander, General Karl von Plettenberg, quickly panicked and ordered the 2nd Guard Division's commander to: "Immediately withdraw everything to the northern bank of the Sambre."[232]

Fortunately, just as Plettenberg's withdrawal order was being received by the troops at the front, yet another FFA 1 observer appeared at Guard Korps HQ. The newly arrived aviator reported seeing the same three columns reported by the previous flight, but admitted being unable to initially make out their composition. The observer then declared that he ordered the pilot to descend to the dangerous altitude of 700 meters (2,300 feet) where he was able to decipher that the columns were mainly comprised of cavalry. The report, delivered at 10:40am, prompted Plettenberg to rescind the retreat order. By holding his forces on the southern bank, Plettenberg enabled X AK to fully deploy its forces and begin the counter-attack that ultimately drove the French back three miles to the line Walcourt—Mettet.[233]

The tactical reconnaissance work performed by FFAs 1 and 21 during the morning of the 22nd was of incalculable value. FFA 21's report identifying the French intent to launch a full-scale counter-attack alerted Emmich in time to smartly deploy both of his divisions to meet the coming assault. The subsequent attack made by both of Emmich's divisions was the determining factor in the German victory of the 22nd that ultimately resulted in Second Army maintaining possession of and later expanding the southern bank bridgeheads. In the Guard Korps' sector, the airmen of FFA 1 kept their corps headquarters fully updated throughout the morning with a flurry of reports on the strength of the French forces opposite the 2nd Guard Division on the southern bank. The section's final report of the morning clarifying that the advancing columns were only cavalry literally saved the day for Bülow's army. Without that report, the Guard's forward troops would have abandoned the southern bank, leaving X AK unsupported and unable to deliver its attacks. Consequently, instead of beginning the offensive on the 23rd from the southern bank, Second Army would have had to spend the day reclaiming the river crossings.[234]

Both of Bülow's left wing flying sections continued to operate throughout the afternoon, keeping the corps commanders updated on the progress of the ground troops. FFA 21 conducted two more flights between noon and sunset, each taking note of X AK's attacks as well as detailing troop concentrations in the French rear-area. Meanwhile, FFA 1 dispatched an aircraft late in the afternoon to reconnoiter the French positions in the area of Mettet directly south of Winckler's 2nd Guard Division. By nightfall both corps HQs (and later Second Army HQ) were equipped with excellent intelligence detailing the French deployment along the line Mettet—Wagnee—Walcourt. These reports were later cited when Bülow and his staff planned the attacks of the 23rd.

Second Army's other flying sections had a minimal impact on the battles of the 22nd. For example, FFA 18's aircraft were tasked throughout the day with reconnoitering west of Charleroi along both banks of the Sambre as X RK and VII AK completed their approach marches. Unfortunately, the section failed to deliver anything of significant importance. Elsewhere, Second Army HQ's FFA 23 provided reports of the situation at Namur as well as French deployments along the Meuse.

The airmen of Second Army's four flying sections were once again informed of the general situation at their pre-dawn meetings on the morning of the 23rd. They were specifically told of Bülow and Hausen's plan to attack French Fifth Army from both directions and were directed to plan tactical flights accordingly to assist with that endeavor. In addition to these tactical flights that were to be performed as circumstance dictated, each of the army's three corps flying sections were given special instructions regarding possible long-range flights. On the army's left wing, the Guard and X AK HQs strongly suspected that the French had once again redeployed during the night. Thus, both corps commands ordered their flying sections to perform reconnaissance flights over the French defensive line as early as possible to facilitate the ground troops' success. Meanwhile on Bülow's other wing, FFA 18 was ordered to perform long-range flights to better ascertain the strength and position of Lanrezac's left flank. With First Army beginning to execute its attack, this task was considered to be of great strategic importance.

Poor weather played a considerable factor in the *Fliegertruppe's* inability to tactically support the ground troops on the 23rd. A thick layer of fog hung over the entire sector at dawn for the third consecutive day, grounding all early flying activity. By 7:30am the fog had dissipated and each of Second Army's flying sections began their day's first flights. The delay unfortunately resulted in the ground troops' attack being launched without any of the updated aerial intelligence that the corps commanders had requested.

Taking off at 7:15am, a DFW B.I belonging to FFA 23 conducted the first sortie of the day. The flight had been sent to the area of Mettet to determine the strength of the French positions in the area. Thick clouds, however, caused the observer to return with nothing definitive to report. FFA 21 sent two aircraft airborne at 8:30am to report on the French deployments opposite X AK. The subsequent reports delivered to X AK HQ described in detail the deployment of French III CA, who stood directly in the path of the corps' advance. According to both reports, French infantry had begun to entrench on the heights between Gomzee and Hanzinelle with strong artillery in support. One observer was able to successfully locate the positions of several massed batteries and included this information in his report.

Despite being assigned to a critically important sector of the front, the airmen of FFA 1 were unable to make a significant impact on the fighting of the 23rd. The actions of the previous two days had taken a heavy toll on the section's aircraft, forcing many to be grounded for maintenance or repair. Consequently, the section only had two airworthy machines on the morning of the attack. Thus, with only two available aircraft, the section struggled to consistently provide updated information to Guard Korps HQ throughout the day.

FFA 1's first flight of the morning took off at 9:00am with orders to reconnoiter the area around Mettet to the Meuse. The Guard's two divisions had expected to link up at some point in the afternoon with the inner wing of Third Army after the latter had forded the river. Quality aerial reports of the Meuse sector were therefore required by Guard HQ to determine where its divisions could open the crossings for Hausen's troops. After zigzagging their way around the assigned

Above: German artillery bombards the fortifications at Namur. (Author's Collection)

sector, the airmen returned at 10:45am with a report of strong French reinforcements moving south from Phillipeville through Florennes. Regarding the area east of Mettet along the Meuse, the observer reported seeing only weak detachments. Unfortunately, the morning's cloud cover had concealed the presence of I CA that was moving directly against the Guard's left flank. As a result, the Guard was unexpectedly drawn into battle and later compelled to divert forces to the east to check the French attacks.

FFA 1 was ordered to send up another aircraft around noon to investigate the situation on the corps' left flank. With the exact strength of the French force still unknown, the airmen were directed to focus their search exclusively around the area of St. Gérard. For 30 minutes, the pilot circled over the assigned location in vain as heavy clouds continued to inhibit the observer's ability to decipher the size of the unknown French force below. Thus, upon his return, the observer was only able to provide a list of isolated French units in front of the 2nd Guard Division, who had by then fully deployed its field and heavy artillery and had begun to force the French back.[235]

FFA 1's limited capabilities seriously affected its impact on the battlefield throughout the rest of the afternoon. The standard aerial doctrine of the time demanded at least one aircraft to remain grounded "at the ready" whenever the ground troops were heavily engaged. The section was therefore only able to have one aircraft airborne at a time during the most important moments of the battle. Consequently, FFA 1 was unable to provide any tactical support to its divisions on the 23rd. Instead, the afternoon consisted of long-range flights along the Meuse to provide the corps headquarters staff with updated reports on Third Army's progress crossing the river. It was confirmed after the final flight of the day returned with a report from Third Army's XII RK HQ, that Hausen's forces would be unable to participate in the fighting on the west side of the river until the following morning.[236]

Tactical reconnaissance sorties were performed with limited impact on Second Army's right wing where X RK renewed its assault against Lanrezac's left. In order to assist the reservists (who had no aviation assets of their own), two Albatros B.Is from First Army's FFA 11 (IX AK) were temporarily transferred under the control of X RK HQ. The aircraft operated from a small, hastily constructed airfield outside the small village of Couillet near the corps headquarters. The first flight took off at 12:15pm after X RK's Chief of Staff, *Oberst* Gottfried Marquard, personally ordered a report on the French deployment opposite the corps' left division (19th RD), who had become seriously engaged around the town of Nalines. An hour later the observer returned with only a general outline of the French deployment in the sector—information that the front line troops already possessed. Additional flights were conducted over the following three hours with minor success as X RK struggled to gain ground. Although these reports kept X RK HQ informed of the conditions at the front, they did little to affect the battle. Indeed, the situation would often change before the airmen's information could reach them.

Operating under the direction of Second Army HQ, FFA 23 performed long-range reconnaissance flights as far as Mariembourg over the French rear-area. Each flight returned with evidence that the French were withdrawing troops towards the south. The first, delivered at 10:25am, described "disorganized columns in retreat." Later that day the section's commander, *Oblt*. Otto Vogel von Falckenstein, piloted the future fighter ace Rudolf Berthold over the front lines where Berthold reported seeing evidence of a French retreat as far south as Philippeville. These reports, however, had little impact on the attitudes of the Second Army HQ staff, who remained committed to the belief that Lanrezac was holding his ground with plans to renew the battle the following day.[237]

A thick blanket of clouds moved into the sector late in the afternoon that grounded each of Second Army's flying sections for the remainder of the day. Although frustrating, the lull in the action gave Bülow's airmen a much-needed break. Each section had been unusually active for three consecutive days, placing considerable stress on the men and their aircraft. For example, FFA 21's *Oblt*. Friedrich von Dalwig had personally performed six flights in less than fifty hours. Similarly, *Oblt*. Ernst von Bornstedt of FFA 1 was airborne five times over the same period.[238] The aircraft were equally as taxed. By late afternoon of the 23rd, the number of serviceable aircraft in FFA 21 was down to three while FFA 1 had been reduced to only two.

Tired and overworked from three consecutive days of activity, the pilots and observers used the time to rest and recuperate while their mechanics feverishly worked to rehabilitate the aircraft. By the morning of the 24th Second Army's *Fliegertruppe* were practically back to full strength. FFA 1's mechanics had repaired two aircraft, thus increasing the section's strength to four (the section's maximum strength after losses). FFA 21 reported having five operational aircraft by the afternoon of the 24th. Each of the army's other flying sections had at least four aircraft available for use on the 24th. Large quantities of fuel and supplies were procured during the night by members from each section in anticipation of performing numerous flights on the 24th. However, once it was discovered that Lanrezac's army had retreated, Bülow's airmen returned to a more normal level of activity. Specifically, each flying section ordered one or two aircraft airborne to locate and monitor Lanrezac's retreat.

Third Army *Fliegertruppe* During the Battle of Namur

Although their ground troops were unengaged, the airmen of Hausen's Third Army were highly active during the first two days of the battle. The army's flying sections had already been employed in the days leading up to the 21st searching around Dinant on both banks of the Meuse. Working with HKK 1, the aviators were able to provide Hausen and his staff with "rough outlines representing the French line of defense above and below Dinant."[239] Indeed, multiple aerial reports on the 20th described strong French forces, complete with artillery, entrenching on the river's western bank. An eleven mile long column was also discovered marching north on the Rocroi—Couvin road. While all the bridges in the sector were reported to be intact, the aviation reports left no doubt as to how the French planned to stop Third Army's advance.[240]

Throughout the 21st, aircraft from Third Army HQ's FFA 22 were ordered to reconnoiter the region west of the Meuse in order to acquire additional details of French Fifth Army's deployment. By the end of the day the section had produced two excellent reports pinpointing the location of three large columns in the area around Philippeville. Hausen's flying section was directed the following day to perform long-range flights from the area of Charleroi to Fumay where they confirmed OHL's suspicions that at least three corps were operating in the area. Later that morning, FFA 22 was compelled to send an aircraft to the south in support of German Fourth Army. In the midst of their battle in the Ardennes, Fourth Army HQ had become seriously concerned about a possible attack against its right flank and had solicited Third Army HQ's help in determining whether these fears were valid. Accordingly, a flight from FFA 22 was ordered to search as far south as Sedan and report on any threats to Third and Fourth Army's inner flanks. Flying below cloud cover at an altitude of 1,600 feet, the observer was able to identify an infantry division accompanied by an artillery column at St. Menges and at least one additional division to the north of Bertrix. As the aircraft traveled toward Sedan and prepared to turn back, they were suddenly shot down by ground fire. The pilot, *Ltn*. Erich Janson, was killed in the crash[241] while the observer—*Ltn*. von Stietenkron—managed to escape and get back to German lines

Above: Saxon Jäger infantryman. (Author's Collection)

where he formally delivered his report to Third Army HQ on the evening of the 23rd. In his memoirs of the campaign, Hausen described meeting Stietenkron and hearing of his remarkably story:

Upon entering the Château de Taviet that evening [23rd], we found a French airplane that had been forced down by the fire of German infantry lying in pieces on the ground. For several days following its fall it had been regarded as part of the spoils of the army cavalry. We were in the midst of rejoicing at the skill of the German infantry fire which had brought such success when we received proof of the equal precision of the French infantry fire. A Ltn. of Hussars, Herr von Stietenkron, presented himself to me, with a wound in his head which was bleeding freely. He also appeared to be suffering from a severe nervous shock. According to his story, during a long distance flight which he had made as an observer on August 22nd with his comrade, Ltn. Jansen, pilot aviator, their plane while at the height of 500 meters, had come within range of some French infantry fire, to which his comrade had fallen victim.

After a forced landing at an unknown spot, probably at Paliseul, he found himself, upon recovering consciousness, in the center of some French infantry at rest. The French soldiers treated him with the greatest indignity and coarseness until the sound of a bugle called them to arms.

When the infantry resumed its march, the Ltn. was kept confined, with the intention, no doubt, of putting him to death. After nightfall, however, thanks to the cover offered by the neighboring woods, he was able to escape from captivity and make his way to the German advance posts. While he was a prisoner, the [written] record of the observations made by him vanished. The destroyed plane and his dead comrade, killed by a bullet in the head, remained where they had fallen.[242]

Elsewhere on the 22nd, Third Army FFAs 24 (XIX AK) and 29 (XII AK) were each ordered to concentrate their searches west of the Meuse to determine the French strength along the river. Ordered to scout the area west of Onhaye and Anthée, an LVG of FFA 24 found large numbers of reinforcements moving north towards the front. First, a five mile long column, with a strong vanguard turning east towards Rosée and Corenne, was observed on the Philippeville–Florennes road. To the south at Mariembourg, supply trains accompanied by infantry were reported marching north towards the battle zone. Finally, at Philippeville, a French military airfield was discovered with 24 tents. Upon landing, the observer's report was immediately delivered by automobile to Third Army HQ at Leignon, who forwarded it to Second Army HQ that evening.

Meanwhile an aircraft of FFA 29 patrolling the area of Mettet located large numbers of French troops moving east toward the Meuse. The report, delivered by *Ltn.* Kurt Völkers to XII AK HQ, described in detail the strength of the French reinforcements being sent to defend the river:

Three columns moving east from Anthee. Lead Column consisting of infantry and horses—5 kilometers in length. Second Column: infantry and significant artillery—12 kilometers in length. 3rd Column: infantry and artillery—5 kilometers in length. Trains attached.[243]

The report was immediately shared with Third Army HQ, who passed it along to OHL that evening. Völkers pilot on the flight was Gerhard Sedlmayr, a civilian who volunteered for service with the

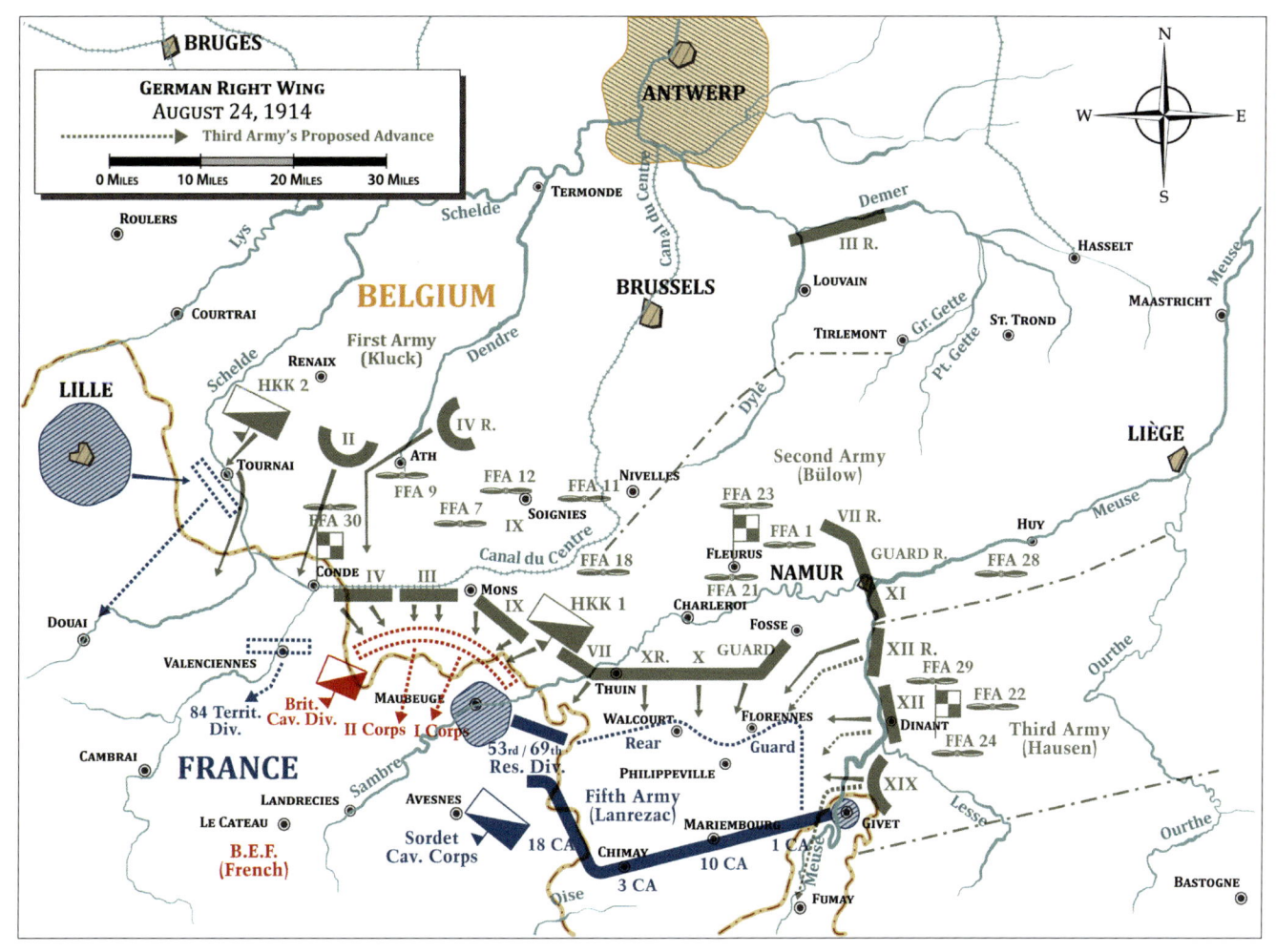

Fliegertruppe upon the outbreak of war. Prior to the war Sedlmayr worked for several aviation firms as a test pilot, earning a reputation as one of the finest pilots in Germany by achieving several national flying records. Together, Sedlmayr and Völkers flew together for the remainder of the campaign.

On the morning of the 23rd the airmen of Third Army's four flying sections were fully briefed of the general situation and were informed of their roles in the day's battle. With an understanding of their army's role in the attack, Hausen's aviators prepared themselves for a day of tactical reconnaissance work assisting the ground troops cross the river and attack the French defenders' flank and rear. Thus, during their pre-dawn meetings, FFAs 24 and 29 supplied their observers with updated information regarding the placement of friendly heavy artillery batteries in preparation to carry out artillery cooperation sorties throughout the morning.

Thick clouds and fog hung over the Meuse sector at sunrise, causing all aerial activity to be delayed. Consequently, Third Army was forced to start their attack at 5:50am without their aviators. The first flight of the day was performed by an Albatros B-type of FFA 22—taking off at 8:25am to report on the effect of the army's preliminary bombardment that had been underway for over two hours. Despite encountering a great deal of smoke and lingering fog, the crew managed to return to Third Army HQ's forward command post at Taviet shortly before 10am with a brief written report describing the French strength along the river's western bank to be weak, as well as significant movements to the south that suggested a retreat had been ordered.

FFA 22's report had the potential to significantly impact the future course of the campaign. At 8:35am, Hausen had received a radiogram from OHL recommending Third Army send troops to cross the Meuse south of Givet in order to cut off French Fifth Army's eventual retreat. Hausen, on his own initiative, had already ordered this to be done. However, with an aerial report now in his possession suggesting a retreat had already begun, Hausen immediately augmented the orders to commit his entire left wing corps—XIX AK—south of Givet to strike the French rear. In his memoirs, Hausen described the effects of his airmen's important message:

Above: Albatros B.II powered by a 120 hp Benz. The design was simple, practical, and robust. The B.II was developed from the B.I by fitting a smaller, two-bay wing. The two-bay wing was cheaper to produce and maintain than the three-bay wing of the B.I and gave better performance. For a time the two types were produced simultaneously before the B.II prevailed. Once it was removed from the front, the B.II was built in quantity as a trainer throughout the war. (Aeronaut)

Upon my return to Taviet, I studied some aviation intelligence of the greatest importance that left no doubt at all about the situation. The enemy operating to the south of the Sambre had commenced his retreat, passing over the Dinant bridge, and evacuating part of his position to the west of the city.

We were able to estimate then, with a good deal of probability, that the resistance on the Meuse, above and below Dinant, would not last very much longer. This resulted in the immediate decision to carry out the operation which already had been confided to General of [the 40th] Division Götz von Olenhusen, not only with the disengaged troops of XIX AK, but with the entire XIX AK, because that operation had within it the germ of great success.[244]

After making this decision, Hausen radioed the following to OHL at 11:45am: "The enemy is in full retreat from Dinant to Phillipeville. Army follows." He simultaenously radioed Second Army HQ: "Enemy in full retreat: Third Army pursues overtaking to the left, army's right wing in direction Houx—Furnaux—Corenne." All necessary orders to divert the army's advance southwest were then promptly dispatched that evening under the false assumption that further resistance would be weak and all of Third Army would have been moved across to the west bank.

Meanwhile, FFAs 24 and 29 spent the morning relocating a number of their aircraft to a combined forward airfield near the village of Loyers in order to better tactically support the army's attack across the river. Both sections immediately had aircraft airborne upon their arrival with orders to specifically assist the 32nd Division's (XII AK) attack at Leffe. Similar to the battles in Lorraine, Third Army's airmen worked closely with the artillery by dropping notes over the side of the aircraft to help destroy troublesome targets that were often well hidden. These artillery cooperation sorties all along the Meuse continued throughout the morning, silencing French guns which had greatly hindered the Saxons' progress.[245]

By midafternoon it had become obvious to the Third Army HQ staff that their initial assessment of the French resistance on the Meuse had been wrong. Nonetheless, despite his army's slow progress, Hausen still believed that French Fifth Army was in the midst of a withdrawal. Thus, he remained committed to his plan of pushing his left wing to the southwest the following morning to either cut off Lanrezac's forces or drive a wedge between them and their neighbor to the south (Langle de Cary's Fourth Army). In order to properly carry out this task, Third Army HQ and its subordinate corps commands needed to be aware of the latest position of the French main body and its rear-guards. Unfortunately, rain had moved into the area and had caused two of FFA 29's flights to abort flights headed toward the French rear-area.

Among Hausen's flying sections, only FFA 24 experienced success reporting on French troop movements during the afternoon of the 23rd. At 12:30pm a message from the section was delivered to corps headquarters stating that the major roads between Givet and Fumay and to the southwest of XIX AK were empty. However, the observer was also able to report a large baggage train accompanied by several battalions of infantry halted at Anthée on the road from Dinant—suggesting Lanrezac's troops were currently being stripped away from the Meuse. A later flight, returning at 6:35pm, found several large infantry columns moving south of Philippeville toward Mariembourg. North of Philippeville, the observer noted a "12 kilometer long infantry column with trains" moving south toward the city. Finally, a "six kilometer long column" was observed marching south at Emerton, just two miles southeast of a major battle involving Second Army's troops at Mettet.[246]

FFA 24's reports, combined with the army's successful crossing late in the afternoon, convinced Hausen to proceed with his plan of turning his army southwest the following morning [24th] in the attempt to overtake and envelop Lanrezac's retreating force. Third Army HQ believed that Lanrezac's command was severely damaged from its prolonged combat with Second Army and could be encircled and destroyed. Thus, in accordance with the directions already received from OHL that morning, Third Army HQ began issuing orders that evening to its corps commanders directing the advance to begin before dawn in order for the army to be in a position to cut off French Fifth Army before they could escape.

Hausen radioed OHL on the evening of the 23rd with a brief report of the day's combat and his intentions for the following day.

On August 23rd, at noon, aerial scouts ascertained that the enemy was making a retreat on the left bank of the Meuse. Hostile rear-guards aggravated the army's crossing of the Meuse until late into the night. In this the enemy was supported by extremely difficult country and innumerable franc-tireurs to such an extent that at 23:00, only weak forces from both corps held the left bank, despite their extreme bravery. On the morning of 24 August, the attack will be renewed and the advance continued.[247]

Third Army HQ dispatched the following day's orders at 2:30am to all subordinate commands. In accordance with Hausen's instructions, the pursuit was to begin before dawn along both banks of the Meuse in a southwesterly direction. Those elements of the XII and XIX AKs that had already crossed the river were instructed to first capture the French stronghold of Onhaye before proceeding southwest. All other commands (with the exception of one division of XII RK screening Namur) were to advance in the direction of Fumay and Rocroi. The objective for the 24th was the line: Franchimont—Fumay. From here, Hausen's forces were expected to continue a rapid advance and then strike Lanrezac's flank and rear.

Around 4:00am Hausen received a visit from a Second Army HQ staff officer that would fundamentally change how Third Army was employed on the 24th. Bülow's messenger informed Hausen that Second Army HQ now estimated that five French corps were in the region and that it "was of the utmost urgency that the Third Army, by an attack in an east–west direction, should support the attack of the left wing of Second Army in the general direction of Mettet."[248] In other words, after three consecutive days of battle, Bülow and the Second Army HQ staff were deeply concerned that their troops would be unable to stop another large scale French counter-attack on the 24th. Indeed, Bülow feared that a strong, coordinated French offensive across his entire front could shatter his entire command and drive its remains back across the Sambre. If that were to occur, the German offensive to quickly end the war in the west would be over and all western army commanders would have to order a retreat. Thus, faced with his neighbor's desperate appeal for assistance, Hausen had no choice but to change plans and immediately direct his army to the west to come to the aid of Second Army. These new orders were quickly dispatched to each of the army's corps commanders at 5:45am.

Hausen had serious reservations about abandoning his plan to direct part of his army to the southwest to pursue the French. Using multiple aviation reports, Hausen had come to the conclusion that French Fifth Army was beaten and preparing to retreat. If Third Army moved southwest on the morning of the 24th, the isolation and destruction of French Fifth Army was a serious possibility. Unfortunately, the older and more conservative Bülow relied solely upon intelligence gathered by the ground troops at the front when he evaluated his army's situation during the evening of the 23rd. Consequently, Second Army HQ was inclined to believe that they were facing five French corps capable of mounting a massive coordinated counter-offensive despite having reports from their own airmen as well as Third Army which confirmed significant withdrawals in Lanrezac's rear-area. As a result, Bülow felt that he required Third Army's assistance to resolve the situation tactically in his army's front before any larger victories could be pursued.[249] Faced with the prospect of what could happen if Second Army HQ's worst fears came to fruition, Hausen felt obligated to order his army to attack directly west towards Mettet on the morning of the 24th—thus forfeiting the Germans' best chance to cut off Lanrezac's retreating army.

Hausen later described the painful decision in his memoirs:

After having made a report to Supreme Headquarters and also informing the neighboring armies of my plans [to send the army southwest], I drew up at Taviet at 2:30am, August 24th, the army order relative to that day's operations. Hardly had that order been dispatched when Major von Fouqué, an officer from the General Staff Headquarters of Second Army, presented himself at 4:00 in the morning and imparted the following information:

"That Second Army was under the impression that it had been engaged on August 23rd with an enemy having a strength of approximately five army corps, occupying a fortified position, which had been successfully attacked by X RK; that in general Second Army's attack had been successful; but that, however, in view of the renewal of the attack proposed for August 24th at daybreak it was of the utmost urgency that Third Army, by an attack in an east-west direction, should support the attack of the left wing of 2nd Army in the direction of Mettet."

This demand for help from Second Army, repeated like the clang of a bell the "instant desire" already heard at 6:30pm on August 23rd, forced upon me the grave necessity of deciding whether to adhere to my original order to Third Army, issued about an hour and a half earlier, or to lend an ear to the proposition of Second Army to march in an east-to-west direction. The short time that remained for me to reflect and decide on this proposition—because the Second Army was to attack at daybreak—made it impossible to engage in any new deliberations either with Supreme Headquarters or Second Army. The report, presented by the agent of Second Army gave rise to the thought at Taviet that not only had the conflicts fought by the Second Army on August 23rd failed to achieve the results envisaged by the cherished schemes of Second Army HQ, but also that there was no cause for rejoicing at all because the success with which the enemy had attacked the X AK. This point of view could not be invalidated by the self-serving declaration of Second Army HQ that: "the attack of the Second Army was generally favorable."

...With a heavy heart, I found it necessary to give myself up to the new reflections at 4:00am on the morning of August 24th. I pondered over the circumstances under which the idea of an offensive toward the southwest had been developed and had achieved so much importance. I examined the strategic perspectives that would be available in the event of the execution of such an offensive... [However,] A tactical check of Second Army made such a success impossible. Then the Third Army, separated from XI AK, faced the danger in turn of being cut off from Second Army by an enemy superior in numbers and of being thrown back onto the Fourth [Army]. The possibility of even more fatal consequences stared us in the face, because the Third Army had penetrated very deeply toward the southwest in a region where the configuration of the terrain, the condition of the routes, roads and bridges would paralyze the freedom of its movements. Consequently, not only was the entire operation of the right wing of the German army compromised, but the success obtained by Fourth Army was also placed in jeopardy.

...For the rest, the manner in which the request for help had been made clearly indicated that the situation of Second Army was of such a nature that it demanded immediate aid by the most rapid means and with the least possible delay. This appeal for assistance, renewed in a form so characteristic of Second Army HQ, was bound to shake our confidence in any cooperation with the latter, following the unexpected blow of the untoward events of August 23rd. But as a condition precedent to the success of the attack projected by Third Army in a southwesterly direction, it was necessary that it be covered, and that the movements of Second and Third Armies planned for August 24th be made to agree from a strategic viewpoint. Therefore, the difficult decision of the Second Army bound the Third to give thought to the designs of the former: to march no further in a southwesterly direction, but instead in an east-to-west direction. We thus subordinated our personal views of the situation in favor of the tactical plans of our neighbor...This concession on the part of Third Army HQ permitted a realization of concerted action in the interest of the Second Army at a critical hour.[250]

As ordered, Third Army began its westward advance at dawn in support of the Guard Korps on Bülow's left wing. After deploying for combat, the troops quickly discovered that the French main body had withdrawn during the night. Upon hearing this news, Hausen ordered his aviators airborne to clarify the situation. FFA 22 was accordingly directed to send aircraft to reconnoiter the roads from Dinant to Philippeville, concentrating along the line Surice—Gomezée in order to determine the direction Lanrezac's army was taking and the strength of its rear-guard. FFA 29 was assigned the area between Yvoir—Beaumont and Dinant—Philippeville with a request to report on all traffic observed on the roads: Philippeville—Mariembourg—Couvin—Chimay. The two sections collectively provided Third Army HQ with five reports between 7:00am and noon that gave Hausen and his staff a complete picture of French Fifth Army's right wing. The first report, delivered at 7:35am by an observer from FFA 22, described retreats across a wide front with large bivouacs established southwest and west of Philippeville. Another FFA 22 report brought to Third Army HQ at 9:00am confirmed that the French main body was in the midst of retreat in a southwest direction through Philippeville. Based upon these messages as well as reports from his commanders at the front, Hausen ordered the axis of his army's advance be redirected back to the southwest at 9:45am in the hopes of maintaining pressure on the retreating French. In the meantime, additional aerial reports continued to arrive throughout the rest of the morning which specifically named the various roads the French were using for the retreat.[251] Unfortunately by that time it had become too late for Hausen's Saxons to reach the French flank and arrest the retreat. Indeed, Third Army was compelled to halt for the evening along a line east of Philippeville. Meanwhile, Lanrezac's main body stood on the evening of the 24th just south of Chimay and Mariembourg with its flank safely anchored against the small fortress of Givet. The French had escaped.

****

Third Army's *Fliegertruppe* performed brilliantly throughout their first major test of the war. With the exception of XI AK's FFA 28, which was detached from Third Army for the siege of Namur, each of Hausen's flying sections played a significant role during the battle. First, Hausen's airmen discovered the build-up and deployment of French troops along the Meuse prior to the army's arrival on the evening of the 22nd. These reports confirmed that the eastern bank of the Meuse was free of French troops, allowing Third Army to complete its approach march and deploy for the assault against the western bank without the fear of a hostile presence directly in their front. Once Third Army became engaged, airmen from FFAs 24 and 29 showed their versatility by conducting extraordinarily difficult artillery cooperation flights over the battlefield as well as vitally important long-range reconnaissance sorties behind French lines. These reconnaissance reports had the potential to fundamentally change the battle and the campaign. Citing his airmen's intelligence, Hausen made the decision to send his army southwest on the morning of the 24th to cut off Lanrezac's troops—a choice that could have possibly turned an ordinary tactical victory into an extraordinary strategic one.

The missed opportunity at The Battle of Namur was the first of several instances during the war's opening campaign where the *Fliegertruppe* had the potential to make a larger impact upon the campaign but was unable to do so due to mistakes made by the army leadership. Indeed, with military aviation still in its infancy, many senior commanders mistrusted their aviators' information. Consequently, many of the *Fliegertruppe's* reports were questioned, minimized, or ignored. For example, while Hausen trusted his aviators and had decided to formulate a plan based upon their reports, Bülow disbelieved his own aerial reports suggesting that a French withdrawal had already begun. As a result, he foolishly requested that Third Army abandon its plans to envelop the French in favor of directly supporting Second Army's left wing. To be fair, with their troops exhausted and badly damaged, Second Army HQ was in an admittedly tough position on the evening of the 23rd. A large-scale French counter-stroke could have indeed split the army in two and ended the German right wing's offensive. However, reports from Second Army's own airmen as well as messages from Third Army HQ had confirmed that such an attack was not forthcoming. Nevertheless, Second Army HQ continued to fear a French strike and requested Third Army come to their aid—eliminating a great opportunity for the German right wing to win a decisive victory in the process.

Although the Battle of Namur had ended in a tactical German victory, the French retreat on the evening of the 23rd ended all possibility of destroying the allied left wing in the initial battle as the Germans had hoped. The German War Plan was predicated upon quick decisive victories in the field, allowing large numbers of troops to be rapidly transferred to the East. Lanrezac's decision to retreat not only saved his own army from certain destruction on the 24th but also prolonged the possibility of a decisive German victory in the West until the German right wing's pursuit could force a second major battle. It therefore became imperative for the German pursuers to quickly bring Lanrezac's defeated army to battle and create the conditions for its destruction—an act that required the assistance of the *Fliegertruppe* and the trust of the army leadership.[252]

Kluck's First Army Engages The BEF

Maintaining contact with Second Army's right flank, Alexander von Kluck's First Army began its turn to the south on the morning of August 21st. Throughout the previous four days, First Army's cavalry and aviators had been employed in a massive search for the BEF but were unable to find any signs of their existence.[253] First Army HQ had learned from newspapers that the English had already arrived on the continent but were unable to determine where they landed or where they planned to deploy in the allied battle-line. First Army HQ was therefore hesitant about turning to the south before first confirming the BEF's position. The risk to the entire right wing's operations was incalculable if the English were able to strike First Army's flank and rear after the army had wheeled south. In his post war history of the campaign, General von Kuhl explained the operational necessity of locating the BEF *before* the army could join Second Army along the Sambre:

Therefore, the First Army considered it better to only advance a small distance on the 21st and to assume a strongly echeloned formation right and left so as to be prepared for all emergencies. It was not yet desirable to form the army facing in a definite direction. It was especially not desired to wheel left [south] too early. We had to remain in a position which would enable us to execute a sufficiently broad sweeping movement. The prescribed turn could not be resumed until we were sure that the English approach did not threaten our flank. The English were to be attacked by surrounding their left flank, throwing them against the French, and cutting them off from the ports. This was already regarded as a necessary condition for further operations. If the design succeeded, it would, at the same time, initiate the surrounding of the French left wing in the most effective manner.[254]

Recognizing the dire need to locate the British, First Army HQ ordered each of the army's five flying sections to maintain a lively aerial presence on the 21st in order to patrol the various areas of western Belgium and locate the English as quickly as possible.

The aviators' search area was divided into two sectors. The first in a westerly direction along the line Termonde—Alost—Audenarde—Renaix and the second from Ath—Mons in the south. High ranking German leaders including Moltke himself believed incorrectly that the English would be found to the west in the vicinity of Lille. The majority of aircraft employed on the 21st were therefore sent to reconnoiter the western sector. Only one flying section, IX AK's FFA 11, was ordered to patrol the area of Mons where the English were in fact concentrating. Thus, despite Kluck's five flying sections combining to perform 15 flights on the 21st, none were able to return any evidence of the BEF's location. Likewise, the army's cavalry corps, HKK 2, was also unable to locate any allied forces during its advance up to the line Grammont—Ath—St. Ghislain. Oddly, a flight conducted by FFA 11 in the area of Mons had unknowingly flown over lead elements of the BEF but was unable to identify them due to heavy cloud cover.

As a result, First Army HQ remained ignorant of the British Army's position for yet another crucial day.

Acting as flank guard, First Army advanced a short distance on the 21st in refused echelon to protect the right wing's line of communication. That evening First Army HQ received a report from OHL detailing the location and strength of all known allied corps in the German right wing's sector, specifically the areas where Second and Third Armies were to attack. When it came to identifying the allied strength opposite First Army however, the report turned to speculation. Regarding the English, Moltke inexplicably believed that a majority of the BEF had not yet landed on the continent.[255] To their credit, Kluck and Kuhl dismissed Moltke's claim and continued to operate under the (correct) assumption that the BEF, in its entirety, had already landed. First Army HQ therefore expected to make contact with the English sometime over the following 72 hours. With time running out and Bülow's army now engaged along the Sambre, the important question was: where?

In accordance with the German General Staff's pre-war hypothesis, First Army HQ believed the BEF had landed at Boulogne and would operate from the direction of Lille. On that presumption, Kluck ordered First Army to move southwest in echelon on the 22nd in order to maintain the ability to fight the BEF, should they appear from the west. As a result, each of First Army's corps were spread out over a large area—a fact that would significantly impact Kluck's ability to concentrate his forces once the army became engaged the following day.

In their pre-dawn orders for the 22nd, First Army HQ ordered its aviators and cavalry to continue their search to the west for the English in the area of Mons to the south and Lille—Tournai. Once again the *Fliegertruppe* were directed to concentrate the majority of its flights towards Lille due to the General Staff's erroneous assumption that the BEF were in the area. Throughout the afternoon, aircraft from FFAs 7, 9, 12, and 30 conducted flights in the area of Ghent—Lille—Tournai. Each of these flights returned reports to their respective HQs reporting the area "to be completely free of enemy troops." Nonetheless, despite not finding anything, these flights were of great value as the German Official History explains:

Importantly, these observations confirmed that the British troops recently disembarked in French ports were not concentrating in the area Courtrai—Lille—Tournai; thus, First Army's flank and rear did not yet appear to be exposed to danger.[256]

As was the case the day before, only FFA 11 was assigned to the area of Mons. The section's first three

> ### A Serious Concern: The German Army's Flawed Liaison Doctrine
>
> Poor communication between ground commanders and their aviators was extremely prevalent throughout the campaigns of 1914. On many occasions the various HQ staffs failed to dispatch adequate instructions to their flying sections. As a result, the airmen remained grounded during decisive moments in the fighting. In other cases, such as with FFA 11 on August 22nd, flying sections dispatched critically important intelligence that never reached the attention of the army commander. In either case, the *Fliegertruppe* were unable to make a more significant impact despite maintaining the capability to do so.
>
> Examples of poor communication such as IX AK HQ's failure to transmit FFA 11's vital message on the 22nd was evidence of the German Army's faulty organizational model at the start of the war. Unlike the French, the Germans had not yet created the position of "aviation staff officer" on the staff of the various army HQs. An aviation staff officer's role was to supervise the aviation assets under the army's command and ensure that they were employed in a manner that best suited the commander's concept of operations. Most importantly, the aviation staff officer would act as liaison between army headquarters and the aviation sections by sending and receiving all aviation communiqués. During the war's first months, however, this duty was delegated to various staff officers who were already heavily burdened with other tasks. These officers understandably prioritized their traditional work, resulting in aviation reports often becoming neglected or even ignored. For example, the German Official History estimates that "about 50% of available air reports never came to the attention of First Army HQ during the Marne Campaign."[259] As we shall see, these lost reports had a significant impact on the outcome of the campaign. Only the existence of a dedicated aviation staff officer would have solved the problem and dramatically improved communication between the army leadership and its aviators. Unfortunately, the lessons learned from these mistakes were not fully recognized until the spring of 1915 when Major Thomsen ordered a *Stabsoffizier der Flieger (Stofl)* be attached to each Army HQ's staff.[260]
>
> When asked after the war about the lost aerial reports during the Marne Campaign, General Kuhl replied:
>
> *"The Army HQ's FFA 12 was able to bring messages the fastest, as it was connected directly to the army commander. Therefore, not only the written messages arrive quickly, but the observers could easily be summoned to explain in person.*
>
> *The Corps FFAs would often times be handled by the Corps HQ's "Ic" (Intelligence services officer). Through him, the Corps HQ could be kept up to date with the rapidly changing situation at the front. However, by the time the Ic would have time to transmit information to Army HQ, the situation at the front had very often been drastically changed and the aviator's information was essentially null and void.*
>
> *These Ic officers of the various corps HQs would spend hours riding to and from the HQs and airfields. These officers were so over-burdened with work, especially during the periods of heavy combat of the Marne battle, that it was easy to see how some aviator's messages were not given appropriate attention.*
>
> *The transmission of messages from Corps HQs to Army HQ was carried out by the General Staff: "the daily news from Corps HQs was at times the job of the Ordinance Corps Officer of First Army HQ. According to the documents from OHL the aviation messages regularly were addressed late in the evening after other matters had been addressed. The telephone connection with the Corps HQs was extremely inadequate. It was very common that interruptions would occur in the middle of a conversation. With these unfortunate conditions, it is clear why some of the messages sent by the Corps Commanders never arrived at First Army HQ. Consequently, aviation messages and other important communiqués throughout the battle were either never received or were overlooked. In either case, important messages were not put to use to assist our forces during the battle."[261]*

flights of the morning were unable to locate any signs of allied forces in the area. Nevertheless, FFA 11's commander, *Hptm.* Helmuth Wilberg, continued to direct his aircraft in the direction of Mons. Finally, shortly after 3:00pm, one of the section's LVG B.I biplanes landed at the airfield with definitive proof of the BEF's location. The observer, *Oblt.* Klein, reported large columns of troops in the vicinity of Bavai marching towards Binche. Heavy automobile and other transport/supply columns were also witnessed moving north towards Mons from Valenciennes. Upon landing, Klein's report was immediately dispatched by automobile directly to IX AK HQ.[257] Unfortunately, perhaps due to IX AK HQ's overburdened staff, Klein's valuable report was never forwarded to First Army HQ. In fact, the only aerial report sent by IX AK HQ on the 22nd was FFA 11's first report of the day stating: "No military activity found in the sector: St. Ghislain—Mons." This left the First Army HQ staff to believe FFA 11 had not found any signs of the BEF on the 22nd— a false assumption that would play a major role in the following day's engagement.[258]

Despite the inexplicable loss of FFA 11's important reconnaissance report, First Army HQ was able to gain valuable intelligence on the 22nd regarding the location of the BEF. At 11:00am HKK 2 transmitted the following message by radio: "Patrol taken under fire on the canal six kilometers east of Mons. Le Roeulx (northeast of Mons) clear. Country clear as far as Escaut." Fifty minutes later another message was received stating: "Patrol of the 4th Cavalry [Division] has positively identified an English squadron at Casteau northeast of Mons." Meanwhile, IX AK HQ radioed at 3:00pm that: "According to declarations made by local inhabitants, Mons is occupied by English Troops." Later that evening, in its nightly report to army headquarters, IX AK confirmed that: "18th Division reports that the passages of the Canal du Centre between Nimy and Ville-sur-Haine are occupied by the British." Likewise, HKK 2's final message of the day bluntly stated: "English at Maubeuge."[262]

Together, these messages convinced Kluck, Kuhl, and the First Army HQ staff that the BEF was to the south in direct support of French Fifth Army's left wing. Additional evidence presented itself in the form of a captured English aviator that had been shot down by ground fire at the village of Enghien. It was soon discovered that the aircraft had taken off from Maubeuge, directly south of Mons. Although this information confirmed that the BEF was to the south, the exact location of its main body remained an open question. Drawing upon FFA 11's initial aerial report stating "no military activity in the sector of Mons," First Army HQ theorized that the BEF was between Maubeuge and Valenciennes—well to the south of Mons where they were actually concentrating. Kuhl, a believer in aviation, trusted his airmen's report declaring the area of Mons to be clear. If his aviators were unable to see enemy troops along the canal, Kuhl assumed the BEF's main force was still to the south, moving up behind a light cavalry screen that HKK 2 and IX AK had reported around Mons. At this point the loss of FFA 11's report detailing large numbers of British troops less than five miles south of Mons weighed heavily. If Kluck and Kuhl had received Klein's report, First Army HQ would have dropped their assumption that their opponent's main strength was well to the south at Maubeuge and would have likely concentrated their forces at Mons to overwhelm and destroy the English forces deployed there.

Believing erroneously that the BEF's main strength was a day's march to the south, First Army HQ did not expect to become seriously engaged on the 23rd. Consequently, Kluck did not issue any attack orders on the evening of the 22nd. Instead, the army's order for the 23rd prescribed a short march to the heights south of the line Mons—St. Ghislain. Meanwhile on the army's left wing, IX AK was ordered to capture Mons itself. III AK on their right was to take St. Ghislain and the heights to the south. Meanwhile, on the army's right wing, IV and II AKs were to turn south and establish contact with III AK's right flank— remaining *en échelon*. Thus, on the morning of the 23rd each of First Army's lead corps moved south expecting to capture Mons, St. Ghislain and the canal that ran through them with little or no resistance.

Under the command of Sir John French, the British Expeditionary Force had landed at Le Havre, Rouen, and Boulogne on August 14th.[263] The BEF was composed of I Corps (1st and 2nd Divisions) under the command of Sir Douglas Haig, II Corps (3rd and 5th Divisions) commanded by Sir Horace Smith-Dorrien, and General Edmund Allenby's cavalry division. To bolster their numbers, an additional infantry brigade— the 19th— was created out of rear echelon troops and pressed into front line service. After landing on the continent, the BEF immediately moved into position in the area between Le Cateau and Maubeuge. According to Joffre's Special Order 15 written on August 21st, the BEF was to move north in the direction of Soignies and take up positions alongside French Fifth Army left in order to assist Lanrezac's attack against the "northern enemy group." It should be remembered that this order was part of Joffre's optimistic plan to split the German line and cut off the right wing with simultaneous offensives using his own center and left. By the evening of the 22nd, however, the situation had changed dramatically. First, the French offensive into the Ardennes had been decidedly defeated, resulting in Third and Fourth Armies' precipitous retreat. Secondly, the original intelligence estimates regarding the strength of the German right wing had been proven to be false. Instead of deploying their main strength south of the Meuse, it was discovered that the Germans had deployed at least eight corps to the north.[264] Thus, after being driven back by determined attacks on the 21st and 22nd, French Fifth Army was now faced with defending the Sambre— Meuse against two converging German armies with large numbers of the Kaiser's troops reported further west moving towards the BEF. In light of these revelations, Sir John French recognized that Joffre's attack orders to the allied left wing commanders were no longer feasible. Therefore, with Lanrezac's Army heavily engaged against superior forces on his right, the British commander decided to hold his positions on the 23rd, covering the French left flank for another 24 hours.

As it happened, the British deployment was entirely ill-suited for defense. Smith-Dorrien's II Corps was thinly deployed along the canal at Mons facing north while Haig's I Corps was massed together to the southeast near the village of Givry. The BEF had moved into these positons on the 22nd under the assumption that the allied left wing's offensive would commence the following day. When the situation changed, the British leadership failed to redeploy any of its forces. Consequently, the II Corps was left alone to bear the brunt of the coming German attack.[265]

Thick fog and rain covered First Army's sector at dawn on the 23rd. In the pre-dawn hours First Army HQ had dispatched orders to each of the army's flying sections directing aerial reconnaissance of the areas to the south and southwest in the attempt to definitively locate the BEF's main body. The poor weather, however, unfortunately caused each of the army's flying sections to remain grounded throughout the morning—leaving Kluck and Kuhl uninformed of the situation in their front. Meanwhile, the army's lead corps had begun their southward advance towards Mons expecting to only encounter light resistance. Then, at 9:20am, First Army HQ received a message from HKK 2 announcing large numbers of troops in the area of Tournai (near Lille). The news of hostile forces on the army's right flank deeply worried Kluck and Kuhl, who began to rethink the possibility of the BEF operating from the direction of Lille. Concerned about a possible British strike against their flank and rear, First Army HQ ordered the army's advance to be temporarily suspended until the situation could be cleared up. With his flying sections grounded, Kluck ordered HKK 2 to investigate and determine if the English were indeed in the area. At 1:00pm he received his answer. The troops previously seen at Tournai were not English, but were elements of the French garrison from Lille. In the meantime, First Army HQ had learned from local inhabitants that up to 40,000 English troops had been seen the day before marching north towards Mons. Together, these two reports confirmed that the BEF was deployed to the south in the direction of Maubeuge and had already established a significant presence at Mons.

After receiving confirmation that their flank was clear, First Army HQ ordered the army's advance to continue. However, unbeknownst to Kluck and Kuhl, IX AK had already become engaged at Mons when the earlier order to halt the advance was transmitted.

Above: Sir John French, commander of the British Expeditionary Force. (Author's Collection)

At 9:00am IX AK's 18 ID had arrived in the area and observed large numbers of English troops in the town defending the canal. Within minutes, the division's artillery deployed and began a 'vigorous bombardment' from the heights to the north. At 9:30, 18 ID's infantry deployed in skirmisher lines and began to advance by bounds against the British defenders. By noon, IX AK had pierced the English center, triggering a British withdrawal. Throughout the battle the Germans skillfully used artillery and machine guns in close support to assist the attacking infantry seize the formidable English positions. By nightfall, IX AK had successfully pushed the English forces back 1.8 miles south of the canal.

Meanwhile the III AK, deployed directly on IX AK's right, was prepared to attack at 10:00am. However, First Army HQ's order to suspend the advance had reached III AK HQ before its divisions began their assaults. The corps therefore remained unengaged for four hours until it was finally given permission to attack at 2:00pm. After finally receiving permission to attack, III AK quickly deployed both of its divisions abreast and attacked the English positions west of Mons between St. Ghislain and Jemappes. The German infantry, advancing by bounds, managed to cross the canal at three points, vigorously driving the British defenders back over two miles. By 7:20pm, forward elements from 6 ID had established themselves on the high ground 1.5 miles south of the canal.

First Army HQ's misconception regarding the BEF's location at Lille had caused Kluck and Kuhl to spread their forces out *en échelon* prior to the battle. Consequently, the IV AK was compelled to make a 15 mile march to the battlefield on the 23rd once it had been finally decided to swing the army to the south. This lengthy approach march resulted in IV AK moving into position alongside III AK's right flank with very little daylight remaining. The corps' subsequent attack was therefore unable to make a substantial impact before darkness descended over the sector. Noted historian Terence Zuber believes that a decisive German victory could have been won on the 23rd had IV AK started their day closer to the field or had III AK simply been allowed to deliver its attack at 10:00am when it was ready to do so.[266] It should be noted that both of these scenarios could have been a reality had *Ltn.* Klein's aerial report from the previous afternoon been successfully forwarded to First Army HQ.

Despite being desperately needed, the *Fliegertruppe* of First Army were unable to play a significant role during the battle at Mons. First, the abysmal weather throughout the morning denied First Army HQ the use of their aviators to clear up the situation at Tournai. Had Kluck's airmen been available, First Army HQ would have gained an understanding of the BEF's position far sooner—resulting in III AK being allowed to deliver its attack and IV AK arriving on the battlefield much earlier than they actually did. As it happened, aerial activity in First Army's sector began around noon when an aircraft of IX AK's FFA 11 took off on a 125 mile long flight, traveling as far south as Guise.[267] The flight had been requested by IX AK HQ to determine whether the bulk of the British Army was deployed in First Army's front or if it was further south. Two hours later, the aviators returned with a definitive answer. While patrolling north of the line Bavai—Maubeuge at 12:30pm at an altitude of just 3,300 feet, the airmen discovered multiple highly condensed bivouacs and other signs of "strong troop concentrations." Over the next 45 minutes, the airmen carefully searched the roads south of Maubeuge only to find "insignificant movements." At 1:50pm, during the return-leg of the flight, the observer discovered "a three kilometer long enemy column marching north towards Mons on the La Longueville—Malplaquet road." Next, two smaller columns of equal size were seen at Genly (less than two miles southwest of Mons) similarly marching toward the battlefield. After returning to FFA 11's airfield east of Soignies around 2:00pm, the observer crafted his report and dispatched it by automobile to IX AK HQ who forwarded its contents to First Army HQ that evening at 6:00pm. Kluck and Kuhl reviewed the report and immediately came to the conclusion that "the bulk" of the BEF was concentrated directly in their front.[268]

Hoping to quickly obtain a decisive victory, First Army HQ ordered the attack to continue at dawn on the 24th. Specifically, the three corps (IX, III, IV AKs) already on the scene were directed to assault the hostile forces in their front, turn their left and throw them back upon fortress Maubeuge. Meanwhile, II AK was pushed south by night march to the town of Condé where they were expected to move into position and overwhelm the BEF's open left flank on the morning of the 25th.[269] First Army HQ was now unmistakably in an all-out offensive mindset; acting in accordance with the German War Plan to envelop and destroy the BEF with the full intention of later moving against French Fifth Army's flank.

The combat of the 23rd had seriously damaged the BEF's II Corps. Together, the IX and III AKs had succeeded in punching a hole in II Corps' center, forcing the English forces on either flank to withdraw in confusion to a line roughly three miles to the south of the canal. BEF HQ originally ordered II Corps to defend their new positions on the 24th. However, when Sir John learned of French Fifth Army's decision to withdraw on the evening of the 23rd, he smartly ordered his entire army to follow suit. Accordingly, the BEF began their retreat at dawn on the 24th— leaving behind a strong rear-guard to hold off Kluck's renewed attacks while the rest of the army completed their escape to the southwest.

Each of First Army's forward corps resumed their attacks against the British line at 5:00am. By mid-morning, German attackers across the front were reporting that their adversaries had already initiated a retreat and were fighting a rear-guard action. Aware of this fact, First Army HQ ordered all subordinate commanders to aggressively continue their attacks as planned, resulting in substantial territorial gains and the destruction of large portions of the British rear-guard before nightfall.

First Army's leadership immediately ordered aerial reconnaissance flights upon hearing of the BEF's retreat. Good weather allowed aerial activity across the entire army's sector to begin at 7:00am and continue uninterrupted until dark. While First Army HQ's FFA 12 was assigned the Mons–Valenciennes region and the territory on the army's open right (western) flank, the corps flying sections were directed to reconnoiter the area south of the army in the direction of each respective corps' advance. At 8:45am, the first meaningful report of the day was dropped by one of FFA 11's LVG B.Is at the eastern exit of Mons where IX AK HQ's forward command post was located. The report read:

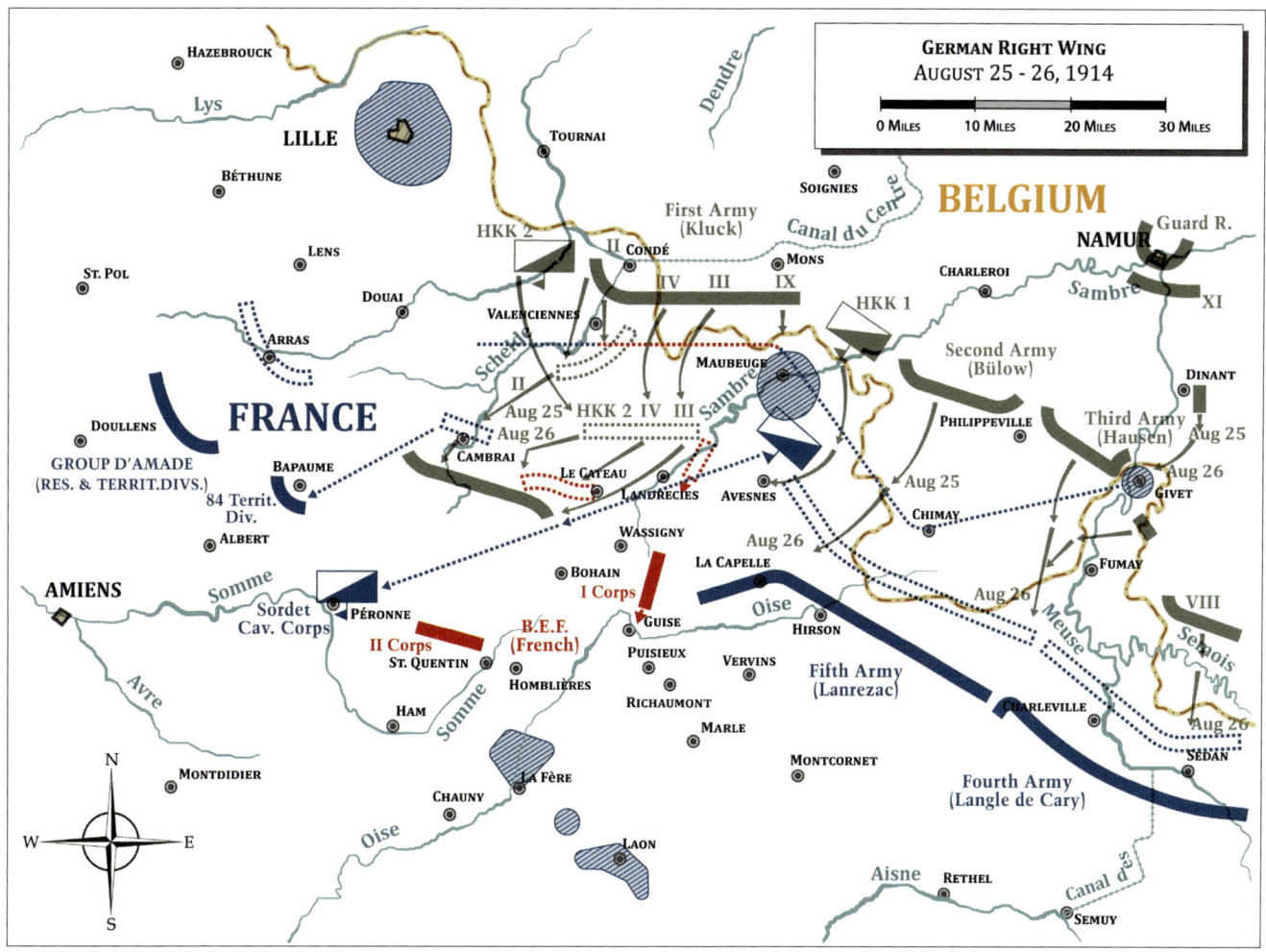

To: IX AK HQ[270]
1. Enemy retiring from the line St. Ghislain—Givry in a westerly and southerly direction.
2. The chief roads employed for the retreat: Givry—Bavai; Mons—La Longueville; Mons—Genly—Bavai; Angre—Sebourg; Audregnies—Angre; Roissin—Bry; Bavai—St. Waast—Jenlain
3. Artillery in position: —
    a) South of the Roman Road near Chateau Poussiere, South of Givry
    b) Immediately west of Nouvelles
    c) Immediately west of Wasmer

This important message confirmed that the British were retreating *away* from Fortress Maubeuge.

Knowledge of the direction of the BEF's retreat was critically important for First Army HQ, who was planning an attack against the English left flank. As Kluck and Kuhl studied the sector's map, they acknowledged two possible directions the English could take. Either the BEF would fall back directly to the south and southeast upon Fortress Maubeuge, or they would turn southwest in the direction of Le Cateau. Seeking a decisive victory, the ultimate question for First Army HQ was: which direction to pursue?

Sixteen flights were ultimately conducted by First Army's aviators on the 24th. Many of them returned with excellent reports describing the size and direction of the British columns in retreat. The majority of these reports arrived at First Army HQ during the evening of the 24th as Kluck and Kuhl were discussing the proper course of action for the following day. Almost immediately, the intelligence from FFA 9's final flight of the day was singled out. The message stated:

Infantry and artillery columns at 4:30 retreating on Bavai—Pont road. At 4:45 baggage and other trains

Above: German infantrymen valiantly charge the enemy's position. (Author's Collection)

*near Bachant driving on Aulnoye-Noyelles road.*²⁷¹

FFA 9's report definitively proved that British columns were marching west past Maubeuge on the Bavai—Aulnoye—Landrecies road.²⁷² Yet another message from II AK's FFA 30 reported French territorial troops on the BEF's left flank retreating towards the west and southwest. In total, no less than five reports were reviewed by First Army HQ which described British columns fleeing towards the southwest. Therefore, based upon this information, Kluck issued orders at 8:30pm for his army to conduct their pursuit in a southwesterly direction.

The situation drastically changed at 2:00am when a radio transmission arrived from III AK HQ. The transmission contained the reconnaissance report produced from FFA 7's final flight of the day. Delivered in person to III AK HQ at 8:30pm, the observer's testimony appeared to contain new information regarding the direction of the BEF's retreat:

To: III *Armee Korps* HQ
1. Enemy columns of all arms retiring along the roads:
   Bellignies—Bavai;
   La Flamenbrie—Bavai;
   Wargnies—St. Waast—Bavai;
   Gommegnies—Bavai.
2. Bivouac with vehicles at the railway station 3 kilometers [1.8 miles] north west of Le Quesnoy.
3. All roads leading from Le Quesnoy to the southwest and south and all roads through Forest Mormal are clear.
4. **General Impression**: **General retirement on Maubeuge.** [emphasis added]
5. The machine was bombarded east of Le Quesnoy by anti-aircraft guns on automobile mountings, but was not hit.

The message from III AK radically changed First Army HQ's perception of the situation. According to the report, the British appeared to now be yielding in an easterly direction through Bavai toward Fortress Maubeuge. Although this conclusion was in direct contrast to numerous other aerial reports, First Army HQ nonetheless made the decision to act on it. New orders for the 25th were promptly created and dispatched by automobile to each corps commander directing their columns to turn to the south with the intent of cutting off the BEF and proceeding against the French left wing.²⁷³

Receiving their new orders at 8:15am, First Army's various corps quickly shifted the direction of their advances. While IX AK covered Maubeuge, HKK 2 was ordered toward Guise, II AK to the west of Le Cateau, IV AK to Landrecies and III AK to Maroilles. By noon however, First Army HQ had received three additional aerial reports confirming that the entire BEF was indeed retreating to the southwest towards Le Cateau and *not* in the direction of Maubeuge as FFA 7's late night message had opined. Each of the flights was conducted by different flying sections, most notably First Army HQ's own FFA 12 who reported:

Long columns of all arms have been observed marching from Bavai on the highway leading to Le Cateau; rear at 9:30 at Bavai, advance guard 1 kilometer northeast of Croix [northeast of Le Cateau]. Numerous smaller columns, isolated companies, squadrons, batteries and automobiles were crossing the region of Selle Brook to the south and north of Solesmes on the roads leading towards the southwest.[274]

Another report delivered by FFA 7 stated:

An artillery battle [rear-guard action] is taking place at Solre-le-Chateau and southwest of Valenciennes; all enemy arms are retreating on the Maubeuge—Le Cateau road; on the same railway line, heavy traffic moves in the direction of Le Cateau.[275]

First Army HQ was immediately forced to admit their error and send a counter-order directing the army to once again change directions back to the southwest. By nightfall the army had re-oriented itself in the correct direction and was prepared to take up the pursuit the following day. Unfortunately, the loss of time caused by First Army HQ's mistake had done irreparable harm. The BEF's left flank was now unreachable. Had Kluck's men simply continued to the southwest as originally ordered, First Army's III and IV AKs (as well as HKK 2) would have been in position to envelop and destroy the BEF's II Corps on the morning of the August 26th.[276] Instead, the Germans would be forced to settle for a less-ambitious attack, initially employing just three brigades of IV AK and HKK 2—a significant difference that would ultimately have a fundamental impact on the outcome of the campaign.

Kuhl's decision to turn to the southeast on the morning of the 25th was a terrible blunder.[277] The change of direction was made based upon a single report that contradicted numerous other aerial reports submitted throughout that afternoon. Furthermore, FFA 7's report didn't necessarily conclude that the entire BEF was retreating southeast upon Maubeuge.

It was merely the observer's opinion based on the units he had seen during the flight which caused him to write the fateful words: "General Impression: General retirement upon Maubeuge." While it was wrong for the observer to offer his "impression" based on limited evidence, the fault for the missed opportunity lies squarely with First Army HQ.[278] Kluck and Kuhl should have never allowed a single aviator's opinion to supersede the corroborated conclusions of at least five other aerial reports. But, as English historian Walter Raleigh points out, it was Kluck's wishful thinking that influenced First Army HQ into making the decision:

...but he [Kluck] says enough to show that he was misled chiefly by his own preconceptions. Hope told a flattering tale, and he seems to have been possessed by the idea that the British army would be tempted into the fortress of Maubeuge.[279]

First Army HQ's actions on the evening of the 24th–25th is another example of how the *Fliegertruppe's* inadequate organizational model diminished the army's ability to be successful. Had First Army HQ been staffed with an aviation staff officer, all of the aerial reports from the 24th would have been gathered and presented to Kluck and Kuhl with a formal conclusion made by the aviation officer himself. The large number of reports describing the BEF's retreat to the southwest suggested overwhelmingly that the British were falling back upon Le Cateau, not Maubeuge. Thus, an aviation staff officer's suggestion based upon all the available intelligence would have undoubtedly been to pursue in a southwesterly direction—resulting in a much more favorable opportunity on the morning of the 26th.

First Army Re-Engages the BEF at Le Cateau

Late in the evening of the 25th, Smith-Dorrien's II Corps retreated through Solesmes on its way to Le Cateau. After a short rest, the retrograde movement was continued through the early morning hours to the line Le Cateau—Inchy—Caudray. In accordance with Sir John French's orders, the retreat was to begin again shortly after dawn. However, the frantic pace of the German pursuit had brought the forward elements of Kluck's army within striking distance of the resting English troops. The proximity of the Germans gave Smith-Dorrien no choice but to stand and fight on the 26th in the hopes of delaying the Germans long enough to continue the retreat under the cover of darkness that evening. Accordingly, Smith-Dorrien messaged BEF HQ to inform them of the situation and his intention to fight on the 26th. Sir John gave his consent but reminded Smith-Dorrien to resume the retreat as soon as possible.

The ensuing battle at Le Cateau began at 9:00am and lasted throughout the day as all available German units were rushed forward to attack British II Corps. Because of First Army's turn to the southeast on the 25th, only three brigades from IV AK were initially available to attack Smith-Dorrien's right and center.[280] Furthermore located on the German right, HKK 2 was directed to attack and turn II Corps' left. These formations were simply not enough, however, to achieve First Army HQ's goal of completely enveloping the British left before they could escape. Thus, the only attainable goal Kluck's troops could accomplish was to simply inflict as much damage upon the British as possible.

Despite being initially outnumbered, the German attackers were able to use their superior combined arms doctrine to break through the British line. At 11:15am, after a morning of hard fighting, IV AK's commander General Sixt von Armin directed all available troops from both of his divisions to act in concert against the British right flank. With the close support of their artillery, Armin's troops methodically pushed the British right wing off the commanding heights located southwest of Le Cateau—making the rest of the British position untenable. Upon learning of the deteriorating circumstances on his right, Smith-Dorrien issued a general retreat order at 2:00pm. On the British left, HKK 2 fought a brilliant dismounted action throughout the day, skillfully using their *Jäger* light infantry and artillery to heavily damage the BEF's 4th Division.[281]

Fighting continued along the front until darkness allowed Smith-Dorrien's men to withdraw. Covered by large formations of cavalry, the BEF retired throughout the evening amidst pouring rain, putting some distance between themselves and their pursuers—effectively saving the BEF from destruction. According to Sir John French, "the condition of the army was pitiful." Indeed, after their costly engagement at Mons, the British had lost a further 7,812 men and 38 guns on the 26th.[282] The crippled BEF was therefore unable to take to the offensive for the foreseeable future.

First Army's leadership learned of the extent of their victory from the initial reports of captured men and matériel. Recognizing that the BEF was now badly damaged, Kluck began to plan the army's next move. While First Army HQ correctly assumed that the English would retreat that evening under the cover of darkness, their direction remained uncertain. In order to maintain contact with the channel ports, it was suspected that the BEF would continue retreating to the southwest. Before acting on that suspicion, however, both Kluck and Kuhl consulted all available intelligence, including all that was gathered by their aviators.

Like the rest of the army, Kluck's airmen had not anticipated a general engagement to occur on the 26th. Indeed, it was well after 9:00am when IV AK HQ recognized the battles raging around Le Cateau were more than just a simple rear-guard action.[283] Thus, the First Army *Fliegertruppe* were not prepared to provide any tactical support to the ground troops. They had planned instead to reconnoiter the roads and to gather updated information on the BEF's location. However, as news of the action at Le Cateau spread, the commanders of FFAs 7 and 9 each modified their plans to concentrate on the area around the fighting. Meanwhile, FFAs 11, 12, and 30 proceeded with their original orders to patrol the regions along both of the army's flanks.

The first meaningful reports of the day were delivered to IV AK HQ at 10:00am and 12:15pm by its own FFA 9. While the initial message provided a general outline of British II Corps' deployment, the latter report focused specifically on the British right wing. Unfortunately, due to the peculiar nature of the battle, these reports had no tactical impact on the fighting. Further north, the aircraft of III AK's FFA 7 had spent the morning relocating to the section's new airfield. Their first patrol of the day took off around noon in the direction of Étreux to ascertain whether there was a threat to the corps left flank as it rushed toward the fighting at Le Cateau. The observer quickly found elements of the unengaged British I Corps withdrawing south between Le Favril and Oisy—confirming that Haig's command was not a threat. At 2:00pm, FFA 7 ordered its second flight airborne. The crew had been ordered to closely monitor the British right wing before III AK's forward units arrived in the area to attack it. It was clear, however, by the time the crew's Albatros B.I had reached the area of Le Cateau that the British right had already been defeated. Proceeding to the south, the observer found signs that a general withdrawal had indeed been ordered. Baggage trains and supply columns of varying sizes were reported all along the Roman Road leading southwest from Le Cateau. Upon landing the observer shared his completed report with the section's commander, *Hptm.* Wilhelm Grade. Recognizing that his section's two flights had collectively discovered the direction of British I and II Corps' retreat, Grade personally delivered his airmen's reports to III AK HQ, who forwarded the information to First Army HQ later that evening.[284]

Poor weather once again played a determining factor for First Army's *Fliegertruppe* when a drizzling rain moved into their sector late in the afternoon causing all aerial operations to be suspended at 4:00pm. Heavy rain and thunderstorms continued to

Above: Jeannin *Taube* A.283/14. Although the Jeannin *Taube* was not the best nor most numerous *Taube* at the front, it was one of the best-known based on its prewar fame. (Aeronaut)

ground all flights over the next 24 hours as the army began its pursuit. First Army's airmen were therefore unable to gather updated information regarding the retreat of the English main body. Nevertheless, FFA 7's excellent reporting on the afternoon of the 26th proved vital in helping Kluck and Kuhl to begin the pursuit on the 27th in the correct direction. According to First Army HQ's orders, the army was to advance to the southwest toward the Somme on both sides of Péronne.[285] Despite not having the use of their aviators Kluck's forward columns maintained contact with the retreating allied forces as the next phase of the campaign began.

Joffre and Moltke React to The Battle of The Frontiers

August 23rd was a disastrous day for the allied left wing. Having already suffered horrendous losses over the previous 48 hours, Lanrezac's Fifth Army had once again been attacked and thrown back by German Second and Third Armies. Later that evening, Lanrezac was faced with the realization that he had been outflanked and that his current position would soon become untenable. Consequently, Lanrezac immediately ordered a general withdrawal in order to save his army from becoming isolated and destroyed by Bülow and Hausen's forces.[286] Meanwhile at Mons, Kluck's First Army was successful in punching a hole in the center of British II Corps' line—compelling Sir John French to order the BEF's retirement. By sunrise the entire allied left wing was hastily retreating. Thus, when troops from German Second and Third Armies found only rear-guards opposing them on the morning of the 24th, the two armies' leadership promptly recognized the need to aggressively pursue their defeated adversaries and quickly re-engage them to achieve the decisive victory prescribed by the German War Plan.

As Kluck's First Army pursued the BEF on the morning of the 24th, airmen from Second and Third Armies were immediately ordered airborne to determine Lanrezac's route of retreat. As mentioned previously, airmen from Third Army's FFA 22 and 29 monitored Lanrezac's retreat throughout the day and successfully kept the army leadership informed of their whereabouts. Drawing heavily upon their aviators' reports, Third Army reverted back to their original orders of marching to the southwest but were unable to catch Lanrezac's fleeing main body. To the west, Bülow's Second Army pressed after the French in a southerly direction.[287] However, Lanrezac's sizeable head start had allowed his army to gain some separation from their German pursuers.

French Fifth Army resumed its retreat before dawn on the 25th, using the cover of darkness to increase the gap between themselves and the Germans. This withdrawal was carried out in concert with the other allied armies who had been similarly defeated and forced to withdraw. Across the front, three French armies, the BEF and the French territorial troops on the English left were all in retreat as Joffre and his staff desperately endeavored to formulate a new operational plan to salvage the situation. Lanrezac later described this dark period in his memoirs:

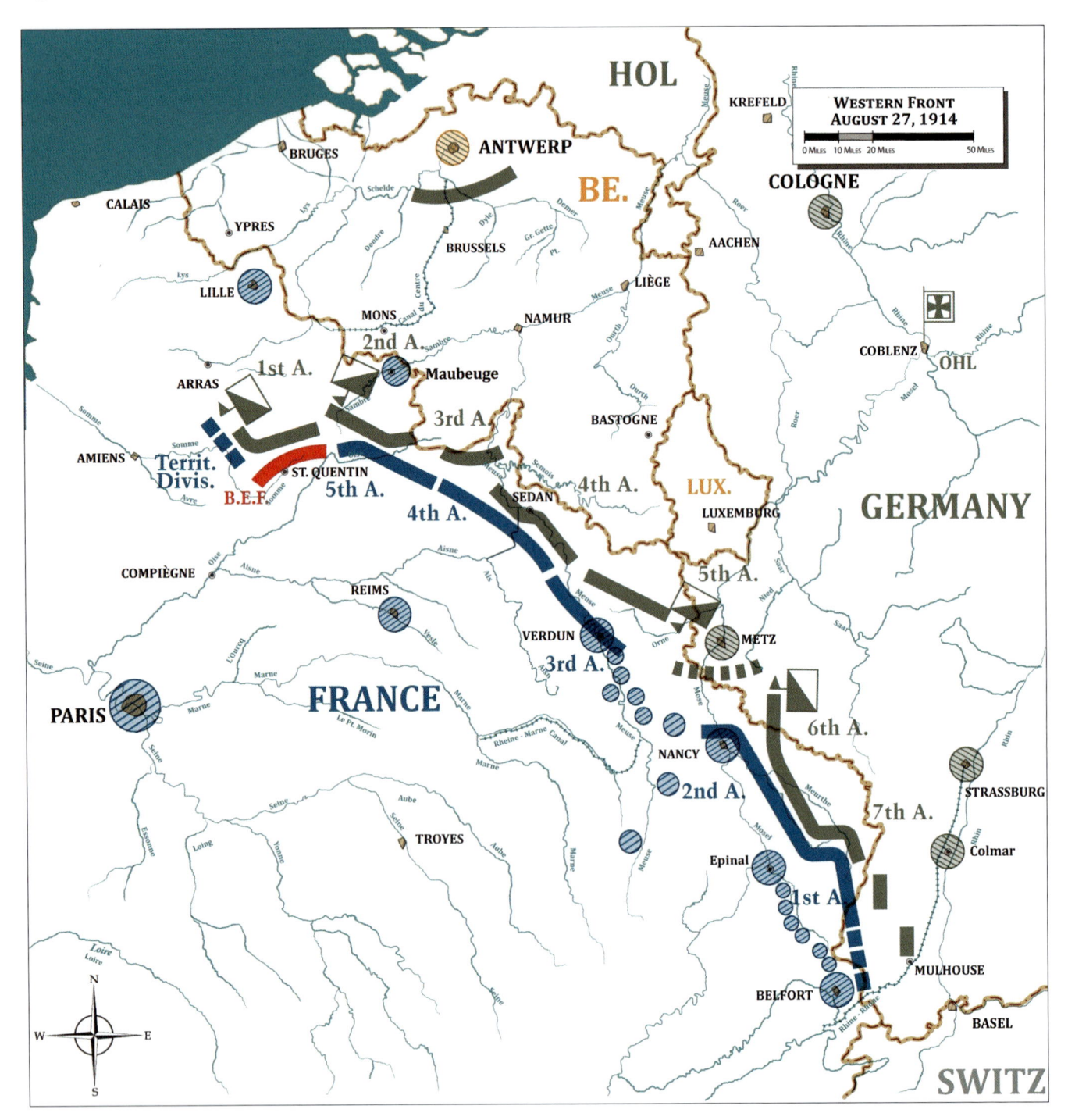

The whole French Army was thus in a most unfortunate situation. It was not only Fifth Army that had sustained a serious defeat. Langle's [Fourth] Army was beaten north of the Semoy and found itself compelled to retreat toward the Meuse, exposing the right of Fifth Army along a depth of more than two marches. Ruffey's [Third] Army was not much better off between Arlon and Thionville and had to fall back on Verdun. The armies of Castelnau [Second] and Dubail [First] after vain efforts to dislodge the Germans from Morhange and Sarrebourg, were compelled to retreat, that of Castelnau on the Couronne de Nancy and that of Dubail behind the Mortange. We were defeated everywhere from the Sambre up to the Vosges!... All our armies, greatly fatigued, had nothing left to do but beat a retreat as quickly as possible in order to avoid total destruction.[288]

Lanrezac's troops continued their retreat throughout the 25th as his rear-guards successfully delayed the German advance. Meanwhile, Hausen's Third Army continued to race south in the attempt to engage and cut off the fleeing French columns. Aviators from both

Second and Third Armies closely watched Lanrezac's forces throughout the day, keeping army leadership informed of their location and route of retreat. Using their aviators' reports, both Army HQs directed their columns to stay within striking distance. The German leadership was therefore extremely confident on the evening of the 25th as Second Army Chief of Staff Otto von Lauenstein's 7:00pm note suggests:

The success of the two day battle at Namur has proven to be very important. This morning, aerial scouts reported that all French corps continue their retreat. The fruits of our victory now fall to 3rd Army, whose route we have opened across the Meuse through our bold offensive across the Sambre; it is marching directly into the flank of the enemy we recently defeated. The success on our right wing against the British does not seem so grand; apparently they have recognized the danger threatening them in time and are in full retreat to the south. But the result of the last three days still stands: the Belgians, British and French are defeated. That is a good start.[289]

From his headquarters at Vitry-le-François, Joffre immediately began working on a new plan of operations. After two full days of discussion and debate with his staff, the French Commander-in-Chief decided upon creating a new army far to the west on his extreme left in order to strike the powerful German right wing's open flank from west to east in a massive counter-offensive. This new army was to be mainly formed from units originally deployed on the French right in Lorraine. Consequently, the plan required the existing armies to continue their retreat in order to buy time for the necessary forces to be transferred by rail to a staging area on the Allied left from where the counter-attack was to ultimately be launched.[290]

Joffre's new strategy was immediately put into effect on the evening of the 25th with the transmission of his "General Instructions No. 2" to each of the Army HQs. The order, which outlined in detail the Allies' new plan of operations read, in part, as follows:

1. *Having been unable to carry out the offensive maneuver originally planned, future operations will be conducted in such a way as to re-construct on our left a force capable of resuming the offensive by a combination of the Fourth and Fifth Armies, the British Army and new forces drawn from the east, while the other armies hold the enemy in check for such time as may be necessary.*
2. *Each of the Third, Fourth, and Fifth Armies, during its retreat, will take account of the movement of the neighboring armies, with which they must remain in liaison. The movement will be covered by rear-guards left at favorable points in such fashion as to utilize all obstacles to halt the enemy's march, or at any rate to delay it by short and violent counter-attacks in which artillery will be the principal element employed...*
6. *Before Amiens, between Domart-en-Ponthieu and Corbie, or behind the Somme between Picquigny and Villers-Bretonneux, a new group of forces will be constituted by elements transported by rail (VII CA, four reserve divisions and perhaps another active army corps) between August 27th and September 2nd.*
   This group will be ready to take the offensive in the general direction of St. Pol—Arras or of Arras—Bapaume.[291]

Joffre's transmission was received by each of his army commanders during the early morning hours of the 26th. By dawn, all five French armies were acting in accordance with the new strategy established by the order. To the south, First and Second Armies were to maintain their fortified positions along the line Toul—Épinal with the expectation to "fix" the German forces currently present in the area and deny them the freedom to be redeployed elsewhere. Ruffey's Third Army was to continue its retreat, seeking to rest its right flank upon Verdun. Next, Langle's heavily damaged Fourth Army was instructed to retreat to the Aisne with orders to continue to the area around Reims. Perhaps most importantly, Lanrezac's Fifth Army was directed toward St. Quentin with strict instructions to keep liaison with both Fourth Army and the British. Finally, on the extreme left, Sir John French's independent BEF was asked to maintain contact with Lanrezac and to retreat southwest in the direction of Péronne. Altogether, Joffre envisioned his left and center armies withdrawing to the line of Amiens—Reims—Verdun. It was in the vicinity of Amiens on the BEF's left where Joffre intended to assemble a new army composed of two active corps and four reserve divisions. This newly composed army was to aggressively strike the German right wing's open flank and roll it up in concert with frontal attacks launched by the French Fourth and Fifth Armies as well as the BEF. News of the British defeat at Le Cateau, however, ultimately compelled Joffre to postpone the counter-offensive and to order the withdrawal to continue.[292]

On the evening of the 25th, Moltke reviewed the reports from his subordinate army commanders and proclaimed a "decisive victory" had already been obtained in the West.[293] This claim was nothing short of delusional. General Tappen, OHL's Chief of the Operations Department, later described Moltke's mindset:

The wholly favorable news that came to us each day and which continued to arrive on the 25th, combined with the great victory obtained by the Sixth and Seventh Armies in Lorraine, gave rise to the conviction at OHL that the great decisive battle on the Western Front had been fought to our advantage. Believing in this "decisive victory" the Chief of the General Staff decided, in spite of objections presented to him, to send reinforcements to the Eastern Front. He believed that the moment had come when, having won a decisive victory in the West, substantial forces might be diverted to the East, in conformity with the general plan of operations, to seek an equally decisive result there.[294]

In accordance with this fantastical view, Moltke ordered two corps (XI AK and Guard RK) from the critically important right wing to be transferred to the Eastern Front where the situation was reported to be rapidly deteriorating. To shore up the situation in the East, Moltke made the decision to relieve the commander of German Eighth Army—*Generaloberst* Max von Prittwitz—and replace him with *Generaloberst* Paul von Hindenburg. Hindenburg and his new chief of staff, *Generalmajor* Erich Ludendorff, ultimately stabilized the situation with their famous victory at Tannenberg *without* the assistance of the troops detached from the western armies' right wing, who were on a train traveling to the Eastern Front at the time of the battle. Consequently, Moltke's misguided transfer of forces needlessly deprived Bülow's Second Army of the critically important Guard RK while reducing Third Army's field strength to just 2½ corps. Furthermore, in addition to these two corps' redeployment, it should be noted that the German right wing's overall strength was already reduced significantly by the necessity to detach troops to besiege the Franco-Belgian fortresses in the rear-area. First, the Belgian Field Army's presence within Antwerp compelled First Army to commit III RK to cover that fortress. Second, Maubeuge's impressive garrison of 50,000 men kept Second Army's VII RK, plus some important heavy artillery batteries, away from the important engagements at the front throughout the remainder of the campaign.[295] Finally, Third Army's 12 RD (XII RK) was used to reduce Fort Charlemont at Givet—keeping the division from the rest of Hausen's Army until the evening of September 7th.[296] In total, whether from Moltke's senseless redeployment or for necessary siege operations, the absence of these forces would ultimately prove to be a deciding factor when the right wing armies re-engaged their French adversaries at the Marne in early September.

The conclusion of the "Battle of the Frontiers" ushered in a new phase of the campaign. These initial battles, which had begun on August 20th at Morhange—Sarrebourg on the German left wing and had ended on the 24th with the Franco–British withdrawal from the Sambre, had admittedly not brought about the "decisive victory" necessitated by the German War Plan. However, the initial reports from the center and right wing commanders suggested that the situation was entirely favorable. Thus, Moltke believed that total victory in the West was at hand. All that was required, in OHL's view, was "an immediate and continuous pursuit of the enemy, who now seemed to be engaged in a general withdrawal to the southwest toward Paris."[297] Moltke's right wing and center armies were therefore expected to constantly harass and pressure the allies and deny them the ability to rest, reorganize or recover from their defeats. In accordance with this view, OHL issued a memorandum to all seven western armies entitled: *General Guidelines for First through Seventh Armies for the Continuation of Operations*. This directive, which effectively established a new plan of operations for the western armies, was dispatched by automobile to each army commander on the evening of the 27th.[298]

General Guidelines for First through Seventh Armies for the Continuation of Operations[299]

The enemy, comprising three groups, has sought to check the German offensive.

On his northern wing, supported by the British Army and by elements of the Belgian Army, before our First, Second and Third Armies, he has adopted an attitude in the main defensive in the region Maubeuge—Namur—Dinant. His plan, which consisted of taking the German right wing in the flank, broke down before the outflanking movement of our First Army.

The central enemy group was assembled between Mezières and Verdun. His left wing took the offensive and moved beyond the cut of the Semoy against our Fourth Army. This attack did not succeed. He then sought to cut off the left wing of our Fifth Army from Metz by an attack debouching from Verdun. This attempt also failed.

A third powerful enemy group attempted to penetrate Lorraine and the valley of the upper Rhine, in order to march towards the Rhine and the middle Main, passing on either side of Strasbourg. Our Sixth and Seventh Armies succeeded in repulsing this effort after hard combats.

All the French active army corps, including two newly formed divisions have already been engaged and have sustained heavy losses; the majority of the French Reserve Divisions have likewise been engaged and are severely shaken. One cannot yet say what value must be given to the capacity for resistance that the Franco-British Army retains.

The Belgian Army is in a state of complete disintegration; there is no longer any question of its ability to take the offensive in the open field. At Antwerp there may be about 100,000 Belgian troops, mobile forces as well as garrison. They are severely shaken and have little capacity for offensive enterprises.

The French, at least their northern and center groups, are in full retreat in a westerly and southwesterly direction, that is on Paris. It is likely that in the course of this retreat they will oppose new and bitter resistance. All information from France indicates that the French Army is fighting to gain time and that it is a question of holding the greatest possible part of the German forces on the French front to facilitate the Russian offensive.

The Anglo-French group of the north and center may, after the loss of the line of the Meuse, offer a new resistance behind the Aisne, with their extreme left pushed perhaps as far as St. Quentin, La Fère and Laon and their right wing established west of the Argonne, about in the region of Ste. Menehould. The next line of resistance will probably be the Marne, with the left wing supported by Paris. It is also possible that forces will be concentrated along the lower Seine.

On the southern wing of the French army the situation is not yet clear. It is not impossible that in order to relieve his northern wing and center, the enemy may resume the offensive in Lorraine. If this southern wing retreats, it will endeavor, resting on the fortified triangle Langres—Dijon—Besançon, constantly to attack the flank of the German armies by debouching from the south or to hold its forces in readiness for a new offensive.

... It is therefore important, by moving the German army rapidly on Paris, not to allow the French time to recover, to prevent the creation of new units and to eliminate so far as possible the country's means of defense.

Belgium will be organized by a general government under German administration. It will serve as the Zone of the Rear for First, Second and Third Armies for supplies of food, and will thus materially shorten our right wing's lines of communication.

His Majesty orders that the German armies advance in the direction of Paris.

The First Army, with 2nd Cavalry Corps under its orders, will move towards the Lower Seine, marching west of the Oise. It will be ready to intervene in the actions of Second Army. It will be further charged with the protection of the right flank of the armies. It will prevent the formation of new enemy units within its zone of action. The elements left behind to besiege Antwerp (III and IX Reserve Korps) are hereby assigned to OHL, IV Reserve Korps will be available to First Army.

The Second Army, with the 1st Cavalry Corps under its orders, will cross the line La Fère—Laon. It will besiege and capture Maubeuge and later La Fère and Laon, the latter place in concert with Third Army. The 1st Cavalry Corps will reconnoiter before the front of Second and Third Armies. It will also give information to Third Army.

The Third Army, crossing the line Laon—Guignecourt will march on Château-Thierry. It will seize Hirson as well as Laon and Fort Condé, the two latter positions in concert with Second Army. The 1st Cavalry Corps, operating before the fronts of Second and Third Armies, will give information to Third Army.

The Fourth Army will march by way of Reims on Épernay. The Fourth Cavalry Corps under the orders of Fifth Army, will also give information to Fourth Army. Siege material necessary for the capture of Reims will be placed later at the disposal of this army. VI Army Corps will pass under the orders of Fifth Army.

The Fifth Army, to which VI Army Corps is henceforth attached, will march towards the front of Châlons-sur-Marne, Vitry-le-François. It will assure the protection of the left flank of the armies by echelons to the rear and to the left, until Sixth Army is in a position to undertake this protection west of the Meuse. The 4th Cavalry Corps will remain under the orders of the Fifth Army; it will reconnoiter before the fronts of Fourth and Fifth Armies and will also report to Fourth Army. Verdun will be invested...

The Sixth Army, with Seventh Army and the 3rd Cavalry Corps under its orders, will have as its mission first, supporting itself on Metz, to oppose an enemy advance into Lorraine and Upper Alsace. The fortress of Metz is placed under the orders of Sixth Army. If the enemy retires, the Sixth Army, with the 3rd Cavalry Corps under its orders, will cross the Moselle between Toul and Épinal and will advance in the general direction of Neufchâteau. The protection of the left flank of the armies will then devolve upon Sixth Army. In this event the Sixth Army will be reinforced by elements of

> the Seventh Army (XIV, XV AKs and an Ersatz Division) …
>
> The Seventh Army will remain initially under the orders of the Sixth Army. If the Sixth Army crosses the Moselle, the Seventh Army will become entirely independent. The fortress of Strasbourg and the fortifications of the Upper Rhine will remain under its orders. The mission of the Seventh Army will be to prevent any enemy breakthrough between Épinal and the Swiss frontier. It is recommended that it constructs substantial defensive works before Épinal and thence to the Vosges, as well as in the valley of the Rhine, connecting them with Neuf-Breisach, and that it place the main body of its forces behind its right wing…
>
> All the armies must act in mutual cooperation and lend each other mutual support in overcoming the various obstacles of the terrain. If the enemy presents powerful resistance along the Aisne and later along the Marne, it may become necessary to have the armies converge towards the south instead of southwest.
>
> It is urgently desirable that the armies move forward rapidly, so as not to allow the French time to re-form and organize new resistance. Accordingly, the armies will report when they can begin their respective movements.
>
> (Signed)
> von Moltke

While it is beyond the scope of this book to critically analyze the merits to Moltke's latest plan of operations, it is important to identify that the August 27th directive outlined the German Commander-in-Chief's overall strategy to conclude the campaign in the West. Despite possessing a significantly weakened right wing, Moltke believed he had the sufficient strength to finish off the western allies. As before, the destruction of the allied armies in the open field remained the primary objective. In Moltke's view, this could only be obtained with an aggressive and relentless pursuit that would allow his right wing to attack and envelop the allied left before it was safely anchored to Paris.[300] To that end, the German center and right wing armies were directed to continue their advance in a southwesterly direction with instructions to turn south once it was time for the final battle to occur.

Second and Third Armies After the Battle of Namur

German Second and Third Armies continued their pursuit of the defeated French forces from the 26th–28th without success. Bülow's Second Army advanced in a southwesterly direction each of the three days, keeping them in almost constant contact with Lanrezac's rear-guards. By the evening of the 28th, the Germans had caught up and made contact with significant French forces along the Oise River between Hirson and Guise.[301] To Bülow's left, the Saxon Third Army continued its southerly advance over the same three day period. This created a gap between Second and Third Armies that gave Second Army HQ cause for concern. Nonetheless, Hausen and his staff desired to maintain the option to assist Fourth Army, which was struggling to cross the Meuse during its pursuit of French Fourth Army. Throughout the 27th, Third Army HQ received requests from both Second and Fourth Army HQs asking for assistance. With OHL's backing, Hausen ultimately decided to turn his columns to the southwest to aid Second Army. However, the morning of the 28th brought desperate pleas from Fourth Army HQ for Third Army's intervention on its right wing. The tone of these requests left Hausen and his staff no choice but to commit their forces toward the south in order to help the left hand neighbor. Consequently, Second Army was left alone to face French Fifth Army on the 29th and 30th.[302]

The Second Army *Fliegertruppe* were highly active during the three day pursuit. Aviators from FFAs 1, 18, and 23 supplied Bülow's headquarters with key information on French Fifth Army's whereabouts throughout the 26th. More specifically, FFAs 1 and 18 kept watch over the French columns in their corps' sector while FFA 23's aircraft were used to establish the location of Lanrezac's left flank. According to these reports, the allied columns were marching southwest in the area of Oisy (south of Landrecies)—Fourmies—Hirson—Guise. That evening, having reviewed the intelligence from his airmen and cavalry, Bülow decided upon moving forward with a plan to "gain the left flank of the enemy by pursuit, overtaking him to the right." In other words, Second Army HQ planned to aggressively push its right wing forward to turn French Fifth Army's left—driving them back to the southeast away from Paris.[303]

Strong winds and heavy rain grounded Bülow's airmen throughout the morning and afternoon of the 27th. Late in the afternoon however, FFA 21's *Oblt*. Behn braved the elements in his LVG B.I to reconnoiter the French positions south of the Oise. Determining the strength of the French troops along the river was significantly important to Second Army HQ, which wanted to rapidly pursue Lanrezac's forces across the river but was unwilling to do so if the southern bank was strongly defended. Flying as low as 2,500 feet to avoid low-lying clouds, Behn's aircraft was repeatedly subjected to heavy ground fire while the observer, *Oblt.* Count von Oriola, closely studied the terrain along the river's southern bank from Etréaupont (four miles west of Hirson) to Guise. After repeatedly circling the area, Oriola noted that the region was almost entirely devoid of French troops. Indeed, only in the vicinity of the sector's most important bridges were some rear-guards discovered. Upon returning to the section's airfield at Barbençon (located far in the rear-area to the east Beaumont), Oriola hastily penned his report, which ended with the phrase: "Conclusion: In the Oise sector there are only weak rear-guards," onto two index cards and presented them to FFA 21's commanding officer—*Hptm.* Franz Geerdtz. Recognizing the overwhelming importance of Oriola's observations, Geerdtz summoned a motorcycle courier to take the report directly to Second Army HQ at Avesnes—a distance of just over 18 miles. Curiously, the messenger inexplicably failed to reach Bülow's headquarters until the following morning.[304] While there has never been a formal investigation into the matter, it is highly likely that the motorcyclist became lost as he attempted to navigate the hostile countryside in the dark by himself. Nonetheless, as a result Second Army HQ remained ignorant of the true French strength along the Oise. Forward elements from the Guard Korps had suggested that there were large numbers of French troops defending the southern bank and had reported to Second Army HQ that "strong opposition was likely."[305] Thus, Bülow ordered his left wing corps on the 28th to "hold itself in readiness" north of the river until his right wing was fully concentrated and had captured St. Quentin.

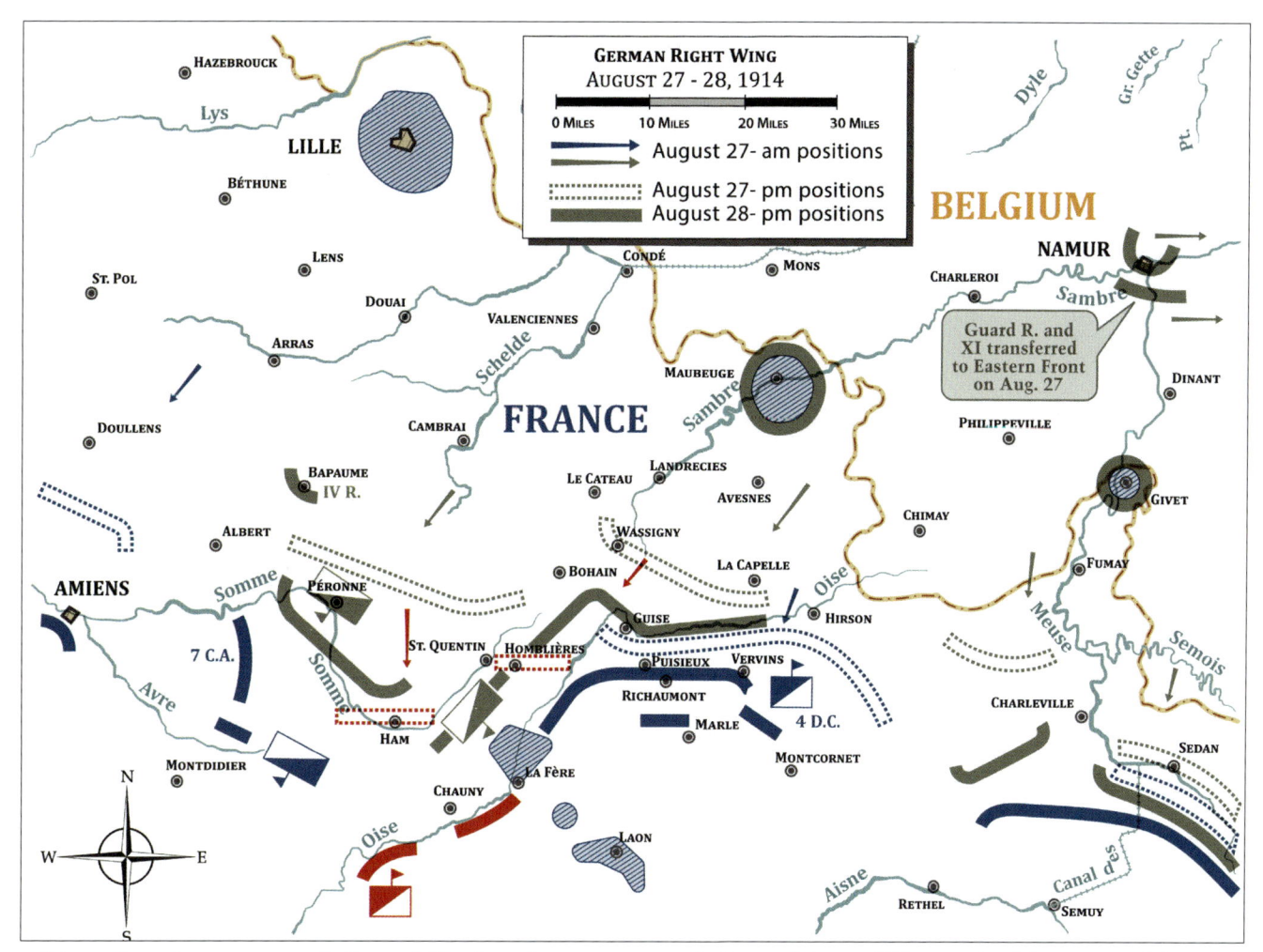

FFA 21's report finally reached Second Army HQ at 8:00am on the morning of the 28th. Within an hour Bülow dispatched a counter-order directing his left wing corps to cross the Oise and take full possession of the southern bank:

Second Army HQ – Order #10
Dispatched from: South Exit of Oisy, 28 August, 9:00am

To: X Reserve Korps, X Armee Korps, Guard Korps
A fliers' report states enemy behind the Oise are weak in numbers. X RK will move forward with its right column over Croix-Fonsomme to Fonsommes with its left column to Martigny-Carotte.

Reconnoitering desired in southerly direction by aviators and cavalry over the Oise sector between Thenelles and Vadencourt.

Reconnaissance results to be immediately reported to Second Army intelligence office at Étreux.

X AK and Guard Korps will seize the Oise sector. X AK will advance through Guise on Courjumelles and through Flavigny on Landifay. Guard Korps will secure its left flank, moving with its right column over Wiege-Faty on Sains-Richaumont and with its left column according to own choosing.

As soon as the Oise sector has been seized, reports are requested to be immediately sent to me at Étreux.

Army HQ today, Étreux.

(Signed)
Generaloberst Karl von Bülow

Bülow's counter-order was promptly dispatched to the army's left wing corps, who took possession of the Oise southern bank after minor combats with the French rear-guard. At the same time, the right wing (VII AK and X RK) moved south of St. Quentin. Heavy rains kept the majority of Second Army's airmen grounded on the 28th. Because the day had been originally planned as a day of rest, the flying sections spent the morning and afternoon relocating to new airfields directly behind the army's front line. On the right wing, FFA 18 moved from Avesnes to a new airfield at Bohain while FFA 23 relocated to Étreux. Meanwhile, FFA 1 and 21's new airfields were established northeast of Guise and La Capelle respectively. These relocations were sorely needed. In the case of FFA 1, the section's airfield had remained at Cerfontaine (west of Phillipeville), some 31 miles behind their corps' divisions.

Unfortunately, the lack of aerial activity on the 28th had kept Second Army HQ completely in the dark as to French Fifth Army's latest movements. Unbeknownst to Bülow and his staff, the delay caused by FFA 21's lost report on the evening of the 27th had allowed Lanrezac to freely redeploy his forces on the 28th in preparation for a large-scale counter-attack that was to be launched the following day. Had Second Army's left wing aggressively pushed across the Oise early in the morning, their lead units would have likely discovered the French army's unusual activity and then reported it to Second Army HQ. Armed with the knowledge that Lanrezac had halted his retreat, Bülow could have prepared his forces to fight a general engagement. Instead, an unsuspecting Second Army was surprised by a powerful French counter-offensive that had the potential to break the army in two.

Above: *Hptm.* Franz Geerdtz, commander of FFA 21 (Courtsey of Clemens Pongratz)

The Battle of Guise-St. Quentin

Joffre's General Order No.2 had placed great faith on the allied left wing's ability to retreat in good order until the time had arrived to deliver the counter-offensive. Together, French Fifth Army, the BEF, and the French territorial divisions on the British left were to avoid making unnecessary contact with the Germans while they continued their withdrawals and bought time for the formation of the new French Army on the British left. Once this new army was created and in position, it was expected to act in concert with the rest of the left wing armies to deliver a devastating blow against German First Army's flank and rear. Joffre understood that the success of this plan was predicated upon each of his left wing armies simultaneously delivering attacks with all the available troops at their disposal. The British defeat at Le Cateau however, had shattered all hope of the BEF being able to participate in the attack, which was originally planned to be launched from the direction of Amiens. This compelled Joffre to postpone his grand counter-offensive and order the retreat towards Paris to continue. However, with German First and Second Armies rapidly closing the gap between themselves and the forces on the allies' extreme left wing, Joffre felt it was necessary to order Lanrezac's Fifth Army to temporarily halt their withdrawal in the area of St. Quentin—La Fère and launch a determined counter-attack against the pursuing Germans' flank. This action would, in Joffre's mind, give the badly damaged BEF as well as the newly formed French Sixth Army a chance to distance themselves from the Germans and reorganize themselves for the coming offensive.

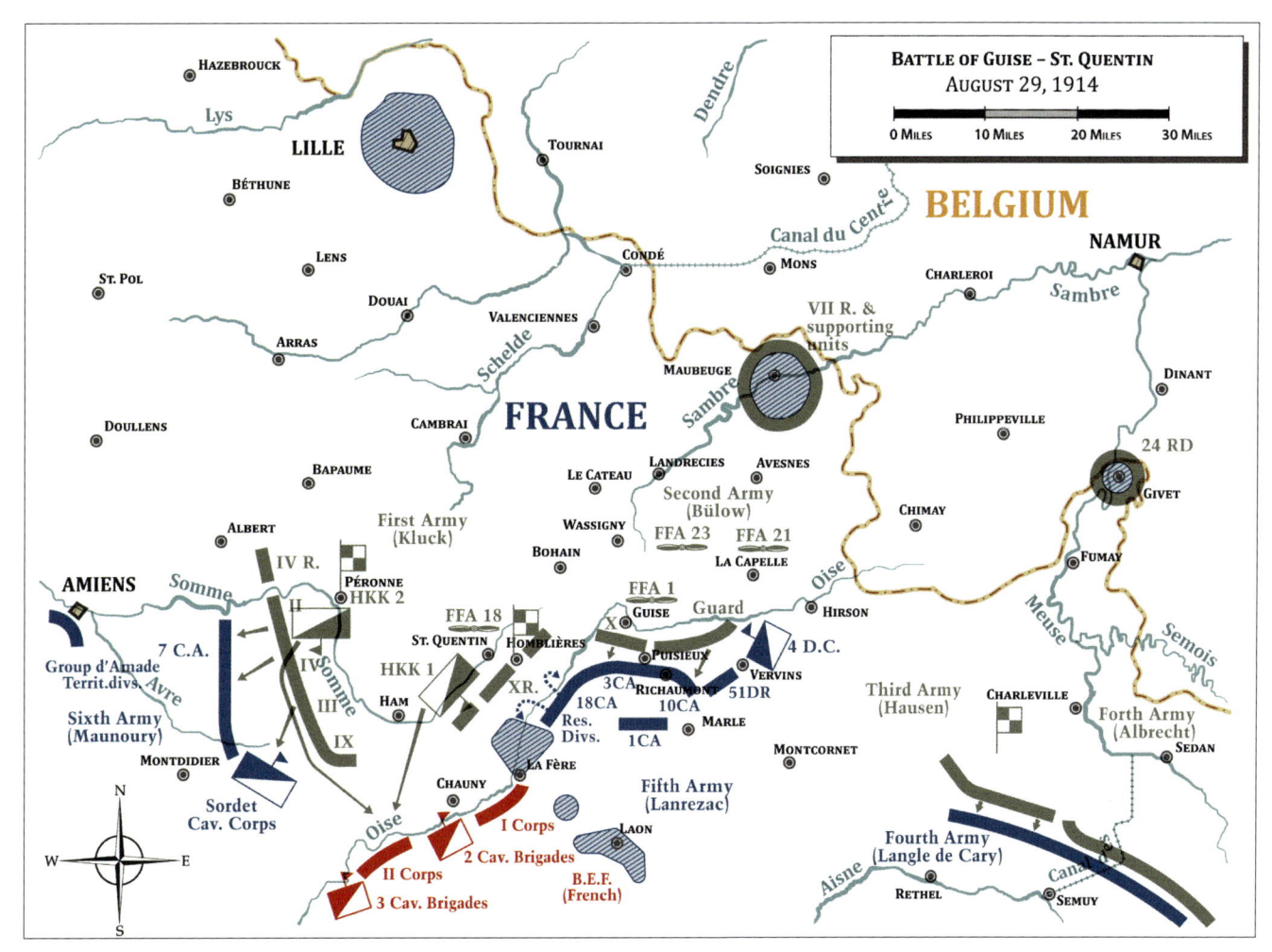

Joffre informed Lanrezac of the new plan by sending a staff officer to Fifth Army HQ at Marle on the afternoon of the 27th. The maneuvers required by the order were incredibly risky, involving halting the retreat, turning all of Fifth Army around, and realigning it to carry out an offensive towards the west. The dangers of executing these movements were increased by the fact of German Second Army's close proximity. In light of these issues, Lanrezac and his staff were strongly against Joffre's plan. Nonetheless, despite disagreeing, Fifth Army HQ faithfully obeyed their superior's wishes and began issuing orders to each of their corps to begin the complex redeployment on the morning of the 28th.[306]

Fortunately for the French, the loss of FFA 21's aerial report had kept the Guard and X AK from pushing south of the Oise on the 28th and making contact with Fifth Army while it was carrying out its redeployment. Lanrezac's forces were therefore able to complete their movements without major incident and were in position by nightfall to launch their attacks on the morning of the 29th.[307] In accordance with Lanrezac's orders, the X CA and a division of cavalry were deployed south of the Oise facing north in order to cover the army's right flank. Meanwhile facing west, the III and XVIII CAs as well as Valabreuge's exhausted reserve division group were to cross the Oise and attack towards St. Quentin. Finally, the I CA was deployed as army reserve in a position to move either north or west as circumstances required.[308]

Lanrezac's counter-offensive began as planned at

dawn on the morning of the 29th. With the element of surprise, the III and XVIII CAs pushed west towards St. Quentin while X CA held its ground opposite the Guard Korps. Heavy fog had prevented aerial activity until 9:00am, leaving the German leadership blind to the scale of the French attack throughout the morning. Bülow and his subordinate commanders had anticipated that the French would continue their retreat on the 29th and were shocked to learn that the army's lead units were facing heavy resistance. At that moment, the army was split into two groups (X RK and VII AK on the right and X AK and the Guard on the left) with a large nine mile gap between them. Located between St. Quentin and Guise, this breach in the German line immediately became a source of concern for Second Army HQ upon receiving word that "X RK was being attacked from a southeasterly direction by forces of some importance."[309]

Without aerial reconnaissance to bring clarity to the situation, Second Army HQ was compelled to manage the crisis without any idea of the attacking French force's strength. First, Bülow quickly recognized the need to close the gap between his army's two wings. Thus, from a hillside near Homblières, the army commander personally redirected VII AK's 13 ID (on its way from Maubeuge to rejoin the army) to deploy on X RK's left and plug the breach. Meanwhile, troops from X AK and the Guard reported making contact with French forces south of Guise. In a series of methodical attacks, these two corps slowly pushed the French back from one position to the next, incurring heavy casualties in the process. At the center of one of these attacks, the 1st Regiment of the Guard, commanded by Prince Eitel Friedrich of Prussia, was able to seize the important heights around Richaumont. During the attack's most critical moment, the Kaiser's second son seized a drum and began beating on it amidst heavy rifle fire to rally his men. With the situation restored, the prince personally led the final attack to capture the heights.

The Guard Korps' successful counter-attack caused the French X CA's commander, *Général* Gilbert Defforges, to send a plea for assistance to Lanrezac:

I am violently attacked on my whole front. They are getting around my right flank. I will hold at all costs. Give me support as soon as possible on my right and my left.[310]

Lanrezac granted Defforges wish by sending the I CA to his sector. The decision could have proved decisive but I CA's commander, Franchet d'Esperey, wasted several hours carefully deploying all of his artillery in preparation for his attack. As a result, d'Esperey's thrust did not begin until 6:00pm. With his bands playing *La Marseillaise,* d'Esperey and his staff personally led the attack against the Guard positions at Richaumont. Within an hour I CA had recaptured the town and had pushed the German lines back to the north. Although the attack was successful, the long delay in deploying the artillery had caused d'Esperey's men to run out of daylight. Consequently, the French were obliged to halt their advance and take up positions along the newly captured heights.[311]

To the west, the III and XVIII CA's attack against Bülow's right wing failed. Before crossing the Oise, III CA's right flank was struck by troops from the X AK, compelling the corps' commander to turn his divisions to the north to meet the threat. Thus, the XVIII CA was left to carry out the attack against Bülow's right by themselves. In a morning of brutally heavy fighting, X RK's heavily outnumbered 19 RD skillfully blunted the French attacks. Later that afternoon, X RK's other division (2nd Guards RD) arrived along with four heavy howitzer batteries. Together, these fresh forces allowed X RK to launch a successful counter-attack against the French left flank. Threatened with encirclement, XVIII CA's commander, *Général* Jacques de Mas Latrie, ordered a general retreat back across the Oise.

The morning's thick fog burned off around 9:00am, allowing flight operations to commence. FFA 1 and 21 on the army's left wing conducted the first flights of the day. Both sections' commanders ordered their airmen to determine the strength of the French forces in their corps' sector and promptly report back to corps headquarters. With the opposing forces heavily engaged in close proximity, the task of carrying out any tactical reconnaissance was extremely difficult. Nevertheless, FFA 1's airman returned a report to Guard Korps HQ at 2:00pm estimating three enemy corps were engaged with X AK and the Guard. FFA 21's report detailed the deployment of French artillery in III CA's sector. On the army's right, VII AK's FFA 18 was unable to provide any assistance because the section was already in road column moving forward to their new airfield at St. Quentin when the heavy fighting began.

During the course of the day's combat, Guard Korps HQ had become concerned of a possible French attack against their left flank, which was left uncovered by the gap that had opened between the Second and Third Armies. Accordingly, FFA 1 sent an aircraft to observe the area south of the corps' left wing. The airmen proceeded to diligently reconnoiter the area Vervins—Montcornet—Marle. The subsequent report was delivered to Guard Korps HQ at 4:45pm stating that the: "roads from Vervins—Montcornet and to the east of this line are free. Large columns observed from Montcornet—Ebouleau [marching in the direction of

Above: Kondor *Taube* Type H photographed at the Kondor factory. It was powered by a 100 hp Mercedes. The factory name on the roof has been removed by retouching on the original photograph. (Aeronaut)

Laon]." This report confirmed that the French were not moving forces against Second Army's left but were in fact marching southwest *away* from the St. Quentin—Guise battlefields.[312]

After spending the morning and afternoon in the field at his forward command posts directing the battle, Bülow returned to Second Army HQ at Homblières around 5:00pm to issue the following day's orders. Both Bülow and his staff believed that the army would have to continue the attack the following morning to attain a favorable decision. However, considering Lanrezac's surprising decision to attack on the 29th, the question remained whether the French would bring up fresh forces during the night to renew the offensive and if so where? This question was critically important for Bülow, who had to decide where to focus his army's *Schwerpunkt* (focal point) for the following morning's attack.[313] While the issue was being debated, two aerial reconnaissance reports from FFA 23 were produced that brought clarity to the situation. The first described "strong troop concentrations in the area of Sons-et-Ronchères—Châtillon—Bois-lès-Pargny and at Crécy. Weaker concentrations at Mortiers and Dercy; also at least one brigade spotted at Marle." The second report, describing road traffic further south, stated:

A long column of artillery and wagons on the road: Bucy—Sissone; a long column of infantry and artillery on the road: Ébouleau—Pierrepont; near Sissonne much military activity was recognized. The roads: Montcornet—Reims and Corbeny—Reims are empty.[314]

Although the reports did not definitively prove Lanrezac's intentions, they led Bülow to believe that the French were willing to accept battle the following morning by possibly bringing forward the various columns observed by his airmen. Second Army HQ therefore dispatched orders for the battle to be continued at daybreak.

Bülow was proven correct. At dawn, elements of I CA, III CA, and X CA attacked the formidable defensive positions taken up by the X AK and the Guard. The French attacks were launched in a disjointed manner that enabled the Germans to repulse each of the attacks with ease. After the initial attack failed, the French began an intense artillery bombardment that was thought to be in preparation for a second attack. The German artillery answered the French cannonade with counter-battery fire—ushering in one of the largest artillery duels of the war's opening campaign. The artillery battle lasted throughout the morning without the second French attack ever materializing. Once the French artillery fire slackened and it had become clear that they were not going to attack again, Bülow's Second Army prepared to launch the attack they hoped would bring the battle to a victorious conclusion.

From his headquarters at Laon, Lanrezac fully understood the danger his army faced on the evening of the 29th. Thus, once the result of the day's battle had become obvious, he telephoned Joffre's headquarters seeking permission to order a retreat. Despite failing to breakthrough Second Army's line, Joffre believed Fifth Army's attack had successfully bought enough time to organize the new French Sixth Army on the Allied battle-line's left wing. Therefore, Lanrezac's request to initiate Fifth Army's retreat at dawn on the 30th was granted. However, due to poor staff work,

Above: German infantry storm Fort Loncin during the Battle of Liège. (Author's Collection)

Joffre's order failed to reach Fifth Army HQ until late in the morning. As a result, meaningless attacks were launched by I, III, and X CAs throughout the early morning hours, resulting in unnecessary casualties. Once the order was finally received, Lanrezac's men disengaged and began a retreat under the cover of massed artillery fire.

After answering the French cannonade with a lengthy preparatory bombardment, infantry from X AK and the Guard attacked the French positions at 1:00pm only to find Lanrezac's rear-guard. After receiving confirmation of the French withdrawal, Second Army HQ ordered a general advance and pursuit. On the army's left, the X AK and Guard Korps met almost no appreciable resistance as they easily threw back the French rear-guard and began to push south after French Fifth Army's main body. Likewise, VII AK and X RK spent the remainder of the day crossing the Oise and fighting rear-guards that were covering French Fifth Army's left wing. Second Army's advance was ultimately brought to a halt due to the troop's exhaustion. This allowed French Fifth Army to slip away late in the afternoon and put some distance between themselves and their pursuers.[315]

For a second straight day, a heavy fog followed by thick haze grounded all aerial activity until early afternoon. The first flights of the day took off after noon with orders to determine the direction of Lanrezac's retreat. Despite the late start, Bülow's airmen were able to successfully determine the direction of the French withdrawal. Performing two consecutive flights, FFA 23's *Ltn.* Ernst Leonhardi confirmed that French Fifth Army was indeed conducting a general retreat. The first report, delivered to Second Army HQ shortly after 2:00pm, stated:

- *Three long columns marching to the south:*
    —At Villers-le-Sec toward Crécy
    —At Monceau-le-Neuf toward Crécy
    —At Marle toward Laon

- *A covering force of two columns was observed advancing (northwest) between Fort de Mayot and Renansart.*

- *In the area Renansart-Origny only weak detachments remain.*[316]

Incredibly, despite similar messages from troops on the ground, Second Army HQ was unwilling to believe Leonhardi's initial report that Lanrezac's entire army was in retreat. Thus, on his own initiative, Leonhardi performed a second flight, bringing back a report at 4:00pm that described the concentration of one division at Renansart as well as a long column retreating from Monceau-le-Neuf toward La Fère.[317] Based upon these two reports, Second Army HQ finally issued pursuit orders to its corps commanders at 4:45pm. Later that evening, Second Army HQ received messages from its corps flying sections which confirmed that the retreat was in the direction of Laon—Crépy—La Fère. FFA 18's communique read: "all roads east of the Oise and north of La Fère, retreating movements in mostly southern and southwestern directions." FFAs 1 and 21 reported the withdrawal of infantry and artillery columns from west of Marle and from Vervins toward the south respectively. Finally, at 8:45pm, another observer from FFA 23 delivered the following message:

In my opinion the retreat is being carried out in good order, columns are rarely broken. At no point were columns next to one another except in the streets of Marle. South of the Serre Sector there are no entrenched works. No rail traffic observed on the lines Marle—Laon and Montcornet—Laon.[318]

After receiving overwhelming evidence from his aviators that the French had begun their retreat, a jubilant Bülow radioed OHL informing Moltke that Second Army had won a "complete victory" and that "the French, in the strength of four army corps and three divisions, were in full retreat." Another message sent to Third Army HQ announced that "strong French forces had been decisively defeated in a two-day battle at St. Quentin and thrown back on and east of La Fère."[319]

While Bülow's communiqués correctly declared the Battle at St. Quentin to be a victory, it was not a "complete victory" as he claimed. Rather, it was an "ordinary" tactical success that did little to change the overall situation. As each army commander was aware, the German War Plan was predicated upon a quick and decisive victory featuring the destruction of a significant portion of the allied armies in the field. Thus, the continuation of the allied withdrawal in the aftermath of the battle was effectively a defeat by postponing another general engagement. Victory in the West was not possible until the German right wing armies could once again draw the Allied left into battle under favorable conditions—a feat that would require the *Fliegertruppe* to furnish the army's leadership with the necessary intelligence to maneuver their respective commands into position to deliver a knockout blow. Fortunately for the Germans, the first month of the war had proven that their aviators were up to that task.

Chapter 2 Endnotes

1. Ernest von Hoeppner, *Germany's War in The Air* (Nashville, 1994), p.16.
2. For an excellent historiography of the development of the German War Plan, see: Terence Zuber, *The Real German War Plan 1904–1914* (Gloucestershire, 2012).
3. Reichsarchiv, *Der Weltkrieg 1914–1918* (Berlin, 1925), Band I, pp.80–82.
4. There are a lot of myths and misconceptions regarding the German War Plan in 1914. Popular histories and standard textbooks alike have often described it as "The Schlieffen Plan" in reference to Alfred von Schlieffen's famous *Denkschrift* of December 1905, which resembled the deployment scheme of 1914. "The Schlieffen Plan," as described by these "popular historians," was centered around a grand encircling movement that would feature the German right wing armies marching past the western side of Paris to the lower Seine River. Paris, according to the myth, was the objective of the plan and therefore the objective of the German offensive in August 1914.
The truth is much more complex. First, Schlieffen's famous memorandum called for a substantially larger force than the Germans had. Second, it was originally written for only a one front war with France. While it did indeed feature a strong German right wing marching through Belgium and around Paris, it was Schlieffen's own opinion that encircling Paris should only be done as a last resort. Indeed, his true intention with the march through Belgium was to draw out and annihilate the French army in the open field. Moltke, as Schlieffen's successor, agreed that the plan's primary objective was the destruction of the French forces. He also used the deployment scheme as a basis to form his own plan. He famously disagreed, however, that an overwhelmingly strong right wing, as Schlieffen prescribed, was necessary given the credible possibility of a French offensive against the German left in Lorraine. Thus, beginning in 1908, Moltke began to reduce the superiority of the right wing from Schlieffen's 7:1 to 3:1 in order to check the perceived threat to his left wing. By 1914, the plan could no longer be reasonably linked to Schlieffen's original memorandum. Nevertheless, neither Moltke nor Schlieffen's concept of operations called for a dogmatic plan for a "march on Paris" as is often claimed. On the contrary, despite the many changes from its

birth in 1905 to its implementation in 1914, the German War Plan remained a flexible document dedicated to the prompt destruction of the allied armies in the field. For an excellent analysis of the plan's evolution, see: Terence M. Holmes, "The Reluctant March on Paris: A Reply to Terence Zuber's 'The Schlieffen Pan Reconsidered,'" *War in History*, Volume 8, Issue 2 (2001), pp.208–232; Holmes, "Asking Schlieffen: A Further Reply to Terence Zuber," *War in History*, Volume 10, Issue 4 (2003), pp.464–479.

5 In this case, the right wing armies were to make a much shallower wheel through Belgium, marching down the Meuse to attack the French main body in the flank and rear.

6 Zuber, *The Real German War Plan 1904–1914*, pp.145–146.

7 Zuber, *The Real German War Plan 1904–1914*, pp.144–145.

8 Mark Osborne Humphries & John Maker (ed.), *Germany's Western Front 1914*, Part 1 (Waterloo, 2013), p.123.

9 Upon the declaration of war, Moltke's title was renamed from "Chief of the General Staff of the Prussian Army" to "Chief of the General Staff of the Field Army," which had the authority over the units of all the German States.

10 Hermann Cron, *Imperial Germany Army 1914–18: Organization Structure, Orders of Battle* (Eastbourne, 2001), pp.13–15.

11 Dirk Rottgardt, *German Armies' Establishments 1914/18*, Vol.1 (West Chester, 2009), pp.18–34; Holger Herwig, *The Battle of the Marne: The Opening of World War I and the Battle That Changed the World*, (New York, 2011) p.46.

12 Cron, *Imperial Germany Army 1914–18*, pp.95–98; Rottgardt, *German Armies' Establishments 1914/18*, Vol.1. pp.34–38.

13 Rottgardt, *German Armies' Establishments 1914/18*, Vol.1. pp.32–33, pp.50–54.

14 Elard von Loewenstern, *Mobilmachung, Aufmarsch und erster Einsatz der deutschen Luftstreitkräfte im August 1914* (Stuttgart, 1939) p.2.

15 Ian Sumner, *German Air Forces 1914–1918* (Long Island City, 2005) p.5.

16 Numbers contained in the official *Frontbestend 1914*. Ferko–UTD, Box 20, Folder 5, "Frontbestend 1914."

17 James McCudden, *Flying Fury*, (London, 1933) p.36.

18 Bergmann served as First Army's Quartermaster/Chief Operations Officer—also known as a: "Ia." Bergmann wanted to place the airpark at an old factory in order to give all the unit's carpenters, machinists and mechanics the ideal setting to perform the major repairs that they were often required to do. Hermann von Kuhl and Walther von Bergmann, *The Supply and Movement of German First Army During August & September 1914*, The U.S. Command and General Staff School Press (Fort Leavenworth Kansas, 1929) p.158.

19 Sumner, *German Air Forces 1914–1918*, p.5.

20 In theory, the additional 13 gas wagons were given to the section to guarantee that a sufficient quantity of gas would always be available at the front. While one column of gas wagons was refilling at the AFP, the other column would be with the unit at the front. Oliver Richter, *Feldluftschiffer* (Erlangen, 2013), pp.5–6.

21 Loewenstern, *Mobilmachung*, pp. 2–4; Imrie, *Pictoral History of the German Air Service*, pp.21–22.

22 The Army officially possessed nine dirigibles at the outbreak of war. Three private Zeppelins were also requisitioned bringing the potential total to 12. However, two of the army's prewar ships and two of the civilian ships were declared unfit for service. John R. Cuneo, *The Air Weapon 1914–1916* (Harrisburg, 1947), pp.12–13.

23 Cuneo, *The Air Weapon 1914–1916*, pp.14–15.

24 The guns were assigned by army as follows: Second Army–1, Fifth Army–1, Sixth Army–1, Seventh Army–2, Eighth Army (Eastern Front)–1. Hoeppner, *Germany's War in the Air*, p.28.

25 The horse-drawn guns were 7.7cm's while the stationary platforms were equipped with the old 9cm field gun. Cuneo, *The Air Weapon 1914–1916*, p.198.

26 The ability to pursue a defeated army and not allow it to recover was the key to a decisive victory in 1914. As we will see, although the Germans were able to win every meeting engagement, it proved very difficult to overcome rear-guards that employed large numbers of quick firing French 75's.

27 Loewenstern, *Mobilmachung*, p.1.

28 Document #82 "Weisung des Generalstabes des Flugzeuge im heeresdienst" can be found in: Kriegswissenschaftliche Abteilung der Luftwaffe, *Die Militarluftfahrt bis zum Beginn des Weltkrieges 1914*, Band III, pp.187–190.

29 The contemporary German definition of "strategic reconnaissance" in this context meant "operational reconnaissance" in today's parlance. Simply put, there were no systematic definitions for strategic, operational and tactical operations. Therefore, to avoid confusion, this text will endeavor to replace the term strategic reconnaissance with "long-range reconnaissance."

30 See: Jack Snyder, *The Ideology of the Offensive-Military Decision Making and the Disasters of 1914* (Ithaca, 1984).
31 Terence Zuber, *The Mons Myth* (Gloucestershire, 2010), p.14.
32 Frank Buchholz, Joe Robinson and Janet Robinson, *The Great War Dawning* (Vienna, 2013), pp.211–215.
33 Terence Zuber, *Ardennes 1914* (Gloucestershire, 2009), pp.50–73.
34 Martin Samuels, *Command or Control? Command, Training and Tactics in the British and German Armies, 1888–1918* (London, 1995), p.82.
35 Fire included machine gun and artillery fire, but principally it meant rifle fire. Zuber, *Ardennes 1914*, p.32.
36 Zuber, *Ardennes 1914*, p.51.
37 Frank Buchholz, Joe Robinson and Janet Robinson, *The Great War Dawning*, p.216.
38 Zuber, *Ardennes 1914*, pp.51–52.
39 Zuber, *Ardennes 1914*, pp.44–48.
40 Zuber, *Ardennes 1914*, pp.25–26.
41 Frank Buchholz & Joe and Janet Robinson, *The Great War Dawning*, pp.233–241.
42 Herwig, *The Battle of the Marne*, p.48.
43 Hug, "Carrier Pigeon Flieger," pp.299–300.
44 As translated in: Peter Kilduff, *Iron Man: Rudolf Berthold: Germany's Indomitable Fighter Ace of World War I* (London, 2012), p.31.
45 The first four sections to become operational were on the left wing where the flying sections had been stationed in peacetime. Thus, with very short distances to travel, these four sections were able to become quickly activated. These were: Sixth Army HQ's FFA 5, Seventh Army HQ's FFA 26 and two corps sections—FFA 20 and 3—belonging to Seventh Army's XIV AK and XV AK respectively. Cuneo, *The Air Weapon 1914–1916*, pp.29–31; Ferko–UTD, Box 20, Folder 5, "Mobilization of the Fliegertruppe."
46 During this period, it was found that a number of the older aircraft—particularly *Taubes*—were unable to safely operate under wartime conditions. In other instances, ground crews were reported having difficulties assembling some of their aircraft, which caused delays in getting the flying section fully operational. These occurrences were rare however, with most FFAs quickly reporting fully operational after the arrival of all their men, machines and equipment. Cuneo, *The Air Weapon 1914–1916*, pp.16–17.
47 Robert A. Doughty, *Pyrrhic Victory* (Cambridge, 2008), pp.22–24.
48 Zuber, *The Real German War Plan 1904–14*, pp.152–153.
49 Sewell Tyng, *The Campaign of the Marne*, (Yardley, 2007) pp.26–27.
50 Zuber, *The Real German War Plan 1904–14*, p.155.
51 Zuber, *The Real German War Plan 1904–14*, p.155
52 Herwig, *The Battle of the Marne* (New York, 2011), p.56.
53 For more on the development of offensive doctrine in the pre-war French Army, see: Douglas Porch, *The March to the Marne: The French Army 1871–1914* (Cambridge, 2003), pp.213–231.
54 Doughty, *Pyrrhic Victory*, p.26.
55 Herwig, *The Battle of the Marne*, (New York, 2011) p.61.
56 Tyng, *The Campaign of the Marne*, pp.362–364; Zuber, *The Real German War Plan 1904–14*, p.155.
57 Tyng, *The Campaign of the Marne*, p.52.
58 Calling the defenses around Liège "formidable" would be an understatement. The construction of the field works in between and around each of the forts had been fully completed in the final days of July. These positions included steel and wooden spikes, barbed wire entanglements and redoubts for both the infantry and field artillery. Telephone lines were also laid to facilitate communication between the troops in the trenches and the garrisons inside the various forts. Clayton Donnell, *Breaking the Fortress Line* (South Yorkshire, 2013), pp.35–39.
59 Herwig, *The Battle of the Marne*, (Random House, 2011) pp.105–108
60 Humphries and Maker (ed.), *Germany's Western Front 1914*, Part 1, pp.96–97.
61 Donnell, *Breaking the Fortress Line*, pp. 43–54; Tyng, *The Campaign of the Marne*, p.53.
62 Herwig, *The Battle of the Marne*, p.113.
63 Some historians have suggested that this decision had the reverse effect—implying that Liège could have longer delayed the German advance if the 3rd Division's infantry was still present. This, however, is a foolish opinion considering the large number of German siege batteries that were ultimately used in the fortress' reduction.
64 A concise summary of the siege at Liège is in: Donnell, *Breaking the Fortress Line*, pp.33–66; For a complete narrative of the battle with a focus at the tactical level see: Terence Zuber, *Ten Days in August: The Siege of Liège 1914* (Gloucestershire, 2014).
65 Donnell, *Breaking the Fortress Line*, pp.57–66.
66 Many popular historians including the authors of the British Official History have opined that the Belgian resistance at Liège significantly delayed and therefore heavily impacted the German right wing's future operations. However, the participants of the

campaign, both Allied and German, say otherwise. In their post war writings, von Kluck, von Bülow, and von Kuhl all say the Belgian resistance at Liège had *no impact* on the course of the campaign. Even General Dupont, the Chief of the French Intelligence Service, was compelled to concur. The simple truth is that the right wing armies wouldn't have even been fully concentrated and ready to begin an advance until the 14th at the earliest. As it happened, First Army crossed the Meuse on the 14th and moved into assembly areas on the Meuse western bank on the 15th–16th while the final forts were being bombarded. On the 18th, the right wing confidently began their advance behind a massive cavalry screen consisting of HKKs 1 and 2. Zuber, *The Real German War Plan 1904–14*, p.155; Tyng, *The Campaign of the Marne*, p.58; Humphries and Maker (ed.), *Germany's Western Front 1914*, Part 1, pp.106–107.
67 Loewenstern, *Mobilmachung*, pp.14–15.
68 Both of Fest.FA Cologne's aircraft were airborne for four hours on their respective flights. Loewenstern, *Mobilmachung*, p.14.
69 The ship's payload comprised of just four 50 kilogram projectiles. It seems, knowing this fact, that the attack was meant to be more psychological than anything else. Cuneo, *The Air Weapon 1914–1916*, pp.14–15.
70 Ernst A. Lehmann, *Zeppelin: The Story of Lighter Than Air Craft* (Charleston, 2015), p.48; Cuneo, *The Air Weapon 1914–1916*, pp.14–15; Humphries and Maker (ed.), *Germany's Western Front 1914*, Part 1, p.102.
71 Tyng, *The Campaign of the Marne*, p.56.
72 Ferko–UTD, Box 11, Folder 3, "FFA 9 notes."
73 Adolf von Schell, *Battle Leadership* (Columbus, 1933), p.21.
74 Ferko–UTD, Box 20, Folder 5, "Frontiers Battles, 1914"
75 Loewenstern, *Mobilmachung*, pp.14–15.
76 Loewenstern, *Mobilmachung*, p.15.
77 Loewenstern, *Mobilmachung*, pp.15–16.
78 Ferko–UTD, Box 20, Folder 5, "Frontiers Battles, 1914"; Box 11, Folder 3, "FFA 18"; Box 11, Folder 3, "FFA 21."
79 Egmont F. Koenig, *The Battle of Morhange-Sarrebourg 20 August 1914* (USCGSC Student Paper Series, 1933), p.7.
80 Herwig, *The Battle of the Marne*, p.76.
81 Karl Deuringer, *The First Battle of the First World War: Alsace-Lorraine*, Translated and edited by: Terence Zuber, (Gloucestershire, 2014), pp.32–33.
82 Herwig, *The Battle of the Marne*, pp.76–77.
83 Seventh Army HQ's plan was extremely significant. By deploying the entirety of his only two active corps for a counter-offensive at Mulhouse, Heeringen was completely abandoning his pre-determined deployment plan which had called for both of these corps to complete their concentrations and take up a defensive stance on the Rhine River's east bank in order to cover Sixth Army's left flank. By prematurely throwing his only two active corps into battle, Heeringen ultimately put Seventh Army at a disadvantage for the more important battles that happened a week and a half later on August 20th. Deuringer, *The First Battle of the First World War*, p.33.
84 *Der Weltkrieg*, I, pp.160–164.
85 Landesarchiv Baden–Württemberg, Generallandesarchiv Karlshrue (hereafter: GLA), 456 F1, Nr. 533, "Aufstellung über Operationen der 7. Armee vom 3.8. bis 16.9.1914 (Reinschrift mit Prüfungsvermerken)."
86 All three of Seventh Army's flying sections had been stationed in peacetime within a very close proximity to their initial deployment areas. Seventh Army HQ's FFA 26 and XIV AK's FFA 20 were both stationed at Freiburg while XV AK's FFA 3 was located at Strassburg-Neudorf. Even Seventh Army's Air Park—AFP 7—was stationed in Freiburg in peacetime. Thus, by being already in the Alsace-Lorraine region, the three flying sections as well as the army's air park were able to quickly move all the necessary equipment and personnel into position to begin active operations. GLA, 456 F5, Nr. 60, "Mobilisierung von Fliegern und Flugzeugen."
87 GLA, 456 F2, "XIV Armee Korps."
88 Loewenstern, *Mobilmachung*, pp.22–23.
89 Fortunately for the Germans, Bonneau's troops neglected to conceal their positions around Mulhouse—specifically east and north of the city. FFA 26's observers took full advantage of this and were able to sketch out the French troops' defensive line. GLA, 456 F5, Nr. 60, "Mobilisierung von Fliegern und Flugzeugen."
90 Hug, "Carrier Pigeon Flieger," p.299.
91 An excellent overview of this entire battle can be found in the German Official History. See: *Der Weltkrieg*, I, pp.160–168.
92 GLA, 456 F1, Nr. 533, "Aufstellung über Operationen der 7. Armee vom 3.8. bis 16.9.1914 (Reinschrift mit Prüfungsvermerken)."
93 Loewenstern, *Mobilmachung*, pp.22–23.
94 Seventh Army's return to its defensive position on Sixth Army's flank was disastrous. First, XIV AK was drawn into combat with French troops in the Vosges mountains. Secondly, XV AK impatiently decided without authorization to cancel their rail transport and march overland—thus further

fatiguing the troops, who were described by Sixth Army Chief of Staff Krafft von Dellmensingen as "highly over taxed and exhausted for a number of days" in the days after their battle in Alsace. See: Dieter Storz, "This Trench and Fortress Warfare is Horrible!" *The Schlieffen Plan: International Perspectives on the German Strategy for World War 1*, edited by: Hans Ehlert, Michael Epkenhans & Gerhard P. Gross (Lexington, 2014), p.146.
95 Doughty, *Pyrrhic Victory*, pp.58–60.
96 Deuringer, *The First Battle of the First World War*, pp.33–35.
97 Loewenstern, *Mobilmachung*, p.29.
98 The troop concentrations were the French IX and XV CAs. Loewenstern, *Mobilmachung*, p.27.
99 Thanks to the airmen's messages, the commanders on the ground were able to deploy reserves to the proper areas in time to check several French attacks. Furthermore, on the 12th, a crew belonging to FFA 1b was able to cooperate with local artillery batteries by landing and providing a sketch of the French positions in the area.
100 Loewenstern, *Mobilmachung*, pp.27–28.
101 Aerial darts were made of iron and were the size and shape of a ballpoint pen. To keep the center of gravity at the point, the upper part had four grooves. These darts were thrown overboard by the handful. Falling through the air, they descended vertically and could pierce a steel helmet. After a few weeks, the darts were no longer dropped because their efficiency was so poor. See: Hug, "Carrier Pigeon Flieger," pp.299–300.
102 Loewenstern, *Mobilmachung*, pp.28–30.
103 GLA, 456 F1, Nr. 533, "Aufstellung über Operationen der 7. Armee vom 3.8. bis 16.9.1914 (Reinschrift mit Prüfungsvermerken)."
104 Storz, "This Trench and Fortress Warfare is Horrible!" p.149.
105 Herwig, *The Battle of the Marne*, p.85.
106 Storz, "This Trench and Fortress Warfare is Horrible!" p.149–150.
107 It is important to note that the actions of the other German armies would have been dramatically different if the French had sent a larger force into Lorraine. In order to cover Fifth Army's attack through Metz, Fourth Army would have been moved southwest to protect its flank and rear. The German right wing armies would have therefore taken a much shallower wheel through Belgium, moving south down both banks of the Meuse. The mere fact that the German High Command had contemplated and temporarily enacted this plan is evidence that the German War Plan was fluid and not the nonsensical "march around Paris in 40 days regardless of how the enemy is acting" Schlieffen Plan that is commonly taught by historians and cited as evidence of sinister German intentions.
108 Dellmensingen was the major decision maker at Sixth Army HQ. As a royal personage, tradition demanded that Rupprecht assume an army command at the start of the war. However, Rupprecht had very little interest in the military. Indeed, Erich Ludendorff famously retorted that the Bavarian Crown Prince was "a solider from a sense of duty," and that "his inclinations were not the least bit military." Nonetheless, while a strong sense of duty and honor ultimately moved Rupprecht to do an excellent job commanding his army group later in the war, it was quite clear in the summer of 1914 that he needed the assistance of a seasoned General Staff Officer to be his chief of staff and the major decision maker at Sixth Army HQ. Dellmensingen fit the role wonderfully by taking charge and making the overwhelming majority of the operational decisions during the opening campaign of the war. Crown Prince Rupprecht of Bavaria, *In Treue Fest: Kronprinz Rupprecht, Mein Kriegstagebuch*, 3 volumes. (Munich, 1929), Volume 1.
109 Storz, "This Trench and Fortress Warfare is Horrible!" pp.150–152.
110 The corps flying sections had to spend the late evening of the 14th and most of the day of the 15th moving through and around traffic jams caused by the army's support columns. By the time the flying sections were prepared to begin operations on the 16th and 17th, Lorraine was hit with a series of heavy thunder and rainstorms that made flying impossible. Egmont F. Koening, *Morhange-Sarrebourg*, pp.8–11.
111 Loewenstern, *Mobilmachung*, p.48.
112 Herwig, *The Battle of the Marne*, pp.89–92.
113 Elard von Loewenstern, *Mobilmachung*, pp.50–51.
114 Deuringer, *The First Battle of the First World War*, pp.116–126.
115 See the French Official History: *Les Armées Françaises dans la Grande Guerre*, 103 vols. [hereafter: FOH], Volume 1 (Paris, 1922–1938), p.257.
116 Loewenstern, *Mobilmachung*, p.54.
117 BayHStA, GK II. AK (WK) 1.8.1914–31.12.1914.
118 Loewenstern, *Mobilmachung*, p.54.
119 Although the XX CA was indeed on French Second Army's extreme left, the corps had protection from a group of reserve divisions that were deployed slightly behind them and to the west. It is unclear whether the observer saw these forces or not.
120 Ferko–UTD, Box 20, Folder 5, "Frontiers Battles, 1914"; Loewenstern, *Mobilmachung*, p.54.

121 Loewenstern, *Mobilmachung*, pp.55–56; Ferko–UTD, Box 11, Folder 5, "Bavarian Aviation, 1914."

122 The fighting in this sector began around 4:00am when VIII CA's 15th Division launched a preemptive attack that struck the I Bavarian RK at Gosselming and the I Bavarian AK at Oberstinzel. Within an hour, German counter-attacks supported by massed artillery, had quickly checked the French attack and drove them back. By 7:00am, the early French gains had been completely erased. Over the following two hours, the German artillery heavily bombarded VIII CA's front while I Bavarian AK and I Bavarian RK's infantry reorganized and prepared to attack. Then, shortly after 9:00am, the German counter-stroke began—systematically pushing the badly damaged and disorganized French troops back from one position to the next until they were expelled from Sarrebourg entirely. FLA 1b was especially active from 5:30–9:00am—first in stopping the French advance and then with the two hour long bombardment that softened up the French defenses before the infantry advance.

123 Loewenstern, *Mobilmachung*, pp.56–57.

124 The section also conducted at least one artillery cooperation flight. Ferko–UTD, Box 11, Folder 5, "Bavarian Aviation, 1914 FFA 1b notes."

125 Second Army's XV and XVI CAs as well as First Army's VIII CA were all badly damaged by the fighting on the 20th. Indeed, of Second Army's three corps, only Foch's XX CA remained combat effective. Herwig, *The Battle of the Marne*, p.94.

126 Deuringer, *The First Battle of the First World War*, pp.126–133.

127 Deuringer, *The First Battle of the First World War*, pp.167–168.

128 Loewenstern, *Mobilmachung*, p.59.

129 GLA, 456 F1, Nr.203.

130 Ferko–UTD, Box 11, Folder 5, "Bavarian Aviation, 1914."

131 Lehmann, *Zeppelin*, pp.48–49.

132 Loewenstern, *Mobilmachung*, p.63.

133 Deuringer, *The First Battle of the First World War*, p.174.

134 Deuringer, *The First Battle of the First World War*, p.177.

135 Deuringer, *The First Battle of the First World War*, p.175.

136 An excellent account of the German side of the fighting in Lorraine during late August and early September can be found in: Storz, "This Trench and Fortress Warfare is Horrible!" pp.156–177.

137 For added security to the group's right flank, Joffre created "The Army of Lorraine" using six reserve divisions and the Verdun garrison. The Lorraine Army was to simply follow the attacking group *en échelon* and protect their flank and rear from any possible sortie out of Fortress Metz.

138 Herwig, *The Battle of the Marne*, pp.137–140; Tyng, *The Campaign of the Marne*, pp.76–82.

139 Zuber, *Ardennes 1914*, pp.87–90.

140 Simon House, *The Battle of the Ardennes August 22*, Thesis Paper, King's College London, 2012. pp.40–45.

141 Zuber, *Ardennes 1914*, pp.80–87.

142 Zuber, *Ardennes 1914*, p.86.

143 Loewenstern, *Mobilmachung*, p.93.

144 Loewenstern, *Mobilmachung*, p.41.

145 Loewenstern, *Mobilmachung*, p.41.

146 Loewenstern, *Mobilmachung*, p.42.

147 Loewenstern, *Mobilmachung*, p.42.

148 Ferko–UTD, Box 20, Folder 5, "Frontiers Battles, 1914"; Loewenstern, *Mobilmachung*, pp.44–45.

149 Loewenstern, *Mobilmachung*, p.45.

150 *Der Weltkrieg*, I, p.305.

151 Zuber, *Ardennes 1914*, p.103.

152 Zuber, *Ardennes 1914*, p.105.

153 Edward S. Johnston, *Study of The German Fifth Army on 21 August 1914, prior to the Battle of the Ardennes*, (CGSS Student Papers Collection 1930–1936); House, *The Battle of the Ardennes August 22*, pp.55–58.

154 Zuber, *Ardennes 1914*, p.108.

155 Tyng, *The Campaign of the Marne*, pp. 82–83. Zuber, *Ardennes 1914*, p.215.

156 Fred During, *A Critical Analysis of the Battle of Virton, August 22 1914*, (CGSS Student Papers Collection 1930–1936).

157 Manfred von Richthofen, *The Red Baron* (South Yorkshire, 2013), p.13.

158 John L. Scott, *Operations on the Right of French Fourth Army, August 22 1914*, (CGSS Student Papers Collection 1930–1936).

159 The officers and men of the Colonial Corps were all veterans of the campaigns in Africa and Indochina.

160 Rex W. Beasley, *A Critical Analysis of the French Colonial Corps in the Battle of the Ardennes with particular attention to the operations of the 3rd Colonial Division at Rossignol*, (CGSS Student Papers Collection 1930–1936).

161 Among the casualties that were killed were the division commander and a brigade commander. The other brigade commander was wounded and taken prisoner. Furthermore, almost all of the division's artillery was captured during the day's fighting. Tyng, *The Campaign of the Marne*, pp.84–87.

162 William E. Bergin, *The French Colonial 5th Brigade at Neufchâteau 22nd August 1914* (CGSS Student Papers Collection 1930–1936). Harry A.

162 Skerry, *German XVIII Reserve Corps at the Battle of Neufchâteau on August 22, 1914* (CGSS Student Papers Collection 1930–1936).

163 Ferko–UTD, Box 20, Folder 5, "Notes on role of aviation during Frontiers Battles, August 1914."

164 Zuber, *Ardennes 1914*, p.141.

165 Hugh H. Herrick, A *Critical Analysis of the Operations of the XVII French Corps in the Battle of the Ardennes, 21 and 22 August 1914*, (CGSS Student Papers Collection 1930–1936).

166 The majority of XVII CA's casualties were sustained during their retreat. Indeed, because of the panic, the corps' officers were unable to martial the number of troops necessary to cover their men's withdrawal. As a result, German batteries were able to follow directly behind their own infantry and freely bombard the fleeing French troops with virtual impunity. Only the onset of darkness later that evening stopped the Germans from decimating the French ranks with pursuit fire. House, *The Battle of the Ardennes August 22*, pp.219–222. Zuber, *Ardennes 1914*, pp.148–151.

167 The best operational study of this engagement, with a focus on how the French missed a significant opportunity to change the course of the battle, forms a chapter in Simon House's excellent study of the Ardennes battles. See: House, *The Battle of the Ardennes August 22*, pp.175–208.

168 Zuber, *Ardennes 1914*, pp.152–160.

169 John H. Carruth, *Study of the Operations of the French XI Corps at the Battle of Maissin, World War, 22 August 1914*, (CGSS Student Papers Collection 1930–1936).

170 House, *The Battle of the Ardennes August 22*, p.74.

171 *Der Weltkrieg*, I, p.316.

172 Fourth Army's final remaining corps—VIII RK—had been deployed far to the rear near Fourth Army HQ at Bastogne and was acting as army reserve. The corps was simply too far away to make any impact on the fighting of the 22nd.

173 House, *The Battle of the Ardennes August 22*, p.74. Grossherzogliches Artilleriekorps, 1 Grossherzoglich Hessisches Feldartillerie-Regiment Nr.25 im Weltkrieg 1914–1918 [FAR 25], (Berlin, 1935), p.27.

174 Loewenstern, *Mobilmachung*, p.67.

175 House, *The Battle of the Ardennes August 22*, p.101.

176 The number of officers and enlisted men available to Poser suggest that many of the enlisted personnel were still busy preparing the section for road column. The type of weapon given to the men of each FFA was the *Mauser Karabiner 98a*—a shortened version of the standard *Gewehr 98* infantry rifle.

177 Loewenstern, *Mobilmachung*, pp.68–69.

178 Zuber, *Ardennes 1914*, p.271.

179 Loewenstern, *Mobilmachung*, pp.68–69.

180 Loewenstern, *Mobilmachung*, p.66.

181 Loewenstern, *Mobilmachung*, p.66.

182 Zuber, *Ardennes 1914*, p.158.

183 Baron Max von Hausen, *Memoirs of the Marne Campaign*, Translated by John B. Murphy, Unpublished Text, (USCGSC–CARL, 1922), pp.132–141.

184 Tyng, *The Campaign of the Marne*, p.93.

185 This operational concept was standard German military doctrine since the days of Moltke the Elder. Indeed, this plan was simply a larger version of Moltke's masterpiece at Königgrätz in 1866. In that battle, the Austrians were "fixed" in place by a number of Prussian frontal attacks. Meanwhile, a separate Prussian force had maneuvered against the Austrian flank and rear. Later, as Chief of the General Staff, Schlieffen taught his subordinates to seek battle under these conditions. Daniel J. Hughes (ed.), *Moltke on the Art of War*, (New York, 1993). Robert T. Foley (Translator and editor), *Alfred von Schlieffen's Military Writings*, (London, 2003).

186 Karl von Bülow, *Bülow's Report of the Battle of the Marne*, Translated by Captain F.G. Dumont (Fort Benning, 1936), Unpublished Manuscript held at USAHEC, p.20.

187 See: Foley (ed.), *Alfred von Schlieffen's Military Writings*, pp.212–213, pp.218–226.

188 Not taking into account HKK 2, First Army was comprised of 320,000 officers and men.

189 Kuhl spent nearly his entire career in the General Staff. From 1897–1905 he worked closely with Schlieffen, becoming one of his most trusted subordinates. In the years that followed, he repeatedly proved his value to Moltke, who opined that Kuhl was "the best man to serve as Chief of the General Staff of an army that I could recommend." Thus, it was only fitting that he was chosen to be the Chief of Staff of the most important army on the Western Front. See: Robert T. Foley, "Hermann von Kuhl," *Chief of Staff: The Principal Officers Behind History's Great Commanders*, Vol.1 (Annapolis, 2008), pp.149–161.

190 Holmes, "The Reluctant March on Paris: A Reply to Terence Zuber's 'The Schlieffen Plan Reconsidered," *War in History*, Volume 8, Issue 2 (2001), pp.208–232.

191 It should be noted that there was a substantial qualitative difference between German reserve divisions and their French counterparts which further adds to this discrepancy. For more detail about the differences see: Bucholz, Joe Robinson,

Janet Robinson, *The Great War Dawning*, pp.161–175.
192 The main bulk of the BEF was moved across the channel between the 12th and 19th. The army was successfully concentrated south of Maubeuge on the 20th, proceeding north until they reached Mons on the 22nd. Edward Spears, *Liaison 1914* (London, 1999), p.31.
193 Thanks to several excellent reports conducted by First Army's airmen during the final days of the Liège siege. See: Loewenstern, *Mobilmachung*, pp.14–16.
194 Zuber, *The Mons Myth*, pp.115–116.
195 Hermann von Kuhl, *The Marne Campaign 1914* (Fort Leavenworth, 1936), pp.29–30.
196 A sixth Division (The 4th Belgian Infantry Division) was garrisoned at Namur. The division would ultimately be separated from the rest of the army for the rest of the campaign and would not be united until late October after the front had stabilized.
197 Kuhl, *Marne Campaign 1914*, p.30.
198 Humphries and Maker (ed.), *Germany's Western Front 1914*, Part 1, pp.136–139
199 Indeed, the Belgian Army was able to carry out three sorties from their positions inside the Antwerp defenses (August 24–26, September 9–13, and 26–27). Although they were defeated, these operations were successful in tying down significant numbers of German troops that would have otherwise been used at the front against the French and British. Moreover, they encouraged the hostile civilian population, who contrary to the laws of war continued to carry out partisan attacks against German officers and men stationed in the rear-area. This ultimately led to the tragedy at Louvain between August 25th–28th when the local population rose up during the Belgian Army's first sortie, which was a thrust in the direction of the medieval city. Many of the Belgian partisans, referred to as '*Francs-Tireurs*,' were members of the civilian militia organization known as the *Garde Civique*. Significant numbers of these civilians began carrying out guerilla warfare contrary to the rules of war established by The Hague Convention—namely that they were fighting without "responsible leaders" or "distinctive military badges plainly visible at a distance." Nor were they openly carrying their weapons as the rules of war required. See: Germany. *The German Army in Belgium: The White Book of May 1915*, Translated by: E.N. Bennett, (London, 1921).
200 Tage Carlswärd, *Strategic Signal Communications with the German Right Wing*, translated by Sigurd N. Ronning (Fort Monmouth, 1933), Unpublished manuscript held at USAHEC, pp.64–68.
201 Ferko–UTD, Box 20, Folder 5, "Frontiers Battles, 1914."
202 Loewenstern, *Mobilmachung*, pp.36–37.
203 Direct quote from First Army HQ Chief of Staff von Kuhl's history of the campaign. See: Kuhl, *The Marne Campaign 1914*, pp.40–41.
204 Loewenstern, *Mobilmachung*, pp.35–36.
205 Humphries and Maker (ed.), *Germany's Western Front 1914*, Part 1, pp.143.
206 Humphries and Maker (ed.), *Germany's Western Front 1914*, Part 1, pp.155.
207 Humphries and Maker (ed.), *Germany's Western Front 1914*, Part 1, pp.155.
208 Humphries and Maker (ed.), *Germany's Western Front 1914*, Part 1, pp.159.
209 Lanrezac was also a serious student of military history, focusing on the life and campaigns of Napoleon. Indeed, he authored a book on the 1813 Battle of Lützen that is still highly regarded to this today.
210 Doughty, *Pyrrhic Victory*, pp.68–72.
211 It should be noted that this plan was based on French intelligence estimates that grossly underestimated the overall strength of the German right wing. For example, the French estimate of August 16 stated that seven German corps (14 divisions) were on the right wing. As it happened, the German right wing was comprised of 24½ divisions (18 active and 6½ reserve). Tyng, *The Campaign of the Marne*, p.359, p.365.
212 Because Fifth Army's approach march was over 60 miles, it is understandable why German aviators had trouble finding any French troops near Namur during the first days of the right wing's advance.
213 Tyng, *The Campaign of the Marne*, pp.101–102.
214 See: Lanrezac, Charles. *French Plan of Campaign and First Month of the War*, Translated by Amico J. Barone, Unpublished manuscript held at USAHEC (1922), pp.67–71.
215 Permission was finally granted by Second Army HQ's Erich Ludendorff, who was present at Guard Korps HQ at the time Winckler's request came in. Humphries and Maker (ed.), *Germany's Western Front 1914*, Part 1, pp.159–160.
216 Lanrezac, *French Plan of Campaign*, pp.93–101.
217 Ian Senior, *Home Before the Leaves Fall* (Oxford, 2012), pp.81–92.
218 Humphries and Maker (ed.), *Germany's Western Front 1914*, Part 1, pp.162–170.
219 Alain Arcq & Achille Van Yprezeele, *Leernes & Collarmont 22 Août 1914*, (Fontaine-l'Évêque, 2013).
220 Humphries and Maker (ed.), *Germany's Western Front 1914*, Part 1, pp.189–197.

221 Belgian civilian *Franc-tireurs* took part in the fighting at Dinant.
222 Hausen, *Memoirs of the Marne Campaign*, pp.158–164.
223 For more on Lanrezac's decision to retreat, see: Lanrezac, *French Plan of Campaign, pp.107–116*.
224 First Army's HKK 2 was in search of the BEF while HKK 1 was conducting its redeployment around fortress Namur.
225 Loewenstern, *Mobilmachung*, p.38.
226 Humphries and Maker (ed.), *Germany's Western Front 1914*, Part 1, p.161.
227 Loewenstern, *Mobilmachung*, pp.37–39.
228 Ferko–UTD, Box 20, Folder 5, "Frontiers Battles, 1914."
229 Ferko–UTD, Box 20, Folder 5, "Frontiers Battles, 1914."
230 Humphries and Maker (ed.), *Germany's Western Front 1914*, Part 1, pp.166–167.
231 Ferko–UTD, Box 20, Folder 5, "Frontiers Battles, 1914."
232 Humphries and Maker (ed.), *Germany's Western Front 1914*, Part 1, pp.168–169.
233 Ferko–UTD, Box 20, Folder 5, "Frontiers Battles, 1914."
234 While it is an overall excellent book, John Cuneo's *The Air Weapon 1914–1916* erroneously states that Second Army's airmen—particularly FFA 1— supplied Bülow with inaccurate information on the morning of the 22nd which prompted the commander to prematurely order an attack across the Sambre that afternoon. Cuneo writes that FFA 1's message regarding the three columns of cavalry had tricked Bülow into thinking his army could achieve "an easy crossing" and that this order fundamentally changed the plans for the right wing's offensive. In reality, FFA 1's message saved the day for Second Army—compelling Bülow to keep his forward units where they stood rather than pulling them back across to the north bank. Without the message, X AK would not have been able to deliver its attack, which shattered multiple French divisions and allowed for large numbers of troops and equipment on both of its flanks to move across river that evening. Indeed, FFA 1 and 21's reports on the 22nd enabled Second Army to deliver its attacks the following morning in accordance with the original plan under more favorable conditions than it otherwise would have been had the flying sections been inactive or had Bülow and his subordinates chosen to ignore them. Cuneo, *The Air Weapon 1914–1916*, pp.42–43.
235 Humphries and Maker (ed.), *Germany's Western Front 1914*, Part 1, p.190.
236 Strangely, this is one of the only instances of using an aircraft to deliver messages between various corps commands during the entire Marne Campaign. General Staff officers had been trained in peacetime to employ aircraft for this purpose, yet it was rarely used. Although there is no explanation for this, one can only guess that other duties had caused the various corps headquarters staff from thinking of using their aviators in this manner. Whatever the reason, the act of neglecting to use their aircraft for this purpose would significantly inhibit the armies of the right wing later in the campaign. Loewenstern, *Mobilmachung*, p.72.
237 Second Army HQ chose to ignore FFA 12's reports and information from Third Army HQ that suggested French Fifth Army was withdrawing its forces and on the verge of a general retreat. Instead, they relied upon reports from the soldiers at the front that suggested five French corps were in the region and that a French counter-offensive on the 24th was a strong possibility. This opinion ultimately prompted Second Army to request Third Army's direct assistance on the 24th—a request that would shift Third Army's axis of advance and eliminate any chance for that army to strike at the French rear.
238 Although the various flying sections at this time in the war had enough aircrews to divide their flights equally, it was rarely done. Commanders preferred instead to send the same aircrews back over the same sector so that the observer could report on any changes that had been made since his last flight. For this purpose, there was a great disparity in the number of sorties performed between crews within each flying section.
239 Hausen, *Memoirs of the Marne*, pp.145–146.
240 Humphries and Maker (ed.), *Germany's Western Front 1914*, Part 1, p.145.
241 Norman Franks, Frank Bailey & Rick Duvien, *Casualties of the German Air Service 1914–1920*, (London, 1999) p.172.
242 Hausen, *Memoirs of the Marne*, pp.179–180.
243 Ferko–UTD, Box 20, Folder 5, "Frontiers Battles, 1914."
244 Hausen, *Memoirs of the Marne*, pp.160–161.
245 The German official history described the artillery's importance: "Only after the German batteries had suppressed the French artillery was the struggle at Leffe decided in the Germans' favor."
246 This column was the remains of the Belgian 4th Infantry Division which had escaped from Namur just before its capitulation.
247 Humphries and Maker (ed.), *Germany's Western Front 1914*, Part 1, pp.187–188.
248 Hausen, *Memoirs of the Marne Campaign*, p.166.

249 Humphries and Maker (ed.), *Germany's Western Front 1914*, Part 1, pp.187–189, pp.195–198.
250 Hausen, *Memoirs of the Marne Campaign*, pp.165–169.
251 Loewenstern, *Mobilmachung*, pp.78–80.
252 In his memoirs, General Lanrezac wrote that the German aviators held a decided superiority over their French counterparts during the Frontiers Battles. Indeed, he wrote, "Germany had organized an air service much better equipped than ours... They also had special planes for their artillery, of which we were totally lacking. This special artillery aviation was so necessary that we had to improvise one for better or worse during the course of operations and which was made very difficult because of the scarcity of both personnel and materiel." Lanrezac, *French Plan of Campaign*, pp.86–87.
253 Lawrence B. Glasgow, *Why Was German GHQ Unable to Locate the BEF Prior To Mons, And Why Was It Surprised to Meet the BEF At Mons?* Unpublished student paper, Command and General Staff School Student Papers Collection 1930–1936. (Fort Leavenworth, 1931).
254 Kuhl, *The Marne Campaign 1914*, p.41.
255 Although the message stated that the British had likely landed at Boulogne and would be employed in the direction of Lille, it went on to claim that: *"There is a tendency here, however, to believe that landings on a large scale have not yet taken place."* Kuhl, *The Marne Campaign 1914*, p.43.
256 Humphries and Maker (ed.), *Germany's Western Front 1914*, Part 1, p.171.
257 Ferko–UTD, Box 20, Folder 5, "Frontiers Battles, 1914."
258 Humphries and Maker (ed.), *Germany's Western Front 1914*, Part 1, p.172.
259 Cuneo, *The Air Weapon 1914–1916*, p.142.
260 Hoeppner, *Germany's War in The Air*, p.40.
261 Bundesarchiv-Militärarchiv, Freiburg (hereafter: BA-MA) RH 61 1223, *Die Fliegerverbande der 1. Armee während der Marneoperationen von Ende August bis zum 9 September 1914*, p.46.
262 Kuhl, *The Marne Campaign 1914*, pp.50–51.
263 J.E. Edmonds, *Military Operations France and Belgium 1914*, Vol.1 (Battery Press, 1996), pp.48–52.
264 Zuber, *The Mons Myth*, pp.112–113.
265 Zuber, *The Mons Myth*, pp.125–126
266 Zuber, *The Mons Myth*, p.161, pp.167–168.
267 Loewenstern, *Mobilmachung*, pp.80–81.
268 Humphries and Maker (ed.), *Germany's Western Front 1914*, Part 1, p.221.
269 Kuhl, *The Marne Campaign 1914*, pp.81–82.
270 Ferko–UTD, Box 20, Folder 5, "Frontiers Battles, 1914."
271 Ferko–UTD, Box 20, Folder 5, "Frontiers Battles, 1914."
272 Indeed, the observer's intelligence was considered so valuable that First Army HQ chose to forward it to OHL later that evening at 11:00pm. Humphries and Maker (ed.), *Germany's Western Front 1914*, Part 1, pp.256–257.
273 Kuhl, *The Marne Campaign*, p.84.
274 Kuhl, *The Marne Campaign*, p.85.
275 Humphries and Maker (ed.), *Germany's Western Front 1914*, Part 1, p.245.
276 Zuber, *The Mons Myth*, pp.200–201.
277 This is undoubtedly Kuhl's greatest mistake of the campaign and in all likelihood, the war.
278 In the words of aviation historian John Cuneo: "the air observer who furnished the erroneous impression violated a cardinal rule of observation aviation. His duty was to confine his report to what he actually saw." Cuneo, *The Air Weapon 1914–16*, p.48.
279 Walter Raleigh, *The War in The Air Vol. 1*, (Nashville, 1998) p.310.
280 It should be noted that some artillery from IV RK was pushed forward around noon to assist HKK 2. Likewise, elements from III AK's 5 ID arrived on the field opposite the BEF right wing at 3:30pm. However, by the time these troops were engaged, the British retreat had already been ordered.
281 The British 4th Division suffered 3,150 casualties while HKK 2 lost roughly 700 in the same sector. An excellent account of HKK 2's actions at Le Cateau can be found in: Zuber, *The Mons Myth*, p.230; Maximilian. von Poseck, *The German Cavalry in Belgium And France 1914*, (Essex, 2008), pp.53–67.
282 What's more, the British only inflicted 2,900 casualties onto the attacking Germans. Zuber, *The Mons Myth*, p.257.
283 Humphries and Maker (ed.), *Germany's Western Front 1914*, Part 1, pp.261–262.
284 Ferko–UTD, Box 20, Folder 5, "Frontiers Battles, 1914."
285 Humphries and Maker (ed.), *Germany's Western Front 1914*, Part 1, p.265.
286 Lanrezac later explained his rationale when he ordered the retreat late in the evening of the 23rd: "To flee was inglorious; yet to have acted otherwise would be to give my army up to complete destruction and would have rendered irreparable the general defeat suffered at the moment by French arms along the entire front, from the Vosges to the Escaut." Lanrezac, *French Plan of Campaign*, p.108.
287 Humphries and Maker (ed.), *Germany's Western*

Front 1914, Part 1, pp.198–201.
288 Lanrezac, *French Plan of Campaign*, p.110.
289 Mark Osborne Humphries & John Maker (ed.), *Germany's Western Front 1914 Part 1*, (Wilfrid Laurier University Press, 2013), p.247
290 FOH Tome 1, Vol. 2, pp. 16–26; FOH Tome 1, Vol. 2, Annex 1, Annex #149, p.125.
291 For the full text of Joffre's 'General Order No. 2' see: FOH Tome 1, Vol. 2, Annex 1, Annex #395, pp.278–280; Tyng, *The Campaign of the Marne*, pp.369–371.
292 Tyng, *The Campaign of the Marne*, pp.129–133.
293 Artur Baumgarten-Crusius, *German Generalship During the Marne Campaign in 1914: Contributions to a Determination of the Question of Responsibility*, Translation from German by US Army General Staff, Unpublished Manuscript held at USAHEC (1922), pp.34–47.
294 Baumgarten-Crusius, *German Generalship During the Marne Campaign in 1914*, p.39.
295 Maubeuge capitulated on September 8th.
296 Tyng, *The Campaign of the Marne*, pp.136–137.
297 Humphries and Maker (ed.), *Germany's Western Front 1914*, Part 1, p.308.
298 Humphries and Maker (ed.), *Germany's Western Front 1914*, Part 1, pp.301–311.
299 Tyng, *The Campaign of the Marne*, pp.371–374.
300 Wilhelm Groener, *Commander Against His Will: Operative Studies of the World War*, Translated by Martin F. Schmitt (Washington D.C., 1943), pp.15–34; Ian Senior, *Home Before the Leaves Fall*, pp.122–124.
301 Karl von Bülow, *My Report on The Battle of the Marne*, pp.41–53.
302 For an excellent and highly detailed analysis of Third Army HQ's predicament on the 27th, see: Constantin Hierl, *Strategic and Tactical Problems for the Study of the Marne Campaign Vol. 1—Studies of the Command of German 3rd Army August 27–29 1914*, Translation (Berlin, 1927).
303 Humphries and Maker (ed.), *Germany's Western Front 1914*, Part 1, p.254.
304 Kurt Hendemann, *Die Schlacht bei St. Quentin 1914: Garde und hannoveraner vom 29. und 30. August. Bd 7b. Schlachten des Weltkrieges, im Auftrage des Reichsarchivs* (Berlin, 1926), pp.7–14.
305 Karl von Bülow, *My Report on The Battle of the Marne*, pp.48–49.
306 Tyng, *The Campaign of the Marne*, pp.147–152.
307 FFA 21's lost report served as another example for the need of an aviation staff officer at each German Army HQ. The absence of an aviation staff officer negated a brilliant performance by *Ltn.* Oriola and *Oblt.* Behn on the evening of the 27th. Together, the crew had made the bold decision to brave the strong winds and rain to reconnoiter the French positions and had successfully discovered and documented French Fifth Army's deployment along the Oise. Had an aviation staff officer existed at Second Army HQ, he would have made inquiries to FFA 21 and the army's other flying sections throughout the evening asking if any sorties had been conducted. With the rest of the army's flying sections inactive, the aviation staff officer would have been focused on obtaining the results of FFA 21's only flight of the day. Thus, it can be reasonably assumed that the missing report would have been delivered before its actual arrival at Bülow's headquarters at 9:00am on the 28th. Had Second Army HQ received the report during the evening of the 27th–28th, the army's left would have been pushed forward south of the Oise at dawn instead of late morning—causing French Fifth Army to become engaged during its redeployment. While it is purely speculative to say what would have occurred next, it is the author's opinion that Second Army would have been better off on the morning of the 29th under this scenario. Whether Lanrezac would have ordered a retreat or had proceeded with an assault toward the German left, it is clear that Second Army would have been able to fight on the 29th under much more favorable circumstances. The loss of FFA 21's report was therefore one of the greatest lost opportunities of the war's first month.
308 Lanrezac, *French Plan of Campaign*, pp.131–139.
309 Karl von Bülow, *My Report on The Battle of the Marne*, p.55.
310 Tyng, *The Campaign of the Marne*, p.154.
311 Senior, *Home Before the Leaves Fall*, pp.146–155.
312 Hh "Deutsche Flieger in der Schlacht von St. Quentin 28. bis 30. August 1914," *Wissen und Wehr*, Jahrgang 1923, Heft 2. (1923) p.147.
313 Humphries and Maker (ed.), *Germany's Western Front 1914*, Part 1, p.378.
314 Ferko-UTD, Box 11, Folder 3, "FFA 23 papers"; Humphries and Maker (ed.), *Germany's Western Front 1914*, Part 1, pp.378–379.
315 Humphries and Maker (ed.), *Germany's Western Front 1914*, Part 1, pp.383–385.
316 Hh "Deutsche Flieger in der Schlacht von St. Quentin 28. bis 30. August 1914," pp.149–150.
317 Hh "Deutsche Flieger in der Schlacht von St. Quentin 28. bis 30. August 1914," pp.149–150.
318 Hh "Deutsche Flieger in der Schlacht von St. Quentin 28. bis 30. August 1914," p.151.
319 Humphries and Maker (ed.), *Germany's Western Front 1914*, Part 1, pp.390–391.

3
Battle of the Marne
September 1914

"Our airmen's efforts greatly assisted the army's tactical situation during the battle [of the Marne] by giving the army leadership outstanding reports that enabled them to make the correct decisions at many different moments of the battle."

General Hermann von Kuhl[1]

First Army HQ Chief of Staff 1914–1915

Despite their inability to achieve the type of victory prescribed by the German War Plan, Moltke's western armies remained in high spirits as they continued to chase the defeated Allied forces toward Paris. Each day German troops across the front vigorously pursued the retreating allied columns, denying them any chance for rest or reorganization. Citing their aviator's long-range reconnaissance reports, German army commanders plotted their troops' advance to remain within striking distance of the retreating allied forces. After struggling for several days to cross the Meuse, the German center's Fourth Army finally gained control of the river's western bank on August 28th, thus facilitating Fifth Army's crossing which resulted in the French Third and Fourth Armies being thrown back in the direction of Reims. On the right wing, Bülow's Second Army had delivered another major blow to French Fifth Army at the Battle of Guise–St. Quentin. Kluck's First Army had likewise defeated the BEF for a second time at Le Cateau and had continued its southwesterly pursuit with the intent to envelop the allied left flank. The overall situation therefore remained highly favorable for the men of the German center and right wing as their advance resumed at dawn on the 31st.

The Right Wing's Pursuit to the Marne

Seeking to envelop the allied left flank, First Army quickly pursued the BEF from Le Cateau towards the southwest. Rain and thunderstorms, however, had grounded aerial activity throughout the 27th, temporarily blinding the leadership. Fortunately, a key aerial reconnaissance report delivered to army headquarters late in the afternoon on the 26th had given Kluck the information necessary to order his army's march in the correct direction the following day, resulting in the continued harassment of the badly shaken British forces. Over the next three days, First Army chased the fleeing allied forces in front of them past St. Quentin in the attempt to reach a position to engage and envelop the allied battle-line's left flank. On the 29th, while Second Army was engaged at St. Quentin, Kluck's II AK overwhelmed the forces of a newly arrived French corps from Alsace (VII CA) that was threatening the army's open right flank. With their line of communications secured, Kluck and Kuhl wheeled First Army to the south the next day (August 30th) in order to be in position to attack French Fifth Army's flank and rear before they disengaged from Bülow.[2]

First Army's *Fliegertruppe* were highly active throughout the advance by maintaining surveillance over the retreating allied columns in their sector. Late in the afternoon of the 28th aviators of II AK's FFA 30 discovered the newly arrived troops from VII CA deploying into positions that threatened the army's right flank. The reports described "intensive railroad traffic at Amiens and Nesle and troop concentrations in the same locality."[3] After receiving their aviator's message, II AK received permission from First Army HQ to turn and attack the threat the following day, resulting in the rout of VII CA at Proyart on the 29th. First Army HQ did not learn, however, of the extent of the victory until the following morning when a crew from FFA 30 landed at Kluck's headquarters near Péronne to report on the battle's outcome—confidently declaring that "the enemy had abandoned their positions and the army's flank was secure."[4] In the meantime, news had arrived from Second Army HQ:

Second Army was attacked yesterday afternoon by ten French divisions at the least, on a front extending from west of Vervins to the region of La Fère. The struggle was a bitter one, but the enemy offensive failed. From the papers of a French corps chief of staff

Above: German postcard illustrating *Taube* reconnaissance aircraft over the front. (Aeronaut)

it was learned that the French had intended to attack St. Quentin while German First Army was to be held frontally by the English and French. At Noyon there is an enemy force of one and one-half brigades. General von Bülow regrets that First Army did not turn [south] facing the Oise in accordance with his expressed desire. The 17th Division [IX AK] will soon be put back under the orders of its army. The enemy appears to be falling back.[5]

With their aviators declaring the army's flank to be clear and Second Army reporting yet another victory, Kluck and Kuhl proceeded to wheel First Army south during the afternoon of the 30th. With good weather, aerial reports poured into army headquarters throughout the day describing in detail the direction of the allied left wing's retreat, including the forces of VII CA which had been mauled at Proyart the day before. All the reports concurred that the Franco-British forces were fleeing along the line Montdidier—St. Just-en-Chaussee with rearguards identified in the area of Bailly (six miles south of Noyon).[6] In addition to standard reconnaissance flights, IX AK's FFA 11 sent an aircraft to bomb the French capital of Paris. Flying in their Albatros B.I (B.180), pilot *Ltn.* Wentscher and observer *Ltn.* Böhmer dropped three small bombs and some propaganda leaflets on the Quai de Valmy in the heart of the city. On the way back to the airfield, Böhmer noted "a large enemy column on the road from Ham to Noyon."[7] Thus, upon reviewing all available aviation reports, Kluck and Kuhl recognized that all the allied forces opposite First Army's right flank were in retreat and therefore no longer posed an immediate threat. First Army HQ accordingly believed that the time had come, as Moltke had suggested in his August 27th directive, to shift the army's axis of advance on the 31st towards the south "in the direction Compiègne—Noyon." By turning their army towards the Oise River, Kluck intended to first push aside any remaining threat from the Franco-British troops of the allied left wing and then strike the flank of the French forces retreating in front of Second Army—thus initiating the final decisive battle that was entrusted to the right wing.[8] Indeed, Kluck's memoirs detail how First Army HQ viewed the situation:

The left wing of the main French forces [Fifth Army] is retreating in a southerly and south-westerly direction in front of the victorious Second and Third Armies. It appears to be of decisive importance to find the flank of this force, whether retreating or in position, force it away from Paris, and outflank it. Compared with this new objective, the attempt to force the British

Above: Wartime Propaganda postcard depicting a *Taube* bombing Paris. (Author's Collection)

Army away from the coast is of minor importance.[9]

After their victory at Guise—St. Quentin, Second Army HQ declared that their troops were unable to immediately begin the pursuit of French Fifth Army. Bülow's army therefore spent the 31st resting, reorganizing, and moving their supply and ammunition columns forward in preparation to continue the advance on September 1st. While the ground troops rested, Second Army's *Fliegertruppe* were busy keeping watch over Lanrezac's retreating columns, which had skillfully managed in 24 hours to put great distance between themselves and their German pursuers. In addition to reporting the French columns' positions, Bülow's airmen observed disorganized French units of varying sizes spread throughout the area, a clear sign of a badly defeated army. Perhaps the most intriguing assessment came from an observer of FFA 23 who reported: "A lack of order is everywhere. Many small and individual

Above: Wartime photograph of the airfield at Stenay, which served as a base for German Fourth Army's Airmen during the latter stages of the Marne Campaign. (Author's Collection)

units observed in retreat completely isolated from one another."[10] The report suggested that Lanrezac's badly mauled army was beginning to disintegrate—a fact that was promptly communicated to both First Army HQ and OHL.

Throughout the campaign, Hausen's Third Army had been placed in the unenviable position of being repeatedly asked to support their neighboring Second and Fourth Armies. Consequently, the overall speed of the Saxon Army's advance was greatly affected as it routinely turned back and forth giving support to both parties as circumstances dictated. By the afternoon of the 29th, Hausen made the choice to take advantage of his army's current position and turn south to offer assistance to Fourth Army on his left. Hausen's decision was based largely upon his aviators' excellent reports regarding the circumstances in Fourth Army's sector after that army had crossed the Meuse. The *Fliegertruppe's* messages plainly illustrated that Third Army was in an excellent position to deliver an attack against the French opposing Fourth Army.[11] Citing this intelligence, Hausen devised a plan to "combine with Fourth Army in an enveloping attack against the enemy's left flank in order to cut off his communications to the west."[12]

On September 1st, in accordance with Hausen's concept of operations, Third Army advanced to the southwest with the objective to envelop French Fourth Army's left flank and push them south into the arms of the advancing German center. Heavy artillery batteries belonging to both XII and XIX AKs were employed on the north bank of the Aisne to assist Third Army's infantry gain possession of the river's southern bank. This enabled the Saxon infantry to successfully capture the river and rout the French rear-guard. During the battle, an aircrew of FFA 24 (XIX AK) personally delivered an extremely detailed reconnaissance report to Hausen outlining the exact positions of French Fourth Army's left wing. Flying

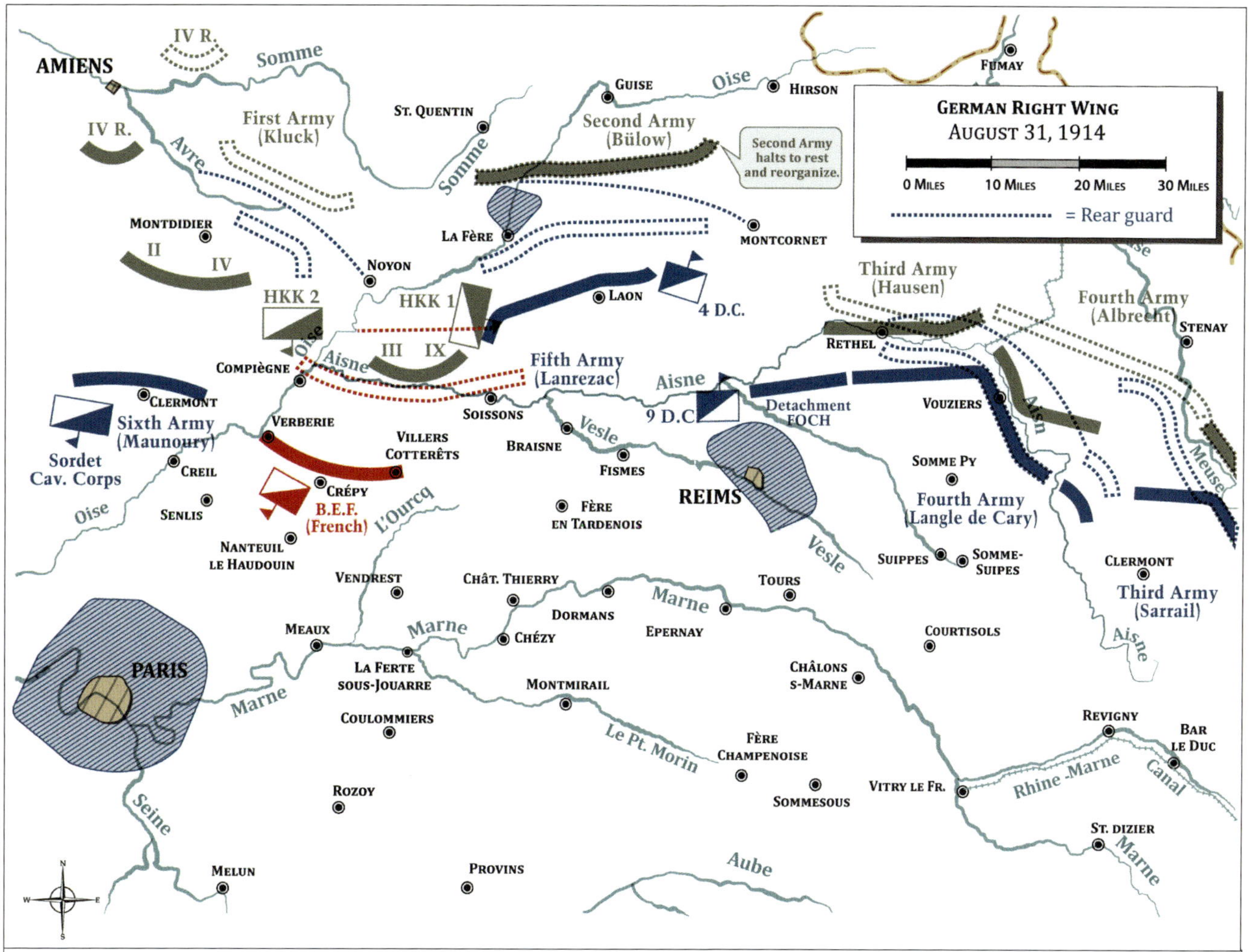
German Right Wing
August 31, 1914
0 Miles
10 Miles
20 Miles
30 Miles
= Rear guard
Second Army halts to rest and reorganize.
Amiens
Somme
IV R.
First Army (Kluck)
Avre
St. Quentin
Guise
Oise
Hirson
Fumay
Second Army (Bülow)
Montdidier
II
IV
La Fère
Montcornet
Noyon
HKK 2
HKK 1
Laon
4 D.C.
Third Army (Hausen)
Fourth Army (Albrecht)
Stenay
Compiègne
III
IX
Fifth Army (Lanrezac)
Aisne
Rethel
Clermont
Sixth Army (Maunoury)
Sordet Cav. Corps
Verberie
Villers Cotterêts
Soissons
Braisne
Vesle
9 D.C.
Detachment Foch
Vouziers
Creil
Crépy
B.E.F. (French)
Fismes
Senlis
L'Ourcq
Fère en Tardenois
Reims
Somme Py
Fourth Army (Langle de Cary)
Meuse
Nanteuil le Haudouin
Suippes
Somme-Suippes
Clermont
Third Army (Sarrail)
Vendrest
Chât. Thierry
Dormans
Marne
Tours
Epernay
Meaux
Chézy
Courtisols
La Ferte sous-Jouarre
Paris
Montmirail
Châlons s-Marne
Coulommiers
Le Pt. Morin
Fère Champenoise
Revigny
Bar le Duc
Rozoy
Sommesous
Vitry le Fr.
Rhine -Marne Canal
Seine
St. Dizier
Melun
Provins
Aube
N
W
E
S

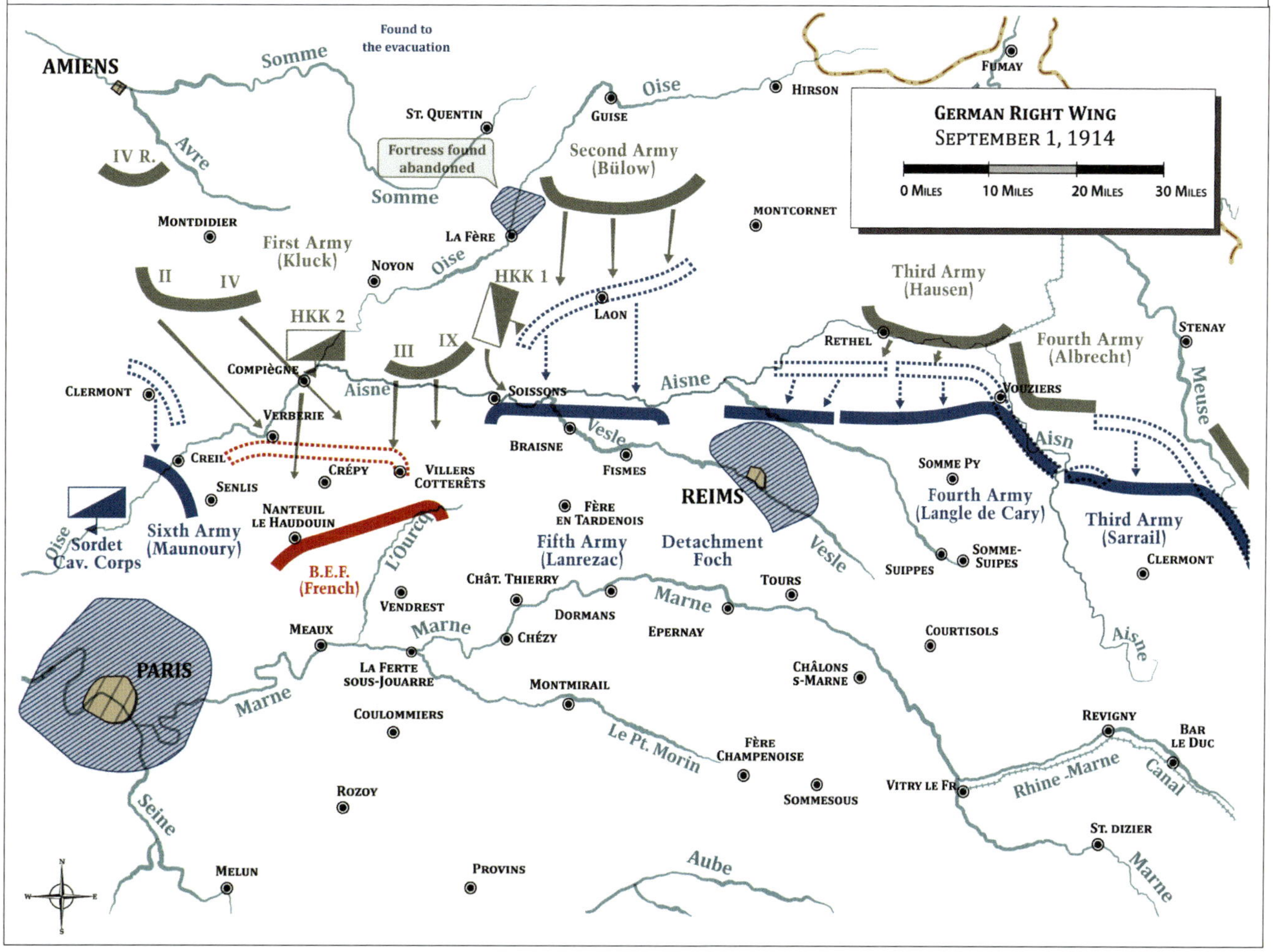
German Right Wing
September 1, 1914
0 Miles
10 Miles
20 Miles
30 Miles
Found to the evacuation
Fortress found abandoned
Amiens
Somme
St. Quentin
Oise
Hirson
Fumay
Guise
IV R.
Avre
Second Army (Bülow)
Montdidier
First Army (Kluck)
La Fère
Montcornet
Noyon
HKK 1
II
IV
Laon
Third Army (Hausen)
HKK 2
Rethel
Fourth Army (Albrecht)
Stenay
Compiègne
III
IX
Clermont
Aisne
Soissons
Vouziers
Verberie
Braisne
Vesle
Creil
Villers Cotterêts
Crépy
Fismes
Somme Py
Senlis
Reims
Fourth Army (Langle de Cary)
Third Army (Sarrail)
Nanteuil le Haudouin
Fère en Tardenois
Sixth Army (Maunoury)
Sordet Cav. Corps
L'Ourcq
Fifth Army (Lanrezac)
Detachment Foch
Suippes
Somme-Suippes
Clermont
B.E.F. (French)
Chât. Thierry
Tours
Vendrest
Dormans
Marne
Meaux
Epernay
Chézy
Courtisols
La Ferte sous-Jouarre
Paris
Montmirail
Châlons s-Marne
Coulommiers
Le Pt. Morin
Fère Champenoise
Revigny
Bar le Duc
Rozoy
Sommesous
Vitry le Fr.
Rhine -Marne Canal
Seine
St. Dizier
Melun
Provins
Aube
N
W
E
S

> ### The Success of Third Army's *Fliegertruppe*
>
> The Saxon airmen attached to Third Army had quickly proved themselves to be some of the best aviators in the German army. On August 23rd at Dinant they had performed several excellent artillery cooperation sorties that assisted the infantry cross the Meuse and attack Lanrezac's weak right flank. Later that afternoon Hausen's fliers brilliantly spotted signs of French Fifth Army's weakened defenses and promptly reported them to Third Army HQ. Believing his airmen's intelligence, Hausen ordered the army to march southwest at dawn the next morning, only to be overruled by a cautious Bülow. Had the attack been conducted as Hausen wished, Lanrezac's army could have been cut-off and destroyed in the western theater's first 'major victory' of the war; a victory that would have been largely owed to the aviator's excellent reporting. As it happened, Lanrezac's army was able to escape encirclement and thus survived to fight another day. Eight days later Third Army's *Fliegertruppe* continued to prove their worth by locating French Fourth Army in its entirety, giving Hausen the ability to confidently change strategies and turn his army south to attack a more opportune target. Once the opportunity had past, however, Hausen's aviators quickly alerted their commander, enabling him to once again switch strategies and keep the army in position to do the most damage. In each case the aviators' contributions greatly assisted Third Army's ground forces—maximizing their potential for success.

at dangerously low altitudes, the observer was able to positively identify the 4th and 9th Cavalry Divisions, Colonial Corps, IX CA and 60th Reserve Division in the area between Vouziers and Reims. With the airmen's valuable report in hand, Hausen understood that it was an impossibility to turn French Fourth Army's left flank. Thus, he was compelled to turn his own army south and join German Fourth Army to throw the French back in a large-scale frontal assault.[13]

Writing home to a family member, *Ltn.* Josef Suwelack—then serving as a pilot in Third Army's FFA 24—described what it was like flying in Third Army's sector during the pursuit from the Frontiers. Suwelack's letters not only highlight how German aviators conducted their work during the war's first campaign, but also give insight into how they viewed their work.

On Sunday [August 30], the Captain and I were flying just north of Rethel and Rheims flying at 6-800 meters altitude. We successfully evaded the enemy's anti-aircraft fire and were able to clearly document French batteries on the heights overlooking Rethel. Half an hour later, our own artillery occupied those very same heights! Our troops proceed with incredible speed and the "red trousers" retreat after every battle. If this continues, we will be in Paris in 4 or 5 weeks! Every night, I sleep through the thunder of the guns. The civilian population here is very scared... From Dinant to Rethel there is hardly a house unscathed. Everywhere we go we encounter a flock of French refugees who travel with all the possessions they can carry. The modern battlefield is a terribly gruesome place... We have found that flying at 2000 meters altitude keeps us very safe while performing our duties. The other day an officer of our section was shot and killed while flying at 700 meters altitude. Flying is a very dangerous task.[14]

Two days later Suwelack wrote home again:

Yesterday (September 2nd) I had to make my first emergency landing in enemy territory 10 km west of Mezieres because my observer had lost his bearings. First, I began looking for German troops, but found none. At the time we were flying at 1,500 meters altitude. After stopping the engine, we heard a terrible cannonade so we knew we had to be in the vicinity of a great battle. The sun blinded us such that we could hardly see. To our right lay a thick forest so I was unable to initially find a place to land. Finally, I discovered a small village located near the forest that had a suitable landing site. After a quick descent, I made a smooth landing in the field I had initially spotted. As I always take a carbine and revolver with me when I fly, my observer and I armed ourselves and went into the village. We quickly came across a group of about twenty young boys, who when they saw our weapons immediately raised their hands and let us go on our way without making a fuss. By that time we were twenty kilometers from our section's field and the sun had already gone down. Nonetheless, we started home in the bright moonlight and eventually got back to the airfield. When we arrived, the airfield had torches lit, but it was hard to tell because there were so many campfires all around. Our current quarters are excellent. The airfield, located on a large farm, was previously used by French airmen. In their haste to retreat, the Frenchmen left the area nicely furnished and in great condition. How nice of them! Meanwhile, the army continues to advance without pause. Hopefully we'll get some rest soon.[15]

As Hausen's Third Army pushed across the Aisne, Kluck and Kuhl's First Army began their southward pursuit of the Franco-British left wing. In accordance with the army commander's wishes, First Army's aviators were airborne at dawn on August 31st to keep watch over the retreating allied columns. Each flying section was assigned a unique sector of the front to patrol, as demonstrated by the following order that was distributed to each section commander shortly before dawn:

Aerial reconnaissance will be made in the following areas:
- II AK [FFA 30]: Reconnaissance of the right flank on the line Clermont—Creil—Senlis.
- IV AK [FFA 9]: Reconnaissance in the area of Compiègne to the southern edge of the woods of Compiègne.
- III AK [FFA 7]: Reconnaissance in the direction of Taillefontaine—Vivières.
- IX AK [FFA 11]: Reconnaissance in the direction of Longpont.
- First Army [FFA 12]: Long-range reconnaissance in the southwest direction towards Paris.[16]

By 1:30pm First Army HQ had a complete picture of the allied deployment in their sector. While the region of the Oise between Noyon, Chauny, Coucy-le-Château and Carlepont was reported "clear of the enemy," several French columns were located marching on Soissons from the north and northeast with no troop movements south of the city—indicating French Fifth Army's location.[17] To the west, multiple strong allied columns (suspected to be English) were spotted marching south from Vic and from Compiègne in the direction of Verberie opposite the army's right wing. In light of this information, First Army HQ ordered II AK on the army's extreme right to push forward cavalry detachments supported by artillery and infantry in motor trucks in the direction of Soissons—Verberie. The leadership of First Army had hoped that this highly mobile mechanized force could make contact and pin down these retreating English forces that had been witnessed from the air. If successful, the remainder of First Army would have the opportunity to catch-up and finish off the resilient forces of the allied left wing, thus giving the army a chance to freely turn toward Lanrezac's force without any danger to their flank. Unfortunately, the mechanized force's herculean effort was ultimately in vain as the allied rearguard successfully held them off in the final hours of daylight.[18]

First Army Flying Sections Available Strength as of September 1st, 1914
FFA 7— 3 aircraft
FFA 9— 4 aircraft
FFA 11— 4 aircraft
FFA 12— 4 aircraft
FFA 30— 3 aircraft

Above: A German infantryman holds a Saxon Regimental standard. (Author's Collection)

Staying aggressive, Kluck decided to continue the chase at dawn the following day—September 1st. Throughout the morning, each of Kluck's lead columns pushed forward with the goal of finally catching and enveloping the allied left wing. However, despite covering a great distance (IV AK marched 24 miles on September 1st) and engaging the allied rearguards at Verberie, Gilocourt, and Villers-Cotterêts, the day ended yet again without a major battle.[19] The reason why was that the allied forces were retreating at breakneck speed, leaving behind significant quantities of matériel which were often spotted from the air. As a result, the Franco-British main bodies slipped away and further increased their distance from Kluck, prompting First Army's leadership to hold a conference to determine their future strategy.

That evening, Kluck and Kuhl summoned the latest aviation reports to determine whether the allied left wing had escaped envelopment or if it was still possible to force them into battle. First Army's flying

Above: German infantry assault the French positions near Soissons. (Author's Collection)

sections had kept watch throughout the afternoon over the retreating Franco-British columns to the south and southwest, reporting on their location as well as the circumstances regarding the various rear-guard battles. An observer from FFA 30 reported:

At La Ferté an enemy infantry division was seen resting at 4:10pm. Strong column retreating on the road between Crépy-en-Valois and La Ferté-Milon. Vehicle parks south of Nanteuil.[20]

Two additional sorties, performed by FFAs 9 and 7, highlighted in great detail the positions of the allied left wing at 1:45pm. According to the reports, the enemy's left flank was located on the line Pont-Maxence—Creil near the town of Senlis. First Army HQ's FFA 12 delivered a similar report. FFA 12's intelligence strongly suggested that the French forces on the BEF's left had successfully escaped from First Army's clutches. This view was further supported by yet another reconnaissance report delivered by IX AK's FFA 11 which stated that, "the enemy are unreachable even if our corps conducts the most strenuous of forced marches."[21]

Provided with seemingly irrefutable proof from his airmen that the Franco-British forces of the allied left wing had escaped, Kluck made the decision to halt the advance and give his army a day of badly needed rest on the 2nd. Indeed, Kuhl's own notes stated that:

No danger from the area Douai—Cambrai—Amiens. The enemy had been dispersed. Catching up with the French appeared impossible. They had gotten away without pursuit. The British will also be difficult to catch. To continue in the present direction makes no sense, the right flank is threatened from the direction of Paris. Consequently, it is now important to halt in order to stage the Army for the coming advance depending on our orders to either continue moving south with protection against Paris or against the lower Seine below Paris. If we continue advancing south and if the French defend the Marne, a flanking attack out of Paris has to be expected.[22]

Thus, First Army HQ informed OHL of this decision by radio with the following communiqué:

Enemy forces in the area of Amiens have fallen back toward the southwest… Fifth French Army has withdrawn its left wing south via Soissons. On this wing, First Army has not been able to catch up with it. Fighting the British at Verberie. On 2

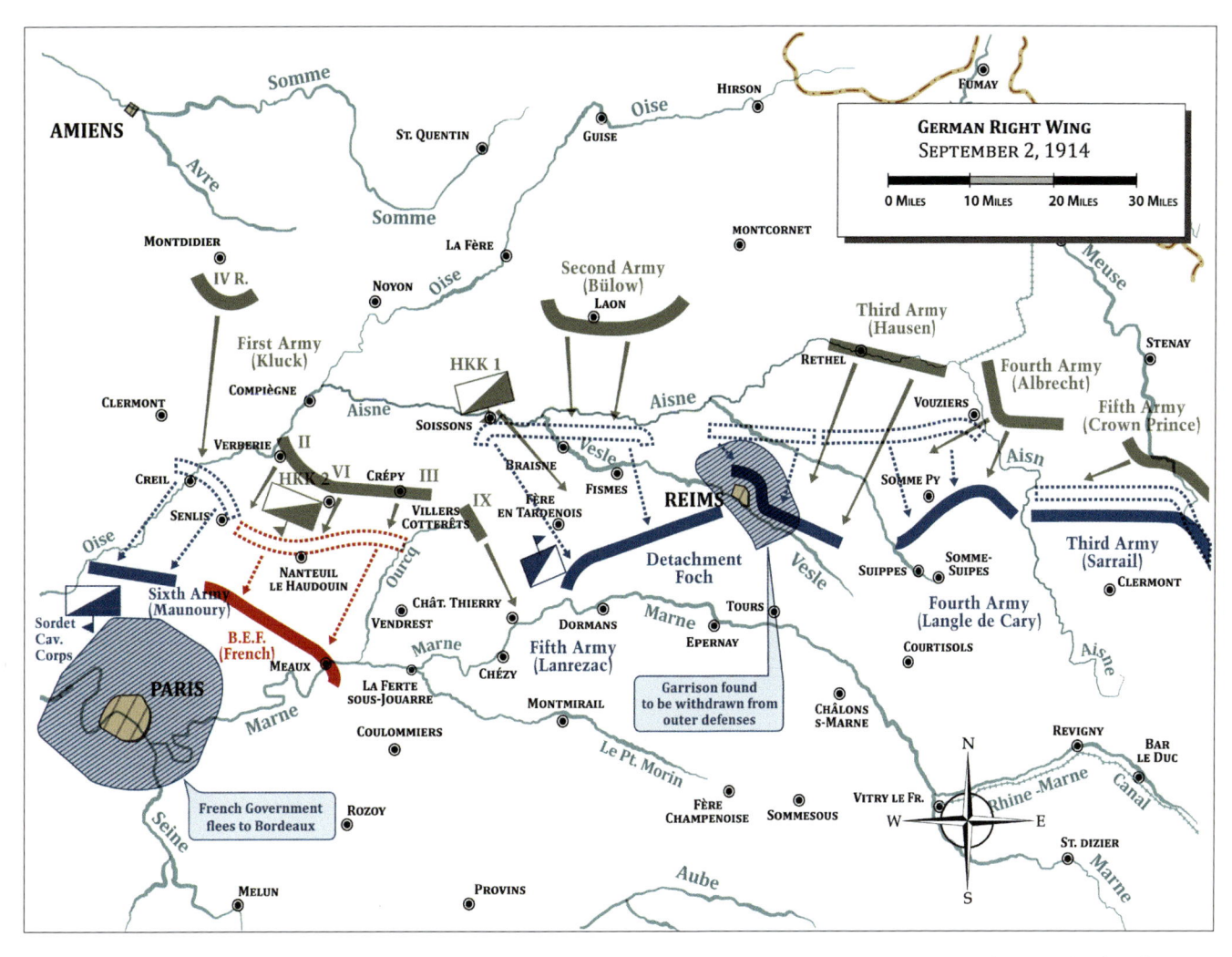

September, First Army intends to deploy along the line: Verberie—La Ferté-Milon so as to be ready for further employment.[23]

Scarcely had the radiogram been sent when a liaison officer from III AK arrived at Kluck's headquarters carrying captured orders that had been found on a British cyclist. The papers detailed the resting positions of the entire BEF for the night of September 1–2. After reading the order, Kuhl realized the entire British Army was within striking distance on the line: La Ferté-Milon—Crépy-en-Valois—Verberie. Kuhl reasoned that an opportunity to finally catch and destroy the BEF should not be passed over. Accordingly, the original order to hold the army in place on the 2nd was scrapped. A new directive, drafted at 10:15pm, called for each of the army's lead corps to turn southwest and participate in a concerted attack against the entire English Army. To avoid any confusion, OHL was sent a revised message stating:

Three English Corps identified immediately in front of First Army. The army will attack tomorrow beyond Creil—La Ferté-Milon so as to be ready for further employment after throwing back the enemy.[24]

First Army attacked as ordered on the morning of the 2nd but ultimately failed to bring about a decisive battle. The English had wisely begun their day's march in the predawn hours and had easily escaped First Army's efforts. Indeed, the only serious combat of the day occurred on Kluck's right at Senlis where II AK became heavily engaged with the BEF's rearguard. To the east, III and IV AKs advanced without incident, leaving the troops in the same strategic situation as the day before.

Located on First Army's extreme left, IX AK HQ received an aerial report early in the morning from FFA 11 confirming that the English had evacuated and would not be brought to battle.[25] Consequently, the corps' commander, *Generalleutnant* Ferdinand von Quast, ordered a two-hour halt to allow his men some rest. In the meantime, Quast received another important aerial reconnaissance report from his own FFA 11 which pinpointed the location of three nearby French corps belonging to Lanrezac's Fifth Army.

CAMPAGNE DE 1914

44   ARMEE ALLEMANDE.   ND. Phot.
L'Empereur d'Allemagne Guillaume II et le Grand Etat-Major. — A droite, le feld-maréchal de Moltke.

Above: Kaiser Wilhelm II and Helmuth von Moltke inspect their troops during an exercise prior to the war. Together, the duo made only one visit to the front during the Marne Campaign. (Author's Collection)

According to the observer's intelligence, the French columns were retreating in a southerly direction on the line: Braisne—Fismes toward the Marne River bridges of Mont-Saint-Père and Jaulgonne (between Château-Thierry and Dormans). The report in its entirety read as follows[26]:

> **To: IX AK HQ**
>
> An enemy column of all arms was observed retreating in disorder on the line Braisne—Fismes in a southerly direction to the Marne River at St. Père and Jaulgonne on the roads:
>
> a) Braisne—Cuiry-Housse—Fère—Beuvares—Mezy.
> b) Braisne—Loupeigne—Fère—Villemoyenne—Gr.Banche.
> c) Fismes—Chéry—Cohan—Cierges—Courmont—LeCharmel—Jaulgone.
>
> Head of column seen at 7:45 at Mont St. Père—Gr.Banche, east of Beuvardes and Jaulgonne. End of columns north of the line Braisne—Fismes. Total combined strength of enemy columns: three army corps.[27]

Quast knew that the hostile forces mentioned in the report represented French Fifth Army's left flank. Although it had previously been declared to be unreachable, the envelopment of Lanrezac's army remained First Army's primary objective. Therefore, with his airmen now reporting the coveted French flank within striking distance, Quast made the bold decision to disregard his previous directives and order a pursuit to the southeast to begin at 1:00pm. Before stepping off, IX AK HQ forwarded FFA 11's report directly to Kluck with an attached note explaining the reasoning behind Quast's decision to change directions toward Lanrezac's flank. Thus, with IX AK already moving southeast, Kluck had an important decision to make. He could either pull IX AK back and reform or turn the entirety of First Army to the southeast in support of Quast's endeavor. After careful consideration, Kluck, Kuhl, and the First Army HQ staff unanimously agreed to take the latter course of action.[28]

Having chosen to support Quast, First Army HQ created a new directive that was dispatched throughout the army at 2:00pm. The order called for the IX and III AKs to advance southeast as quickly as possible with the objective to catch and engage the

Above: Support troops erect telegraph lines behind the German front. These primitive methods of communication, combined with great distances, ultimately caused significant delays in the transmission of important messages from senior officers in the rear to the commanders at the front. (Author's Collection)

newly discovered French columns north of the Marne River. After making contact, these two corps were directed to attack and hold the French columns on the river's northern bank while awaiting the arrival of Second Army (which remained a full day's march behind First Army after halting to rest on the 31st). The remainder of First Army was ordered to shift to their left and prepare to move *southeast* the following morning.

Thanks to their commanders' initiative, the IX and III AKs were already marching southeast when the new army order was received. As a result, the two corps were able to make great progress despite only having a half-day to advance. Marching almost 20 miles, IX AK ended their day outside Château-Thierry while III AK, following *en échelon*, reached the area of La Ferté-Milon. However, despite the great distances covered, minimal contact with French rearguards was made. Kluck's two left wing corps were consequently obliged to renew their advance the following morning.[29]

At 9:45pm First Army HQ's order for September 3rd was dispatched amongst the various corps headquarters. The order specifically detailed the leadership's operational strategy. According to the order, IX AK was to attack southeast into the flank of the recently discovered French columns retreating in front of Second Army. The III AK was ordered to follow *en échelon* in the direction of Château-Thierry and "attack the enemy forces crossing the Marne." However, both corps' liaison officers were told that "in case the enemy is no longer reached, the two corps will clear toward the west of the road: Soissons—Château-Thierry on which the right wing of Second Army was marching." In other words, if the French escaped across the Marne, First Army was to not pursue. Indeed, the liaison officers charged with delivering First Army HQ's order were told that "crossing the Marne is not regarded as likely and will be considered only in exceptionally favorable circumstances."[30] Meanwhile on First Army's right wing, IV AK was to advance into the region of Crouy while II AK was to slip to the left upon Nanteuil. Finally, deployed in deep echelon to the rear, the IV RK was to follow the army's southeastern movement by marching on Senlis, covering the army's open right [western] flank.

First Army's entire plan was predicated upon Second Army taking advantage of IX AK's attack and destroying Lanrezac's troops before they could safety reach the Marne's southern bank. As it happened, Bülow's army had crossed the Vesle River on the

2nd.[31] However, despite accelerating the pace and making good progress, Second Army remained far behind Lanrezac's weakened columns. Consequently, Kluck's First Army, located an entire day's march ahead of Bülow, represented the only German force capable of catching French Fifth Army and bringing it to battle under favorable conditions—an undeniable fact that was seemingly ignored by Moltke when he formulated an entirely new concept of operations for his two right wing armies later that evening.

Moltke's New Concept of Operations

From his new headquarters in a small red schoolhouse far to the rear in Luxembourg, Moltke was in the midst of terribly mismanaging the German pursuit after the victories along the frontiers. In stark contrast to Joffre, who commandeered a touring car and professional driver to constantly move about the battlefront, Moltke remained at his headquarters throughout the campaign—leaving to visit the front on only one occasion. As a result, the German General Staff Chief became increasingly detached from the realities of the front as the days passed and the distance to the front lines increased. He therefore made poor decisions based upon old news that had to be radioed from the various army HQs at the front back to OHL in Luxembourg—a technological necessity that caused delays in transmission of up to twenty hours.[32] Moltke's latest in a series of these poor decisions occurred on the evening of September 2nd as Kluck's army prepared their attack into Lanrezac's flank north of the Marne.

Moltke and his staff had foolishly concentrated the bulk of their attention during the critically important days of August 31st–September 2nd on the battles occurring along the German center. Consequently, OHL remained largely unaware of the fluid situation on the right wing in the days immediately following the Battle of St. Quentin.[33] Nevertheless, despite the lack of information, Moltke decided to establish an entirely new concept of operations based on several assumptions.

Moltke's new strategy was sent out to First and Second Army HQs at 10:40pm in the form of a General Directive stating:

The plan of the High Command is to drive the French back in a southeasterly direction, cutting them off from Paris. First Army will follow the Second in echelon and will protect the flank of the German forces.[34]

Moltke's directive clearly illustrated just how out of touch with reality the German General Staff Chief had become. Written without any regard for the actual positions of the armies mentioned, the order was impossible to successfully implement as written. It therefore confused both Kluck and Bülow and rendered them unprepared for the campaign's next phase.

At the time of the order's transmission, First Army remained a full day's march *ahead* of Bülow's Second Army. If Kluck were to act exactly as ordered, First Army would need to stand fast for 24–48 hours while Second Army moved ahead of them. In the meantime, the nearby allied forces would escape across the Marne and any remaining opportunity to successfully execute the directive's primary objective of "driving the French back in a southeasterly direction and cutting them off from Paris" would be forever lost. As one of the General Staff's most aggressive officers, Kuhl agreed that the directive's primary goal of pushing the allied forces southeast away from Paris should be the right wing's top priority. He also recognized the simple truth that First Army was the only force capable of accomplishing this feat. Thus, Kuhl interpreted the order for First Army to follow the Second *en échelon* to be a mistake that could only have been written from Moltke's inaccurate perception of the relative positions of the two armies. Therefore, First Army HQ chose to act in the spirit of the directive by ordering First Army's advance against French Fifth Army's left flank to proceed. By aggressively continuing their advance, First Army would be accomplishing the directive's principal aim—pushing the enemy columns to the southeast away from Paris.[35]

September 3: First Army Crosses The Marne

Quast's IX AK continued south during the night of the 2nd–3rd to the bank of the Marne River. This aggressive move took the French completely by surprise and allowed Quast's lead columns to seize the bridges at Chézy-sur-Marne and Château-Thierry and push across to the river's southern bank prior to dawn. News of IX AK's impressive progress reached the First Army HQ staff at La Ferté-Milon later that morning as Kluck was preparing to move closer to the front to personally observe the situation.[36] Thus, upon receiving word that part of their army had already crossed the Marne, Kluck and Kuhl were forced yet again into making a crucially important decision: either commit the bulk of First Army south of the river and open themselves up to a flank attack from Paris or draw IX AK back and allow the enemy forces to freely escape. As they debated the issue, Kluck and Kuhl received several excellent aerial reports that clarified the situation and allowed them to confidently make a decision.

The first of the reports (delivered in person by an

observer from FFA 12) arrived shortly before noon. Speaking directly to a First Army HQ staff officer, the airman gave a detailed description of the retreating allied columns' location:

Enemy witnessed falling back on both sides of the river. Many columns of varying sizes also observed throughout the area. To the west, large columns witnessed from Meaux towards the south and moving toward Coulommiers.[37]

The second report, performed by IX AK's FFA 11, arrived at First Army HQ just minutes after the first. Having been radioed directly from IX AK HQ, FFA 11's report described in detail a major combat at 10:00am on the heights southeast of Château-Thierry near the Marne's southern bank. According to the message, the observer had personally witnessed German infantry capture the French defensive positions, causing the French infantry to flee in disorder. The flight's observer also reported seeing large numbers of German troops moving through Château-Thierry preparing to reinforce the units already engaged in the battle. Therefore, based upon what he witnessed, the airman could confidently declare that IX AK had successfully caught the French rear-guard. Furthermore, the observer reported seeing an enemy that was showing signs of disintegration:

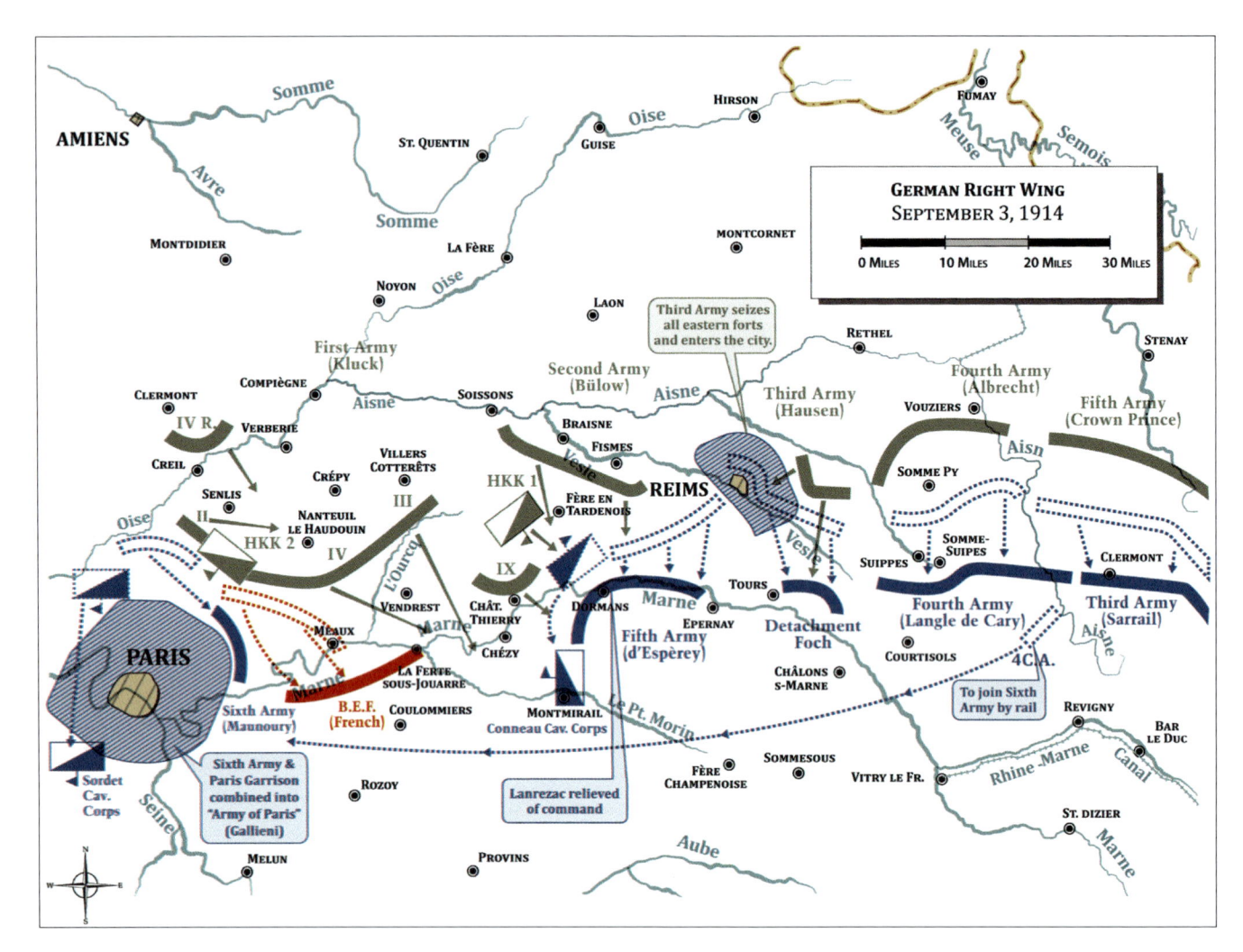

All north-south roads crossing the Marne are crowded with enemy columns of all arms and vehicles interspersed with infantry fleeing in disorder. To the south of the battlefield, the road Château-Thierry—Montmirail is clear of enemy troops. At 10:30am large bivouacs were seen to the south and southeast of Montmirail. Further south, large columns were witnessed retreating in disorder from Montmirail to Montenils and to the east on the roads from Mézy-Moulins to Condé-en-Brie and from Condé-en-Brie to Artonges.[38]

The report concluded by stating that French Fifth Army's forces were falling back in the face of IX AK's advance to positions directly in the path of III AK's line of advance. Thus, with their aviators reporting that the French were fleeing in disorder and with Moltke's new General Directive still fresh on their minds, Kluck and Kuhl made the decision at 1:00pm to order two additional corps (III and IV AKs) across the Marne to assist IX AK. Together, these three corps would attempt to carry out the High Command's wishes of pushing the French back to the southeast away from Paris. Meanwhile, II AK, IV RK, and 4 KD would be kept north of the Marne as a flank-guard to protect the army from any possible attack from the direction of Paris.[39]

In his post-war memoirs, Kuhl described First Army HQ's rationale for making the decision.

IX AK had already crossed the Marne as early as September 3rd and had compelled the French left wing to fight. According to the information furnished by Second Army [via their aviators], the enemy was hurled back in complete disorder. Was it not the duty of First Army to profit by its situation as an advanced echelon? Were we to neglect the last opportunity of overtaking the enemy and let escape the reward of our unspeakable efforts?

Übergang über einen Fluß.

Above: German infantry cross the river ahead of their main body. (Author's Collection)

But in crossing the Marne the army went contrary to the letter of the order from OHL. We were perfectly aware of that fact. The right flank of First Army was to be covered toward Paris by its own echelon formation. Such a measure promised to be sufficient against the enemy forces beaten on the Somme and on the Avre. On the part of the English, there was scarcely any offensive to be feared. There still remained, however, a danger to our right flank (from Paris). We took this chance in order to strive for a great goal that seemed possible of attainment. It was a bold decision at which First Army HQ had arrived. The dice were thrown; the Rubicon had been crossed.[40]

Quast's IX AK continued their advance immediately upon receiving the new army order. Acting aggressively, the corps' two divisions were able to attack the French columns south of the Marne and push them back to the southeast in accordance with the spirit of Moltke's directive. By nightfall IX AK had reached the heights north of Courboin (just west of Condé-en-Brie). To the west, III and IV AKs reached the Marne in the vicinity of Charly and La Ferté-sous-Jouarre while II AK and IV RK remained north of the river near Nanteuil-le-Haudouin covering Paris.[41]

Meanwhile, FFAs 7, 9, and 30 had been ordered to keep watch over the retreating allied columns south of the Marne opposite the army's right wing.[42] Flying a 160 mile flight, FFA 9's *Oblt.* Kleine (pilot) and *Oblt.* Blumenbach (observer) spotted hostile forces in the area between Coulommiers and Montmirail retreating in good order in a southerly direction under the cover of a strong rear-guard consisting of artillery and cavalry. Blumenbach also reported substantial military traffic boarding a train at the rail station at Coulommiers. A later flight performed by FFA 9 discovered significant troop concentrations on the road Othis—Dammartin and at Dammartin itself.[43] At the same time, II AK's FFA 30 performed several flights in the direction of Paris in order to observe any potential threat to First Army's right flank. The section's first report was delivered to II AK HQ shortly after 11:00am:

At 10:40am—1 infantry division resting on the march southwest and south of Dammartin. Enemy artillery south of Dammartin. A group of artillery spotted on the move from Villeneuve to Dammartin.[44]

The second report stated:

At 5:30pm—the area Senlis—Creil—Luzarches—

Above: Pontoon train in road column. These formations were vital in allowing the German Army to cross France's numerous rivers during the period of open warfare in 1914. (Author's Collection)

Dammartin—Nanteuil was free of the enemy. However, several enemy battalions were seen moving east near Villeron (along the railway Paris—Creil).[45]

FFA 30's reports were of great operational value. They proved that the French were sending forces from Paris to strengthen their left wing—an unmistakable sign of a major redeployment. With these reports in hand, Kluck and Kuhl would have undoubtedly been aware of the fact that there was a growing threat to their army's vulnerable right flank and would have reconsidered their decision to cross the Marne as they did, thus fundamentally changing the course of the campaign. Incredibly, despite their undeniable value, the two reports inexplicably never reached First Army HQ. Consequently, Kluck and Kuhl remained unaware of the growing threat to their right flank and continued with their plan to send III, IV, and IX AKs across the Marne.[46]

The loss of FFA 30's reports was perhaps the greatest example of how the lack of an aviation staff officer negatively impacted German operations during the war's first campaign. Without an aviation staff officer or at least a standard doctrine of liaison, each FFA's reports within an army were handled differently, often resulting in haphazard treatment of potentially vital information. From documents found in the *Reichsarchiv*, it has been discovered that all of FFA 30's reports during the campaign were handled by II AK HQ's "Ia" staff officer: Major Rudolf Mengelbier.[47] A corps HQ's "Ia" (1st General Staff Officer) was arguably the most overworked member on staff. He was responsible for updating the corps staff and divisional liaison officers of the tactical situation and the intentions of the corps commander, writing operational orders, and overseeing communications with Army HQ as well as neighboring corps HQs.[48] Therefore, taking his workload under consideration, it is likely that Mengelbier never got around to reviewing the aviation reports on the evening of the 3rd and thus never forwarded them to First Army HQ.[49]

In an interview with the *Reichsarchiv* after the war, Kuhl reflected upon the impact of FFA 30's lost reports:

Our aviator's work was considered to be of great operational significance. The prevailing view at First Army HQ and at OHL at the time of September 3rd was that the army's right flank was not in danger. I believed that defeated French forces that were

Vernichtung französischer Artillerie in den Kämpfen zwischen der Aisne und Oise.

Above: French artillery batteries participate in a rear-guard action to cover the allied retreat from the Oise to the Aisne Rivers following the Battle of Guise. (Author's Collection)

engaged in the area of Amiens were in fact on our army's right flank. However, I believed that these forces were not capable of offensive action and were simply retreating with the other Franco-British forces. These enemy forces were expected to continue their retreat and seek refuge inside Paris, but as they had already been engaged and defeated, they were considered to be of no threat to our army's flank. The aviator's messages of the 3rd that discovered enemy forces with artillery on our army's right flank advancing from Paris—Dammartin would have greatly altered First Army HQ's mindset. If I had received these important messages, I would have adopted immediate counter-measures to ensure the safety of our army's right flank. At once, I would have stopped the army's southern advance and would have dispatched reinforcements to the army's flank to resist any attack from the direction of Paris.[50]

Kuhl's statement clearly illustrated the value of the two lost reports. Their absence caused First Army to advance across the Marne and unknowingly expose their flank to a growing threat. Had the reports been delivered, it is reasonable to conclude that First Army HQ would have halted their pursuit of French Fifth Army and would have redeployed their main strength to the army's right wing.[51] Had this happened, the subsequent deployment would have resulted in a much more favorable scenario for First Army as events later proved. Thus, the loss of FFA 30's aerial reports served as one of the most critical mistakes of the campaign and arguably the war.[52]

September 3rd also saw Bülow's Second Army execute a spirited pursuit of Lanrezac's regiments to the banks of the Marne. Second Army's aviators flew long-range sorties throughout the day over the retreating French columns, keeping Bülow and his staff fully informed of their location and condition. Witnessing the effect of IX AK's attack, an observer of VII AK's FFA 18 reported to Second Army HQ: "in the area of Épernay—Château-Thierry the enemy is retreating in total disarray." An attached sketch drawn by the observer showed the head of the French column on the line Orbais—St. Martin-d'Ablois (southwest of Épernay) at 12:30pm. The sketch further indicated the presence of "large disorganized masses of infantry, artillery, and vehicles at Condé-en-Brie." FFA 23's Rudolf Berthold also observed the panicked French retreat during a sortie on the 3rd:

Every day I have flown reconnaissance missions. First we followed the enemy incessantly to St.

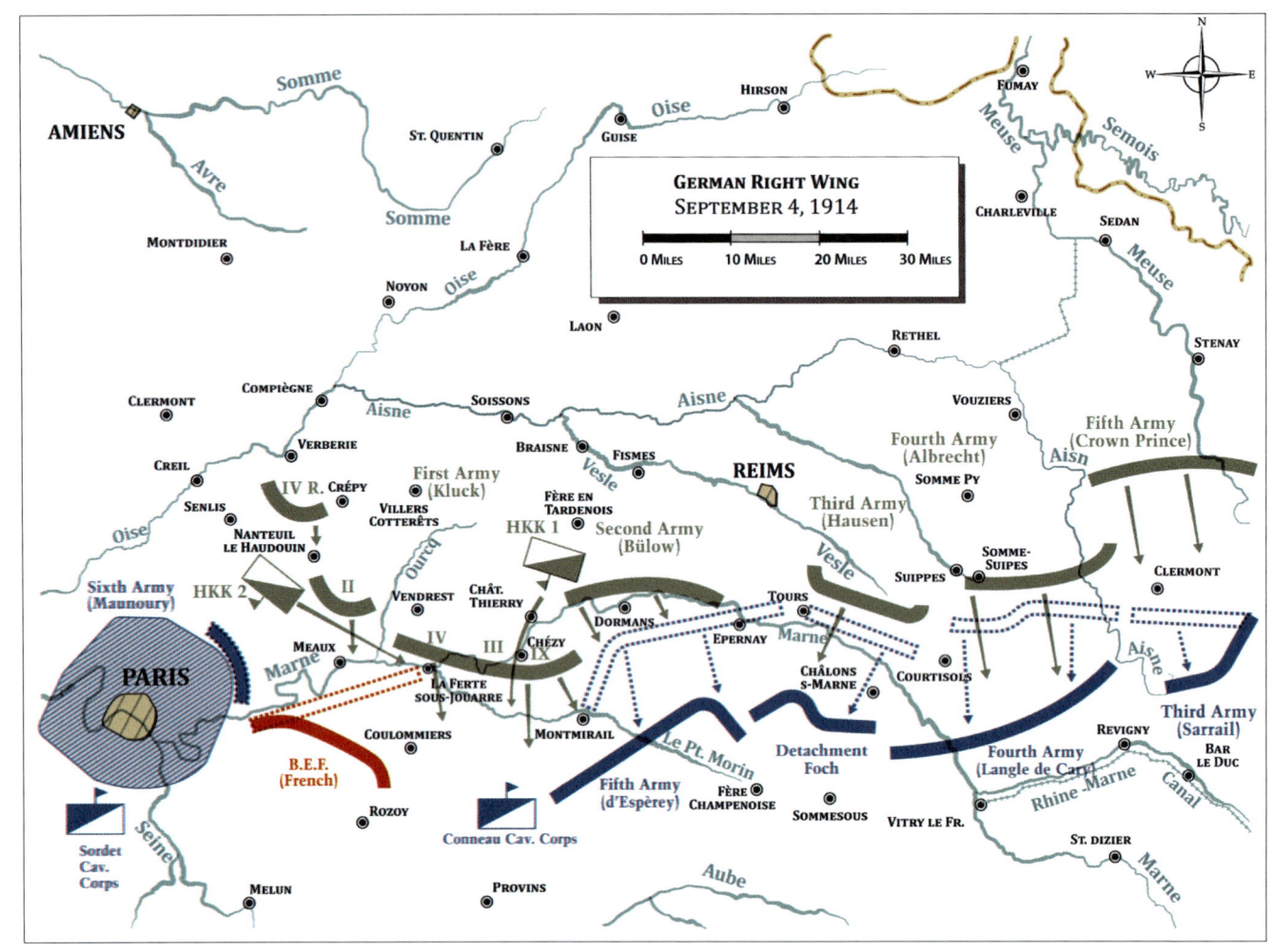

Quentin. They were on forced marches... and I was amazed by the relatively good order displayed by the retreating French columns... But, the soldiers were scarcely across the Marne River when there was no longer any restraint: without discipline they threw away their weapons and knapsacks and fled into the countryside... I went down to 100 meters' altitude and wrote and sketched what I saw... The Army Command's order was clear: relentless pursuit![53]

That evening Second Army's Chief of Staff, *Generalleutnant* Otto von Lauenstein, dispatched a message to Kuhl at First Army HQ stating: "Second Army has closely pursued the enemy today up to and beyond the Marne. Enemy also pouring back south of the river in complete disorder. Marne bridges in part destroyed." Second Army's order of the evening on the 3rd read in part: "...Enemy beyond the Marne is retreating to the south and southeast. Only small detachments are holding this evening in front of X AK. Pursuit beyond the Marne will be continued tomorrow." Thus, with these communications the coordinated pursuit across the Marne by First and Second Armies was assured.[54]

To the east, Hausen's Third Army advanced toward the Marne in the direction of Tours—Châlons with the intention to seize the fords in that sector. However, Hausen's forces met significant French resistance and were unable to reach the river as originally planned. As a result, the Saxon Army ended the day to the north of Tours—Châlons with orders to renew their advance the following morning.

Nevertheless, despite having failed to reach the Marne, Hausen's army still managed to score a significant victory on September 3rd when the XII RK captured the fortified city of Reims without a fight. The quick and bloodless capture of the important French city was primarily the result of the work done by the Third Army *Fliegertruppe*, who had kept close watch on it throughout the day. Indeed, shortly after noon an airman from FFA 22 landed at Third Army HQ to alert Hausen of large numbers of French troops seen withdrawing south out of the city. Hausen deduced from the airmen's report that Reims had been vacated and directed XII RK to conduct a surprise attack against the fortresses' eastern front. As ordered, XII RK's 23 RD quickly dispatched its two brigades to capture the fortresses' outer defenses. Meeting no resistance, the attacking infantry captured their objectives without a single shot fired. Continuing on, the Saxon reservists entered the city's interior and declared Reims a "captured city" shortly after sunset—a victory that would allow Third Army to continue their advance without the need to detach a large covering force to cover its lines of communication.[55]

September 4: The Pursuit Continues

The following day, September 4th, each of the three German right wing armies continued their advance south of the Marne. In Third Army's sector, Hausen's Saxons successfully crossed the Marne and pushed as far south as possible, compelling Third Army HQ to declare the following day "a day of rest" for the entire army. To the west, Second Army advanced south of the Marne to the line: Pargny-la-Dhuys—Épernay without making contact with hostile forces. To Bülow's right, First Army's IX AK threw back French forces in the direction of Montmirail, thus continuing their trend of pushing back French formations to the southeast as Moltke's directive prescribed. Located on IX AK's right, III AK defeated a French rearguard and reached its objective of the day near Montolivet. Meanwhile, IV AK reached the village of Rebais without incident. In covering towards Paris, II AK came to a rest in the elbow of the Marne northeast of Meaux while IV RK remained deployed *en échelon* in the area of Nanteuil-le-Haudouin.[56]

The flying sections of all three right wing armies were busy throughout the 4th keeping watch over the retreating allied columns. Third Army's airmen continued their excellent reconnaissance work delivering reports directly to army HQ that described in detail "the enemy's abandonment of Châlons-sur-Marne and the immediate environs of the city." It was further reported that the French forces in their sector showed signs of withdrawing from the region south of Esternay with two French corps ending the day far to the south along the line: Sézanne—Fère-Champenoise. Meanwhile, Second Army's *Fliegertruppe* similarly spent the day updating their army commander on the position of French Fifth Army's retreating columns in their sector. However, Second Army's distance from the French prevented the airmen's reports from changing the operational situation. Thus, the day ended without any combat in Second Army's sector.[57]

The majority of the sorties performed by First Army's fliers on September 4th were to the south and southeast over the Grand Morin River sector. As a result, First Army HQ remained fully informed of the general position of French Fifth Army and parts of the neighboring BEF. FFA 12 delivered the first report of the day at 12:15pm which declared: "enemy forces

Above: General Ferdinand von Quast, commander of IX Armee Korps. (Author's Collection)

retreating in haste from Montmirail. Intermittent and partly disorganized columns south of First Army on the roads from Montmirail to Esternay. Dust clouds also noticeable on all roads south of Esternay."[58] FFA 11 produced a report later that afternoon describing an artillery battle that had been witnessed at Montmirail between forward elements of IX AK and a French rearguard:

I observed the enemy's artillery firing in several groups from the heights southwest of Montmirail to cover the retreat of the rest of the enemy's forces.[59]

With FFA 11's report in hand, Kluck and Kuhl had verification that IX AK had caught and reengaged French Fifth Army—an encouraging sign that the enemy could be brought to battle the following day. Shortly thereafter, II AK HQ radioed that its FFA 30 had found "large [British] bivouacs west of the line Coulommiers—Rebais near Villers, Aulnoy, Coulommiers, and Mauperthuis with large columns departing southwest from Coulommiers."[60] In the

Above: Wartime photograph of a *Taube* reconnoitering the area around Paris. (Author's Collection)

meantime, FFA 7 delivered another report confirming that "2 enemy columns were observed moving south on the road Montmirail—Rieux—Tréfols." Together, these reports gave First Army HQ a complete picture of the situation to the south of their attacking left wing. This enabled Kluck and Kuhl to plan the following day's operations accordingly—a testament to the hard work of their aviators, who had been performing long-range patrols almost non-stop since the Battle of Mons ended on August 23rd.

Although Kluck's aviators had successfully clarified the Franco-British deployment to the south, the situation to the west remained unclear. The loss of FFA 30's reports on September 3rd had caused First Army to blindly continue its advance without any idea of the growing threat to their vulnerable flank. Most members of the First Army HQ staff believed the flank to be safe, but the lack of a report confirming this was unsettling. Thus, desperate for definitive intelligence, First Army HQ sent the following message to II AK HQ prior to dawn on the 4th: "*Aerial reconnaissance to be conducted in the direction of Dammartin—Paris—Choisy with particular focus on the area of Coulommiers.*" Amazingly, despite receiving a direct order to send their airmen west to the army's right flank, II AK HQ directed all aerial reconnaissance to be conducted to the south. As a result, all reports produced by FFA 30 on the 4th illustrated the BEF's deployment—news of little strategic value.

Nevertheless, a second opportunity to learn of the situation on the army's right flank presented itself later that afternoon when one of FFA 9's Taubes strayed off course and accidently located French troop concentrations in the area of Dammartin (northeast of Paris). In his report, the flight's observer, *Oblt.* Roos, described what he witnessed:

Significant enemy deployment taking place on the line Villeron—Chennevières—Épiais-lès-Louvres. One infantry regiment located south of Le Blanc-Mesnil. Supply trains moving out from Paris towards the east. Large camps witnessed in the eastern sections of Paris full of large and small tents.[61]

Upon landing, Roos' report was immediately dispatched to IV AK HQ in order to be forwarded to First Army HQ. Incredibly, the staff officer that received the message disbelieved the localities mentioned in the report based upon FFA 9's assigned search area. Questioning the report's accuracy, the staff officer tragically chose not to forward Roos' intelligence report.[62] As a result, First Army HQ remained uninformed of the enemy's growing strength on their open right flank despite having specifically ordered aerial reconnaissance flights in that direction the night before. Once again, the absence of a dedicated aviation officer on staff at army headquarters was seriously damaging the German chances for success.

That evening, at their new headquarters near the village of Rebais, Kluck and Kuhl carefully analyzed the situation and began to make plans for the following day—September 5th. Both generals concurred that an envelopment of French Fifth Army was no longer feasible. It was also agreed that the further First Army drove south, the harder it would be to cover their flank towards Paris. Finally, it was acknowledged that the army had reached the limit of its endurance and badly needed a day of rest. Thus, an obvious question presented itself: should First Army, with its IX and III AKs having made contact earlier that day, continue south on the 5th or should the order be given to stand fast? With First Army HQ remaining ignorant of FFA 9's important discovery regarding significant enemy activity around Paris and with the army's left wing already on the southern bank of the Marne, the decision was ultimately made for the three left wing corps to continue their southerly advance for one more day for the express purpose to push the allies back and clear space for the German right wing armies to rest and reorganize for the next phase of the campaign.[63]

Earlier that afternoon OHL had received two radiograms sent by First Army HQ the previous day during the army's southward advance across the Marne. The messages informed Moltke that First Army was unable to comply with OHL's General Directive of September 2nd and had crossed the Marne *ahead* of Second Army to drive the French away from Paris. This was the moment that Moltke first learned of First Army's position south of the Marne with its right flank fully exposed to a possible attack from Fortress Paris. Unbeknownst to Kluck and Kuhl, Moltke held evidence that the French had been transferring large numbers of troops from other sectors of the front towards Paris on the allied left wing in obvious preparation for a counter-offensive. However, OHL's great distance from the right wing (as well as Moltke's relaxed command style) had prevented this information from ever reaching the commanders in the field. Consequently, Kluck and Kuhl had made their decision to cross the Marne on September 3rd without adequate knowledge of the allied forces' strength in the area of Paris—knowledge that OHL possessed before September 3rd.[64]

Moltke's knowledge of the French redeployments towards Paris had been acquired exclusively through the work of the *Fliegertruppe*. On September 2nd, 3rd and 4th, aviators from the Third, Fourth, Fifth, and Sixth Armies reported large numbers of French troops being moved by rail away from their respective sectors to the west in the direction of Paris.[65] Recognizing their importance, each army commander forwarded this intelligence directly to Moltke at OHL for his consideration. Upon reviewing the reports, Moltke was able to predict Joffre's intent to relocate significant numbers of troops in order to launch a counter-offensive out of Paris against the German right wing. Thus, when the German Commander finally learned of Kluck's decision to cross the Marne on the afternoon of the 4th, he understood that decisive action had to be promptly taken to avert a disaster.

Moltke Changes Strategy

After learning of the French redeployment and the serious danger it posed to his own right wing, Moltke decided to formulate an entirely new concept of operations for the seven western armies. He believed that the Franco-British forces were retreating southward toward the Seine River with the objective to draw in their German pursuers and expose them to an attack from the direction of Fortress Paris. Moltke also understood that the right wing armies did not possess enough troops to continue a southward advance and cover their flank toward the French capital. In short, the Germans' initial offensive had reached its *Clausewitzian* 'culmination point.' They could proceed no further. A major change in the German Army's operational strategy was required. Although redeploying large numbers of troops to the right wing to resume the offensive would have been preferable, Moltke recognized that this would take too much time to complete and was therefore not practical. Thus, after careful consideration, Moltke decided to turn First and Second Armies into a giant flank guard with Second Army being assigned the sector between the Seine and the Marne while First Army held the area between the Marne and the Oise. Together, these two powerful armies would check any allied sortie from Paris, giving the forces of the German left and center the ability to launch an offensive of their own.

According to Moltke's new concept of operations, the campaign was now to be determined by the armies of the German left and center. On the left, the Sixth and Seventh Armies were to utilize newly available siege batteries to attack and breakthrough the French line along the Moselle between Toul and Épinal. At the same time, the two center armies (Fourth and Fifth) were to turn southeast and strike the French between Verdun and Vitry-le-François while the Saxons of Third Army were to attack south towards Troyes and drive a wedge through the center of the French front—penetrating deep into their rear and threatening their line of communications. Finally, First and Second Armies were to hold defensive positions and check any allied efforts out of Paris with carefully coordinated counter-attacks.[66]

Moltke's new General Directive, describing in detail the commander's new plan, was dispatched shortly after midnight on the 5th to each of the seven western armies. The order read in part:

> The enemy has eluded the enveloping attack of the First and Second Armies and has succeeded, with part of his forces, in gaining contact with Paris. Reports from the front and information from reliable agents lead furthermore to the conclusion that he is transporting towards the west forces drawn from the line Toul—Belfort and that he is also proceeding to withdraw forces from before our Third—Fifth Armies. It is therefore no longer possible to roll up the entire French army towards the Swiss frontier in a southeasterly direction. We must rather expect to see the enemy transfer numerous forces into the region of Paris and to bring up new forces in order to protect his capital

and threaten the right flank of the German armies.

The First and Second Armies must therefore remain before the eastern front of Paris. It will be their mission to oppose offensively any enemy effort coming from the region of Paris and to lend each other mutual support in these operations.

The Fourth and Fifth Armies are in contact with important enemy forces. They must endeavor to drive these forces without respite in a southeasterly direction, which will have the effect of opening the passage of the Moselle to the Sixth Army between Épinal and Toul...

The Sixth and Seventh Armies will initially conserve their present mission of holding the enemy forces which are before their front. They will advance as soon as possible to attack the line of the Moselle between Toul and Épinal covering themselves towards these fortresses.

The Third Army will march on Troyes—Vandœuvre; as circumstances dictate it will be employed either to support First and Second Armies beyond the Seine in a westerly direction or to participate in a southerly or southeasterly direction in the operations of our left wing.

Accordingly, His Majesty orders:

I. First and Second Armies will remain facing the eastern front of Paris to oppose offensively all enemy attempts starting from Paris: First Army between the Oise and the Marne... Second Army between the Marne and the Seine. It is recommended that the two armies keep the main bodies of their forces far enough away from Paris to be able to retain sufficient liberty of action for their operations. 2nd Cavalry Corps will remain under orders of First Army and will transfer one division to 1st Cavalry Corps. 1st Cavalry Corps will remain under the orders of Second Army and will transfer one division to Third Army.

The mission of the 2nd Cavalry Corps will be to observe the northern front of Paris between the Marne and the lower Seine as far as the coast. Distant reconnaissance beyond the Lille—Amiens railroad line will be undertaken by the aviation of First Army.

1st Cavalry Corps will observe the southern front of Paris between the Marne and the Seine below Paris; it will reconnoiter in the direction of Caen, Alençon, Le Mans, Tours, and Bourges and should receive the aviation necessary for this purpose.

The two cavalry corps will destroy the railroad lines leading to Paris as near to the capital as possible.

II. Third Army will march on Troyes. A division of cavalry will be transferred to it by 2nd Cavalry Corps. Reconnaissance is ordered towards the line Nevers—Le Creusot; aviation necessary for this mission should be attached to it.

III. The Fourth and Fifth Armies advancing resolutely towards the southeast will open the passage of the Moselle to the Sixth and Seventh Armies...

IV. The mission of Sixth and Seventh Armies remains unchanged.

(Signed) Moltke[67]

Moltke's new concept of operations represented a complete abandonment of the original German War Plan which had prescribed all major offensive operations to be conducted by the right wing. According to the new stratagem, the two right wing armies were to relinquish the initiative and hold their ground while the armies of the left and center carried out a converging attack against the allied right wing, which was expected to be significantly weakened after the redeployment of forces to Paris. Perhaps the most damning part of Moltke's latest brainchild, victory was dependent upon the left wing armies breaking the virtually impregnable fortress line between Toul and Épinal—a plan that was ultimately doomed to failure.

Due to its great distance from Luxembourg, First Army HQ did not receive Moltke's order until 7:00am on the morning of the 5th—several hours *after* the army's order of the day to continue south had already been dispatched to the troops. Consequently, First Army's lead columns spent the morning hours of the 5th marching south towards the Seine, further exposing themselves to the French forces gathering around Paris before new orders could be drafted to pull them back. During that time however, Kluck's army was able to successfully push back large numbers of allied troops through Sézanne, Esternay, and Provins towards the south, forcing the enemy in several instances to hastily abandon their bivouacs and vehicles out of fear of capture.[68]

To ensure his new directive was properly executed, Moltke dispatched a member of his staff, *Oberstleutnant* Richard Hentsch, to First and Second Armies on the morning of 5th. After a long and hazardous drive to the front, Hentsch arrived at

First Army HQ at Rebais at 2:00pm. Upon his arrival, Hentsch informed Kluck and Kuhl of the bleak situation. Indeed, according to Kuhl's notes Hentsch bluntly stated the following:

The situation is bad. Our left wing armies are held up in front of Nancy—Épinal and in spite of heavy losses is not progressing a single step. Verdun is cut off. To the west of Verdun, Fourth and Fifth Armies are executing a sweeping movement in order to deliver a flank attack upon the French stationed behind the front: Verdun—Toul. There too the advance is very slow. Large numbers of troop movements are taking place from the French right wing toward Paris...[69]

Listening intently, Kluck and Kuhl received Hentsch's assessment with total bewilderment. Throughout the campaign the two generals had been led to believe that the German left and center armies were victorious and making substantial gains each day—successfully tying down the French forces in their respective sectors and not allowing any major redeployment to occur. Instead, according to Hentsch, the left and center armies had been held in check, creating the conditions for large numbers of French troops to be transferred to Paris on First Army's right wing. Puzzled by the fact that they were just now learning of these redeployments, Kluck and Kuhl instantly recognized the danger First Army was in and promptly took action.

Thus, acknowledging the seriousness of the situation, Kuhl immediately created a plan to move First Army to safety north of the Marne. Kuhl discussed the specifics of the plan with Hentsch, who had remained at First Army HQ throughout the afternoon. After receiving a detailed briefing, Hentsch declared the plan to be in conformity with Moltke's directive and emphasized several times that the movement was to be executed "calmly and that there was no particular hurry." This suggestion was based on the reasonable assumption that the allies were still in the process of retreating and redeploying and would not be able to launch a unified counter-offensive against the German right wing for at least two days.[70] However, unbeknownst to Hentsch and Kuhl, the allied counter-attack had already started to the north in IV RK's sector and would soon spread across the entire front the following day. As a result, First Army would have no chance to march to the northern bank of the Marne in the 'slow and methodical manner' that had been suggested by Hentsch. Instead, Kluck's army would be compelled to rapidly redeploy its forces north of the Marne to defend its flank and rear in the decisive engagement of the campaign.

Joffre's Counter-Offensive: The Plan

Although the attack had been defeated, French Fifth Army's counter-offensive on August 29th at Guise—St. Quentin was successful in buying the French enough time to transfer additional forces and create a new army—the Sixth Army—on the allied left wing in accordance with Joffre's revised plan of operations. Allied troops across the front were obliged to continue their retreat after the engagement at St. Quentin in order to reorganize and concentrate their strength for the coming counter-offensive. Thus, the withdrawal continued into the first days of September as the allied leadership waited for an opportunity to strike. That moment finally presented itself on September 3rd when French aviators discovered that German First Army had crossed the Marne and presented its flank to the newly created French Sixth Army and the large garrison inside Fortress Paris. Joffre and the allied leadership agreed upon learning of Kluck's movements that the time for the allied counter-offensive had finally arrived.

At 10pm that evening, after spending most of the day finalizing the details of the plan with his staff, Joffre dispatched his *General Instructions No. 6* to each of his subordinate army commanders. The new directive outlined exactly how the anxiously anticipated allied counter-offensive was to be conducted.

> **General Instructions No. 6**
> I. It is desirable to take advantage of the exposed position of the German First Army and to concentrate against it the strength of the Allied left armies.
>
> All dispositions will be taken during the day of September 5th with a view of launching an attack on the 6th.
>
> II. The dispositions completed on the evening of September 5th will be:
> a) All available forces of the Sixth Army northeast of Meaux ready to cross the Ourcq, between Lizy-sur-Ourcq and May-en-Multien in the general direction of Château-Thierry.
>   All available elements of the Cavalry Corps that are in the vicinity will be restored to the command of Sixth Army for this operation.
> b) The British Army, established on the front Changis—Coulommiers facing east, ready to attack in the general direction: Montmirail.
> c) Fifth Army, contracting its front slightly on the left, will establish itself on the general front Courtaçon—Esternay—Sézanne, ready to attack in the general direction from south to north.
>   Conneau's Cavalry Corps will maintain liaison

Above: Michel-Joseph Maunoury, commander of French Sixth Army. (Author's Collection)

> between the British Army and Fifth Army.
> d) Ninth Army (General Foch) will cover the right of Fifth Army, holding the boundaries of the Marshes of St. Gond and moving a part of its forces forward on the plateau north of Sézanne.
>
> III. The offensive will be taken by the respective armies on the morning of September 6th,
>
> (Signed)
> J. Joffre[71]

Each allied army had a unique role to play in the coming operation. In accordance with Joffre's *General Instructions No. 6*, the counter-stroke was set to begin at dawn on the 6th. Located on the extreme left, General Michel-Joseph Maunoury's Sixth Army was to aggressively attack German First Army's flank and rear. Meanwhile, French Fifth Army (now under the command of Franchet d'Espèrey after Lanrezac was relieved on September 2nd) was ordered to turn around and attack their familiar adversary, German Second Army, while Sir John French's battered BEF attacked on their left. To the east, the newly formed French Ninth Army (created when French Fourth Army was split in two) was likewise ordered to turn around and attack the closest hostile units in their sector: German Second Army's left wing and Hausen's Third Army. On their right, French Fourth and Third Armies were directed to halt their retreats and dig in—thereby holding the German forces in their sector while the counter-attacking allied left wing sought victory. Finally, the French First and Second Armies were to continue to use the formidable defensive positions of the French Fortress line to repulse the inevitable attacks that were to be delivered by the armies of the German left wing.[72] Joffre envisioned that the allied armies would win a great victory. Working together, the allied armies would wrest back the initiative and eventually take the war to German soil. Unfortunately for Joffre, the actions of a single German corps commander the day before the battle would ultimately cost him and the Entente the element of surprise and a chance at that highly anticipated victory.

Von Gronau Saves First Army

In accordance with Joffre's *General Order No. 6*, Maunoury's Sixth Army moved east on the morning of the 5th out of Fortress Paris towards the Ourcq River. From this position, Maunoury's forces were to conduct its attack at dawn the following morning against German First Army's exposed right flank. In order to keep the element of surprise, Maunoury was directed not to engage the Germans on the 5th, using the day instead to simply move into position and prepare for the following day's attack. Regrettably for Maunoury and the Entente, the French advance had been spotted by German cavalry who promptly alerted IV RK's commander, *General der Artillerie* Hans von Gronau. With his cavalry reporting a sizeable enemy force advancing in his direction, Gronau decided on a preemptively attack. Accordingly, IV RK's two divisions (7th and 22nd RDs) were ordered to move west to the heights near Monthyon and prepare for battle.[73]

At 1:00pm, after quickly seizing the hills and deploying their artillery, Gronau's men opened fire on Maunoury's lead columns, thus officially starting The Battle of the Marne a full day before Joffre had intended. Under heavy fire, Maunoury's men frantically deployed for battle, then launched a series of failed assaults against the heights held by Gronau's forces. The battle that ensued continued unabated throughout the afternoon all along IV RK's front from St. Soupplet in the north to Penchard in the south.

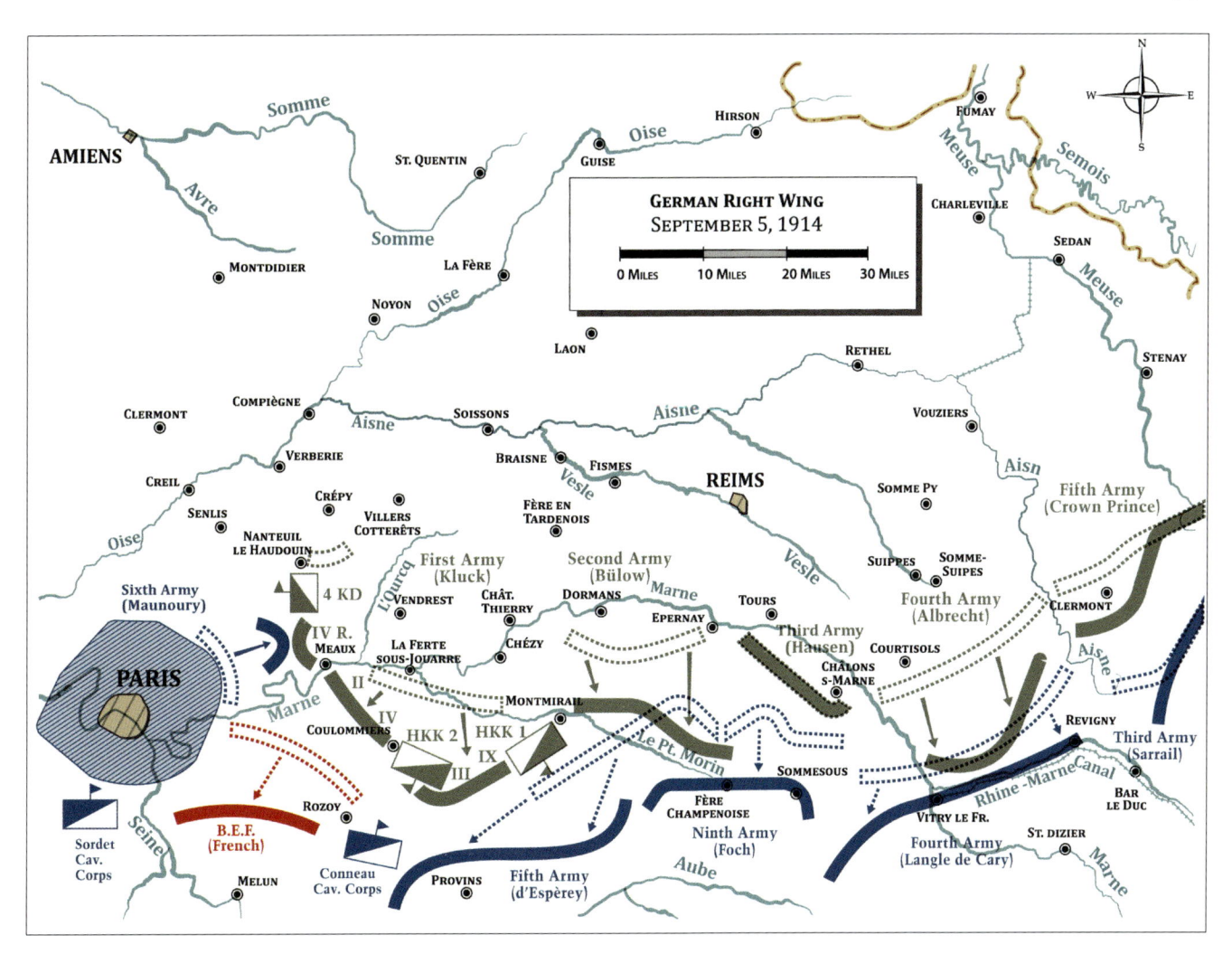

While the French attacks were uncoordinated and poorly supported by artillery, Gronau and his staff were seen constantly moving around the battlefield, guiding IV RK's reserves and coordinating its artillery fire for maximum effect. As a result, the outnumbered German reservists were able to turn back the French advance, keeping First Army's flank and rear protected as darkness brought the day's desperate fighting to a close.

Despite winning the day, Gronau and his staff smartly understood that the French would reinforce themselves and continue their attacks the following morning. Late in the afternoon Gronau's cavalry had reported that two unengaged French columns were seeking to maneuver around IV RK's open right flank—threatening the small understrength reserve corps with encirclement. Thus, Gronau responded by ordering his troops during the night to fall back to a more defensible position behind the Thérouanne River. From here, the IV RK would hold its ground while awaiting reinforcements.[74]

First Army's aviators had been heavily active throughout the 5th by flying sorties to the south and southwest ahead of the army's advance. On Kluck's left wing, FFAs 7 and 11 produced reports that kept First Army HQ informed of French Fifth Army's whereabouts as they retreated further south beyond Provins and Sezanne.[75] First Army HQ's FFA 12 performed several long-range reconnaissance flights to the south in order determine the extent of the enemy's redeployment toward Paris. The subsequent report informed Kluck that major enemy rail activity was sighted at 11:40am over the area Nogent-sur-Seine—Romilly. Indeed, the observer's message concluded that, "the enemy's activity suggests overwhelmingly that a major redeployment is taking place to the west."[76] During the same flight, the observer also noticed a RFC airfield north of Melun with 32 aircraft congregated wingtip to wingtip. Meanwhile, FFA 9 dispatched its aircraft to the southwest, locating the BEF's main body west of Rozay at Tournan.

Finally, on the army's right-wing, II AK's FFA 30 was charged with reconnoitering the area around Paris after having failed to follow orders and do so the day before. In order to cover a larger area of the important sector, the section's commander, *Hptm.*

Vor dem Feind in guter Deckung.

Above: A German field gun operating with the assistance of the battery's observation wagon. (Author's Collection)

Felix Wagenführ, ordered all flights to be performed south of Paris in the morning and north of the city in the afternoon with each flight covering a unique area of the sector. Although the morning sorties produced several noteworthy reports regarding the location of allied troop concentrations as well as heavy rail traffic moving through the French capital to the north, the section's final flight of the afternoon produced the day's key discovery. While searching the area around Meaux at 5:45pm, the flight's observer, *Ltn.* Viktor von Knobelsdorff, noticed a large battle between French Sixth Army's lead columns and the IV RK. Knobelsdorff took note of the French force's strength and position and informed his pilot, *Ltn.* Hans-Carl von Ruville, to continue along their previously agreed route. Flying in their personally assigned Albatros B.I, Ruville and Knobelsdorff quickly finished their patrol and returned to FFA 30's airfield at Pommeuse. Upon landing, Knobelsdorff hastily penned his report and dispatched it via touring car to II AK HQ. The report read:

Airplane #: Albatros B.65.14
Pilot: *Ltn.* von Ruville
Observer: *Ltn.* von Knobelsdorff

Pommeuse,
5 September 1914
7:00pm

To: II AK HQ
1. 5:45pm—IV RK heavily engaged in battle northwest of Meaux against an enemy force from the direction of Paris.
2. Cavalry regiment with infantry and a few vehicles advancing on the road Lagny—Ferrières as far as at Saint Georges.
3. North and south Touran and south Presles and at Courquetaine—light troops witnessed—<u>an enemy army corps</u>, partially in bivouacs, partially in movement in these towns.
4. Weak numbers of infantry at Chaumes.
5. North and south Rozy enemy division in bivouac. Columns of vehicles along the route Nesles—Bornay, start of the road to Fontenay. Many vehicles and fighting troops in Rozoy, partially in movement heading west.
6. Time of observation: 5:45—6:15pm.

(Signed)
von Knobelsdorff[77]

Recognizing the report's importance, the II AK HQ staff immediately telephoned its contents to First Army HQ for Kluck and Kuhl's consideration.

First Army's command team first learned of IV RK's battle late in the afternoon when a message from

Gronau arrived at army headquarters. The message was sent to inform the army's commander that IV RK's cavalry had detected a French force attempting to maneuver into First Army's rear and that the corps' two divisions had been ordered to engage them. The brief message read:

First Army HQ:
1. In the forest Montgé—Cuisy there is enemy cavalry which prevents our own from getting a view of the situation. Behind this forest, more enemy cavalry and infantry have been seen.
2. The 7th Reserve Division will advance to the attack in several columns by way of Cuisy-Montgé upon St. Mard in order to drive the enemy back.
3. The 4th Cavalry Division will take part in the attack by advancing through Marchemoret in the direction of Dammartin.
4. The 22nd Reserve Division will be held in readiness between Barcy and Monthyon.
5. IV RK HQ staff will be with the 22nd Reserve Division

(Signed)
von Gronau[78]

Kuhl initially believed the enemy forces described in Gronau's message to be numerically weak and considered their presence to be a minor delaying action aimed at slowing the German advance as it passed Paris. The general's attitude quickly changed, however, when FFA 30's report arrived later that evening describing the enemy's strength in IV RK's sector. Concerned by the report, Kuhl was immediately presented with a message from IV RK's chief of staff that had been transmitted shortly after the day's battles had concluded. The report confirmed FFA 30's observations by stating that the corps' two divisions had successfully engaged a large French force north of Meaux, but was required to fall back to stronger defensive positions after sunset to avoid becoming enveloped the following morning. Fully understanding the seriousness of IV RK's situation, Kuhl immediately telephoned II AK's commander, *General der Infanterie* Alexander von Linsingen, and ordered him to immediately awaken his men and begin marching north to the aid of IV RK. As directed, the men and animals of II AK began moving shortly after midnight. Marching continuously throughout the night, Linsingen's two divisions arrived in IV RK's sector just in time to assist Gronau's men repulse French Sixth Army's renewed attacks at dawn on the morning of the 6th.[79]

Had Gronau not acted in the manner he did, French Sixth Army would have been in a better position to launch its attack on the morning of the 6th against First Army's flank and rear. Instead, thanks to Gronau's actions, First Army HQ learned of Joffre's intentions and was able to recall II AK north of the Marne—a decision that ultimately saved IV RK and First Army from disaster when Maunoury's entire army aggressively resumed their attack the following morning as part of Joffre's grand counter-offensive.

Above: Alexander von Linsingen, commander of II Armee Korps. (Author's Collection)

The Eve of Battle

Although IV RK's battle on the 5th had alerted Kluck and Kuhl of the impending attack against their flank, the two generals remained totally unaware of the impending attack that was to be delivered by the other allied armies. The continued retreat of the Franco-British forces to the south on the 5th had deceived the German leadership into thinking that no counter-offensive from that direction would occur the following day. Consequently, First Army's two left wing corps, III and IX AKs, were given orders to rest on the 6th while the remainder of the army began their movement north of the Marne to counter French

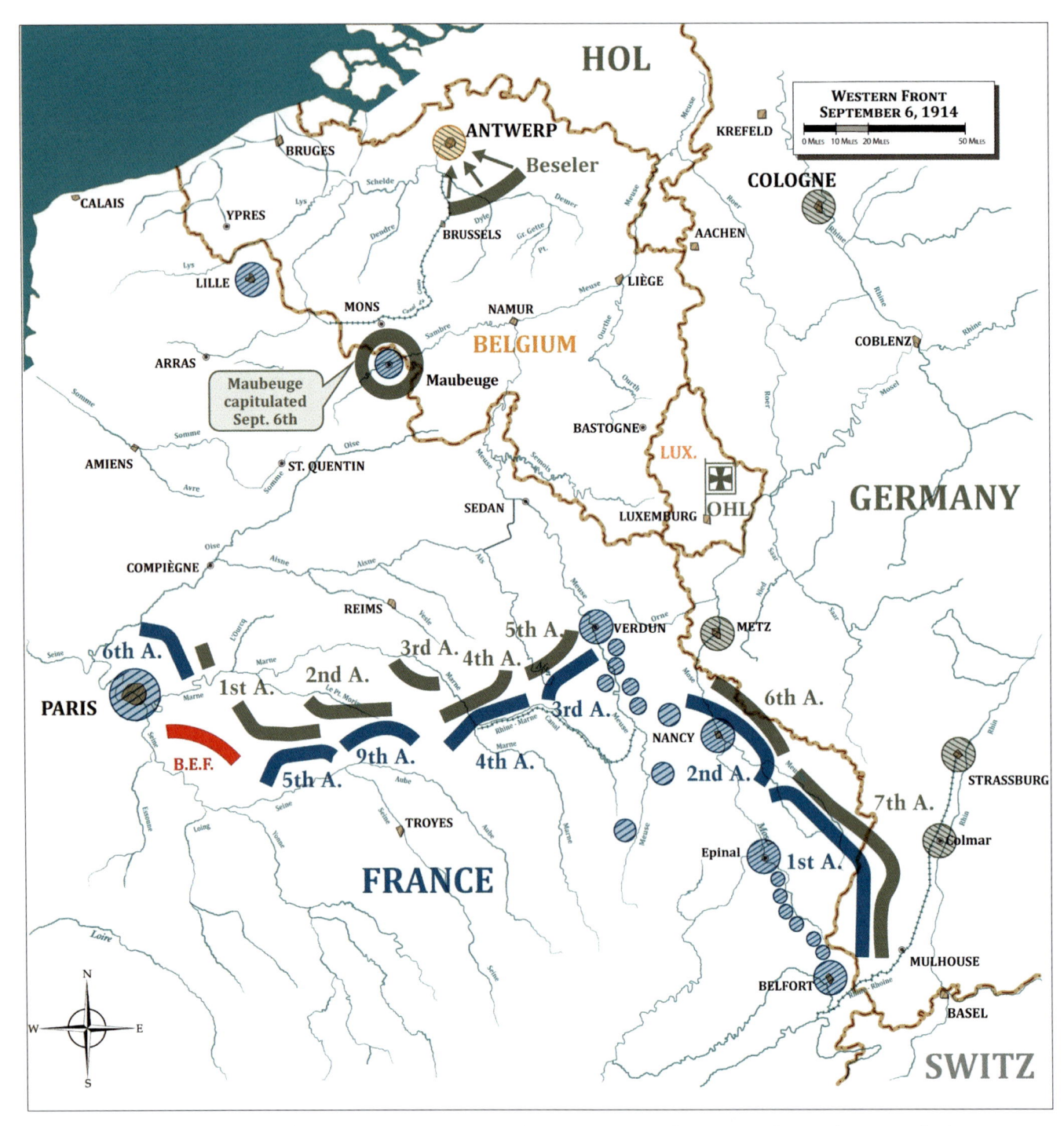

Sixth Army's sortie out of Paris. German Second Army HQ similarly believed that the allied forces in their sector were incapable of offensive action on the morning of the 6th. As a result, the officers and men of Bülow's army expected no major contact with the enemy on September 6th and had planned to mostly use the day to rest and reorganize.

Meanwhile, allied troops throughout the front were informed of Joffre's counter-attack plan. For secrecy, the specifics of the plan were only known by the various Army HQs and had been kept from the Franco-British troops themselves until the evening of the 5th when each army commander dispatched the attack orders amongst their men along with the following appeal from Joffre.

As we engage in the battle upon which the safety of our country depends, all must remember that the time for looking backward has passed; every effort must be devoted to attacking and driving back the enemy. Troops that can no longer advance must hold ground won at any cost and die in their tracks

Above: Louis Franchet d'Espèrey, the newly appointed commander of French Fifth Army following the firing of Lanrezac on September 3. (Author's Collection)

Above: Infantryman belonging to the Guard Korps. (Author's Collection)

rather than retreat. In the present circumstances, no weakness can be tolerated.

J. Joffre[80]

Joffre's dramatic message had the desired effect upon his army. The serious losses, numerous hardships, and constant fatigue of the past month were all quickly forgotten. The minds of the allied soldiers turned instead to preparing for the coming counter-offensive that was to begin at dawn against the German right wing. If successful, the Entente's forces would turn back the German invader and reclaim France's sacred soil. If the attack failed however, the war would be lost. Thus, with perhaps the next 100 years of European history at stake, the campaign's decisive battle was finally underway.

September 6

After weeks of anticipation, French Fifth and Ninth Armies finally launched the counter-offensive at dawn on the 6th in concert with French Sixth Army's renewed assaults on the Ourcq. Located on the extreme right of Joffre's attacking force, Ferdinand Foch's Ninth Army attacked from the area of Fère-Champenoise against the left wing of Bülow's Second Army. To the west, Franchet d'Espèrey's Fifth Army aggressively assaulted Bülow's right wing as well as two corps (III and IX AKs) composing First Army's left. Further west, Sir John French's BEF began a northeasterly advance towards Coulommiers and German First Army's right wing. However, the English had begun the day significantly behind their French allies and were therefore not expected to be a factor in the initial stage of the attack.

Possessing the advantage of numerical superiority, the first hours of the battle were an overwhelming French success.[81] Joffre's forces boldly attacked along a 50 mile front and made some impressive initial gains against the unsuspecting German troops opposing them. By mid-morning frantic messages from the frontlines were reaching the various German corps and army headquarters describing a violent enemy attack and requesting reinforcements. Completely surprised, the German leadership was forced to take quick and dramatic action to stop the allied advance.

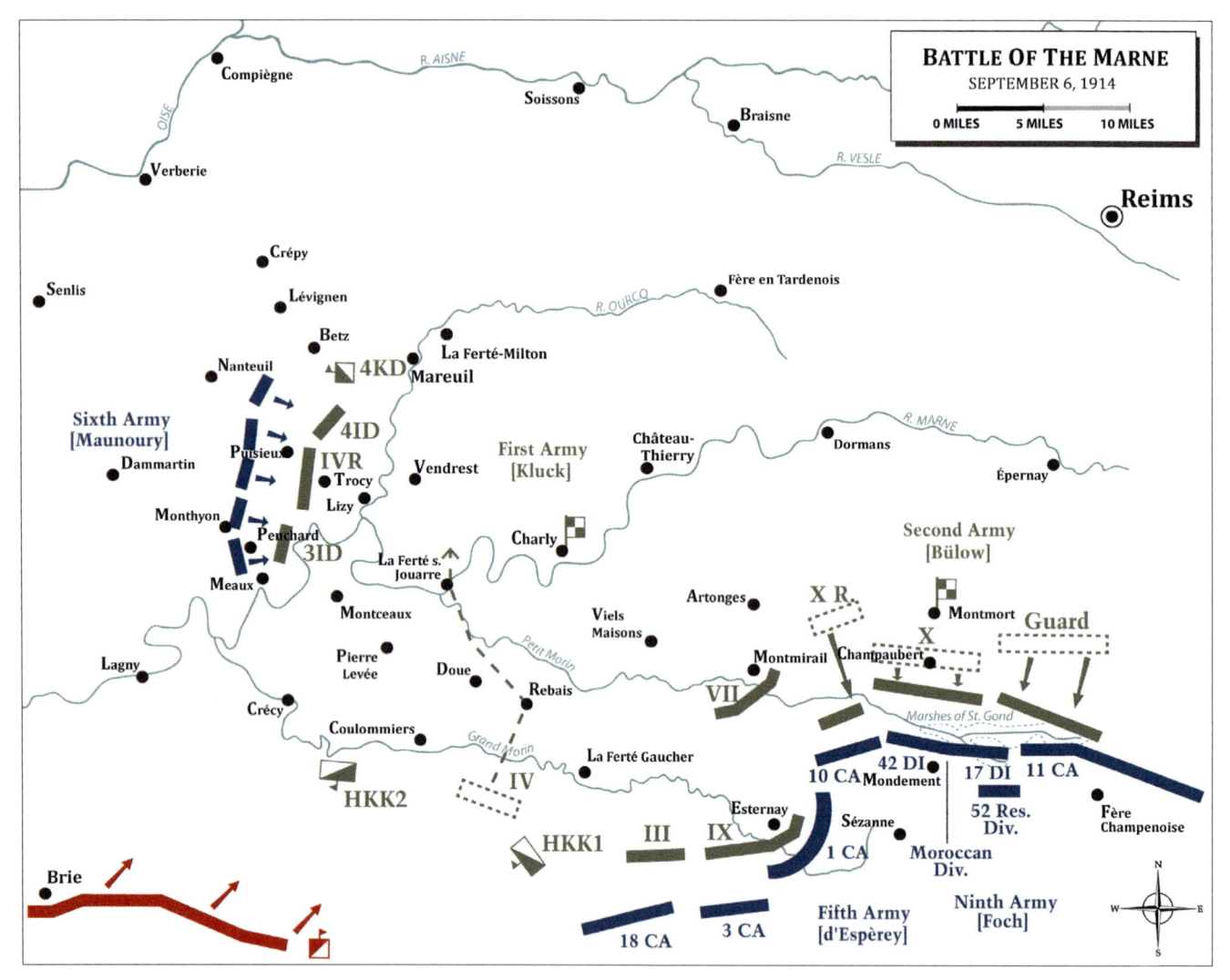

At their headquarters in the village of Montmort, Bülow and the Second Army HQ staff were shocked to hear reports of a large-scale French attack. They had anticipated no combat on the 6th and had planned to use the day to begin quietly swinging the army to the west to face Paris in accordance with Moltke's latest directive. The reports of heavy fighting and the sounds of cannon fire, however, caused Second Army's leadership to quickly reevaluate their situation. At that point it was still unknown whether the Allied attack was indeed a counter-offensive or merely a strong rear-guard action to assist Joffre's forces retreat across the Seine. Thus, it was imperative that Bülow and Second Army's corps commanders quickly determine the intentions and strength of the attacking enemy force. To that end, with the entire army under assault, Second Army HQ ordered all flying sections airborne in order to give the leadership an accurate representation of what was occurring at the front.

Bülow's request for "immediate aerial reconnaissance in a southerly direction" was received by each of Second Army's corps FFAs between 9 and 10am. Hearing the gunfire themselves, Bülow's airmen were already aware of the unusual activity at the front. Extraordinarily thick cloud cover hung over the army's sector, however, causing several FFAs to initially send a single aircraft airborne to determine if aerial reconnaissance was in fact feasible. By noon each of the aircraft had returned with bad news—the cloud cover was too thick to conduct any meaningful surveillance. Throughout the rest afternoon, Bülow's aircraft and aviators remained on alert, waiting for the cloud cover to dissipate.

Without any reports from his aviators, Bülow initially believed the allies' actions to be a ploy designed to facilitate their main body's retreat across the Seine. Second Army HQ held this belief as late as noon when the following order was dispatched amongst Bülow's corps commands:

The French Fifth Army has already got the larger part of its forces behind the Seine. Non-contradictory reports indicate troop displacements from Romilly—Nogent-sur-Seine toward the west; only covering troops are left north of the Seine. Vigorous pursuit at all cost for the purpose of annihilating these forces and destroying the section of railway is required. Hence, continuation of the pursuit.[82]

As ordered, Second Army's various columns advanced south toward the sound of the enemy's cannon fire—believing it to be a rear-guard. However, without the use of their aviators, Bülow's forces unknowingly stumbled into the teeth of French Fifth and Ninth Armies' advance. To the east on Second Army's left wing, the X AK and Guard Korps made contact with French Ninth Army at the northern boundary of the "Marshes of St. Gond"—a broad region of impassable swamplands that possessed only five small roads and three narrow foot-paths.[83] After a quick and powerful attack, Bülow's left wing successfully pushed their French adversaries back across the marshes. Staying aggressive, the German leadership agreed to press forward in order to seize the heights south of the swamps that commanded the entire region. However, because the roads were poor and the swamps themselves were impassable, both the X AK and Guard each sent one division around either side of the marshes to take the heights on the other side.

Marching around the swamps' western perimeter, X AK's 19 ID quickly met stiff resistance on the Poirier Crest—the most important terrain feature in the area.[84] Despite taking the crest, 19 ID was unable to proceed further and open a crossing for the rest of X AK. By nightfall, the division had entrenched with plans to resume the attack the following morning. Meanwhile at the eastern end of the marshes, the Guard Korps' 2nd Guards ID also failed to open a crossing. Acting as Second Army's flank guard, Guard Korps HQ ordered the division to advance slowly and cautiously. As a result, they made little progress against weak opposition. During the advance, Guard Korps HQ received a note from their own FFA 1 which stated that the cloud cover had broken and an aircraft had been sent airborne to reconnoiter near the marshes opposite Second Army's left flank (which was left uncovered by Third Army's decision to halt their advance on September 5th). At 5:15pm an LVG B-type dropped a note at corps headquarters warning of "strong reserves behind the enemy front." Deeply concerned about his flank, Plettenberg halted his corps' attacks in favor of waiting for Third Army's arrival the following morning. The remainder of the 2nd Guard Division's evening was therefore spent moving

Above: Ewald von Lochow, commander of III Armee Korps. (Author's Collection)

around the marshes to link with Third Army's right wing in preparation for the following day's attack.[85]

News of X AK's battle in the swamps quickly reached X RK on Bülow's right wing. X RK's new commander, *General* Johannes von Eben, had been initially ordered to bring his two divisions (19 RD and 2nd Guards RD) to the area Montmirail—Le Gault in preparation for the army's turn toward Paris the following day, but the sounds of cannon fire to the east demanded action. Acting on his own initiative, Eben turned his forces east in order to attack into the left flank of the French forces engaged with X AK in the marshes. By noon, X RK had switched directions and crossed the Petit Morin River. Before reaching their objective, however, Eben's left division (2nd Guards RD) unexpectedly made contact with a different force, French Fifth Army's X CA who was acting as Ninth Army's flank guard. In an extremely disorganized and bloody battle, Eben's men pushed back the French forces five miles before sunset ended the day's hostilities.

French Fifth Army's main attack violently struck First Army's left wing (III and IX AKs), which was positioned six miles ahead of Second Army. Operating under the false assumption that the French were continuing their retreat across the Seine, both of Kluck's left wing corps commanders had promised their men a day of rest as they covered First Army's redeployment north of the Marne. Consequently, the troops were caught completely unprepared when d'Espèrey's artillery opened fire with their preliminary bombardment. Nevertheless, the officers of both III AK and IX AK HQs remained calm and ordered their men and artillery to deploy and prepare for an attack.

In the meantime, III and IX AK HQs requested aerial reconnaissance in order to ascertain the enemy's deployment and intentions. Was the enemy's artillery fire just a demonstration to cover their army's retreat across the Seine or were they conducting a preliminary bombardment as part of a larger offensive? At 10:00am, one of FFA 11's LVG B-types returned with the answer. Peeking through gaps in the thick cloud cover, the flight's observer, *Ltn*. Schwab, successfully discovered several large French columns rapidly advancing to the north towards IX AK—an unmistakable sign that French Fifth Army had halted their retreat and begun a counter-attack.

Upon landing, Schwab was driven from FFA 11's airfield at Leuze to IX AK HQ to deliver his report in person. Schwab's report read:

Airplane #: LVG B.255   Leuze, 6 September 1914
Pilot: *Ltn*. Wehrig   10:00am
Observer: *Ltn*. Schwab

To: IX AK HQ

1. Strong enemy forces (1–1½ Army Corps) witnessed at 9:30am between Seine and Esternay. The enemy is advancing north and will take positions on the line Villouette—Escardes.
2. East of the railway of Esternay (near the woodland of Villenauxe) there was no view due to cloud cover.
3. Sketch of enemy forces witnessed attached.
4. Flight broken off at 9:40am due to intermittent engine failure over Villenauxe.

(Signed)
Ltn. Schwab[86]

Schwab's report was immediately brought to Quast, who was seen outside IX AK HQ anxiously awaiting news from the front. Reviewing the report's accompanying sketch, Quast quickly concluded that the 'strong enemy forces' were likely the I and III CAs that composed French Fifth Army's center. Thus, the report revealed that d'Espèrey's army, in its entirety, had halted its retreat and was now *advancing* against the German battle-line.

Staying true to his reputation as one of the army's most aggressive commanders, Quast ordered a preemptive attack against French Fifth army's left and center. Accordingly, IX AK's two divisions advaned south from Esternay towards the approaching French columns. By noon Quast's forces were fully engaged along a front from Courgivaux to Châtillon. At Courgivaux, IX AK's right division (18 ID) pushed back Charles Mangin's 5 DI two miles to the village of Escardes. Meanwhile to the east at Châtillon, the 17 ID battled inconclusively with the entirety of I CA until darkness brought the day's fighting to a close.[87]

Quast's attack was supported on the right by III AK, who had similarly learned of French Fifth Army's advance through the work of their aviators (FFA 7). The report, delivered to III AK HQ at 11:15am, read:

Airplane #: Euler B.172/13   Villiers-les-Maillets,
                              6 September 1914
Pilot: *Ltn*. Engwer   11:00am
Observer: *Ltn*. Fink

To: III AK HQ

1. Column of infantry seen at 10:05am. Head of column at Chevru, end at Jouy-le Châtel.
2. 1 Infantry regiment seen at Amillis.
3. A similar column of infantry, artillery and cavalry on the road Rozoy—Coulommiers. The head of the column was seen 3 kilometers southwest of Coulommiers at 10:20am.
4. See attached sketches for general location of enemy troops.

(signed)
Ltn. Fink[88]

III AK's aviators had discovered the XVIII CA on the extreme left of French Fifth Army as well as elements of the advancing BEF. Thus, with the French position known and IX AK fully engaged on their left, III AK HQ made the decision to attack.

III AK had begun the day with orders to "conduct a retirement to the north in accordance with OHL's new concept of operations" and had already started its movement when Fink's reconnaissance report arrived. As a result, it took over two hours for the corps' two divisions to turn around and deploy for battle. In order to cover the corps' redeployment, III AK's commander, General Ewald von Lochow, ordered "every available gun" to shell the French positions. Thanks to the German artillery arm's excellent mobility, III AK was able to quickly place over 150 artillery pieces for the cannonade. The French quickly responded, however, with over 200 field guns of their own, resulting in the campaign's largest artillery duel. Late in the afternoon, after hours of constant bombardment, the artillery fire diminished, allowing three French divisions to successfully attempt an advance.[89] Outnumbered and holding ground of little value, both III and IX AK's withdrew to strong defensive positions along the Grand Morin River.

Thick cloud cover hung over III and IX AK's sectors throughout the remainder of the afternoon, causing FFAs 7 and 11 to remain inactive. Aerial reconnaissance towards the west, however, was possible and was conducted by the aviators of First Army's right wing. These sorties were performed in order to keep First Army HQ informed of the allied forces actions during the right wing's withdrawal north of the Marne. The first of these reports, delivered to First Army HQ at 7:00am by FFA 12, alerted Kluck and Kuhl that Joffre's force had halted their retreat and that a counter-offensive had begun.

Airplane #: Albatros *Taube* A.70   Charly,
                                       6 September 1914
Pilot: *Ltn*. Haupt   7:00am
Observer: *Hptm*. Petri

To: First Army HQ

1. Between 6:00 and 6:30am—enemy establishing trenches south of Pezarches.
2. Enemy column from Marles (end) to Lumigny (head).
3. Enemy forces observed at a road fork 3 kilometers [1.8 miles] south of Lumigny.
4. Enemy column observed from La Houssaye (end) to Crèvecoeur (head).
5. Enemy column halted north of Chapelles.
6. Concentration of troops witnessed north of Tournan.
7. Enemy column observed at road fork southeast from Courquetaine (end) to Retal (head). Troop concentrations to the east of the road fork. Short train at railway station at Guignes.
8. Enemy troop concentrations southeast of Courtomer.
9. Another enemy column on the road Rozoy—Vaudoy in the area south of Voinsles. Length: 2½ kilometers [1½ miles].
10. Enemy column seen from La Jariel—Vaudoy. Head of the column on the road south of Vaudoy, end at La Jariel.

(Signed)
Hptm. Petri[90]

Kluck and Kuhl had expected the allied forces directly to the south of First Army (part of French Fifth Army and the BEF) to use the day to continue their retreat across the Seine. FFA 12's report quickly disproved this belief by confirming that the allied troops in First Army's sector had not only stopped their retreat, but had taken to the offensive. Using the report's information, Kuhl dispatched a message to HKK 2 ordering them to prepare for a delaying action. By delaying the advancing allied columns, Kluck's cavalry would buy time for the army to redeploy north of the Marne where IV RK was facing a numerically superior force capable of striking the army's rear. Thus, as ordered, Kluck's cavalry held off the slowly advancing forces of Sir John French's BEF, enabling IV AK to rejoin II AK and IV RK north of the Marne.

Desperate for additional intelligence, First Army HQ ordered FFA 12 to immediately dispatch three aircraft on long-range patrols. Taking off together at 10:45am, each aircraft was sent in different directions (west, southwest, and south) in order to give First Army's leadership a complete picture of the allied deployment. At noon, the first of the three aircraft (LVG B.#256) returned from the Ourcq front with the following report—written and delivered by *Ltn*. Menzel.

Airplane #: LVG B.256   Charly,
                                       6 September 1914
Pilot: *Ltn*. von Kleist   12:00pm
Observer: *Ltn*. Menzel

To: First Army HQ

1. Infantry column (spread over 3 kilometers [1.8 miles]) south of the road Monthyon—Barcy. Column appears to be advancing upon Barcy (observation: 11:35am).
2. Artillery on hills 113 and 89 firing in a northeasterly direction from prepared positions (11:37am).
3. Rail traffic from Paris—St. Mard to Nanteuil (3 trains). Troop embarkations at the St. Mard rail station (11:38am).
4. Infantry column (estimated strength: 1 battalion) seen resting north of road Ève—Lagny-le-Sec (11:40am).
5. An enemy column seen on road Lagny-le-Sec—Silly-le-Long. Head of column at Silly-le-Long at 11:43am. End located at Le Plessis-Belleville.
6. One column observed on the road Bargny—Betz—Acy-en-Multien. Head of Column at the northern edge of the forest located south of Betz. End at Bargny.

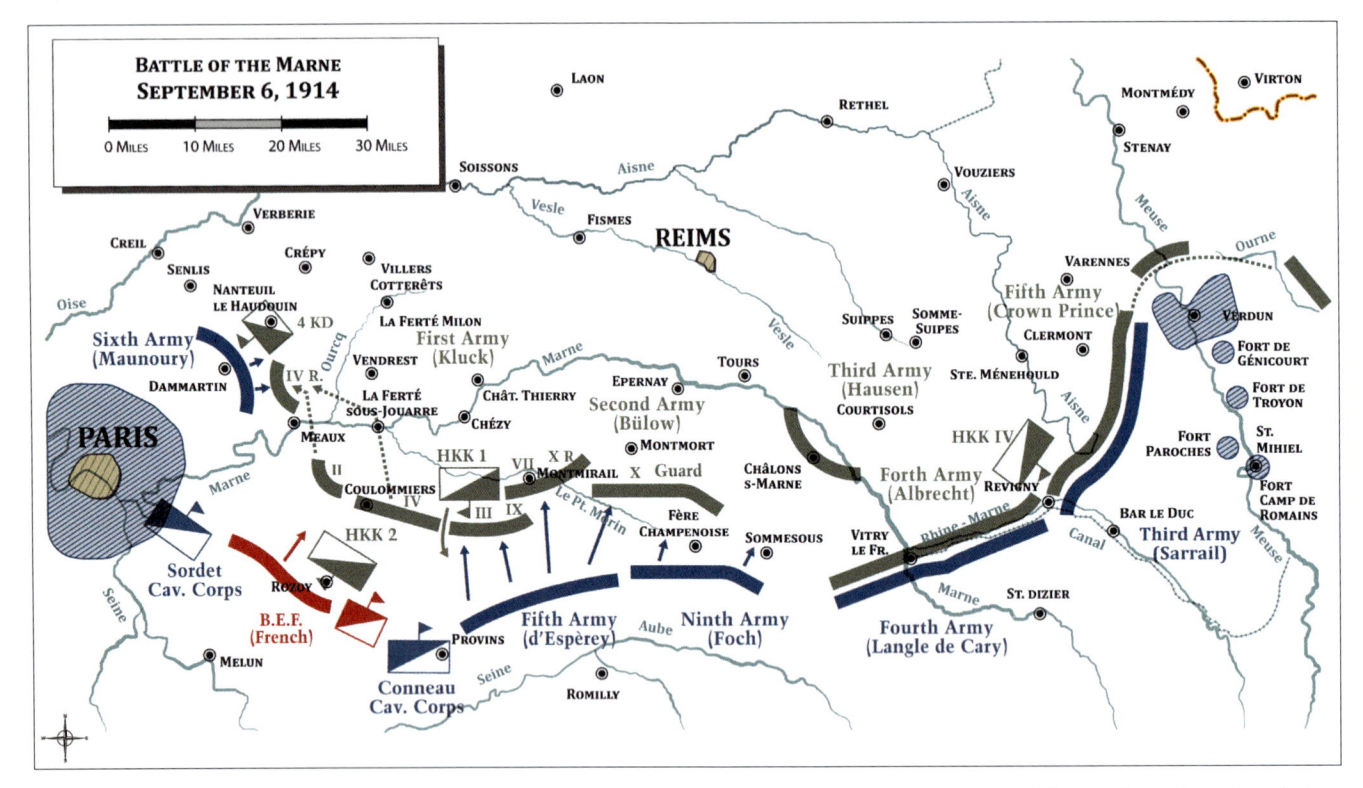

7. Another column was seen marching to Antilly from the wooded area: Bois du Rois; head of column west of Antilly. End of column could not be seen due to dense fog.

(Signed)
Ltn. Menzel[91]

Menzel's report proved that French Sixth Army had been substantially reinforced since first making contact with IV RK the day before. It became clear to both Kluck and Kuhl that Joffre and the allied leadership intended to seek a decision on the Ourcq. Thus, recognizing the need to reinforce the Ourcq front as quickly as possible, Kuhl dispatched an order to IV AK HQ requesting an artillery regiment be rushed ahead under close cavalry escort to IV RK and II AK's aid.

Meanwhile, FFA 12's second aircraft arrived at First Army HQ with pleasant news. The aircraft had been sent southwest to observe the location and condition of the BEF. After a 90 minute patrol the flight's observer, Oblt. von dem Hagen, reported that the English were not yet in position to threaten First Army's operations on the Ourcq. Thus, Hagen's report gave Kluck and Kuhl confidence in knowing that, for the moment, the Ourcq front's flank and rear was secure.

Several hours passed without word from FFA 12's third aircraft, which had been sent south to III and IX AK's sector. Finally, at 4:30pm, word from the crew arrived. Their message was sent from Second Army HQ where the aviators had been forced to land due to poor weather. Despite the region's thick clouds, the flight's observer managed to observe significant numbers of French troops advancing towards IX AK. Accordingly, the following message was relayed to Kluck and Kuhl from Second Army HQ:

Strong enemy forces observed while patrolling the line La Ferté-Gaucher—Provins—Esternay. Enemy columns seen south and southwest of Esternay advancing in the direction of our troops. To the east, the enemy was also seen advancing. No enemy troops were witnessed moving south. All observed enemy forces were being employed to attack in a northerly direction. The enemy's retreat has been halted. Solid bank of clouds prevented further exploration of the enemy's deployment or return to (home) airfield. Forced landing made at Montmirail as a result of the weather. Will attempt return flight at dawn.[92]

FFA 12's final report of the day, relayed through Second Army HQ, confirmed what Kluck and Kuhl had already suspected: the entire allied left wing had stopped their retreat and were now launching a coordinated counter-offensive. The pressing question was: what should First Army's leadership do about it? With the help of his aviator's report, Kuhl had an answer.

First Army's *Fliegertruppe* had performed brilliantly on September 6th. Each of the aviation reports had combined to give Kluck and Kuhl a complete picture of the Franco-British movements in their sector. To the south, French Fifth Army was moving against III and IX AKs as well as the right wing of Bülow's Second Army. To the southwest, the BEF was timidly advancing into the gap that was created after II and IV AKs withdrew to the Ourcq. In the west, a reinforced French Sixth Army was attacking Gronau's IV RK, now supported by Linsingen's II AK (who had arrived on the Ourcq shortly after dawn), in the attempt to envelop and encircle First Army's open right flank.

Citing their aviators' reports, Kluck and Kuhl quickly devcloped a plan to neutralize all three threats. To combat French Fifth Army, III and IX AKs were ordered to withdraw to the northern bank of the Grand Morin River. From there, the two corps were to cover Second Army's flank. Accordingly, both corps were temporarily placed under the command of Second Army HQ, which was closer and better suited for the task. Next, the BEF's advance was to be opposed by the combined strength of HKKs 1 and 2. Together, the two cavalry corps were to fill the gap around Coulommiers [henceforth referred to as "the Coulommiers Gap"] and delay the English advance until First and Second Armies achieved a decision in their respective sectors. Finally, along the Ourcq Front, IV AK was to concentrate its strength as quickly as possible to assist IV RK and II AK defeat French Sixth Army's offensive out of Paris. Indeed, it was believed that IV AK's arrival would give the Germans a numerical and qualitative advantage along the Ourcq—allowing for the engagement there to be fought to a conclusion.

Kuhl correctly recognized during the afternoon of the 6th that the Ourcq Front was ultimately where the battle was to be decided. This realization was largely the product of *Ltn.* Menzel's reconnaissance report. Menzel's intelligence revealed that the French had shifted more troops to the sector during the night of the 5th–6th, thus revealing Joffre's intent to envelop First Army's right wing. First Army HQ had the presence of mind, however, to dispatch II AK to the sector on the night of the 5th. As a result, II AK's two divisions were able to give Gronau's IV RK immediate support when the day's fighting began shortly after dawn. Together, II AK and IV RK repulsed all of French Sixth Army's attacks throughout the afternoon of the 6th. By sunset the fighting in the sector had died down with each side reinforcing and preparing to renew the battle the following day.[93]

Kluck and Kuhl both knew that additional forces were necessary to achieve victory on the Ourcq. Accordingly, First Army HQ ordered IV AK to execute a forced march during the night in order to be sent into the battle at dawn on the 7th. Keeping Moltke informed of the battle's progress, First Army HQ radioed the following message to OHL at 10:45pm:

II AK and IV RK north of the Marne are engaged in violent fighting southwest of Crouy against powerful enemy forces attacking from Paris. IV AK will come into action there tomorrow. III and IX AKs covering the flank of Second Army west of Montmirail which to the east of this point is attacking powerful enemy forces toward the south.[94]

The arrival of IV AK along the Ourcq was believed to be enough to stabilize the front and prevent a French breakthrough. Thus, First Army HQ was relatively optimistic on the evening of the 6th despite their difficult situation. With the assistance of their aviators, Kluck and Kuhl knew the Allies' operational intentions and had taken the necessary steps to counter them. Feeling confident, the two generals went to bed on the evening of the 6th believing IV AK's arrival could secure their flank. The events of the next 24 hours, however, would ultimately prove this belief to be in error.

September 7

Late in the evening of the 6th at his headquarters near the village of Montmort, Bülow and the Second Army HQ staff met to determine the following day's operations. The meeting quickly turned sour, however, as Bülow learned more of his army's dire situation. After quizzing each of the corps' liaison officers it became clear that the troops were too exhausted to execute the wheeling movement to the Seine as Moltke's orders had outlined. Even more concerning was the situation on the army's right wing where First Army's III and IX AK had been temporarily left under Bülow's command. News from HKK 1, which was now working with HKK 2 in the Coulommiers Gap, confirmed that French Fifth Army's left wing and the BEF together posed a major threat to Second Army's right wing. Considering the troop's exhausted state and the vulnerability of his flank, Bülow made the decision to withdraw his right and center behind the protection of the Petit Morin River. OHL was notified of the decision by radio at 2:00am with the following message:

In view of the contact with the enemy at Rozoy, the III and IX AKs are retiring behind the Petit Morin. Left wing of Second Army [X AK and Guard Korps] remains [on the] offensive. Support by the Third Army with all available forces immediately necessary.[95]

By pulling his right wing back 9–12 miles, Bülow had bought time for First and Third Armies, as well as Second Army's left wing, to blunt Joffre's counter-attack.

Above: Friedrich Sixt von Armin, commander of IV Armee Korps (Author's Collection)

Bülow's order reached the right wing corps at 3:00am. By sunrise the III and IX AKs as well as the X RK were safely deployed along new defensive positions on the high ground north of the Petit Morin. Extending X RK's line was VII AK's 13 ID, who had spent September 6th unengaged near Montmirail as the army's flank guard. Together, this force was charged with holding off French Fifth Army and protecting Second Army's vulnerable right flank until a decision could be gained elsewhere along the front.

Bülow believed the battle in his sector would be decided to the east in the Marshes of St. Gond where Second Army's left wing was to link with Hausen's Third Army and attack towards the southwest. The objective of this push, in Bülow's mind, was to breakthrough the French positions, turn west, and roll up French Fifth Army's flank. Owing to Third Army's day of rest on the 5th, Bülow and Hausen both believed their forces had sufficient strength for victory to be attainable. Furthermore, aerial intelligence performed by Third Army aviators on the 6th suggested that the French forces in their sector to be weak and susceptible to defeat if subjected to vigorous attacks. Consequently, Second and Third Army HQs ordered their troops to attack at dawn. Thus, the decision in Second and Third Armies' sector was pinned upon the hope that Bülow's left wing and Hausen's right could combine to break the enemy's positions and roll up their flank before Second Army's right wing was overwhelmed on the Petit Morin—a plan that also depended upon First Army winning its battle on the Ourcq.[96]

Kluck and the First Army HQ staff moved from Charly to Vendrest at 7:00am to meet with their corps commanders and determine an appropriate strategy for the day's impending battle. The main topic of discussion was the proper employment of *General* Friedrich Sixt von Armin's IV AK, who had arrived in the Ourcq sector during the early morning hours after a long night of marching. IV AK's arrival lengthened the battle-line's right wing considerably, enabling the Germans to return to the offensive and launch an enveloping attack. Kluck and Kuhl both consented to the idea, believing to hold numerical superiority on the front's right wing. Thus, the plan agreed upon was to hold firm on the left wing while the reinforced right wing attacked and enveloped the French flank.[97]

As Armin's troops and artillery were moving into position for the attack, First Army's *Fliegertruppe* began the day's first flights. Each flying section was assigned a specific area of the front to cover according to orders dispatched by First Army HQ at 6:00am. In the west, FFAs 9 and 30 were respectively assigned the northern and southern halves of the Ourcq Front. Next, III AK's FFA 7 was ordered to send all their aircraft southwest to report on the BEF's progress against the two cavalry corps, who were covering the gap left by the redeployment of II and IV AKs. Working with Second Army, FFA 11 was to send its aircraft due south in the direction of French Fifth Army. Finally, First Army HQ's FFA 12 was to dispatch aircraft to the Ourcq and to the southwest in the direction of the English in order to keep the army's commander fully updated of the situation in both of the important sectors.[98]

The first of these important reports was forwarded from III AK HQ to First Army HQ shortly before noon. The message, written by FFA 7's *Ltn.* Fink, read:

Enemy column spotted at 10:00am marching at Pezarches towards Coulommiers. At 10:05am bivouacs witnessed northeast of Tournan. South of the Marne there are no enemy forces west of La Ferté-sous-Jouarre.[99]

Fink's report was outstanding news. Kluck and Kuhl had begun the day very concerned about an aggressive English advance into the Coulommiers Gap between the Ourcq Front and Second Army. Indeed,

First Army could face annihilation if the two cavalry corps defending the gap were overwhelmed before victory on the Ourcq could be secured. However, Fink's report confirmed that the BEF was advancing very slowly into the gap and would not be a threat to breakthrough for at least another 24 hours. In other words, thanks to Fink's report, First Army HQ knew they could proceed with the enveloping attack on the Ourcq without having the fear of an English move against their rear.

First Army HQ's elation over Fink's report was quickly negated, however, when the first aerial report from the Ourcq Front arrived bearing bad news. The message was brought directly to Kuhl by FFA 9's *Ltn.* Rohdewadd (pilot) and *Oblt.* Fulda (Observer), who had chosen to land at First Army HQ's airfield to deliver their report as quickly as possible. Fulda's report read:

FFA 9   Vendrest,
                                   7 September 1914
Pilot: *Ltn.* Rohdewadd   10:00am
Observer: *Oblt.* Fulda

To: IV AK

1. Enemy troop concentrations at Nanteuil. (Strength: 1 infantry division). Rail traffic from Paris at Nanteuil rail station.
2. Enemy's defensive positions on their left wing are crude and weak, heavier in the center by St. Soupplets. Near Meaux on the right wing their enemy defenses are also very strong.
3. Road Mareuil—Betz—Crépy-en-Valois and the area east and north of the road are free of the enemy. Road Senlis—Crépy also free of enemy troops.

(Signed)
Oblt. Fulda[100]

Fulda's reconnaissance revealed that the French had brought up fresh reinforcements during the night and were in the process of redeploying even more by rail from Paris. It could therefore be assumed with certainty that Joffre, in the decisive battle that he was seeking, was devoting his main effort to the envelopment of First Army's right wing. Thus, IV AK's arrival and planned envelopment attack would not be decisive as First Army HQ had hoped. On the contrary, the recently discovered French reinforcements gave Maunoury's Sixth Army the numbers of troops and artillery pieces necessary to attempt a decisive attack of their own. Recognizing this fact, First Army HQ had a seriously important decision to make. Should First Army act defensively and fall back to close the gap with Second Army or should Kluck recall III and IX AKs from protecting Second Army's flank along the Grand Morin to throw back the reinforced French Sixth Army offensively?

After examining both options carefully, Kluck and Kuhl audaciously decided to remain on the offensive. Orders were therefore written for III and IX AKs to be immediately withdrawn from Second Army's flank for use on the Ourcq Front's right wing in a large-scale enveloping attack. Second Army HQ was notified of the decision at 11am with the following message:

Intervention of III and IX AKs on the Ourcq urgently needed. Enemy being greatly reinforced. Request that corps be sent in direction of La Ferté-Milon and Crouy.[101]

Bülow received Kluck's message with total bewilderment. Second Army HQ had started the day believing that two corps would be left on the Grand Morin under their control and had planned their operations accordingly. Consequently, if the two corps were to be detached, Second Army's right wing would be left dangerously weak and vulnerable to envelopment. Nonetheless, Bülow was compelled to comply with Kluck's request and the two corps began their long journey to the Ourcq shortly after 1pm. In the meantime however, Second Army's operations would have to be adjusted to avert disaster.

First Army HQ's plan was not without its problems. Due to the great distance (70+ miles) between the Grand Morin and the Ourcq, Kluck could not expect the attack to be delivered until the morning of the 9th. Therefore, First Army's battered forces on the Ourcq would have to hold off the numerically superior French Sixth Army by themselves throughout the remainder of the 7th and all of the 8th while awaiting the arrival of III and IX AKs to turn the tide. Furthermore, the two corps' departure from the Grand Morin increased the gap between First and Second Armies to 31 miles. If the BEF, who was slowly advancing into the gap, was able to successfully split First and Second Armies, the campaign and perhaps the war would be lost. Thus, the two cavalry corps defending the gap were notified of the situation and ordered to delay and prevent a breakthrough at all costs. German victory on the Marne was now dependent upon the two cavalry corps holding the Coulommiers Gap until victory could be obtained on the Ourcq.[102]

While undoubtedly controversial, First Army HQ's decision to remain on the offensive was the correct choice to make. Indeed, recalling the III and IX AKs for an enveloping attack on the Ourcq was Kluck's only option that gave the Germans an opportunity to win the engagement and avoid a prolonged two front war. This opinion was first presented in Kuhl's memoirs when describing First Army HQ's situation and reasoning for making the decision:

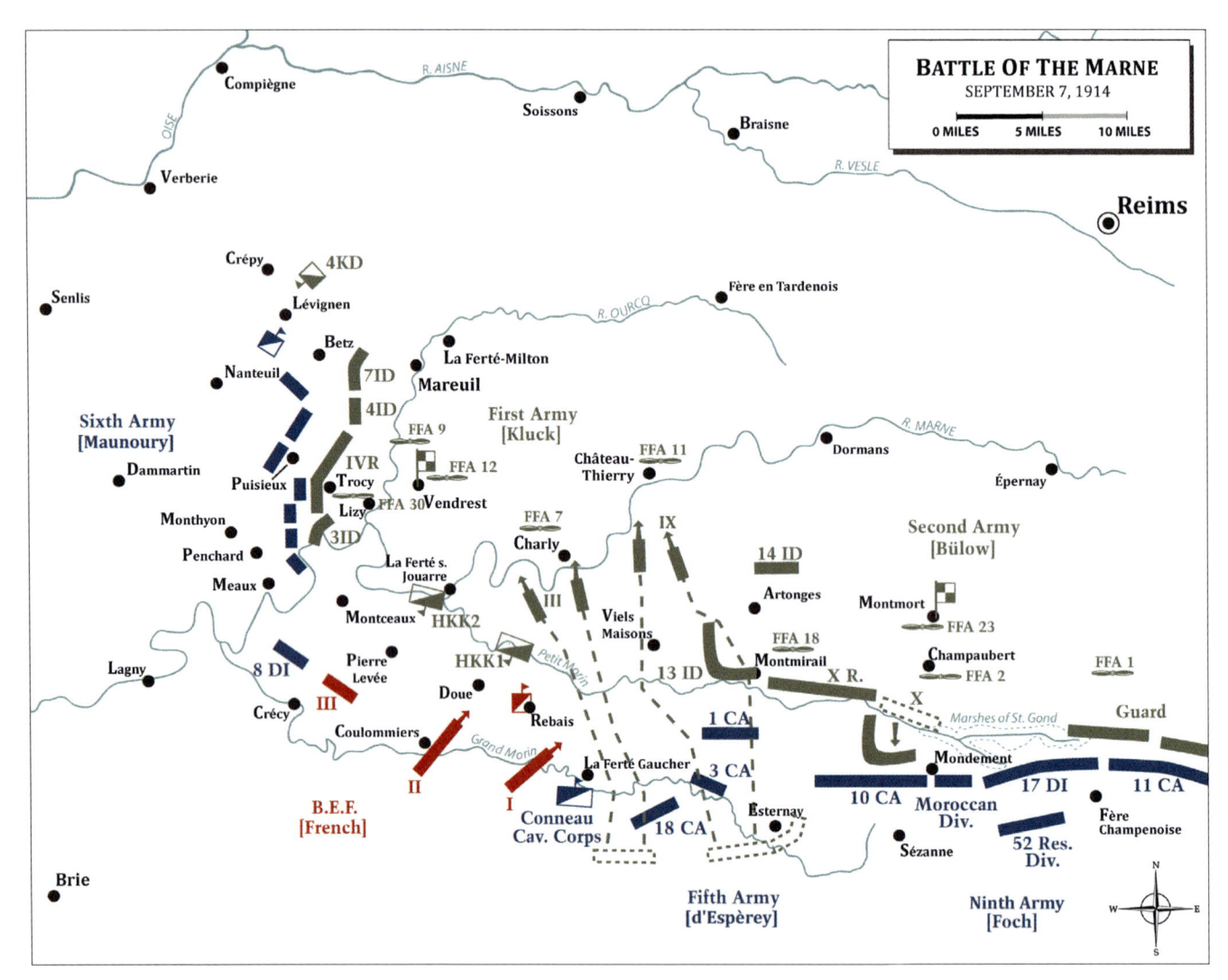

If First Army were to be thrown back upon the eastern bank of the Ourcq, toward which at the time its columns and trains were still crowding together; there would inevitably result the greatest danger not only to First Army but to all the German armies. The fate of the whole Battle of the Marne was at stake.

The army command considered once more whether the task of the army could be accomplished defensively. A defensive position behind the Ourcq could only make our situation worse; compressed in the angle between the Ourcq and the Marne, the army would have fallen back into an unfavorable tactical situation without escaping envelopment. To draw the army back in the midst of combat, say with the left wing upon Château-Thierry, was extremely risky. After a little while the army would have found itself confronted again with the same situation. With the cooperation of the English army henceforth assured, the enemy was able to initiate his envelopment movement more effectively.

Nothing but the offensive offered a possible solution. The enemy had to be thrown back. It was the best way of protecting the flank. But to that end it was necessary call up the III and IX AKs. This was the sense of the decision reached by General von Kluck.

To be sure, the gap which already existed between the flank of First and Second Armies was thus increased and the flank of Second Army, which First Army was charged with protecting endangered. We were going on the assumption, however, that the English, after their repeated defeats, heavy losses and continued retreat since the Battle of Mons, would hardly be in a position to pass at once to the vigorous offensive.[103] It appeared to us to be possible to hold them, at the latest on the Marne, until a decision could be obtained on the Ourcq. If Second Army continued, as we expected, to wheel to the right and to make progress between the Seine and the Marne in the westerly direction, the gap was sure to become narrower. The course of the fighting up to the evening of September 6th, both as regards Second Army and III and IX AKs had given us no cause for concern.[104]

Recognizing the offensive as their best option, Kluck and Kuhl adopted additional measures to increase their chances of success. First, HKK 2's two cavalry divisions (4 and 9 KDs) were ordered to retire to the north to better cover the Ourcq Front's flank and rear. Next, senior officers attached to the corps already on the Ourcq were informed of III and IX AK's impending arrival and were given an appeal to stand firm, regardless of cost, and await the arrival of the two corps. Finally, III and IX AKs were given instructions to push themselves as far as possible during their forced marches in order to come into action as soon as possible.

Both the III and IX AKs spent the remainder of the 7th marching northwest towards the Ourcq Front. By the end of the day III AK had reached Charly—La Ferté-sous-Jouarre while IX AK rested for the evening in Chézy—an impressive distance of 37 miles covered since daybreak.[105] After a brief rest, the two corps were to strike camp at 2:00am and resume their march. Lochow's III AK was to take the route from Montreuil-aux-Lions and La Ferté-sous-Jouarre upon Mareuil and Crouy while Quast's IX AK was to advance by way of Château-Thierry upon La Ferté-Milon. Thus, the two corps were to force march another 37 miles on the 8th before taking a short rest and then ultimately going into action on the Ourcq Front's right wing on the morning of the 9th. Certainly, it was a bold plan reminiscent of Davout's heroic forced march at Austerlitz.[106] The pertinent question was: could the rest of the German line hold until First Army's attack was delivered?

In the meantime, Second Army HQ immediately took action to shore up their vulnerable right flank. Richthofen's HKK 1 was ordered to fall back north of the Grand Morin into strong defensive positions that would allow them to delay the BEF's advance into the gap between First and Second Armies. Furthermore, VII AK's 13 ID on the extreme right of the army's battle-line was directed to extend itself to the west on a vast frontage from Fontenelle to Montmirail.[107] The troops of the 13 ID and X RK on their left were then ordered to dig in on the heights overlooking the Petit Morin. From these formidable positions, the three divisions were expected to hold off an enemy far superior in numbers and defend the army's flank and rear for a considerable length of time. Finally, VII AK's second division (14 ID) was kept behind the right wing as Bülow's army reserve—ready to shore

Above: A romantic depiction of a German cavalryman defending his position from behind a river. Operating from behind the Petit Morin and later the Marne, both HKKs 1 and 2 were instrumental in delaying the BEF's progress towards the breech that had opened between First and Second Armies. (Author's Collection)

up any failing sector of the army's defenses. In the mind of Bülow, each of these decisions combined to temporarily secure Second Army's vulnerable right flank and buy time for the German forces attacking elsewhere along the front to win a major victory. The key to the German strategy, however, continued to depend on the ability of the two cavalry corps in the gap to hold firm until that victory was achieved.

Fortunately for Bülow, the allies failed to attack or even threaten his army's right wing during the 7th. Indeed, having missed German Second Army's withdrawal to the Petit Morin, d'Espèrey's badly damaged Fifth Army spent most of the morning reorganizing and preparing for a German attack. By noon, English and French aviators had reported that German Second Army's entire right wing had redeployed to the heights north of the Petit Morin. Upon receiving the reports, d'Espèrey immediately telephoned each of his corps commanders, demanding that the army reach the Petit Morin by nightfall. Shortly thereafter, d'Espèrey received a request for

Above: *Jäger* light infantry scout the allied positions. (Author's Collection)

assistance from Foch's Ninth Army on his right, who was engaged with Second Army's left wing as well as elements of Hausen's Third Army. As it happened, Foch's forces had suffered terrible losses and were in danger of collapse. Thus, in order to stabilize the situation, d'Espèrey was compelled to send his X CA to Foch's left wing. Although the reinforcements bolstered French Ninth Army's position, d'Espèrey's men were unable to advance to the Petit Morin as quickly as planned out of concern that their own flank was left vulnerable to attack.[108] As a result, the troops of X RK and 13 ID remained unengaged throughout the entire day. The German official history of the battle described the action on Second Army's right wing during the afternoon of the 7th as follows:

Having arrived on the north bank of the Petit Morin, the 13 ID, 19 RD, and 2nd Guards RD dug in under peacetime conditions; for far and wide not an enemy was to be seen and the day came to an end without an enemy infantryman appearing or even artillery fire falling anywhere on the German positions.[109]

To the southwest, the two cavalry corps holding the gap between Kluck and Bülow's armies used their elite *Jäger* light infantry and generous allotment of machine guns to slow up the BEF and keep them at least one day removed from Bülow's vulnerable right flank. By evening, Richthofen's HKK 1 was installed in defensive positions on the heights north of the Petit Morin with instructions to continue the delaying action the following day. Thus, the situation on the army's right wing remained entirely favorable on the evening of the 7th.

Despite the lack of combat, German commanders of Second Army's right wing remained convinced throughout the morning that an attack was imminent. As a result, VII AK HQ ordered an aircraft of its own FFA 18 to fly south and reconnoiter French Fifth Army's deployment in order to determine when and where the attack would occur. At 12:15pm, after a brief flight over the enemy's positions on the Grand Morin, the flight's observer delivered the following handwritten note to the VII AK HQ staff.

Enemy positions in the sector largely unchanged since yesterday. Enemy force, estimated strength three divisions, remains south of the Grand Morin in the area north of Sancy—Cerneux. One enemy corps witnessed in column on the line Aulnoy—Neuvy. Column's lead elements have begun crossing the Grand Morin at Neuvy. Two columns, each one division in strength, observed moving north from Esternay. First column on the road Esternay—Champguyon. Second column witnessed departing Esternay through Viviers. These two columns are the only enemy forces moving north of the Grand Morin.[110]

The contents of the observer's message were immediately forwarded to Bülow at Second Army HQ for his consideration. Bülow and his staff deduced from the aviator's observations that the French would not engage Second Army's right wing until the morning of the 8th. Furthermore, reports from HKK 1 of the BEF's slow progress into the gap suggested that the English involvement would continue to be a minor factor. Therefore, believing his right to be temporarily secure, Bülow ordered VII AK's 14 ID (acting as army reserve) to move from behind the army's right wing towards the left to take part in the combat in the Marshes of St. Gond. This decision to redeploy an entire division from the highly vulnerable right wing would later have tremendous repercussions on the outcome of the battle.

Bülow's plan had been to hold defensively with his right while the left wing, supported by Hausen's Third Army, broke the French line around the Marshes of St. Gond and wheeled toward Paris to threaten French Fifth Army's lines of communication. Third Army's cooperation however, was not guaranteed. The Saxon Army was already understrength and spread dangerously thin. Hausen recognized that a strong French attack was likely to split his army and sever the entire German battle-line in two. Thus, before committing any of his forces to an offensive in support Second Army's left wing, Hausen ordered aerial reconnaissance be conducted to determine if his army was at risk of being assailed. Three aircraft from Third Army HQ's FFA 22 were therefore sent airborne shortly after 9:00am with orders to reconnoiter deep behind French Ninth Army's front line. Upon returning, Hausen's aviators delivered favorable news. The French forces opposed to Third Army were split into two groups connected only by a weak cavalry screen. The left group was concentrated near the Marshes of St. Gond opposite Bülow's Second Army while the right was attacking to the east toward German Fourth Army. Thus, Hausen knew he was free to support Second Army on his right without the fear of an enemy attack against his army's weak center.

In his memoirs, Hausen described the critically important task given to his aviators and the subsequent decisions that were made as a result of their reports.

During the forenoon of September 7th, the Headquarters of the Second Army informed me that because of a strong attack emanating from Paris it found itself forced to withdraw the III and IX AKs and the X RK behind the line of the Petit Morin, but that on the contrary the left wing of the Second Army would maintain the offensive, and that therefore the immediate assistance of all the disposable forces of the Third Army was urgently needed. While this request for assistance coincided with my point of view that the enemy intended to continue his offensive of September 6th along the entire front, it was necessary, however, for me to consider what might happen if the enemy took the offensive against the Third Army. Orders were issued, therefore, for aerial reconnaissance to explore the country south of the sector assigned to the Third Army and also the routes Fère Champenoise—Plancy—Méry and Vitry-le-François—Brienne. This reconnaissance was made with three aircraft for a depth of 200 to 250 kilometers [125-155 miles], and from its results it was possible for me to form before noon the following idea of the enemy's situation: That he would continue the offensive already commenced in a northerly direction with—

The right wing of his group of armies, opposed by Second Army, proceeding in the general direction of Lenharrée, east of Fère Champenoise, while

the left wing, which was advancing against Fourth Army, reinforced by some troops debarking from the railroad, marched toward the line Vitry-le-François—Sompuis.

Between the two attack groups, there was one cavalry division with some strong artillery and a few infantry detachments…Behind this cavalry screen, there did not seem to be any other troops arriving along the route Mailly—Sommesous, because the roads leading to Arcis-sur-Aube and to Vinets, the heights of the two sides of the Seine between Romilly and Troyes, as well as the network radiating to the north and northeast of Troyes, apparently were bare of French detachments.

It was otherwise along the roads leading from Brienne to Vitry-le-François. The traffic activity of the railroad from Nogent-sur-Seine to Troyes via Romilly and from Vendeuvre to Troyes, and the accumulation of matériel in the railroad stations of Romilly, Troyes and Brienne, clearly showed that the adversary, with his east group greatly reinforced by these additional men and supplies, intended to advance via Brienne toward the north.

I concluded:

First, that the 32 ID, engaged alongside the 2nd Division of the Guard, ought to be able to arrest this new push of the enemy with the aid of the 23 RD.

Secondly, that tentatively the piercing of the center of Third Army was not immediately to be feared; and

Thirdly, that the XIX AK, engaged with the 23 ID and the VIII AK [Fourth Army], would find itself obliged to assume the defensive instead of attacking.[111]

After receiving the aviation reports, Hausen ordered his army's right wing (32 ID with 23 RD in support) to conduct an attack against the eastern end of the Marshes of St. Gond. Spearheading the assault would be Second Army's 2nd Guards ID. However, a large portion of the Guard's artillery was not available for the attack. Consequently, the attacking Germans were unable to properly support their infantry with adequate artillery fire, prohibiting Bülow and Hausen's combined force from making any significant gains on the 7th. By nightfall the attackers dug in where they stood with plans to resume the attack the next day.[112]

Second Army's attacks on the marshes' western perimeter were far more successful. Supported by artillery, Emmich's X AK attacked at first light with an objective to push the French back and secure safe passage through the swamplands for their ammunition and supply columns. After failing to throw the enemy in their sector back in the morning, Emmich's men were reinforced and were ultimately successful in occupying a strong position beyond the marshes by nightfall. From this position, Bülow expected X AK to breakthrough the French center the following day.[113]

In order to attack in greater strength on both sides of the marshes, the German leadership had been compelled to leave the marshes themselves lightly defended by a small infantry detachment supported by artillery. By late afternoon, the French had recognized the German deployment scheme and had decided to attempt a breakthrough to split Second Army's left wing in half. Accordingly, a small force of three infantry companies was quickly gathered to assault the German positions across the marshes. After marching single file over the marshes' causeways, the French attackers reached the northern bank where they temporarily expelled the German infantry detachment from the town of Aulnizeux. Within an hour, hastily summoned German reinforcements forced the small French detachment from the village and pushed them back to the marshes' southern boundary. Although it had been repulsed, the French attack compelled Bülow to use VII AK's recently arrived 14 ID to plug the gap north of the marshes.

Aviators belonging to Second Army's left wing corps attempted several tactical flights throughout the day on both sides of the marshes. The subsequent aerial reports were of little help however, as the fighting on both sides of the marshes was largely unrecognizable from the air. Thus, recognizing the futility of further flights, the commanders of both FFAs 1 and 21 discontinued flight operations around 3:00pm.[114]

Under direct orders from Bülow, Second Army HQ's FFA 23 performed two flights late in the afternoon deep behind the allied lines. Both of the aircraft returned late in the afternoon with detailed reports of French Fifth and Ninth Armies' deployment. According to the first report the French forces around the Marshes of St. Gond were numerically weak and spread thin. Although the observer reported several large artillery groups in the area, it was his opinion that a strong attack could break the French line. The aviator's conclusion was based from the observation that the roads behind the French positions were free of troops—suggesting that all available reinforcements had already been moved forward into the line. The second report, much more sobering in tone, informed Second Army HQ of French Fifth Army's position opposite Second Army's vulnerable right wing. After reading the observer's message, Bülow concluded that the bulk of d'Espèrey's Fifth Army had crossed the Grand Morin and were briskly marching north towards his army's right wing. Thus, by sunrise a numerically superior force would almost assuredly be engaged with Second Army's vulnerable right wing.

That evening Bülow and the Second Army HQ staff used FFA 23's reports as a basis to formulate their plans for the following day. Citing the reports, Bülow declared his army's right wing too weak to hold against French Fifth Army's anticipated attacks. Second Army HQ therefore ordered X AK on the eastern end of the marshes to fall back to the Petit Morin where it could lend support to the forces of the army's threatened wing. On the western end of the marshes, Second Army's Guard Korps was ordered to continue the offensive and work with Third Army to breakthrough the French lines and carry the attack into their rear. The plan for Bülow's Second Army therefore remained the same—hold firm defensively while the Guard Korps and Third Army as well as First Army obtained the campaign winning victory in their respective sectors.

Meanwhile, at his headquarters in the town of Châlons-sur-Marne, an optimistic Hausen and the Third Army HQ staff evaluated their situation. Early in the morning the army command had received a message from OHL stating that the French were

Above: German and French infantry enter into hand-to-hand combat during the Battle of the Marne. (Author's Collection)

"launching a decisive attack by rushing forces of sufficient importance from Paris to menace the right flank of the German army." Throughout the afternoon the Saxons received multiple reports from the right wing armies that confirmed OHL's statement by describing the increased ferocity of the attacks from the direction of Paris against First Army and the heavy enemy resistance opposite Second Army. Hausen and his staff therefore deduced that if the enemy was in great strength against First and Second Armies that they must be weak in Third Army's sector. Therefore, Hausen aggressively ordered his forces to prepare for an energetic attack to break the French line and relieve pressure on the neighboring First and Second Armies. In his history of the campaign, Hausen described the situation and his reasoning for planning the bold attack, which was set to begin before dawn the following day, September 8th.

...The Third Army, however, was able to maintain itself unswervingly on the entire front which had been achieved by it, and to defend this front successfully against all attacks directed against it by superior French forces, while the Second Army was forced to retire to the line of the Petit Morin before a very strong offensive coming from the region of Paris. This retreat gave rise to the supposition that the French High Command—as had been announced by the German Supreme Headquarters—had succeeded in "rushing from Paris some forces of sufficient importance to menace the right flank of the German army." If that was the case—and the retreat of the Second Army seemed to indicate that it—I did not see how it could be possible for the adversary to have a numerical superiority along the entire front.

From that point on it seemed to me that a new and energetic attack on our part was the best method of reconnoitering the enemy situation in order to break through his position where it appeared to be a little weak, and in this manner to parry the attack which the French were making with superior numbers against the right flank of the German armies [First and Second Armies].

It seemed to me that it was necessary to act immediately, not only because of the menace directed against those two armies, but also because Third Army—remaining on the site of the battle of September 7th—was still in the immediate proximity

Above: A group of German 77mm field guns come into action during battle. (Author's Collection)

of the enemy.

Inspired by our experiences of the combats of September 6th and 7th, we therefore resolved to take the initiative ourselves in order to realize better the tactical dispositions and also with the idea of avoiding so far as possible the attacks of French infantry and the action of French artillery. To accomplish this, an assault against the nearest army before dawn on September 8th was indicated, this attack to be pursued with the bayonet as far back as the French batteries.

With this end in view, and having in mind the relationship with the two neighboring armies, I requested of the headquarters of the Second Army the cooperation of the 2nd Infantry Division of the Guard, and also obtained from the headquarters of the Fourth Army the promise of the cooperation of the VIII AK. In conformity with our understanding and in line with the considerations already mentioned, I issued an order to Third Army at 6:00 on the evening of September 7th in which these considerations were textually reproduced.[115]

Thus, after reviewing all the available intelligence, Hausen correctly theorized that the best response to the Allied attacks against First and Second Armies was to seek a rapid decision in his army's sector. However, Third Army was already weakened and had suffered heavy casualties from the French field artillery. The attack was therefore ordered to begin at 3:00am under the cover of darkness in order to secure the advantage of surprise and minimize the impact of the French artillery. The details of the attack plan were sent to OHL at 9:15pm, whereupon Moltke radioed back his approval shortly after midnight. With OHL's consent secured, Hausen's forces moved into position and prepared for the attack that would ultimately shatter the French line and bring the Germans one step closer to victory.[116]

As the reader will recall, Kluck and Kuhl originally ordered their forces on the Ourcq Front to remain on the offensive throughout the afternoon of the 7th. First Army HQ's plan was to throw the French back and buy time for the III and IX AKs to complete their forced marches to the battlefield. Accordingly, at 12:15pm, the recently arrived IV AK on the Ourcq Front's right wing was given the order to attack while the depleted forces of the left and center held defensively.

Unfortunately, necessity had caused both the II and IV AK's to deploy their divisions piecemeal along the Ourcq Front. When II AK arrived on the morning of the 6th, it deployed its two divisions on either flank of Gronau's command. Likewise, IV AK was compelled to detach a brigade (8 ID's 15 IB) to bolster Gronau's men in the center of the German line while the rest of the corps moved into attack positions on the front's extreme right. Thus, as directed, IV AK's 7 ID along with II AK's 4 ID deployed into attack formations and began their advance with the support of several grand batteries of field artillery. This attack collided with the French 61st Reserve Division on the extreme left of French Sixth Army's battle-line—throwing the French reservists back in great confusion. The extent of the victory, however, was not fully recognized by the attackers, resulting in a lost opportunity to exploit their tactical success. By sundown, the forces of IV AK moved back to stronger defensive positions knowing the offensive could not be safely resumed until III and IX AK's arrival on the morning of the 9th.

Meanwhile, the French attempted several large-scale attacks against First Army's left wing near the village of Étrépilly which guarded the approaches to a series of commanding heights known as the Trocy Plateau. With orders to hold defensively, the German infantry in the sector spent the morning digging in at the base of the plateau while large numbers of German artillery unlimbered in advantageous positions on the heights behind them. Consequently, the forces of the German left were already well established in strong defensive positions by the time the French attack had gotten underway. Nevertheless, the local French commander hastily ordered his men forward, perhaps believing that the men of the First Army's Ourcq Front had been weakened after the two previous days of combat. Taking advantage of their position on the plateau, the German artillery assisted the defenders repulse the French attacks with heavy losses. Further north, the French 63rd Reserve Division launched a similar attack against the German center, held mostly by the men of Gronau's IV RK. On the heights behind Gronau's divisions rested several groups of batteries totaling more than 50 guns. With a commanding view of the valley below them, the German field guns decimated the attacking French division, causing many to flee in panic. By nightfall, all major fighting had come to a close with the opposing positions virtually unchanged.[117]

In the air, First Army's *Fliegertruppe* played a major role in the battles on the 7th by supplying the army leadership with numerous valuable tactical reconnaissance reports throughout the afternoon. The first of these was written by FFA 9's *Oblt.* Roos and delivered to IV AK HQ on the front's right wing at 1:30pm. Roos' flight had been ordered to reconnoiter the positions north of the attacking German right wing in order to ensure there was no danger to the army's flank as it advanced. After scouting the positions on the army's flank, the airmen were to continue to the French rear-area directly opposite IV AK to determine the strength of the enemy reserves being moved into the sector. Taking off at 12:30pm, Roos' Taube #A.173 quickly flew over the assigned areas and promptly returned to the section's airfield. Ten minutes after landing, Roos produced the following handwritten report—his second of the day—in person to IV AK's Chief of Staff, *Generalmajor* Leo von Stocken.

To: IV AK
Area to the north of Betz free of the enemy. Crépy-en-Valois and surrounding environs also clear of enemy troops.

A munitions column, 1 kilometer in length, witnessed on the road Silly-le-Long—Nanteuil-le-Haudouin. Baggage trains also witnessed further south.

An enemy infantry division observed at Brégy marching towards Nanteuil. Munitions columns parked west of Chèvreville. Limbered enemy artillery moving east from Chèvreville. Additional enemy artillery deployed in positions west of the line Fossé-Matin—Douy-la-Ramée.[118]

By observing large numbers of French troops moving north towards Nanteuil-le-Haudouoin, Roos confirmed First Army HQ's theory that the French were changing their strategy from breaking through the Ourcq Front close to Meaux to enveloping it from the north. The report also admitted that no substantial hostile forces were north of IV AK's positions, thus allowing the German troops to continue their attack without fear of envelopment.[119]

Operating from a forward airfield near the village of Trocy-en-Multien behind the German center, II AK's FFA 30 provided information that greatly assisted First Army repulse the French attacks. The first flight took off at 12:30pm under orders to take note of French Sixth Army's front line strength. Accordingly, one of the section's three AEG B.Is flew along the entire width of the French line from Penchard in the south to Betz in the north—drawing heavy ground fire throughout the flight. Once the aircraft was over Betz, the observer, *Ltn.* Knobelsdorff, ordered his pilot to turn around and fly back down the entire length of French Sixth Army's artillery and reserve line. Finally, after continuing all the way to the Marne, the airmen returned to the airfield. Upon landing, Knobelsdorff hastily penned a report that was forwarded to *General*

Above: Dug in, German infantry supported by two machine guns look to repulse an enemy attack. (Author's Collection)

Linsingen—the overall commander of the Ourcq Front.

To: II AK HQ
Strong enemy forces, estimated strength three infantry brigades, on the line Chambry—Barcy. Several field batteries witnessed unlimbered in close support.

Two infantry regiments observed advancing east of the road Barcy—Marcilly.

A large enemy force, estimated to be at least one division, with significant artillery support in the vicinity of Puisieux.

An estimated three brigades seen advancing east on the line Bouillancy—Villers-Saint-Genest. This force had already made contact with elements of our right wing at the time of the observation, 12:55pm.

Enemy infantry, estimated strength one regiment, moving east on the road Nanteuil-le-Haudouin—Villers-Saint-Genest. Large numbers of artillery and other vehicles being moved from various areas of the front to the north to support the enemy forces in *the area of Villers-Saint-Genest and to presumably extend their line to the north. Few infantry reserves in the enemy's center in the area between Brégy and Bouillancy. Several large artillery groups employed together around Puisieux and Barcy. Area west of Meaux to Annet-sur-Marne completely free of enemy troops.*[120]

Bravely produced by repeatedly flying over the enemy's front lines at low altitude, Knobelsdorff's report informed Linsingen and the First Army leadership of French Sixth Army's entire deployment scheme. In the south, Maunoury's army was reported to have roughly three divisions supported by massed artillery. Meanwhile, the French center was revealed to be numerically weak and spread thin with a generous allotment of field batteries to compensate. The French left was observed to be attacking with three brigades on a narrow frontage. Finally, in the rear-area, all available reserves were seen being moved to the north in support of the army's left wing, thus leaving the center and right without any reserves of their own. Citing the report's findings, Linsingen ordered his advancing right wing to quickly move through Betz and envelop the enemy's left flank. Linsingen accordingly dispatched a note to Gronau's headquarters stating that: "the French center is reported by our aviators to be weak. If these troops attempt an assault, General Gronau has formal permission to launch a counter-attack."

At 3:00pm the weak enemy center comprised of the French 63rd Reserve Division attacked Gronau's troops. Equipped with Knobelsdorff's intelligence, Gronau understood that the attacking French force was too weak to break the German line. He therefore ordered his reserve companies forward to prepare for a counter-attack. As the doomed French attack stalled, Gronau's men turned over to the attack to pursue the retreating enemy. Within minutes, the orderly French retreat turned to panic as German rifle and machine gun fire poured into the backs of the fleeing infantry—leaving a gaping hole in the French line. Only Colonel Robert Nivelle's heroic act of moving several field batteries forward under fire saved the French center from a total collapse. Once unlimbered, Nivelle's 75's checked the German advance and allowed the defeated French infantrymen to reform behind them. That evening Gronau's men fell back to their previous positions at the base of the plateau, remaining in total command of the sector.[121]

It appeared that Kluck's new plan to use the recently recalled III and IX AKs in a large scale attack to envelop and destroy French Sixth Army was going well for First Army. This objective, however, required reliable long-range aerial reconnaissance of French Sixth Army's rear-area to determine the strength and position of Maunoury's reserves. With that information in hand, First Army's attacks would be better planned and have a better chance of success. Therefore, Kluck ordered FFA 12 to perform two flights of French Sixth Army's rear during the afternoon of the 7th. The first took off from the section's airfield at Vendrest at 1:40pm with orders to initially fly southwest to confirm that the III AK was not being menaced while on the march from the Petit Morin, then turn west and scout the allied positions. 70 minutes after taking off, the Albatros B.I returned to the airfield with the following report.

III AK confirmed on the march. Lead elements of the east column seen entering Nogent-l'Artaud at 1:55pm. 2:00pm–Infantry and vehicles of III AK's west column observed at Bussières; head of the column entering Bois de Moras.

Crossed over the Marne at Villenoy at 2:15pm. Area south of the Marne between Jouarre and Villenoy free of troops. 2:20–Ammunition columns and other vehicles spotted at Iverny moving east. Enemy infantry, estimated one regiment, marching east through Montgé in the direction of St. Soupplets.

2:25–Troops detraining at rail station at Le Plessis-Belleville. Significant ground fire in this locale. 2:30–Enemy cavalry, estimated two divisions, with artillery and vehicles in the environs of Nanteuil. The line Nanteuil—Crépy-en-Valois is free of enemy forces.

2:40pm–Observed infantry column moving west on road Villers-Cotterêts—Crépy-en-Valois in the area of Vauciennes. Flew over column at altitude of 2,500 feet and received no ground fire. <u>Strongly suspect the column to be German.</u>[122]

The observer's notes confirmed First Army HQ's suspicions that the enemy was continuing to be reinforced from Paris. Kuhl nevertheless remained optimistic about his army's chances after reading that the French had yet to seriously reinforce their left. If the enemy's left remained weak at the time of the attack, success would be assured. However, Maunoury still had time to redeploy his forces. Thus, with plenty of daylight and the importance of the enemy's left wing well known, Kuhl ordered a second flight be conducted—this time concentrating specifically on the area of the front where III and IX AK's were to make their attack.

First Army HQ's request arrived at FFA 12's airfield shortly after 3:45pm. After a quick pre-flight briefing, pilot *Ltn.* Haupt and observer *Oblt.* von dem Hagen climbed in their Taube (a replacement aircraft provided to the section by AFP 1) and took off for French Sixth Army's left wing. Leaving at 4:05pm, the airmen slowly climbed to cruising altitude then crossed over the German right flank to begin their assignment. Flying an outdated aircraft into the wind, Haupt and Hagen patrolled the French airspace at the slow speed of only 55 miles per hour. This allowed Hagen to make out the enemy's deployment and deliver the following report to First Army HQ in person.

Airplane #: Albatros *Taube* A.70   Vendrest,
                                    7 September 1914
Pilot: *Ltn.* Haupt   6:30pm
Observer: *Oblt.* von dem Hagen

To: First Army HQ

1. Weak enemy forces seen west of Thury-en-Valois at 4:25pm.
2. Roads Thury-en-Valois—Betz and Betz—Crépy-en-Valois were free of enemy forces at 4:30pm.
3. Road from Senlis to Crépy-en-Valois was also free of the enemy. Area observed at 4:45pm.
4. Roads from Senlis to Dammartin and road of

Senlis-Le Plessis-Belleville free of enemy. Enemy forces were seen at the railway station at Le Plessis-Belleville. Vehicles of the column were seen moving southeast in the direction of St. Soupplets.
5. Railway of Le Plessis-Belleville-Compans without traffic.
6. Enemy's deployment: <u>Weaker in the following areas:</u> north, south and west of Ognes, west of Oissery and west of St. Pathus, west of Marchémoret, west of Montgé and north west of Monthyon. <u>Strong in the following areas:</u> South of Montgé and southwest of St. Soupplets.
7. Strong numbers of cavalry and an enemy airfield seen southwest of Compans. Weaker forces west of the line Claye-Souilly.
8. Strong artillery seen firing from Hill 125 north of Villers-St. Genest and south of Betz.
9. 3 Battalions of infantry and several artillery batteries seen at 5:45pm south of Maquelines [northwest of Betz].
10. Strong infantry column at 5:50pm crossing the road north of Thury-en-Valois—Betz.

(Signed)
Oblt. von dem Hagen[123]

Hagen's superb report quickly proved that French Sixth Army's leadership had realized their error and begun shifting their strength to their left wing. In fact, in just the short time Hagen was over French territory, the enemy's left wing had been reinforced and extended considerably. The village of Betz on the extreme left of the French line went from being lightly held at 4:25pm to having two regiments of infantry with considerable artillery support in place at 5:50pm when the aircraft was on its way back to German airspace. Thus, with additional troops detraining behind the enemy's left wing at Le Plessis Belleville, it could only be assumed that the area around Betz would be further reinforced during the night. Kuhl understood that these movements represented the enemy's desire to launch an enveloping attack of their own the following morning using a freshly strengthened left wing. Therefore, he ordered his army's right wing to abandon the gains it had won earlier in the day and withdraw to stronger defensive positions to await the coming attack. Thanks to Hagen's report, Kluck and Kuhl remained one step ahead of the French and were able to buy time for III and IX AKs to complete their forced marches to deliver what they hoped would be the battle's winning blow.

The riskiest part of First Army's current operational plan remained the potential for disaster to the south in the Coulommiers Gap between First and Second Armies. The redeployment of III and IX AKs from Second Army's right wing had dangerously widened the breech to 31 miles, which spread the two cavalry corps defending the sector very thin. If the BEF were able to successfully break the cavalry screen and split First and Second Armies, the results could be disastrous. Either the First Army would be encircled and destroyed or Second Army would be enveloped and rolled up with an attack from west to east. Either way, an allied success of that scale would undo every gain the Germans had thus far achieved. Thus, Kluck and Kuhl's hopes were dependent upon the two cavalry corps holding up the BEF until III and IX AKs attack was delivered. First Army HQ therefore needed as much information regarding the BEF's progress into the gap as possible. To that end Kuhl ordered FFA 12 to conduct another flight over the gap to report upon the British Army's latest progress.

First Army HQ had already received a report from FFA 7 regarding the BEF's location as of 10am that morning. Kuhl recognized, however, that it was entirely possible for the British army to pick up speed as the day progressed. Thus, FFA 12 dispatched a flight piloted by *Ltn.* Schwarzenberger (*Oblt.* Ritter serving as observer) to reconnoiter the area around Meaux near First Army's vulnerable left flank and then further south to determine the BEF's location. Taking off shortly after 12:15pm, the flight was sent along the route Charly—Meaux—Tournan—Coulommiers—La Ferté-sous-Jouarre—Charly. After the flight, Ritter personally delivered his report to First Army HQ. Ritter's observations confidently declared that there were no British units in the area immediately south of Meaux. Nevertheless, despite the report, First Army HQ received messages from the front which deeply worried the headquarters staff. Positioned near Meaux on the Ourcq Front's extreme left wing, the 3 ID sent messages to First Army HQ complaining of destructive artillery fire from the direction of Meaux. It was then assumed by the First Army HQ staff that the British had indeed arrived on the battlefield at Meaux and were about to attack First Army's left flank.[124] In order to test this assumption, First Army HQ ordered FFA 12 to immediately dispatch another aircraft towards Meaux. After an hour long flight, the flight's observer (*Ltn.* Pfähler) reported at 2:30pm that there were no new allied forces opposite the army's left wing. Finally, at 4:00pm, FFA 12 once again ordered Schwarzenberger and Ritter airborne to investigate the BEF's latest location and whether there are any new formations emanating from Paris. Taking off from FFA 12's airfield at Charly, Ritter and Schwarzenberger flew a large clock-wise circle along the route: La Ferté-sous-Jouarre—Coulommiers—Pézarches—Villeneuve—Lagny—Claye-Souilly—Meaux, then finally landing at the section's new airfield at Vendrest from where FFA 12 was to continue its operations the following day. Unfortunately for First Army HQ, Ritter was unable to identify any large size formations during the flight. Consequently, First Army HQ spent the evening of the September 7–8 still worried about the safety of the army's flank. In his post-war history of the campaign, Kuhl explained the somber mood at First Army HQ:

On the evening of September 7 the situation of the First Army was regarded at Vendrest as more unfavorable than it was in reality. The accounts we had received of the conditions of the struggle in the vicinity of Varreddes caused us great uneasiness. I remember distinctly that we were expecting at any moment to see the English arrive in the elbow of the Marne, at Trilport, in the back of the 3rd Division. The success of Sixt von Armin's group is not mentioned in the army order of the evening, which merely states that the army held firm on the line: Antilly—Puisieux—Varreddes. I recall the strong impression produced by the report of the chief of staff of IV Corps when he arrived at Army Headquarters at Vendrest during the night of 7-8th. The war journal of the IV Corps states that doubts were expressed at headquarters whether it would be possible to hold out. The losses were heavy, the enemy artillery seemed superior. The troops were exhausted. The picture presented by the chief of staff was far from encouraging. But the commanding general of the corps had issued orders to stand firm, regardless of cost, and await the arrival of III and IX Corps. The army command could only approve that decision. In the evening large bivouacs were reported in the vicinity of Nanteuil-le-Haudouin, Silly-le-Long, St. Soupplets and westward. He awaited the morrow anxiously. When would the English appear on the Marne?[125]

As Kuhl's personal account suggests, First Army's leadership were seriously alarmed on the evening of the 7th. FFA 12's inability to definitively locate the BEF's main columns, in addition to the latest reports from the other flying sections of new French forces concentrating around Nanteuil opposite First Army's right wing, were both serious problems. Moreover, the German troops deployed along the Ourcq Front had been marching non-stop for weeks and had already fought two or three days of bitter combat against French Sixth Army's fresher divisions which were well supported by field artillery.[126] Thus, First Army HQ was deeply concerned that their front might not hold another day until the arrival of III and IX AKs.

At a council of war held that evening at Vendrest, Kluck and Kuhl used their aviators' intelligence reports to establish an effective plan for the following day. Kuhl recognized from FFA 12's 6:30pm report that French Sixth Army was too strong to be attacked until the morning of the 9th when III and IX AKs were fully concentrated. Indeed, it seemed likely that the French would use their reinforced left wing to try and envelop the German northern (right) flank the following morning. Therefore, a decision was made to have the army dig in along strong defensive positions at the base of the area's various plateaus with all available artillery placed on the heights behind them. Meanwhile to protect the army's flank and rear from any potential British attack from the south, Kluck pulled his army's left wing back to the line Étrépilly—Congis facing the town of Meaux. The rest of the front was ordered to resist any French attacks and buy time for III and IX AK's to begin their offensive. Further south, the two cavalry corps deployed along the Marne were given similar instructions to "protect the left flank of the First Army from any threat from the Grand Morin and Coulommiers and assist the left wing of the Ourcq forces near Trilport with the fire of its artillery."[127]

Back at OHL, some 200 miles from the front, Moltke remained surprisingly detached considering the gravity of the situation. Despite knowing that his right wing had been heavily engaged for over 48 hours, the German Chief of Staff neglected to move his headquarters forward or dispatch a representative to the front. Instead, the right wing's operations were left to the individual army commanders, each of whom had a unique interpretation and solution to the problem currently facing them. At Third Army HQ, Hausen and his staff prepared to use their forces to attack French Ninth Army's weakest positions before dawn—hoping to break the French positions then turn west to attack French Fifth Army's rear. Meanwhile, Bülow had issued orders for his X AK to withdraw behind the Petit Morin to strengthen Second Army's threatened right wing while the Guard Korps participated in Third Army's attack. Finally, on the far right of the battle-line, Kluck's First Army had entrenched in defensive positions with orders to hold at all costs throughout the 8th while III and IX AKs moved into their attack positions. Thus, the German plan continued to depend upon the cavalry corps and Second Army holding the Coulommiers Gap until both First and Third Armies' attacks were successfully delivered. However, with little to no communication between the armies and no guiding hand from OHL, it remained uncertain whether the three commanders' uncoordinated strategy would yield success.[128]

September 8: Third Army's Sector

Beginning at 2:00am, troops of Hausen's Third Army,

Above: Hans von Kirchbach, commander of XII Reserve Korps and acting commander of Third Army's right wing during the Battle of the Marne. (Author's Collection)

supported by the Guard Korps, moved into their attack positions to the east of the Marshes of St. Gond opposite Foch's Ninth Army. The combined force was under the overall command of *General der Artillerie* Hans von Kirchbach—commander of XII RK. In accordance with Third Army HQ's orders prescribing an attack before dawn, Kirchbach scheduled the advance to begin at 3:30am. In order to maintain surprise and reduce the likelihood of friendly fire, Kirchbach's troops were ordered to begin the operation with fixed bayonets, rifles unloaded and breechblocks secured in their bread pouches.[129]

As ordered, the troops of the 32 ID, 23 RD, and the 2nd Division of the Guard moved out at 3:30am and began to advance silently by moonlight through the region's distinct vineyards and marshes. Within 15 minutes the attackers made contact with the men of Foch's first line, who were still asleep with their rifles stacked. Realizing their advantage, Kirchbach's men sounded the trumpets, shouted *Hurrah!* and aggressively pushed their assault deep into the French second line. Completely surprised, the 21 and 22 DIs of French Ninth Army's XI CA were swept back by the German assault. This caused the 18 DI that was deployed in reserve immediately behind them to break and flee without firing a shot. By 6:15am, all four of XI CA's divisions had recoiled six miles from their original positions, leaving the marshes' southern exits entirely in German hands. Continuing on, Kirchbach's men captured the villages of Lenharrée, Sommesous, Connantray, and the important road juncture at Fère-Champenoise. By the time the attack petered out around midday, Kirchbach's troops had routed five French divisions and advanced the German line more than three miles, effectively annihilating Foch's right wing in the process.[130]

Shortly after 6:00am, under the direction of Third Army's Chief of Staff Ernst von Hoeppner, aircraft belonging to FFAs 22 and 29 were sent airborne to monitor the progress of Kirchbach's attack. FFA 22's flight was ordered to reconnoiter the French rear-area to determine the strength of Foch's local reserves and to report whether any large formations were being moved against Kirchbach's command. FFA 29's aircrew was similarly directed to fly along the Somme River[131] on the line Sommesous—Écury-le-Repos to report on the progress of Kirchbach's advance and to, if possible, provide tactical assistance for the ground troops. At 7:30am FFA 22's aircrew presented themselves to Hoeppner to deliver the following report.

Great success! Crossed over the Somme at Normée at 6:35am. Town and environs entirely in our possession. Large numbers of enemy troops west of the line Haussimont—Normée observed fleeing in disorder. Our forces are seen pursuing everywhere in great strength. No effective enemy artillery fire observed. Weak numbers of enemy infantry along the road Fère-Champenoise—Connantray. No enemy reserves south of this line. Enemy infantry, estimated strength one brigade, supported by large artillery group observed in the area west of Fère-Champenoise.[132]

About 45 minutes later, FFA 29's report was delivered to Third Army HQ by touring car from the section's airfield.

Our troops have violently thrown back the enemy and taken control of Sommesous, Lenharrée, and Normée. Lack of an enemy presence along the Somme enabled our patrol to be performed at low altitude without receiving any ground fire. Large numbers

Above: German infantry penetrate the French lines early in the morning of the 8th. (Author's Collection)

of enemy dead were seen throughout the sector on the line Sommesous—Normée. Friendly artillery batteries currently being established on the river's southern bank in the area west of Lenharrée. Our advance is progressing west of the line Lenharrée—Écury without interference from the enemy. Similar results to the west where our infantry is advancing south in the direction of Fère-Champenoise.[133]

Hoeppner immediately passed the aviators' good news along to Hausen. Both reports had confirmed that Kirchbach's attack was a major success. Everywhere along the front French Ninth Army's troops were observed to be hastily retreating. Thus, the situation appeared entirely favorable for Third Army. At 10:30am Hausen's liaison officer to Kirchbach's headquarters sent the following message which further put the aviators' findings into context:

The 2nd Guard Infantry Division has thrown the French back in disorder. 32 ID has captured Lenharrée with some of its infantry detachments. A group of its artillery is now in position on the south bank of the Somme. 23 RD has pushed the enemy out of Sommesous and is currently establishing liaison with 32 ID for an organized attack in the direction of Connantray. Thus far, we have confirmed 22 captured artillery pieces, numerous machine-guns and prisoners.[134]

By midday, Kirchbach's victorious yet exhausted formations were compelled to come to a halt along the heights just south of the line Connantray—Fère-Champenoise for rest and resupply. Without any reserves, Hausen was unable to further exploit the favorable situation. Instead, Kirchbach's divisions were to dig in where they stood with orders to hold their newly won positions until nightfall. Late in the afternoon elements of French Ninth Army's shattered XI CA attempted a feeble counter-attack against Kirchbach's troops which were easily repulsed with heavy losses. Once it was apparent that the counter-attack had failed, Foch ordered his forces to withdraw another 1.2 miles to the high ground on the line Ognes—Mailly, effectively bringing the day's combat in that sector to a close.[135]

Hausen and his staff were extremely pleased with the attack's results. French Ninth Army's right wing had been crushed and the marshes were finally under German control. With the marshes secured, Kirchbach's command could now complete their wheel to the west and relieve pressure on Second

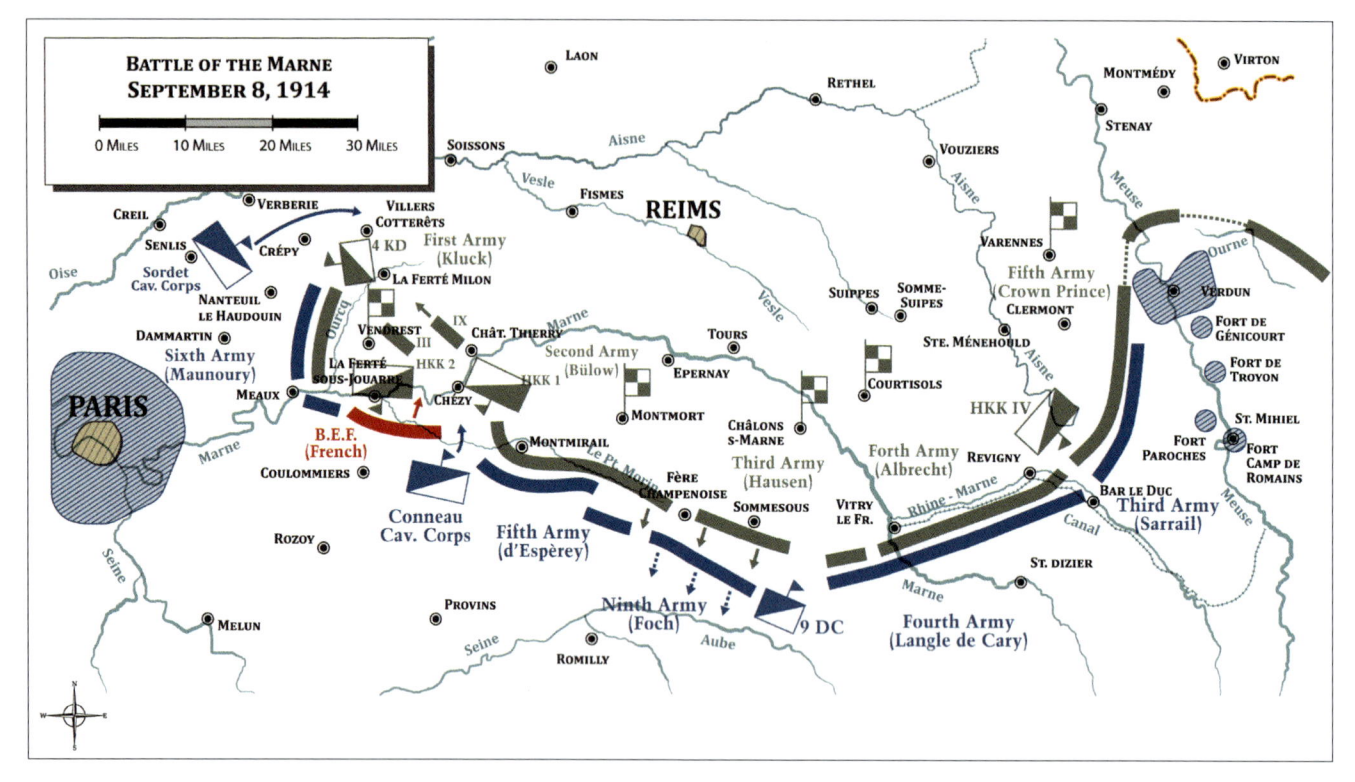

Army's vulnerable right—who was now facing the brunt of French Fifth Army. Furthermore, Third Army received even more good news when the relatively fresh 24 RD arrived on Third Army's right wing late in the afternoon from their previous assignment at the siege of Givet. This substantially strengthened Kirchbach's command and allowed for the renewal of the offensive the following morning with the objective to roll up the remainder of French Ninth Army and threaten Fifth Army's rear. Thus, with the arrival of a fresh division, Third Army HQ eagerly anticipated the morning of September 9th to renew the attack and complete a major victory.[136]

September 8: Second Army's Sector

Upon hearing of Kirchbach's success, General Emmich and the X AK HQ staff ordered their 20 ID to attack due south to retake the ground west of the marshes. Advancing at noon, the depleted division made good progress and ended the day on the Poirer Crest on the marshes' western bank. Further east, General Fleck's 14 ID advanced directly through the marshes, eliminating all nearby pockets of French infantry and securing the swampland's southern bank.[137] Finally, on the army's far left, the Guard Korps participated in Kirchbach's night attack, allowing the 2nd Guards Division to take the important city of Fère-Champenoise late in the morning. By sunset, Second Army's left wing had secured both ends of the marshes and were in position to destroy the remainder of French Ninth Army's left wing the following day.

The aviators of Second Army's left wing were relegated to a purely tactical role throughout the 8th in order to keep the corps leadership well informed of their attacks' progress. Operating from an airfield at Bergères-lès-Vertus on the extreme left of the army, FFA 1 dispatched its aircraft in short-range patrols in support of 2nd Guard Division's attack on Fère Champenoise. The events of the campaign had taken a serious toll on the section, leaving them with the strength of only three serviceable aircraft—two LVG B-types and a replacement Rumpler *Taube*. With no more reserve aircraft available at AFP 2, the section was forced to cut back on the number of flights they could perform each day. Thus, the section's commander, *Hptm.* Jasper von Oertzen, sanctioned only two flights on the 8th; one flight in the late morning to report on the initial attack's progress and the second in the late afternoon to determine the enemy's strength on the corps' right flank.

After a 22 minute flight, the morning patrol returned with a positive report illustrating the corps' victorious advance through Normée and Écury-le-Repos. The section's afternoon patrol produced a report that, among other things, illustrated the locations of several troublesome French batteries that were firing into the flank of the 1st Guards Infantry Division. Using the report, the Guard Korps' artillery was able to silence the French field guns late in the afternoon, thus enabling the German infantry to continue the advance and capture the city of Fère-Champenoise.[138]

To the west, X AK's FFA 21 dispatched four flights throughout the day in support of 20 ID's advance. Like FFA 1, the section had suffered throughout the campaign, losing three of its original six aircraft. However, the section had managed to procure two replacement machines and was thus operating at nearly full strength. Flying a DFW B.I, the section's first flight took off at 9:30am from their airfield at Champaubert with orders to determine if a French assault was imminent. After patrolling the sectors of the X CA, 42 DI, and the Moroccan Division, the flight's observer was able to happily report upon landing that there were no signs of impending offensive action.

At 10:00am, while the first aircraft was still on patrol, FFA 21's commander, *Hptm.* Franz von Geerdtz, received an order from X AK HQ to "send an aircraft southeast across the Marshes of St. Gond to determine the progress made in Third Army's sector." Indeed, X AK HQ had received word of Kirchbach's initial successes, but desired further clarification before going over to the offensive themselves. Accordingly, one of the section's Albatros B.Is piloted by *Oblt.* Behm took off at 10:15am in the direction of Lenharrée with *Oblt.* Oriola serving as observer. Some 90+ minutes later, Oriola delivered the following report in person to X AK HQ.

To: X AK HQ

1. Enemy troops observed retreating in disorder in the direction of Fère Champenoise along the road Morains—Fère Champenoise.
2. Enemy artillery, supply columns and other vehicles seen fleeing into Fère Champenoise.
3. German infantry, estimated one brigade, advancing southwest in pursuit of the defeated enemy force on the line Écury—Normée. Additional German infantry, perhaps one division, in echelon on their right.
4. Lenharrée in German possession. No French resistance seen along the Somme. Friendly artillery batteries currently being established on the Somme's southern bank.
5. Strong enemy forces, over one division in strength, retiring in good order south of the road Connantray—Fère Champenoise.
6. Weak French Infantry in positions to the north and east of Fère Champenoise.
7. Strong mixed enemy force on the line Connantre—Bannes facing east.
8. Large enemy artillery group south of Bannes seen firing on advancing friendly troops.
9. Weak enemy detachments holding the southern exits of the Marshes of St. Gond.

Above: Otto von Emmich, commander of X Armee Korps. (Author's Collection)

10. Mixed enemy force, estimated one brigade, on the heights north of the line Oyes—Soizy.[139]

Oriola's excellent report correctly described French Ninth Army's precarious situation. To the south, Kirchbach's pre-dawn attack had effectively annihilated its right wing. Consequently, a large portion of its center was now bent at an angle with their backs to the marshes. This in turn caused the left wing to become over extended and vulnerable to attack. Recognizing this, Emmich immediately ordered his 20 ID, supported by the corps reserve, to make their attack against the Foch's left wing on the western end of the marshes.

Over the course of 20 ID's advance, X AK HQ ordered two flights. These took off at 2:30pm and 5:15pm with orders to report any rear-area movements behind Foch's left wing. Both flights were quickly performed without incident, giving Emmich's staff an accurate representation of the situation. By nightfall, the 20 ID

Above: German infantry take up positions behind a river. Second Army HQ skillfully deployed their right wing behind the Petit Morin, keeping their flank secured throughout the battle. (Author's Collection)

had retaken the crucially important Poirier Crest and had held off a French counter-attack to drive them off.

To the west, Second Army's right wing remained on the defensive behind the Petit Morin's northern bank expecting an attack from Franchet d'Espèrey's Fifth Army that never came. Unbeknownst to the Germans, d'Espèrey's army was in terrible condition and was unable to conduct a large-scale frontal assault. Thus, with few options, d'Espèrey ordered his right and center to dig in while his left endeavored to maneuver around Second Army's open flank. The only serious action throughout the day therefore occurred in 13 ID's sector on the extreme right of the German line.

After a relatively uneventful morning consisting of sporadic artillery duels, the French began a series of attacks around noon against the German right at Montmirail. Taking advantage of smartly placed machine-guns and heavy artillery, the outnumbered German defenders easily checked each of the French assaults with heavy losses to the attacker. By 5:00pm the French had abandoned their efforts at Montmirail and were seen moving troops even further north around the 13 ID's flank. Unable to extend his line any further, the local German commander sent a message to Bülow describing the enemy's latest movements, warning that envelopment was imminent. In the meantime, the sun had set, bringing the day's fighting to a close with the German positions threatened but still fully intact.[140]

Operating from a forward airfield at Janvilliers, VII AK's FFA 18 was charged with performing all of the aerial reconnaissance duties for Second Army's right wing during the 8th. The section had two aircraft airborne at dawn in anticipation of a major French attack. The section had established liaison with both X RK and 13 ID HQs and had sent the two aircraft to each unit's sector of the front to quickly determine the size and deployment of the attacking enemy force. When the airmen returned roughly 45 minutes later stating there were no signs of an enemy assault, the section's commander, *Hptm.* Ernst von Gersdorff, personally went airborne on a long-range flight to appraise the situation. Acting as the flight's observer, Gersdorff took note of the enemy's front-line strength along the line Viels-Maisons—Montolivet—Morsains—Le Gault-Soigny—Charleville—Lachy, then doubled back over French Fifth Army's rear-area by way of Lachy—Champguyon—Montolivet—Hondevilliers. Gersdorff's report revealed that French Fifth Army was still in the process of preparing for an attack and appeared to be unable to begin any

Infanterie in vollem Feuer. Waldgefecht.

Above: Troops located on Second Army's right wing await an attack from the French during the afternoon of the 8th. (Author's Collection)

assault until mid-afternoon at the earliest. This report, delivered to both X RK and 13 ID HQ's shortly before noon, was of great importance. The troops of the German right wing had been held in their forward defensive positions exposed to violent artillery fire throughout the morning expecting an attack. Gersdorff's report, however, prompted the various right wing commanders to move back significant portions of their infantry to positions in the rear outside the range of the French artillery. This move saved countless numbers of needless casualties, as the French never attempted an attack on X RK's front throughout the entire day. As a result, the X RK suffered only 10 KIA and 96 WIA on the 8th.[141]

FFA 18's five LVG B-Types remained grounded throughout the early afternoon awaiting a change in the tactical situation that would warrant further flying activity. That change occurred at 2:00pm when a representative from 13 ID HQ arrived at the section's airfield requesting aerial reconnaissance in the area of Montmirail. The liaison officer had informed Gersdorff that the French had made several unsuccessful attempts to cross the Petit Morin in the area, causing the division's commander, *Generalleutnant* Kurt von dem Borne, to anticipate an attempt further north against his division's flank.

Borne therefore requested a flight be conducted to determine the French intentions. Shortly thereafter, an LVG B.I was sent airborne with orders from Gersdorff to make a detailed assessment of the French strength north of Montmirail. After a 50 minute flight, the airmen landed back at the airfield and proceeded by touring car to 13 ID HQ to deliver their report. The airmen informed 13 ID's leadership of the enemy's latest movements, which in the observer's opinion greatly threatened the current German position. The observer specified that the French had moved a large number of infantry, perhaps a regiment, with artillery in support, around the German stronghold at Montmirail to the area of Marchais. Furthermore, several batteries of French 75's were seen being deployed on Hill 207, a terrain feature that dominated the German positions. Thus, it was only a matter of time before the French infantry, supported by the newly placed batteries, would overpower the German defenses.[142]

Without any mentionable reserves, all 13 ID HQ could do was send a message forward to the local commander in the area of Marchais alerting him and his men of the coming attack. By 5:00pm the veracity of the aerial reconnaissance report was confirmed when a message from a regimental HQ near the

Above: Built in the 16th Century, Château Montmort served as Second Army HQ throughout the Battle of the Marne. (Author's Collection)

front arrived at 13 ID HQ describing a lively artillery bombardment followed by several major infantry attacks in the area of Marchais. Although these initial attempts had been successfully beaten back, it appeared that the German positions would eventually be either pierced or outflanked. Therefore, 13 ID HQ heeded the regimental commander's warning and sent the aforementioned message to Second Army HQ informing them of the situation.

Shortly after nightfall the French began a series of frontal attacks around Marchais. Firing from their commanding positions on Hill 207, multiple batteries of French 75s heralded the attack with a relentless preliminary bombardment that quickly silenced all German counter-battery and machine gun fire. As soon as their artillery fell silent, hordes of French infantry belonging to *Général* de Maud'huy's XVIII CA launched a massive assault under the cover of darkness along the line Marchais—Fontenelle against 13 ID's center and right wing. Despite greatly outnumbering the Germans, the French attack was easily repulsed with accurate rifle and machine gun fire. Undeterred, de Maud'huy's infantry quickly rallied and charged the German line again, entering into hand-to-hand combat at several points. In the confusion that followed, several companies of French infantry rushed past the German positions undetected through a gap in the line. Upon realizing their good fortune the French infantrymen took up firing positions and fired into the backs of the local German defenders, causing them to flee in panic and confusion. Terror quickly spread down the line, leaving a large gap in 13 ID's center. A local French brigade commander quickly noticed the gap and rushed two regiments forward to claim the valuable ground before it could be recovered. Thus, with his center broken and the French now firmly installed on the heights north of Marchais, von dem Borne ordered a general retreat to the line Montmirail—Artonges, about three miles east of the division's previous position.[143]

Bülow and the Second Army HQ staff followed the events of the 8th from the beautiful Renaissance Château Montmort, located seven miles away from the nearest combat and nearly 15 miles from the action on the army's right wing. The aging army commander had arisen that morning with an overtly pessimistic view of the situation, believing only a major victory by Third Army could save his vulnerable right wing from a catastrophic defeat. By mid-afternoon the good news had arrived of Kirchbach's success, prompting Second Army HQ to order X AK to support the attack with an assault west of the Marshes of St. Gond (an action that had already begun under X AK HQ's initiative). Nevertheless, despite Third Army's gains, Bülow still greatly feared a disaster on his right wing and therefore considered it imperative that X AK push well beyond the marshes by the end of the day. X AK's presence south of the marshes would threaten French Ninth Army with encirclement and force them to retire further south—allowing Kirchbach's forces to continue their wheeling maneuver to face west. From these positions a unified force of Kirchbach's divisions, the Guard Korps and X AK would threaten French Fifth Army's flank and rear. This would undoubtedly compel the French to draw reinforcements from their left, relieving pressure on Second Army's right wing in the process.[144]

In order to better follow X AK's progress, Second Army HQ ordered their own FFA 23 to perform a tactical reconnaissance flight over the western side of the marshes. The section had spent the day exclusively performing long-range flights over the Coulommiers Gap reporting on the BEF's advance but had also smartly kept an aircraft "at the ready" for a tactical reconnaissance sortie should the need arise. As a result, the flying section was able to have an aircraft airborne within 10 minutes of receiving Second Army HQ's order. After a brief 30 minute patrol, the flight's observer reported that X AK was only making negligible progress and would therefore not reach the positions south of the marshes that Second Army HQ had assigned them. Bülow's pessimism immediately returned. In his mind the army's right wing would not be safe unless his left and Third Army had joined hands and wheeled to the west beyond the marshes. Thus, desperate to further improve the situation, Bülow sent the following appeal to Third Army HQ asking for Kirchbach to renew the offensive before sunset:

1st Guards Infantry Division already at Fère Champenoise. Vigorous action by Kirchbach's three divisions, right wing Connantre, immediately necessary. Enemy is currently attempting to envelop Second Army's right wing. No reserves available.[145]

Bülow's request was asking the impossible. Third Army's forces had been engaged since 3:00am and were in desperate need of rest and a hot meal. Furthermore, Second Army's Guard Korps remained pinned down by local French counter-attacks around Fère Champenoise throughout the afternoon. Thus, with his men under attack and completely exhausted, Kirchbach was simply unable to comply with Bülow's wishes. Third Army HQ therefore responded to Bülow's request in the negative, reminding him in the same message that Kirchbach's group had just been reinforced with a fresh division (24 RD) and would resume the offensive at dawn.

As FFA 23's observer predicted, X AK's advance ended far short of their goal south of the marshes. Nevertheless, the situation remained favorable for the Germans as the sun set on the 8th. West of the swamps, X AK had succeeded in recapturing the Poirier Crest and the heights that dominated the area to the southeast. To the east, Kirchbach's combined force had annihilated French Ninth Army's right wing and was poised to split what remained of Foch's command the following day. Most importantly, Second Army's right wing had remained practically unengaged throughout the day, remaining firmly entrenched on the heights overlooking the Petit Morin—albeit with 13 ID now withdrawn three miles to the east.

In spite of all the successes enjoyed around the marshes, Bülow continued to fear an enemy breakthrough on his right. Consequently, when a message arrived at Second Army HQ shortly after sunset describing an enemy breakthrough somewhere between X AK and X RK, the army commander became very concerned. Within minutes, the entire Second Army HQ staff was driving to the front to appraise the situation. Upon arriving to the area in question, Bülow and his staff located *Generalleutnant* Max Hoffman, commander of 19 ID, who had just returned from inspecting the front lines. After exchanging salutes, Bülow explained the reason for his visit and then asked for details of the enemy's breakthrough. A confused Hoffman replied that the French had indeed attempted an attack but had been easily repulsed with great losses. The division commander added that the matter was never in doubt and that he couldn't fathom how such a message could have been sent. After surveying the local defenses, Bülow returned with his staff to his headquarters at Montmort generally pleased with his army's situation and eager to renew the attack with his left the following day.[146]

As Second Army HQ's various touring cars pulled back into the grand driveway at Château Montmort, Bülow was surprised to find OHL's *Oberstleutnant* Hentsch waiting for him. After a brief greeting, Hentsch announced that he "had been sent by OHL to become acquainted with the situation at the various Army HQs and to bring their further steps into agreement with the views of OHL."[147] Bülow, relieved at the thought of the Supreme Command finally taking a more active role in operations, quickly assembled several key members of his staff then sat down with Hentsch for one of the most fateful meetings of the war.

The Mission of *Oberstleutnant* Hentsch

Before further discussing the events that occurred at Second Army HQ on the evening of the 8th, it is imperative to first put Hentsch's mission into proper context. As a member of Moltke's staff, Hentsch began the day in Luxembourg expecting to carry out his regular duties as the director of OHL's Intelligence Section. By sunrise, however, Moltke had called him into an emergency meeting to discuss the latest news from the front. Throughout the previous evening OHL had received numerous reports that clarified the vulnerable position and condition of the right wing armies.[148] These recently transmitted messages belatedly informed Germany's Supreme Command of First Army's battle on the Ourcq, Second Army's struggles around the Marshes of St. Gond, and of the 30+ mile gap that separated the two armies. The latest transmission, intercepted between Richthofen's HKK 1 defending the gap and Bülow's Second Army HQ, read:

(Our) position on the Grand Morin from Villeneuve to Orly has been broken. The Cavalry Corps is retiring slowly behind the Dolloir.[149]

Glancing over his map, Moltke recognized that the Franco-British forces had begun to enter the gap between First and Second Armies. Thus, in order to formulate the appropriate following course of action, Hentsch and several other important members of OHL were summoned to the operations office for the aforementioned emergency conference.

The hastily assembled meeting was attended by Moltke, Hentsch, *Oberst* Wilhelm von Dommes (Chief of OHL's Political Section) and *Oberstleutnant* Gerhard Tappen (Chief of OHL's Operations Department). Although no minutes of the meeting were kept, later testimony revealed that Moltke and Hentsch were both extremely pessimistic about the situation on the right wing, believing a withdrawal of the right flank was already necessary. Tappen and Dommes on the other hand were much more confident and eventually succeeded, after a lengthy discussion, in changing Moltke's views. Thus, by the end of the meeting it was decided that a general withdrawal of the right wing was not yet justified and that the armies concerned should maintain their present positions.[150] Nevertheless, it was unanimously agreed that closer liaison between OHL and the right wing armies was necessary—especially if a general withdrawal ever became necessary. Moltke therefore selected Hentsch to visit the front, appraise the situation, and coordinate the rearward movements of the armies *if a retreat was ordered*.

Considerable controversy regarding the exact wording of Hentsch's orders and their true meaning continues until present day. Since Moltke failed to put the orders in writing, historians have had to debate the issue ever since the war's conclusion. At the center of the discussion is whether Hentsch was given the power to judge for himself if a general withdrawal was necessary and order one in the name of OHL or if he was to merely coordinate the rearward movements of the armies if a retreat had already been initiated by an army commander. In 1917, on Hentsch's request, an official military inquiry headed by Erich Ludendorff looked into the matter. Ludendorff's report read in part:

Colonel Hentsch, then Lieutenant-Colonel and Head of a Section on the Staff of the Chief of the Staff of the Field Army, on the 8th of September, 1914, at Great Headquarters, received verbal instructions from the Chief of the General Staff of the Field Army (Moltke), to motor to the Fifth, Fourth, Third, Second, and First Armies (a round trip of some 400 miles) and bring back a clear idea of the situation. In the case that rearward movements had already been initiated on the right wing, he was instructed to direct them so that the gap between the First and Second Armies would be again closed, the First Army going, if possible, in the direction of Soissons.

Lieutenant-Colonel Hentsch was therefore authorized, in the specified circumstances, to give binding instructions in the name of the Supreme Command.[151]

The court's conclusion, drawn using the testimonies of Moltke, Dommes, Tappen, and Hentsch, was that Hentsch's mission was to gather information and take action to *coordinate* the right wing's retreat with "full power of authority" (*Vollmacht*) in Moltke's name *only if* the right wing's predicament had made such a move necessary. In other words, Hentsch indeed had the authority to *direct* a withdrawal if such an action had already been ordered.

At 11:00am Hentsch, along with *Hptm.* König and *Hptm.* Koeppen of the OHL staff, left Luxembourg headed for Fifth Army HQ. In the event of a mechanical breakdown the officers took two cars, Hentsch and König in one and Koeppen in the other.[152] Before departing, Koeppen wisely asked if he could proceed with his car directly to First Army HQ to evaluate the situation there while Hentsch and König inspected the other armies—suggesting the two rendezvous at Second Army HQ that evening to exchange information. Hentsch rejected the proposal, refusing to either allow Koeppen to proceed directly to First Army or simply begin the tour of the front

Above: The driveway at Château Montmort, where Bülow and Hentsch first greeted one another on the evening of the 8th. (Author's Collection)

with the right wing, which was the main source of OHL's anxiety and the purpose of the trip.

After a three hour drive, Hentsch's two car motorcade arrived at Fifth Army HQ in the town of Varennes. The situation throughout the army's sector was considered excellent. In a brief yet happy conversation, Crown Prince Wilhelm informed Hentsch and König that his army would storm the Forts of Troyon and Les Paroches in the next 48 hours, bringing the German Center group closer to their goal of breaking through the French positions and driving them east in the direction of the attacking German left wing. After leaving Varennes, Hentsch travelled 35 miles to Fourth Army HQ at Courtisols, arriving at 4:15pm. Fourth Army HQ's Chief of Staff, Walther von Lüttwitz, gladly reported that the army had pushed back the French and would commence an enveloping attack along the Rhine-Marne Canal the following morning. Hentsch used Fourth Army HQ's telephone link with Luxembourg to inform Moltke of the highly favorable situation regarding the two center armies, assuring his commander that "there is no urgency (in the center)."[153]

Next, Hentsch drove the eight short miles to Third Army HQ, located in the city of Châlons. Meeting with Third Army Chief of Staff Ernst von Hoeppner at 5:45pm, OHL's liaison team quickly learned firsthand of the heroic performance of Kirchbach's pre-dawn attack and of the great territorial gains that were achieved as a result. After quizzing Hoeppner regarding Third Army's plan to continue the assault the next morning, Hentsch radioed the following to OHL: "Situation and conception [of operations] at Third Army is entirely favorable."[154] Thus, after visiting his first three armies, the situation looked very promising. Indeed, Hentsch's reports regarding Fifth–Third Armies illustrated a situation in the German Center considerably better than what OHL had supposed that morning.

Wishing to not waste any time, Hentsch and his entourage promptly said good-bye to the Third Army staff then departed for Montmort for their fateful meeting at Second Army HQ. Arriving at 7:45pm, Hentsch was unsettled to find the shafts of the headquarters' wagons pointing north (a sign of a planned withdrawal) and the headquarters staff frantically packing up claiming that the French had broken through the army's center. To make matters worse, Bülow and all of the high-ranking members of the army staff were gone, apparently tending to the crisis at the front. Thus, Hentsch could find no one that could either confirm or deny if the rumors of the French breakthrough were factual. Finally, after an agonizingly suspenseful 30 minute wait, Bülow and the Second Army HQ staff returned from the front bringing the news that there had been no breakthrough.

Hentsch quickly greeted the army commander, explained his presence, and requested a meeting to discuss Second Army's general situation. Bülow cordially agreed, asking his chief of staff, Otto von Lauenstein, and *Oberstleutnant* Arthur Matthes (Second Army HQ's Operation's Officer) to join them. Before the conference could begin, however, Bülow excused himself and went to his quarters, leaving Hentsch alone with Lauenstein. While the two waited for Bülow's return, Hentsch and Lauenstein began discussing the events of the day and the overall situation. First, Lauenstein optimistically boasted about the Guard Korps' successes and the subsequent gains made by X AK late in the afternoon west of the marshes. Upon being questioned about the army's right wing, the chief of staff admitted to its vulnerable deployment but stressed that it was holding. Indeed, he explained in detail the enemy's failed attempts to puncture their line in the locale of Montmirail as well as the latest news of a French attempt to envelop the army's flank. Regarding the cavalry of HKK 1 holding the gap, all that was said was that they had fallen back north of the Marne sometime during the afternoon. Hentsch was not pleased with the report, expressing OHL's unfounded opinion that First Army was unable to ward off French Sixth Army's attacks from Paris and that large numbers of 'enemy formations' would soon be in the Coulommiers Gap between the First and Second Armies. Thus, it was Hentsch's uninformed opinion, without first examining the situation at First Army HQ, that the right wing's situation was hopeless.

After hearing Hentsch's report, a panicked Lauenstein rushed out of the room and told Matthes that:

He just learnt from Hentsch that First Army's situation was much more serious than they had been led to believe. According to Hentsch First Army might not be able to turn away the French advance from the direction of Paris nor prevent the enemy from penetrating the gap between the 1st and 2nd Armies. In OHL's opinion circumstances might occur which would mean taking into account the possibility of having to retreat behind the Marne. Despite being reminded of the potential disastrous consequences of such a move, Oberstleutnant Hentsch said that if the enemy broke through in great strength between 1st and 2nd Armies than no other course of action remained. He added that OHL held the view that a punctual and voluntary withdrawal of the right wing armies would not be nearly as fatal as an enemy breakthrough, which would take First Army in the rear and completely wipe it out. If this were to happen, then naturally all of the armies would then be forced to retreat.[155]

Next, Lauenstein recounted Hentsch's pessimistic report to Bülow (who had just returned from his personal quarters), disclosing OHL's opinion that First Army was defeated and no longer able to check the enemy forces emanating from Paris. According to all accounts Bülow's attitude and demeanor immediately changed upon hearing his chief of staff's report. Indeed, he went from being confident and in good spirits to deeply concerned, stating openly that the army's situation was now "extremely serious and even dangerous."[156]

After he had become acquainted with what had already been discussed, Bülow gathered Hentsch, Lauenstein, Matthes, König, and Koeppen into Lauenstein's office to begin the formal meeting. Once everyone was settled, Bülow opened the discussion by summarizing Second Army's situation, repeating once again that the situation was "extremely serious." "After a month of campaigning," said Bülow, "the army has lost a considerable part of its combat value and is no longer able to deliver the final decisive blow that the situation requires." When asked by Hentsch to elaborate, either Lauenstein or Matthes interjected that the army was "burnt to a cinder" (*Schlake*). The over embellished term "cinder" had a clearly noticeable effect upon Hentsch, who henceforth used the term as a basis for all future decisions made in the name of OHL during his trip.

Bülow finished his monologue by criticizing First Army HQ and blaming them entirely for the German right wing's current situation. In nonobservance of Moltke's directive of September 2nd, Bülow opined, Kluck had crossed the Marne ahead of Second Army, which opened their flank to an attack from Fortress Paris. As a result, First Army was attacked under unfavorable circumstances and was compelled to recall III and IX AKs from Second Army's right wing—creating a large gap between the two armies. Then, citing an aviator's information forwarded from First Army HQ, Bülow revealed that, "he had been informed that enemy columns, brigades or divisions in strength, were on the march into the breech and Second Army had no reserves left to attack the enemy or hold him off."[157]

The aerial report mentioned by Bülow was the product of a sortie performed by First Army HQ's FFA 12. Kuhl had ordered the flight to monitor the BEF's progress into the 'Coulommiers Gap' in order to remain fully aware of the possibility of a breakthrough into First Army's rear. Indeed, at 8:40am, FFA 12's commander, *Hptm*. Detten, had taken off with with his pilot, *Ltn*. von Kleist, for a long-range patrol along the line Vendrest—Meaux—Coulommiers—La-Ferté-Gaucher—Saâcy-sur-Marne—Vendrest. About 90 minutes later the duo returned and delivered the following report:

Airplane #: LVG B.75 Vendrest,
 8 September 1914
Pilot: *Ltn*. von Kleist   10:10am
Observer: *Hptm*. von Detten

To: First Army HQ

1. An enemy group observed advancing at 9:00am at [Crécy]-La-Chapelle towards La Haute-Maison. Head at the southern edge of La Haute-Maison. The large column was 3½ kilometers long. An enemy cavalry group, about 4 squadrons, was observed at Les Houis and Courte Soupe [environs of Pierre-Levée] at 8:55am.
2. Enemy [English] division on the march at 9:10am from Boissy-le-Chatel to Doue. Head of column at Doue, end of column at Boissy-le-Chatel. Enemy artillery, approx. 3 batteries, observed north of Doue at Hill 184 in firefight with [rear-guard of] III AK south of St. Cyr [Morin], which crossed the bridge around 9:15 with its end at La Ferté-sous-Jouarre.
3. Enemy group marching from La Ferté-Gaucher through Rebais towards St. Cyr. Head of column at 9:35am was 3 kilometers north of Rebais, end near Hill 180 south of Rebais. Troop encampments were seen near Rebais.
4. The roads from Esbly to Coulommiers, from Coulommiers to both Meaux and La Ferté-sous-Jouarre, and from Rebais to La Ferté-sous-Jouarre are free of the enemy.

(Signed)
Hptm. von Detten[158]

While finding the areas between Meaux and Coulommiers free of enemy troops, Detten had observed three large British columns quickly approaching the Petit Morin with the potential to split the German cavalry screen holding the gap along the Marne. Indeed, it appeared that the BEF was aware of the gap and had shifted the direction of its advance due north towards the two German cavalry corps. Kuhl recognized this and immediately sent a message to HKK 2 alerting them of the threat, informing them that they were to be reinforced as well as ordering the Marne line to be held at all costs. Shortly thereafter, First Army HQ forwarded the valuable information to Montmort for Second Army's consideration. The radiogram, received partly in garbled form at 1:45pm, read:

Airmen confirm branching-off of two enemy columns, apparently a division, through Rebais and Doue to the north. A third column is advancing from La Haute-Maison to the northeast. The Marne line is to be held at all costs...[159]

After disclosing the contents of First Army's message to Hentsch, Bülow summarized the situation as follows: "The enemy has two alternative courses left open to him, either turn against the left wing of First Army or against the right wing of Second Army. Because of our lack of reserves, either movement might lead to catastrophe." Hentsch agreed that First Army's situation was "desperate." Furthermore, he believed that Kluck's army was completely unable to hold the Marne and throw back the British forces marching into the gap. Thus, with First Army seemingly locked in an unwinnable stalemate with the forces attacking from Paris and with no means of stopping the British advance into the breech, the situation appeared ripe for disaster. Hentsch therefore believed that there was no other option for the German right wing other than a "voluntary concentric withdrawal of First and Second Armies to close the gap," adding shortly thereafter that he "had the full authority to order this if necessary in the name of OHL." Bülow responded that because "a breakthrough had not yet become a reality," he believed the current situation did not warrant a general withdrawal of both armies. Instead, according to Bülow, Second Army should be left where it was in order to continue the attack with its left wing while First Army disengaged and moved east, thereby closing the gap. Hentsch demurred, claiming (correctly) that First Army was incapable of executing such a move while fully engaged. At that moment, an aid rushed into the room bringing Bülow the news that 13 ID had been compelled to retreat from the area of Marchais after the French penetrated their line during Maud'huy's night assault. Second Army's right wing was now bent back at an angle facing northwest—widening the Coulommiers Gap even further. Thus, having made the decision earlier in the battle to send VII AK's other division, 14 ID, to the left wing, Second Army HQ had no choice but to abandon Montmirail and pull back their right wing even further to a line behind Margny. From this position, the army's right wing would be temporarily safe from envelopment while the crisis was resolved elsewhere along the battle-line.

The announcement of the retreat of Second Army's

right wing had a devastating impact on Hentsch, who immediately declared that a withdrawal of both armies was a necessity the moment allied forces crossed the Marne River. Bülow and his staff agreed, vowing to continue the attack the following morning with their left wing, yet promising to disengage and retreat in the direction of Fismes the moment they received word that any allied troops reached the Marne's northern bank.

Having reached a decision, the group moved to the next room and had dinner.[160] Afterwards, Hentsch radioed the following cryptic message to OHL, "situation at Second Army serious, but not desperate." Incredibly, Moltke's representative saw no reason to elaborate further or give any indication of what had just been decided. Next, Hentsch, König, and Koeppen retired to their rooms for the night, neglecting to use their time more wisely and proceed directly to First Army HQ. OHL's liaison team elected instead to leave for Kluck's headquarters at dawn the following morning where a final decision would then be made regarding the right wing's future strategy.

Although First Army's situation was at the center of discussion and therefore factored greatly into the decisions made during the Hentsch–Bülow conference, it is important to remember that there was no one present at the meeting who knew the truth regarding that army's circumstances. Regrettably, Hentsch did not send Koeppen directly to First Army HQ at the start of his journey that morning; nor did he choose to immediately leave Montmort after his fateful talk with Bülow. If he had, he would have discovered the situation at First Army was entirely different than what had been originally assumed. Indeed, while Bülow and Hentsch held a pessimistic view on the evening of the 8th based on false assumptions, First Army HQ was highly confident in victory.

September 8: First Army's Sector

In accordance with their aggressive operational strategy, Kluck and Kuhl were obligated to hold First Army together defensively throughout the 8th in order to buy time for the III and IX AKs to complete their forced marches to the Ourcq battlefield. Although III AK was anticipated to arrive ahead of IX AK later that afternoon and could immediately join in the battle, Kuhl wished to employ both corps' infantry together in concert for the final *coup de grâce* against Maunoury's French Sixth Army. The decision was therefore made to hold defensively across the Ourcq Front on the 8th and then launch III and IX AKs' enveloping attack at dawn on the 9th. The details of the plan were plainly illustrated in the official army order for the 8th:

First Army

1) II & IV AKs and IV RK have maintained their positions today on the line Antilly—Puisieux—Vareddes. Strong enemy forces bivouacked this evening at Nanteuil, Silly le Long, and St. Soupplets and to the west of those areas. Fresh enemy reinforcements came into action this afternoon at Betz.

Our aviation reports that weaker forces are opposed to us along the lower Grand Morin, and about a division south of Coulommiers.

2) Second Army is currently locked in battle on the line Montmirail— Fère-Champenoise.

3) II & IV AKs and IV RK are to remain in group formation as before under the command of General Linsingen. The enemy has fought the battle mainly with his strong force of heavy artillery on his right southern wing and in the center.

It will be necessary to hold the positions gained and entrench there. The left wing at Vareddes will have to bend back during the night into a more favorable position. On the right wing the offensive will be pressed forward on the arrival of reinforcements.

4) III AK will start at 2am from Montreuil and La Ferté-sous-Jouarre and march by Mareuil and Crouy so as to come into action on the right flank of the group under General Sixt von Armin, north of Antilly. It is desirable that artillery with cavalry should be sent on ahead.

5) IX AK is ordered to march at 2am from the south of Château-Thierry then to La Ferté-Milon.

6) HKK 2 (without the 4th Cavalry Division) is to protect the left flank of the army towards the lower Grand Morin and Coulommiers; it will also operate from about north of Trilport against the enemy's artillery in position north of Meaux.

7) Army HQ is to remain at Vendrest.

8) A battalion of the infantry brigade of IV RK marching from Brussels and a battalion of the 2nd Grenadier Regiment arrived this evening at Villers-Cotterêts and are attached to the group under General Sixt von Armin.

(Signed)
Kluck[161]

With parts of the army severely damaged and fatigued from 2–3 days of uninterrupted combat, First Army HQ began the 8th with some concerns regarding the army's ability to hold defensively for another day.[162] Composing the center of the battle-line, Gronau's IV RK had been heavily engaged without break since the morning of the 5th. The corps had suffered horrendous losses, especially in officers, and was feared to be on the brink of collapse. With no reserves available, Linsingen (Commander of the Ourcq Front) ordered additional artillery to be concentrated in the center of the front on the heights behind Gronau's men. For additional firepower, III AK's artillery was pushed forward ahead of the corps main body. Thus, with a significant increase in artillery, it was believed that the First Army's weakened line would be able to hold for another day as the plan required.

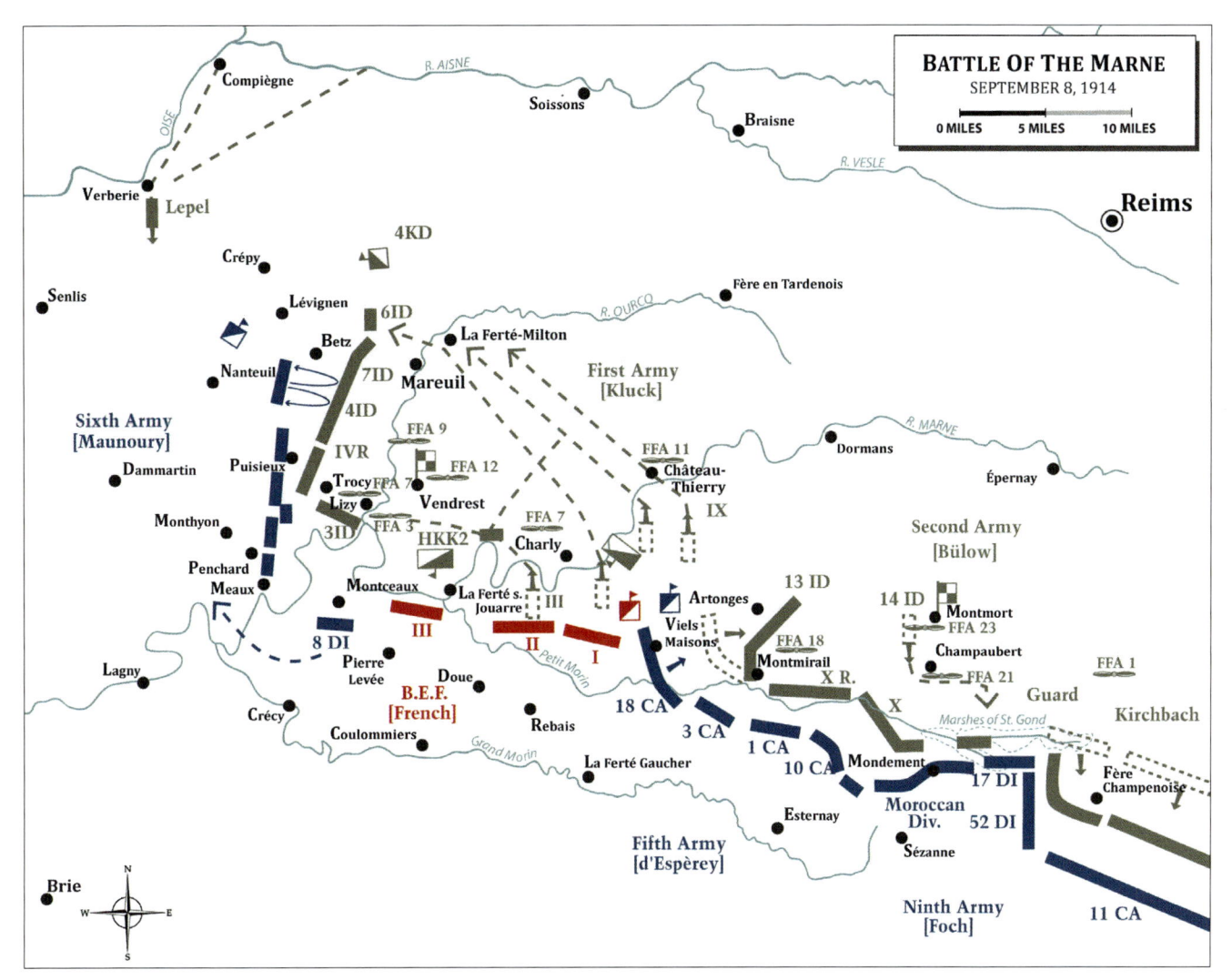

First Army's *Fliegertruppe* had a critically important role in the defensive operations of the 8th. At their pre-dawn meetings, the airmen of each section were informed of the army's overall plan and were assigned a unique area of operations for the day's sorties. All of the army's corps flying sections (including those of the two corps that were en route to the battlefield) were ordered to perform tactical flights over the Ourcq Front. FFAs 9 and 11 were to operate north of the line Nanteuil—Crouy while FFAs 7 and 30 conducted their flights south of this line. Meanwhile, First Army HQ's FFA 12 was instructed to focus all of its flights to the south to check the progress of the BEF's advance into the Coulommiers Gap. The redeployment of the artillery behind the army's left and center was also addressed. The airmen of FFAs 7 and 30 were each directed "to send a liaison officer to the local artillery commanders and establish a system for artillery cooperation."[163] By mid-morning this had been accomplished and each section had an aircraft prepared to exclusively conduct artillery cooperation sorties.

The morning of the 8th brought calm winds and a cloudless sky, enabling the airmen to be airborne at first light. The French had begun the day's combat with an enormous preparatory bombardment concentrating against the vulnerable German center. Several groups of massed batteries, including rare heavy caliber guns from Fortress Paris, were all employed with the intent of softening up the German positions before an

Aufmarsch deutscher Truppen bei Meaux am 8. September 1914.

Above: German artillery rush forward to help stabilize the situation along First Army's center during the fighting on the 8th. (Author's Collection)

infantry assault could be delivered. On the ground, German troops were often unable to determine the exact location of the French batteries in order to begin effective counter-battery fire. Thus, at 6:00am, about 30 minutes after the French cannonade had begun, FFA 30 dispatched its first flight of the day toward the sound of the artillery fire. The flight's observer, *Oblt.* Niemöller, directed his pilot (*Gefreiter* Schauenburg) prior to taking off to fly as quickly as possible across the breadth of the French artillery line from Chambry to Brégy and then return home. The observer had recognized the urgent need for information and constructed the flight plan accordingly to be as speedy as possible. At 6:35am Niemöller returned to FFA 30's airfield at Ocquerre with a full page of notes along with multiple sketches. Not wishing to waste any time with a written report, the observer was quickly driven to II AK HQ, whereupon he delivered a verbal report directly to Linsingen. Niemöller told the commander that while flying over the French gun line, he had counted no less than "26 manned enemy batteries engaged in the bombardment"—a number that left no doubt of the French leadership's intention to breakthrough the German center. "Between Douy-la-Ramée and Fosse Martin," the observer continued,

"there is a very strong enemy artillery group of 14 batteries concentrating all their fire on the area of Trocy. This artillery fire is in preparation of an infantry attack that appears will be delivered against the German positions in the area of Puisieux [IV RK]."[164] About that time II AK HQ received multiple reports from the area of Trocy attesting to the severity of the bombardment, stating that the village was fully engulfed in flames. Drawing upon these statements, Linsingen believed the army's center would be subjected to an attack that it could not repulse. The commander accordingly sent a message to First Army HQ describing the situation and requesting reinforcements.[165]

Over at First Army HQ, Kluck and Kuhl were not surprised to hear of the enemy's latest attempt against their center. In fact, the two generals had strongly suspected that French Sixth Army's leadership would recognize Gronau's IV RK in their army's center as the German weak point and would choose to attack that area in strength. The morning's violent preliminary bombardment recognized by Niemöller and described in Linsingen's report simply confirmed the Generals' suspicions that the French were indeed going to attempt a breakthrough against the IV RK in the area of Trocy. First Army's operational plan however was fully dependent upon the army's left and center holding their ground until III and IX AKs could complete their forced marches to the right wing. Therefore, accepting Linsingen's opinion that the center would not be able to hold without reinforcements, Kluck saw no alternative but to order III AK to detach one of its divisions (5 ID) to the area of Trocy. Although this decision reduced the strength of the force that was to be employed in the battle-winning envelopment attack the following morning by more than 25%, the presence of the 5 ID behind the German center guaranteed the safety of First Army's left and center until the morning of the 9th when Kluck's forces could finally return to the offensive.

By 9:00am the French bombardment had peaked in intensity, prompting the infantry of the 56th and 63rd Reserve Divisions to begin their advance. At the same time, 5 ID's artillery (rushed ahead of the corps' main body) arrived on the field. Galloping at the trot, the recently diverted batteries quickly unlimbered for battle amidst burning farm carts and ammunition wagons.[166] Within minutes, FFA 30's designated artillery cooperation aircraft (operated once again by Schauenburg and Niemöller) was airborne to assist their artillerist comrades in repulsing the coming attack. Flying at dangerously low altitudes, Niemöller immediately located several French batteries that were having an effect on the German infantry. Next, he sketched/took note of the enemy's location and then promptly dropped his notes over the side of the aircraft behind a group of friendly batteries. Repeating this process for over an hour, Schauenburg and Niemöller, working in concert with artillery spotters on the ground, silenced the French forward batteries—forcing them to withdraw out of range. Without any artillery support of their own, the infantry of the two attacking French reserve divisions were easily repulsed and driven back to their starting positions.[167]

While circling over the French lines, Niemöller spotted and took note of the location of French reserves not yet committed to action. As the French infantry attack began to wane, the local French commander ordered these reserves forward. Almost immediately, Niemöller recognized what was occurring and dropped another note at a forward command post stating: "Limbered artillery in the area of Forfry and northwest of Etrepilly. An infantry brigade also seen marching up from a reserve position on the road St. Soupplets—Marcilly." Upon receiving this note, the local German commanders shifted their reserves to successfully check the French advance—demonstrating once again the excellent potential for cooperation between the *Fliegertruppe* and the army.

Above: A German field gun prepares to go into action during a battle with the enemy. (Author's Collection)

Once it was clear that the center was safe from a breakthrough, First Army HQ turned its attention to the northern end of the field where it was believed that the outcome of the battle would ultimately be decided. Kuhl had received an aviator's report at 9:00am from IX AK's FFA 11 suggesting French Sixth Army was moving all available reserves to the northern end of their line. The message, sent directly to Kluck's Headquarters, stated:

At 7:45am on the line Nanteuil—Silly-le-Long, our airmen witnessed a French infantry division advancing against the extreme right of the German line. The roads Nanteuil—Crepy, Senlis—Crepy, and Nanteuil—Betz were free of enemy troops. Accompanying sketch shows the progress of the enemy division between 7:45am and 8:15am.[168]

Although it only mentioned one division, FFA

Above: Wartime postcard depicting two *Taubes* departing for the front. (Author's Collection)

11's brief message raised Kuhl's suspicions enough to immediately order FFA 12 to conduct a long-distance flight of French Sixth Army's left (northern) wing. If Maunoury was indeed seeking a decision on the northern end of the battlefield, Kuhl wished to know about it.

Taking off in their replacement *Taube* five minutes after the order was received, FFA 12's *Ltn.* Haupt (pilot) and *Oblt.* von dem Hagen (observer) performed a two hour flight that totally clarified the situation for First Army's leadership. Delivered to First Army HQ at 11:30am, the report read:

Airplane #: Albatros *Taube* A.70 Vendrest,
8 September 1914
Pilot: *Ltn.* Haupt   11:30am
Observer: *Oblt.* von dem Hagen

To: First Army HQ

1. 9:15am—Strong enemy forces in the area of Maquelines coming from the west.
2. 9:30am—Additional strong forces advancing from Boissy-Fresnoy toward the northwest.
3. 9:40am—Strong forces and columns at Nanteuil marching northeast.
4. Large numbers of infantry spotted at 10:00am between railway and road halfway between Nanteuil and Le Plessis-Belleville.
5. The area to the north, west and south of Dammartin is free of the enemy.
6. The area to the north, west and south Crepy-en-Valois is free of the enemy.
7. Weaker reserves spotted on the line Bouillancy—Monthyon. Stronger reserves are west of Brégy and southeast of St. Soupplets. Enemy cavalry and aircraft observed in the area west of St. Soupplets.
8. At 10:45am the forces mentioned in numbers: #2 and #3, approx. 1 division, at Lévignen.
9. Friendly troops pass through Mareuil at 11:00am.
(Signed)
Oblt. Oskar von dem Hagen[169]

Hagen's report clearly illustrated the deployment and intentions of French Sixth Army. Despite attempting a breakthrough against the German center earlier in the day, Hagen's reconnaissance clearly showed that the Maunoury's primary intention was to envelop the German right flank. A second flight, performed by FFA 9's *Oblt.* Ross, confirmed Hagen's findings. Ross' message read in part:

Above: German infantry holding a strong defensive position in anticipation of an attack. (Author's Collection)

Not only the enemy cavalry corps but also strong enemy infantry and artillery observed north of the line Nanteuil—Antilly. Whether this is for defense of the coming German attack that is to be delivered or to envelop our line in a pre-emptive action is unknown.[170]

Reading the reports, Kuhl undoubtedly recognized that Maunoury was committing all his available reserves to the northern wing for offensive action. Accordingly, the army chief of staff sent a set of orders to his right wing informing them of the situation and prescribing certain measures be taken to strengthen the line.

Acting on Kuhl's orders, the leadership of the German right wing fortified their positions, deployed the majority of their reserves, and dispersed additional ammunition in preparation for the enemy's coming attack. Furthermore, the artillery of III AK's 6 ID, sent under cavalry escort ahead of their infantry, were deployed early in the afternoon amongst the existing field batteries. Together, these actions solidified the German right wing and enabled its commanders to confidently await the French attack that, thanks to the *Fliegertruppe*, they knew was coming.

The French assault against First Army's right wing was easily defeated. Deadly effective rifle and machine-gun fire from the well entrenched German defenders cut down the French infantry, forcing their advance to come to a halt as the survivors sought cover. Next, using their superior range, German artillery firing from the Etavigny Plateau silenced and drove away the supporting French field batteries, causing the unprotected French infantry of 7 DI and 61 RD to fall back and return to their lines without ever having threatened the important German flank.[171]

During both of the day's major actions, First Army's *Fliegertruppe* played a significant role in repulsing the French attacks. In each case, the airmen recognized the enemy's intentions, located their main strength, and promptly reported it to First Army's leadership, allowing them to employ the appropriate counter-measures in order to soundly defeat the French assaults. Specifically, the artillery cooperation efforts of FFA 30 in the area around Trocy were of incalculable value. The German batteries that were in the area had been the focal point of the three and a half hour French preliminary bombardment and were badly damaged. The artillerists had sustained significant casualties with many of their guns and ammunition wagons destroyed, thus seriously reducing their combat effectiveness. Nevertheless, once it became

Above: Assisted by their airmen, German artillery operating from the Trocy Plateau drive back columns of attacking French infantry. (Author's Collection)

clear the French assault was underway, FFA 30's artillery cooperation aircraft was ordered airborne. Working in concert with the batteries on the ground, the aircraft helped silence many of the enemy's most troublesome batteries—ending the French effort to split First Army's center. Without the *Fliegertruppe's* assistance, one can easily argue that the French attack would have either succeeded, or would have been checked with substantially greater losses for the already heavily damaged and fatigued defenders of the German center. These actions combine to serve as compelling evidence that the tactical support given by First Army's *Fliegertruppe* on the 8th was the best performed by airmen on either side in a single day throughout the entirety of the Marne Campaign.[172]

In addition to their excellent tactical work, Kluck's airmen performed several key long-range reconnaissance sorties during the 8th. The most noteworthy flight of the day was the aforementioned report of the Marne sector performed by FFA 12's *Hptm.* Detten which was repeatedly cited at the Hentsch–Bülow discussions at Second Army HQ that evening. Delivered to First Army HQ at 10:10am, Detten's report showed that the BEF was rapidly approaching the German cavalry screen defending the Coulommiers Gap along the Marne River. First Army HQ understood that the gap between First and Second Armies had to be held at all costs for success to be obtained the following day. Otherwise the English would be able to cross the river and strike into the rear of First Army. Thus, at 11:20am, Kluck reluctantly ordered IX AK to send two brigades of infantry and two regiments of artillery to HKK 2 to strengthen the defenses along the Marne between La Ferté-sous-Jouarre and Nogent l'Artaud.[173] *Generalleutnant* Quast, IX AK's commander, received the order at 1:00pm during his march to the Ourcq. However, contrary to First Army HQ's orders, Quast decided due to the depletion of his effectives to leave only a single infantry regiment and one artillery regiment for service in the breech. This group ultimately combined with the army reserve of the same size to form the 'Kraewel Brigade,' which was sent to the Marne and placed under HKK 2's command that evening. This force, created as a result of an aviation report, would ultimately play a major role in delaying the BEF's advance into the gap the following day.

Kuhl recognized the importance of Detten's observations and promptly radioed them to Bülow at Second Army HQ, who famously cited them during

Above: Two German airmen fight off French cavalry shortly after being shot down. (Author's Collection)

his meeting with Hentsch later that evening. Also mentioned in the communiqué was First Army HQ's decision to reinforce the gap with troops from IX AK. Unfortunately, due to poor communications, the radiogram was received in garbled form and Second Army HQ failed to receive any part of the message regarding reinforcements being sent to the breech. Consequently, all the participants at the Hentsch-Bülow meeting believed that the forces defending the gap were much weaker than they really were.[174]

In addition to Detten's sortie, there were several other important long-range reconnaissance flights that occurred on the 8th. Around mid-day, a flight performed by FFA 7 discovered a French division, the 8 DI, moving from their current position on the BEF's left flank to the rear—apparently to be redeployed to the northern end of the Ourcq battlefront. This discovery gave First Army HQ the confidence to continue operations knowing that their army's left wing would not be struck with an attack from the direction of Meaux. Indeed, the airmen's report concluded that the area between Meaux and Lizy-sur-Ourcq was "completely free of the enemy."[175]

Later that afternoon FFAs 9 and 12 each sent out aircraft to inspect French Sixth Army's northern wing. FFA 9's flight was ordered to document the French deployment along the line Crépy-en-Valois—Nanteuil le Haudouin. Unfortunately, the airmen were flying too low and their Gotha Taube was brought down by ground fire behind French lines in an area just north of Crépy. After skillfully landing the aircraft, both the pilot (*Oblt.* Kleine) and the observer (*Oblt.* Blumenbach) were able to evade capture and walk back to German lines.[176] Meanwhile, FFA 12's flight, performed once again by Detten and Kleist, flew the lengthy route Villers-Cotterêts—Crépy-en-Valois—Verberie—Senlis—Nanteuil-le-Haudouin—Dammartin—Vendrest with specific instructions to survey the road network leading out of Paris toward the Ourcq. Upon landing at the airfield at 6:45pm, exactly 90 minutes after taking off, Detten once again drove the short distance to Kluck's Headquarters to deliver his report to First Army's First General Staff Officer—*Oberst* Walter von Bergmann. Detten first commented on seeing several extremely encouraging signs that French Sixth Army was beginning to unravel. The majority of forces witnessed, including large numbers of infantry and cavalry, were falling back under the protection of field batteries to the region of Nanteuil. Furthermore, in Maunoury's rear, there were no troop movements observed from the direction of Paris—indicating that the last of the French reserves had already been committed.

Surprisingly, the most important component of

Detten's report dealt with an observation of *German* troops. While over the area of Verberie, the airmen witnessed a large column moving towards the Ourcq Front from the direction of Compiègne along the line Lacroix-Saint-Ouen—Verberie. This force, which Detten suspected was German after it had refrained from firing on him, was later identified as IV RK's 'Lepel Brigade'—a unit that had been detached from Gronau's command earlier in the campaign for security duties in Brussels. Kluck and Kuhl knew the brigade was on its way to reattach itself to the army, but had no knowledge of exactly when or where they would arrive. Thanks to Detten's report, however, the army leadership not only knew that the brigade was close-by, but that it was marching directly against the rear of French Sixth Army's northern flank—the exact spot that First Army HQ was planning to strike the following morning. Thus, with the knowledge of the Lepel Brigade's proximity and advantageous position, Kluck hastily dispatched a message under cavalry escort to the brigade's leadership notifying them of the situation and ordering them to push forward so that they could participate in the army's envelopment attack the following morning.[177]

That evening, after all of the day's major fighting had come to a close, Kluck and Kuhl held a brief discussion with their staff to review the day's events and evaluate the army's situation. Everyone present agreed that the day could scarcely have gone any better. On the Ourcq, Maunoury's Sixth Army had unsuccessfully attempted two major assaults against the German center and right wing, sustaining significant numbers of casualties in the process. In accordance with their commander's wishes, First Army had absorbed the attacks and held their positions throughout the day, buying time for the reinforcements of III and IX AKs to complete their forced marches. Furthermore, multiple aerial reconnaissance reports proved that the vast majority of the French reserves had been employed in the two failed attacks, suggesting that Maunoury's army was severely weakened and susceptible to collapse if it was attacked on the 9th as planned. Meanwhile to the south, the cavalry divisions of Marwitz' HKK 2 had successfully delayed the advance elements of Sir John French's BEF along the Marne. The arrival of the Kraewel Brigade later that evening further boosted HKK 2's strength, giving Marwitz the ability to confidently continue his mission on the 9th. Finally, perhaps most importantly, after an almost superhuman forced march of 74 miles in less than 48 hours, the men of IX AK and III AK's 6 ID had arrived on the army's right wing ready to deliver the planned envelopment attack against the enemy's left (northern) flank the next day. Thus, taking all of these factors under consideration, the situation on the evening of the 8th was considered by First Army HQ to be extremely favorable as the army's rear remained protected and a large number of reinforcements had finally arrived to deliver the battle's knockout blow against the French flank and rear.

In his memoirs Kuhl described the confidence at army headquarters on the evening of the 8th:

At First Army HQ, still in the course of the evening, we listened to the following radiogram from Third Army to the Second: "Fighting here going well. Heights south of Sommesous carried." We also learned that Maubeuge had capitulated. For our part, we were convinced of having acquired on the Ourcq superiority over the enemy. The victory seemed assured for September 9. The enormous tension we had experienced was beginning to relax in the expectation of the imminent decision.[178]

Unfortunately, due to poor decision making by Moltke and Hentsch, no one else on the evening of the 8th knew the truth regarding First Army's encouraging circumstances. Had Moltke moved OHL closer to the front earlier in the campaign or had Hentsch chose to send Koeppen directly to First Army HQ that morning, the officers at the important Hentsch-Bülow Conference would have understood that the situation on the right wing was far more encouraging than they suspected and could have formulated a plan according to the realities at the front rather than based on their fears and invalid assumptions.

By the evening of the 8th the German right wing was on the precipice of victory at two separate locations—the sectors of First and Third Armies. Kluck's First Army had assembled a large force on their right wing prepared to deliver a large enveloping attack against the flank and rear of Maunoury's weakened Sixth Army. These newly arrived forces were to strike and roll up the French flank, driving them toward Fortress Paris. Elsewhere, the forces of Hausen's Third Army, together with Second Army's left wing, had crushed a significant part of Foch's Ninth Army and forced it back from the marshes. After these impressive gains, Hausen's troops had been reinforced and were poised to continue the attack on the morning of the 9th with the expectation of breaking through what remained of Foch's line—threatening French Fifth Army's lines of communication as they moved against Bülow's weakened right. In other words, on the evening of the 8th the German right wing was in position to deliver two separate attacks at different areas of the front against badly damaged opponents with excellent prospects of success. The obvious weak point, however, was Second Army's right wing and the

Coulommiers Gap between Kluck and Bülow's forces. Although it was admittedly troubling, the events of the next day proved that the reinforced cavalry defending the gap, acting in concert with elements of First Army's redeployed left wing, could have held the gap long enough for First and Third Army's assaults to be delivered.

As it happened, Bülow and Hentsch had already agreed upon an extremely pessimistic operational plan for the 9th—believing First Army was unable to stop either the French forces attacking from Paris or the English moving towards the Marne to the south. It was agreed, citing First Army's "predicament," to hold Second Army in place and attack on the left with Third Army while Hentsch met with First Army's leadership. Once the BEF crossed the Marne, however, the offensive was to be stopped and all three armies of the right wing were to begin a retreat. This course of action, doomed to failure from the start, is a classic example of how critically important operational intelligence and communication between neighboring commands can be. If the major players at the Hentsch conference all knew the truth of First Army's situation, it is highly likely that the plan that would have been agreed upon would have involved allowing First and Third Armies to proceed with their attacks. Unfortunately for the Germans, both Hentsch and Bülow remained unaware of First Army's situation on the night of the 8th and as a result agreed upon a plan that ultimately placed September 9th, 1914 as one of the darkest and most controversial days in German military history.[179]

September 9

In the predawn hours of the 9th, Hentsch and his two aides prepared to finally depart Château Montmort for their visit to First Army HQ. Before leaving however, Hentsch held one final meeting with Lauenstein and Matthes amidst the estate's magnificent gardens. As the officers greeted one another, a panicked Lauenstein said that he believed Second Army was unable to hold its current position unless First Army disengaged on the Ourcq and retreated *eastward* to close the gap and reestablish contact with Bülow's uncovered right flank—a course of action that was suggested by Bülow without success the night before. Hentsch once again responded that he considered the move impossible under the present circumstances. Nonetheless, he vowed to travel to First Army HQ to see if such a move was possible. If it wasn't, Hentsch declared that he would initiate First Army's retreat. Thus, seemingly out of options, the three men decided that Second Army should either begin its retreat as soon as they learned that First Army was *not* moving east to close the gap or when they received confirmation that the allies had crossed the Marne.[180]

Having reached an agreement, the three officers set out to put their plan into effect. First, Lauenstein promised that he and Matthes would immediately discuss the matter with Bülow and obtain his consent to begin the army's withdrawal. In the meantime, Hentsch was to leave for First Army HQ to advise Kluck and Kuhl of the situation and coordinate their retreat so that the breech between First and Second Armies would finally be closed. Together, according to their plan, First, Second, and Third Armies would spend the morning disengaging from the enemy. By afternoon each of the armies would be falling back to their new positions along the Aisne. In other words, the monumental decision to begin the German right wing's withdrawal had been made before First Army HQ had been consulted or any of the day's major combat had begun.

Second Army Initiates the Retreat

Despite their leaders' pessimism, the officers and men of Bülow's Second Army awoke on the morning of the 9th believing the day would bring about the resumption of the battle and a successful conclusion of the promising offensive on their left wing. To that end Second Army HQ ordered the Guard Korps, 14 ID, and 20 ID on the army's left wing to renew the attack against Foch's Ninth Army south of the Marshes of St. Gond. While these divisions were to continue their assault, Second Army's right wing was ordered to hold its position defensively. Indeed, after the dramatic late night breakthrough at Marchais compelled the 13 ID on Bülow's extreme right to withdraw to Artonges, the army leadership had ordered the entire right wing to fall back with them to form a more compact defensive line. As a result, the right wing's new position faced west, running from Bannay through Janvilliers with its right flank resting on the village of Margny—a deployment that temporarily guaranteed the army's safety from envelopment for at least another day.

The left wing's offensive was preceded by an intense preliminary bombardment concentrated against the French positions on the heights of Mont Août, an impressive hill that is easily the highest point in the sector. French artillerists had used the heights throughout the battle as a highly effective artillery spotting platform. Indeed, during the afternoon of the 8th, French guns firing from the heights inflicted heavy losses upon the Guard Korps' 1st Guards Division, causing their commander (Hutier) to halt the advance. Consequently, the Guard Korps lost an excellent opportunity to gain additional ground and inflict more damage on French before nightfall. Thus, with the French artillerists' performance still fresh on

Above: German infantry audaciously attack a French battery. (Author's Collection)

their minds, the officers of the Guard Korps ordered the preliminary bombardment to be focused on the troublesome batteries deployed on the heights as well as the supporting infantry dug in around them.

Second Army's cannonade lasted three hours, silencing the French guns while driving all nearby supporting infantry to the rear. After their artillery had ceased firing, both divisions of Plettenberg's Guard Korps began their assault. Advancing across open ground with all four of its regiments in line, 1st Guards ID easily captured Mont Août, giving the Germans an ideal platform to view the entire battlefield. Several field batteries advancing with the division quickly unlimbered on the hill and opened up with effective pursuit fire into the backs of the fleeing French troops in the valley below—causing their orderly withdrawal to quickly turn into a chaotic rout. To the east, the 2nd Guards ID, advancing with the Saxons of Third Army, captured Connantre—pushing the French behind the River Maurienne. Meanwhile, positioned on the Guard Korps' right, 14 ID attacked from the marshes' southern exits, achieving respectable gains including the village of Allemant while the French forces opposed to them moved east to check the Guard Korps' seemingly more threatening advance. Further west on the attacking wing's right flank, X AK's 20 ID attacked the Moroccan Division holding the critically important chain of ridges at Mondement. If the village and the series of hills around it could be seized, German artillery would be able to subject French Ninth Army to destructive enfilade fire. After a series of staff errors, troops of the 20 ID managed to drive the Moroccans defending the area out of the village, enabling two battalions to establish themselves in the town's fortified château while the remainder of the division moved up to consolidate their gains in order to further continue the offensive.[181]

By noon the left wing's attack had achieved all of its initial goals. The Guard Korps' two divisions, as well as four divisions of Third Army's right wing, had driven back French Ninth Army and captured key territory in the process, including the sector's most important terrain feature at Mondement. From the recently acquired heights of Mont Août and Mondement, German batteries were expected to quickly obliterate the remains of French Ninth Army in the valleys below them. *Oberstleutnant* Dietrich, a member of the Guard Korps who was present on the left wing during the fighting of the 9th, described the situation at noon as follows:

The enemy came down in a hasty retreat from Mont

d'Aout toward the west and southwest; French batteries were getting away at a rapid pace and were subjected to shrapnel fire from three sides. The victory of the left wing of Second Army, obtained in conjunction with the right wing of Third Army, and the thrust through the army of Marshal Foch, were clearly discernable and perhaps already effected.[182]

Thus, after a quick pause to reorganize and bring up their artillery to the new positions, each of the attacking divisions was directed to resume the attack around 1:00pm, whereupon they expected to bring the action in their sector to a decisive conclusion.

Suffering from heavy losses, damaged aircraft, and low fuel, FFAs 1 and 21 could each muster only two flights during the morning to support the left wing's attack. FFA 21 dispatched its first flight at 7:45am with instructions to fly across the breadth of the French positions opposite both of X AK's divisions. Taking off from the section's airfield at Champaubert, the airmen chose to fly a counter-clockwise route taking them north to south over the French positions. First, the observer noted the French movements opposite 19 ID suggested that there was no impending attack. Indeed, artillery fire in the area was reported to be extremely light, suggesting the French troops in the sector (X CA) were too badly damaged to take up any offensive action. Next, the airmen flew over the area of Mondement, which 20 ID was in the midst of attacking. The observer quickly noted the strength of the enemy troops defending the area then returned to Champaubert to deliver his report. According to FFA 21's diary, the report was delivered verbally to a member of the X AK HQ staff, who passed the information down to the relative commands.

FFA 21's second flight of the morning occurred several hours later after the left wing's attack had broken through the first line of French resistance. Taking off around 11:00am, the airmen were ordered to proceed directly to the area of Mondement to determine the French dispositions in the sector. By then, German infantry had driven the Moroccans out of the village and had established themselves in the château and surrounding area. In response French artillery heavily bombarded the area in the attempt to force the Germans out before reinforcements arrived. Once he was over Mondement, FFA 21's observer took note of the enemy's cannon fire and sketched a map of the positions of the closest French batteries. After returning to Champaubert, the observer sent his report along with each of the sketches to X AK HQ for the staff's consideration.[183]

Operating from Bergères-lès-Vertus, FFA 1's two flights were conducted exclusively in support of the 1st Guards ID's assault around Mont Août. The first, performed from 7:45–9:00am, documented the deployment of French artillery and supporting infantry around the heights during the German preliminary bombardment. After writing the enemy's location on paper, the airmen circled around and, on their own initiative, conducted over 30 minutes of valuable artillery cooperation work—dropping notes over the side of the aircraft at the site of local artillery batteries. As the cannon fire ceased and the Guard Korps' infantry began their advance, the section dispatched its second flight of the day with orders to keep the corps leadership informed of the attack's progress. Flying one of the section's remaining original LVG B-types, the aviators circled over the battle-area from 10:00–11:30am, recording the German successes and chronicling with great care the deployment of the newly established enemy positions that were to be attacked later that afternoon. Guard Korps HQ ultimately received the observer's report by touring car at 1:00pm, where its contents were considered during planning for the resumption of the offensive that afternoon. Unfortunately, the hasty decisions made that morning at Second Army HQ rendered the successes on the army's left wing irrelevant—ruling out any possibility for the attack to recommence.[184]

At 8:30am, as the attacks south of the marshes were just getting started, Lauenstein and Matthes met with Bülow to discuss the agreement that had been made with Hentsch earlier that morning in the château's gardens. Lauenstein bluntly told the army commander that the existence of 'the gap' gravely endangered both First and Second Armies and with First Army unable to move east, there was nothing left to do except pull back *all three right wing armies* to close the gap and avert disaster. Bülow demurred. Having not yet heard from First Army HQ regarding the situation in their sector, Bülow remained reluctant to order a general retreat. Besides, an aviation report delivered directly to Second Army HQ at 8:00am by an observer of FFA 18 had brought good news from his army's right wing:

There is no combat in the sectors west and northwest of Montmirail. The roads Montmirail—La Ferté-sous-Jouarre and Montmirail—Château-Thierry are free of enemy troops.[185]

Therefore, with the situation on his right wing under control and no allied troops reported moving northwest of Montmirail towards the breech, Bülow felt entirely unjustified in ordering a general withdrawal at that time. Instead, the army commander resolved to keep the agreement of the previous evening—a retreat would be ordered only when allied troops were reported across the Marne.

Above: Rudolf Berthold photographed while on leave in Germany in October/early November 1914. Best remembered today for his 44 victories as a fighter pilot and award of the *Pour le Mérite*, his reconnaissance sortie of 9 September 1914 was crucial to the Marne campaign. (Aeronaut)

Bülow, Lauenstein, and Matthes remained together at Second Army HQ in order to stay informed of any news from the gap or First Army. At 9:00am, Second Army HQ intercepted a message from HKK 2 to First Army HQ which described strong columns marching easterly from La Ferté-sous-Jouarre. The radiogram was considered by some at Second Army HQ to be confirmation that the BEF had finally reached the Marne River, prompting new requests to immediately initiate the retreat. Bülow remained unconvinced and ordered the army to hold its positions while awaiting further confirmation whether the British had reached the Marne's northern bank. Finally, at 10:00am, FFA 23's *Ltn.* Rudolf Berthold delivered the news that allied troops were indeed north of the river—triggering the retreat of Second Army and eventually the entire right wing.

Berthold's September 9th reconnaissance is unquestionably the most famous *Fliegertruppe* sortie of the campaign. Throughout the first five weeks of the war, Berthold had repeatedly proved himself as FFA 23's most capable observer—flying numerous missions, surviving a forced landing behind enemy lines, and delivering multiple outstanding reconnaissance reports of great value. As a result of his well-earned reputation, the section's commander, *Oblt.* Otto von Falckenstein, selected Berthold to conduct reconnaissance of the Coulommiers Gap along the Marne on the morning of the 9th. The flight had been specifically ordered by Matthes at Second Army HQ to determine if the BEF had reached or had already crossed over the river.

After receiving their briefing, Berthold and his pilot, *Ltn.* Otto von Brachtenbrock, climbed into their DFW B.I and took off at 8:40am headed for the front. The flight plan was simple: Montmort—Château-Thierry—La Ferté-sous-Jouarre—Montmort. Berthold's orders were to document the composition, strength and location of all enemy columns moving into the gap and then deliver the results directly to Second Army HQ upon his return. Brachtenbrock and Berthold completed their flight as ordered in a little over an hour, landing back at FFA 23's Montmort airfield around 9:45am. Berthold hastily jumped out of the aircraft, penned his observations to paper, and was driven to Château Montmort where he presented himself to Bülow shortly after 10:00am to formally deliver his report. Unfortunately, the observer's message was full of bad news:

FFA 23 Montmort, September 9 10:00 am

Pilot: *Ltn.* Bachtenbrock
Observer: *Ltn.* Berthold

To: Second Army HQ

1) 5 enemy columns advancing in a northern direction between Montmirail—La Ferté-sous-Jouarre.

Column 1: On the road Viels-Maisons—Chézy-sur-Marne. Head of column at Chézy at 9:10.

Column 2: On the road Sablonnières—Nogent l'Artaud. Head of column at Nogent at 9:10.

Column 3: On the road Boitron—Pavant. Head of column at Pavant at 9:10.

Column 4: On the road Orly-sur-Morin—Nanteuil-sur-Marne. Head of column at Nanteuil at 9:15.

Column 5: On the road Saint-Cyr-sur-Morin—Saâcy-sur-Marne. Head of column at Saâcy at 9:15.

2) Two large troop encampments witnessed at Replonges east of Hondevilliers and at Bussières.
3) Area between Château-Thierry—Montmirail—Condé-en-Brie is free of the enemy.

(Signed)
Ltn. Berthold[186]

The observer's report deeply impacted Bülow. The army commander quizzed his aviator on the strength and composition of the columns, the status of the area's bridges, and the tenacity of the German resistance in the sector. Berthold answered as best he could. However, he could only speculate as to how long the German defenses in the area could hold. What was certain was each of the five enemy columns was very strong. Together, they possessed the potential of shattering the cavalry screen and turning against First Army's rear. Thus, despite having not yet heard from First Army regarding their current situation, Bülow felt compelled to order a general retreat for his army, compelling the other right wing armies to do the same.[187]

Rudolf Berthold's service on the 9th personified the work of the *Fliegertruppe* throughout the fateful days of the Marne Campaign. Ignoring the stresses from conducting dozens of sorties over five weeks of non-stop campaigning, the airman stoically performed his assigned task and promptly delivered an excellent report outlining the exact location of the Franco-British forces menacing the gap. For his actions on the 9th Berthold became the second individual of Second Army to receive the Iron Cross 2nd Class, receiving it directly from Bülow during a special ceremony on September 13th.[188]

Five minutes after making the decision to start his army's retreat, Bülow dispatched the following message to First and Third Armies:

Aviator reports four long columns advancing towards the Marne. Heads as of 9:00am at Nanteuil-sur-Marne, Citry, Pavant, Nogent-l'Artaud. Second Army initiates retreat, right wing Daméry [Sic].[189]

According to the newly written retreat order, Second Army's vulnerable yet still unengaged right wing was to immediately begin a retrograde movement to the north towards the Vesle. Incredibly, the 13 ID (whose withdrawal the night before had caused great upheaval at Second Army HQ) received Bülow's general retreat order with contempt. The division had received substantial reinforcements during the night and had already completed digging into their new positions.[190] General von dem Borne, the division's commander, later said that, "the order to retreat burst like a bomb in the midst of the leaders and the troops. It was no easy matter to make the order acceptable to the troops by referring to the repulses undergone at another point. It could certainly not be explained from our own position." Borne went further and said he believed his forces could have held the position throughout the 9th *and* the 10th against any enemy attack, giving First and Third Armies the time necessary to complete their promising offensives.[191] Whatever the case, Borne's division as well as those of the X RK on his left were ordered to withdraw at once in a northerly direction.

While the right wing was to start their withdrawal immediately, Second Army HQ concluded that their left should continue with its attack until 4:00pm in order to create sufficient space to safely disengage and begin the retreat. Consequently, the vast majority of the officers and men of the attacking wing's three divisions remained unaware of Bülow's decision as they resumed the offensive against the shattered formations of Foch's Ninth Army. These attacks ultimately made little progress as news of the Bülow's retreat order reached the left wing's division commanders before any subsequent advances occurred. In some cases, the order was ignored. Indeed, *Generalleutnant* Paul Fleck, commander of 14 ID, only obeyed the order after having personally received confirmation from Lauenstein at Second Army HQ that it was authentic.[192] Thus, by late afternoon, all of Second Army was retreating under the cover of a rear-guard.

Although the British columns mentioned in Berthold's message were certainly threatening, they had still not yet crossed the Marne in large numbers. As historian Holger Herwig suggests, Bülow's decision "did not correspond to the situation on the ground." Several other historians have accurately pointed out that the army commander, despite being frustrated by a lack of communication with First Army, could have easily dispatched an aircraft from his own FFA 23 to First Army HQ to obtain a situation report regarding the action in Kluck's sector before making a decision regarding a general retreat. Considering the distances involved, the flight would have taken 30 minutes to reach First Army HQ. Thus, within 90 minutes or less Bülow could have been in possession of all the facts before making a decision. If he had, he would have discovered that Kluck and his staff were already discussing using First Army's left wing to check the advance of 'the five columns' and were in the process of delivering a major blow with their right against the flank and rear of French Sixth Army—a situation that warranted under the circumstances the *holding of Second Army in place* until the attack ended and the situation became clearer.[193]

As it happened, Second Army HQ was entirely

Above: Remaining both bold and aggressive, Alexander von Kluck and the First Army HQ staff brilliantly diffused multiple crises throughout the battle, giving their troops a chance to deliver a knockout blow on the 9th. (Author's Collection)

unaware of First Army's condition throughout these critical hours of the 8th–9th and as a result ordered the retreat at a time when it was not yet justified. In fact, Second Army's retreat order was dispatched and initiated just as First Army's grand envelopment offensive had reached its first objective of the day—pushing the French troops on Maunoury's northern wing back in disorder. Thus, by prematurely ordering the retreat before checking with First Army, Bülow cheated himself and Germany of the opportunity to win the battle or at least create better conditions to carry out a withdrawal. First and Third Armies were attacking with success and Second Army's entrenched right wing remained largely unengaged and unthreatened. In his history of the campaign, Kuhl brilliantly summed up the right wing's circumstances on the 9th. As he points out, the winner of the battle would be the side that possessed the will to do so:

So, on September 9, a crisis existed on both sides. The situation hung by a thread. Whoever was to resolve the crisis to his own advantage had to possess the stronger nerves. To be sure it was a great risk to carry the battle through to a decision. But the prize was worth it. We should have had the audacity. It certainly could not have turned out worse after four hard years of war.[194]

Close examination of the combat in First Army's sector on the 9th validates Kuhl's point. Analyzing the situation on the ground compared to the risks-reward involved, it is plain that the Germans should have, as Kuhl put it, "had the audacity to carry the battle through to a decision." A successful conclusion to First Army's attack would have thrown French Sixth Army back to Fortress Paris, enabling the Germans to pause and reinforce itself before continuing operations as they saw fit.[195] Even if, on the other hand, Kluck's offensive failed to annihilate or drive the enemy back into Paris, the strategic situation would have been better suited to initiate Hentsch's "concentric withdrawal" on the night of the 9th *after* the attack had been delivered rather than the afternoon *during* the confusion of a large offensive. However, to Germany's detriment, Bülow never fully understood what was actually occurring in First Army's sector, which was the western theater's most important point.[196]

First Army Turns to the Offensive

After being surprised and attacked in the flank on September 5th, then forced to conduct three days of costly defensive combats, Kluck's First Army was finally prepared on the morning of the 9th to strike a decisive blow against French Sixth Army's flank and rear. The First Army HQ staff had masterfully managed the crisis and redeployed their entire army (while constantly engaged in heavy combat) so that they continually had sufficient numbers of troops at the correct location to check the French attacks. By the 9th, owing to the heroic forced marches of the III and IX AKs, Kluck's army had assembled three divisions on the right wing, achieving a decisive numerical and qualitative superiority at the key area of the front.

III and IX AK's arrival necessitated a reorganization of the army command structure for the action on the 9th. This new system, established according to Kluck's army order of the 9th, divided the army into three groups—each with a unique role in the coming offensive. First, the newly arrived troops on the right wing—IX AK's 18 ID and 19 ID, III AK's 6 ID, and the 4 KD—were all grouped together under Quast to deliver the main assault against the French flank. On its way to rejoin First Army from an assignment in Brussels, the Lepel Brigade was to take advantage of its location *behind* the French front lines to work together with Quast's forces by attacking through the village of Baron into the French rear-area. Located on Quast's left was the 'Sixt von Armin Group' composed of the 16 IB, 7 ID, and 4 ID. As First Army's center group, Armin's forces were to advance along side Quast's men and take part in the attack "as circumstances would permit." Finally, the left wing group consisting of the 15 IB, 7 RD, 22 RD, and 3 ID (with 5 ID held behind in reserve) was placed under Linsingen with instructions to hold its positions defensively—maintaining liaison with the attacking forces on their right as well as HKK 2 on their left.

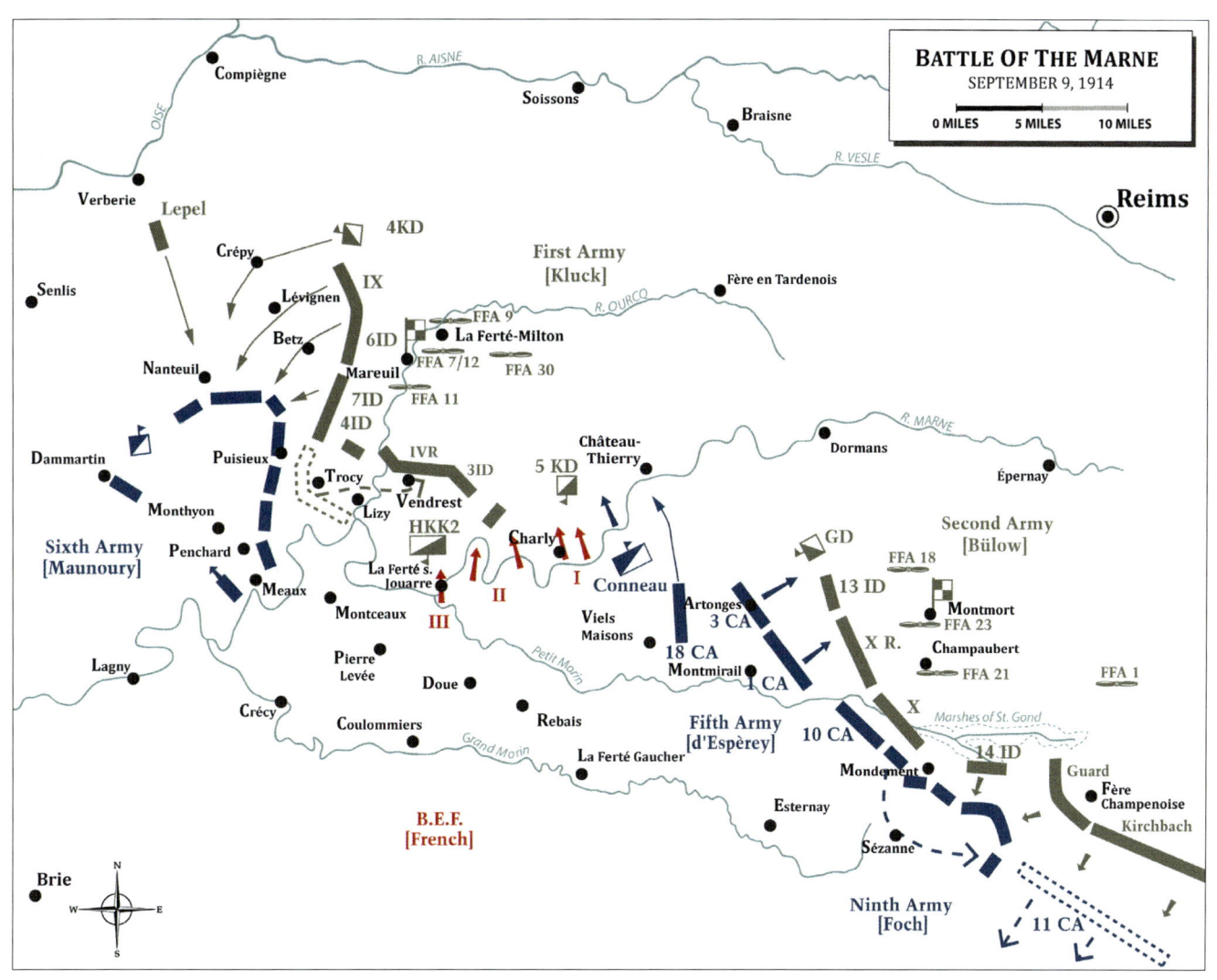

Wishing to be as close to the fighting on the northern wing as possible, First Army HQ was moved north early in the morning to the village of Mareuil directly behind Quast's attacking divisions. From there, Kluck and Kuhl could monitor the right wing's advance and quickly make any necessary adjustments via telephone as the attack progressed. To further assist with command and control, Kuhl dispatched a number of staff officers from First Army HQ to the various wing and corps commands to act as couriers. Regarding aerial reconnaissance, Kuhl ordered all of the corps flying sections to assist with the offensive by offering tactical and artillery spotting flights in their assigned sectors.[197] Specifically, FFAs 9 and 11 were each ordered to conduct operations in support of the troops of the attacking northern wing. Although FFA 9 was traditionally assigned to Armin's IV AK, the section was badly damaged with only three Gotha *Taubes* available on the morning of the 9th. Therefore, First Army HQ wisely decided to move the flying section to the right wing where it could supplement the services of FFA 11, which had five outstanding LVG B-types. Meanwhile FFA 7, with five functional Albatros B.Is, was to perform its flights in the services of Armin's center group. FFA 30 (with four airworthy aircraft) would continue to assist Linsingen's defensive left wing. Finally, First Army HQ's FFA 12 was tasked with maintaining constant surveillance

> **First Army Airfields September 2–9, 1914**
>
> FFA 7
> Sept. 2: Boursonne
> Sept. 3: Coupru
> Sept. 4: Viels Maisons
> Sept. 5: St. Martin du Boschet
> Sept. 6: Villiers-les-Maillets
> Sept. 7: Bézu
> Sept. 8: Beauvoir ou Beauval
> Sept. 9: La Ferté-Milon
>
> FFA 9
> Sept. 2: Villers (south of Crépy-en-Valois)
> Sept. 3: Hervilliers
> Sept. 4: Hervilliers
> Sept. 5: Choisy
> Sept. 6: Crouy
> Sept. 7: Crouy
> Sept. 8: Crouy
> Sept. 9: Précy-à-Mont (NW of La Ferté-Milon)
>
> FFA 11
> Sept. 2: Fontenoy
> Sept. 3: Neuilly-Saint-Front
> Sept. 4: Neuilly-Saint-Front
> Sept. 5: Leuze
> Sept. 6: Leuze
> Sept. 7: Château-Thierry
> Sept. 8: Château-Thierry
> Sept. 9: Mareuil (SW of La Ferté-Milon)
>
> FFA 12
> Sept. 2: Compiègne
> Sept. 3: La Ferté-Milon
> Sept. 4: La Ferté-Milon
> Sept. 5: Rebais
> Sept. 6: Charly
> Sept. 7: Vendrest
> Sept. 8: Vendrest
> Sept. 9: La Ferté-Milon
>
> FFA 30
> Sept. 2: Villers-St.Frambourg
> Sept. 3: Nanteuil-le-Haudouin
> Sept. 4: Trilport
> Sept. 5: Pommeuse
> Sept. 6: Beauvoir ou Beauval
> Sept. 7: Beauvoir ou Beauval
> Sept. 8: Ocquerre
> Sept. 9: Dammard

over the Coulommiers Gap in order to keep Kluck and Kuhl informed of the BEF's progress toward HKK 2 and the Marne River.[198]

With all the necessary preparations in place, First Army's right wing completed a short approach march to the battlefield to begin their attack at 7:00am. From Antilly in the south to Gondreville in the north, the 6, 17, and 18 IDs (supported on the extreme right by a detachment of reservists and the 4 KD) formed a shallow arc facing southwest. The group's objective was to turn French Sixth Army's flank and reach the important town of Nanteuil-le-Haudouin located 11 miles from their starting positions. From there, Quast's divisions, supported by the 'Lepel Brigade' on its right and Armin's group on the left, were to roll up the French line and drive its debris into Fortress Paris. Quast, however, remained uncertain of the exact location of Maunoury's flank, which undoubtedly had changed since the last aerial reconnaissance flight had been performed the night before. Thus, with his vision blocked by the area's wooded terrain, the commander ordered the advance to initially move slowly until an updated aerial reconnaissance report could be obtained. As ordered, the attacking wing began their methodical advance shortly after 7:00am, initially seeing no signs of the French.[199]

The crews of FFAs 9 and 11, like the men of III and IX AKs, were eager to help deliver the battle's decisive blow. At their pre-dawn meetings, they were informed First Army HQ's plan and the *Fliegertruppe's* anticipated tactical role in support of Quast's attack. Because their assigned divisions were expected to advance quickly, artillery cooperation opportunities were considered highly unlikely. Therefore, all attention was to be made in making detailed surveys of the enemy's deployment, particularly the location and strength of all forces operating around Maunoury's open northern flank. At 9:00am, the first opportunity for assistance presented itself when an order from Quast arrived at FFA 9's airfield at Précy-à-Mont (Northwest of La Ferté Milon). The order stated "immediate aerial reconnaissance is requested to locate the French flank." Within minutes, one of the section's two remaining Gotha Taubes (pilot *Feldwebel* Tarnack with observer *Oblt.* Pelzer) took off for the front lines. At 10:05am, after flying the route Précy-à-Mont—Bouillancy—Fresnoy—Crépy-en-Valois—Précy-à-Mont, Pelzer delivered his report to a member of Quast's staff. According to the airman's reconnaissance the French had abandoned all of their positions east of Nanteuil and were falling back in great haste in a southwesterly direction. Furthermore, all French artillery fire on the northern wing was reported as being "severely diminished," leaving Quast to correctly assume that the French were in the process of moving their guns back to a new defensive line. Pelzer finished his report admitting that he was unable to locate and define the position

Above: An excellent example of an early war German airfield. (Author's Collection)

of French Sixth Army's flank due to their withdrawal. Nevertheless, the observer opined that the flank was likely in an area south of Nanteuil.[200]

Pelzer's reconnaissance was invaluable. After spending the morning tediously moving forward, Quast's group had halted at 10:30am to consolidate their initial gains and bring up the artillery to cover the next phase of the advance. Unfortunately, the sluggishness of the advance, resulting in negligible contact with the enemy, was becoming problematic for First Army, who desperately needed to force a decision by the end of the day. However, without the knowledge of the French deployment, Quast was unwilling to risk placing his brigades into road column to hasten the advance. Only the arrival of Pelzer's report changed the situation. With his airmen confirming no French troops directly in front of his divisions, Quast ordered the march to speed up, informing his officers that the French were far off withdrawing in panic. By noon, IX AK had moved through Bargny and Levignen and had begun to establish themselves on the heights east of Boissy-Fresnoy and Villers-Saint-Genest. From there, French Sixth Army's positions could finally be seen. Attack orders were promptly written, directing the offensive to begin around 1:00pm once Quast's entire group was fully concentrated and its artillery unlimbered.

Meanwhile to the south, the morning was surprisingly more eventful on the army's left wing, which had been given a purely defensive role under the direction of Linsingen. After four consecutive days of combat, the divisions under Linsingen's command continued to hold the formidable Trocy and Étavigny Plateaus overlooking the French positions. With the right wing's offensive finally being delivered, it was reasonably assumed after the failed attempts of the previous day that French Sixth Army would not attempt any further frontal attacks in the sector. Therefore, the heavily battered divisions of IV RK and 3 ID, with the fresh troops of 5 ID in reserve, anticipated a relatively uneventful morning in their sector. To the east, the combined forces under the command of Marwitz' HKK 2 anxiously defended the Marne River's northern bank. Using aerial reconnaissance intelligence from the previous afternoon, First Army HQ had alerted Marwitz of the BEF's impending attempt to cross the river. Marwitz consequently deployed his two divisions abreast facing south on the line La Ferte-sous-Jouarre—Montreuil-aux-Lions in order to defend First Army's rear. The sorely needed reinforcements of IX AK's Kraewel Brigade, who had arrived the previous evening, were placed east of the

Above: Operating along a tree line, a German field battery opens fire on nearby allied formations. (Author's Collection)

cavalry in order to extend and improve the defensive line. Despite these measures, the BEF's leading infantry brigade was able to cross the river at 8:15am behind a large cavalry force. These initial crossings, however, were carried out at Nogent-l'Artaud and further east, which did not immediately threaten Kluck's Army. Accordingly, HKK 2 radioed Kuhl at 9:15am stating disingenuously, "all quiet at our front." Shortly thereafter, First Army HQ overheard a radiogram from HKK 1's Guard KD to Second Army declaring, "strong infantry and artillery had crossed the Marne at Charly." Greatly concerned, First Army HQ ordered FFA 12 to immediately send an aircraft to the Marne to sort out the conflicting reports.[201]

As First Army HQ's aviation asset, FFA 12 was exclusively used for long-range reconnaissance duties. On the morning of the 9th, the airmen of the section were informed of the army commander's desire to stay informed of the BEF's position in relation to the Marne. Due to cloud cover, the section's first flight of the day, which had taken off at 7:40am, returned without a report. Undeterred, a second attempt was made at 9:30am. At 10:45am, the observer presented himself to a member of First Army HQ staff reporting "strong columns observed marching north of Rebais in the direction of the Marne." When quizzed about seeing any forces north of the river, the observer answered in the negative. At that moment, however, a car from II AK HQ arrived with an aerial reconnaissance report from their own FFA 30. The report, written by *Oblt*. Niemöller, illustrated in much better detail the location and strength of the allied forces moving towards the Marne opposite HKK 2. Niemöller's message was therefore brought directly to Kühl for his immediate consideration.

Airplane #: Euler B.61.14   Dammard,
9 September 1914
Pilot: *Gefreiter* Schauenburg   10:05am
Observer: *Oblt*. Niemöller

To: II AK HQ

1. Enemy column marching on the road St. Cyr—Saâcy—Méry. Head of column at Méry. A second column likewise in approach from Bussieres to Nanteuil. Head of column at Nanteuil-sur-Marne. Units pushed as far forward as Bézu-le-Guéry and Le Limon. <u>Total Strength is at least 1 corps.</u>
2. Bridging east of Chamigny.
3. Artillery seen at Rougeville (south of Saâcy). Infantry regiment spotted to the east of Les

Pavillons.
4. One column with vehicles advancing at Jouarre. Reserves seen at Sept-Sortes (southeast of La Ferté sous Jouarre) and to the south. Strength of reserves: 1 infantry battalion and 1 field artillery battery.
5. Cavalry with attached infantry (not strong) observed 7 kilometers east of Meaux at Montceaux-lès-Meaux. Another weak force, strength one infantry company, seen at Fublaines. Two batteries observed north east of Trilport.

(Signed)
Oblt. Niemöller[202]

Kuhl immediately understood upon reading Niemöller's report that First Army's left flank and rear was in serious trouble. According to the report, at least one corps was observed crossing over to the Marne's northern bank with significant reinforcements following close behind. Furthermore, it was understood based upon intercepted messages from HKK 1, that there were additional allied formations crossing the river further to the east. Thus, it appeared obvious that Sir John French's BEF had crossed the river in full force and was poised to take First Army in the rear. While Kuhl discussed the matter with the headquarters' staff, a radio message arrived from HKK 2 confirming the aviator's observations stating, "Strong infantry and artillery crossing the Marne at Charly." Before the message could elaborate further, the sender abruptly ended the communication with "I must leave immediately."[203]

The situation was now at crisis point. Even if Quast's attack were successful, it would be for nothing if First Army's left wing was enveloped and destroyed. First Army HQ had no choice but to act quickly. Glancing at a map, Kuhl immediately recognized a solution:

The situation on the Marne appeared, then, to be growing menacing. The right wing of Second Army having fallen back and the English having crossed the Marne, the situation of our left wing on the Ourcq to the north of Congis was becoming untenable. Behind this whole wing there were only two substantial bridges over the Ourcq at Lizy and Crouy. A retreat behind the deeply indented Ourcq could be fatal if it were undertaken at the last moment and under the pressure of the enemy. Accordingly the left wing had to be drawn back at the opportune moment. But the attack on the right wing was to be continued all the more vigorously. Against the English too, relief was to be afforded only by an attack.[204]

According to Kuhl's plan, Linsingen's left wing was to be pivoted back to face south to check the threat of the advancing BEF—allowing for the right wing's attack to continue. The new directive, brilliantly conceived within minutes, is an excellent example of why Schlieffen thought of Kuhl as a "great captain of the future." Instead of forfeiting the advantage or initiative to the enemy, Kuhl sought to retain both by improvising and acting offensively. While Quast's group was to continue their attack, Linsingen's group would move into their positions, covered by the relatively fresh 5 ID that was being held in reserve behind the left wing.

At 11:30am Kuhl's plan was put into action. Linsingen was instructed by telephone to withdraw his divisions back behind the Ourcq and to order the 5 ID to attack the British through the village of Dhuisy. Linsingen's staff quickly drew up the orders for each division and dispatched them to the appropriate commanders. These instructions reached the front line troops at 1:00pm and were executed immediately. Although potentially hazardous, the withdrawal was performed without any interference from French Sixth Army's badly damaged right wing. In a report, *Hptm*. Schutz of the First Army HQ staff described the scene late in the afternoon.

During all this time the French percussion shell and shrapnel fire continued to burst at the same spot as before. No enemy force was visible. When no one was left at Beauvoir au Beauval, I continued to May-en-Multien and climbed to the top of the very high church tower. In the clear evening sky the Eifel Tower was visible to the southwest. The French artillery fire had completely ceased. Despite the excellent view over the whole surrounding country, nothing was seen of the enemy, not even a cavalry patrol.[205]

The lack of interference from the French, illustrated just how badly mauled Maunoury's army really was. By 3:00pm Linsingen's move had been completed and the army's flank was secure from the BEF, who was being checked by the combined efforts of 5 ID, the Kraewel Brigade, and HKK 2, thanks in large part to FFA 30's hastily delivered report.

Hentsch Finally Arrives at First Army HQ

After making the decision to back-pivot the army's left wing to face south, Kluck and Kuhl returned their attention back to Quast's group on the right wing. At 12:34pm, a report arrived from *Hptm*. Bürhmann (First Army HQ's liaison officer to Quast) describing the accelerated speed of the advance during the late morning and Quast's plan to begin the first formal attack of the day on the French positions near Nanteuil-le-Haudouin. Next, Bürhmann announced that the Lepel Brigade was heavily engaged at Baron with excellent prospects of soon striking the enemy in the rear. First Army's command team was therefore pleased with the situation, believing the badly damaged French Sixth Army had spent the morning retreating in panic and were on the verge of being annihilated. At that moment, around 12:30pm, Hentsch and his aides arrived at Kluck's headquarters to inform the First Army HQ staff of the decisions that had been made that morning at Château Montmort.

Forced to take a long detour because of the heavily clogged roads, it took Hentsch five hours to complete the drive from Second Army HQ at Montmort to Kluck's headquarters at Mareuil-sur-Ourcq. Hentsch and Kuhl greeted one another in the street, instantly entering into meaningful conversation while Kluck remained several hundred feet up the road at his command post. Kuhl spoke first. He informed OHL's representative that Quast's group had already thrown back Maunoury's left wing and was in the midst of carrying out an assault outside Nanteuil with excellent prospects of success. Next, Hentsch was told of the aerial reports detailing the BEF's advance across the Marne and of First Army HQ's response to pivot the left wing back to check the threat. Hentsch, undoubtedly surprised as to how well in hand matters were, requested the discussion continue in Kuhl's office.

After moving indoors, Hentsch, Kuhl, and *Oberst* Walter von Bergmann (First Army Deputy Chief of Staff) began a formal discussion. Kuhl reiterated the details of Quast's attack, confidently and accurately depicting what had thus far transpired. Hentsch, believing for over 24 hours that First Army was doomed, was shocked to find out that First Army was on the verge of a major success. Bergmann later noted that: "The surprise was plain to see on his face. I remember clearly that he appeared flabbergasted as his preconceived opinions were contradicted by this [Kuhl's] depiction."[206] Next, after hearing Kuhl's report, Hentsch asked if First Army would be able to come to the aid of Second Army the next day—suggesting to Kuhl and Bergmann that Bülow's situation was critical. Kuhl and Bergmann both replied that the state of the army and the distance to Second Army ruled out First Army's immediate support. Hentsch coldly replied that, "there was nothing left to do but to order a general retreat." Taking out a charcoal pencil, Moltke's representative proceeded to draw the lines of retreat for each army on Kuhl's staff map. To this he added, "Seventh Army is currently being brought up to Saint Quentin from Alsace. This new army will help stabilize the situation and allow offensive operations to resume."[207]

Kuhl and Bergmann vigorously protested Hentsch's proposal. For nearly an hour the two officers attempted in vain to convince Moltke's representative of the superiority of First Army's right wing and the possible gains resulting from its being allowed to complete the assault. Addressing the potential danger of the BEF's advance into 'the gap,' Bergmann reminded Hentsch that the army's left wing was now facing south and that the 5 ID was already executing an approach march to attack the nearest column. When he saw that Hentsch was not budging, Kuhl spoke up to remind Hentsch that to halt a large body of troops such as Quast's in the middle of attack, then execute an about face and retreat, would be extremely difficult and risky—especially considering their level of fatigue after the incessant forced marches and combat of the previous 72 hours.[208] Thus, if a retreat were indeed necessary, Kuhl appeared to be of the opinion that it would be better to do so that evening *after* Maunoury's troops were attacked and thrown back.

Hentsch would not relent. He revealed the fact that he had already come to an agreement with Bülow for Second Army to begin its retreat the moment the BEF reached the Marne—regardless of any input from First Army. When a puzzled Kuhl asked why Bülow would make such a decision, Hentsch responded that "the decision to retreat has been a bitter pill for old Bülow to swallow." Nevertheless, Hentsch told his colleagues that the decision was a necessity because Second Army was "burnt to a cinder"—repeating the over embellished claim that had been made by either Lauenstein or Matthes during dinner the previous evening. Writing in his memoirs, Kuhl later described the effect of Hentsch's story and its impact on his outlook of the situation.

These words remain firmly fixed in my memory...as far as I was concerned it was decisive. If Second Army had been defeated and indeed 'burnt to a cinder' and their retreat already taking place, then First Army could no longer remain standing in isolation where it was. It could only complete the battle on the Ourcq if Second Army were to remain in place.[209]

Sensing that both Kuhl and Bergmann were now beginning to entertain the idea of a retreat, Hentsch revealed that he had been given full authority in Moltke's name (*Vollmacht*) to initiate and coordinate the retreat and therefore demanded that it begin immediately. Kuhl later explained what occurred next.

The interview [with Hentsch] lasted a long time. I opposed in the most vigorous manner the proposition to retreat and insisted over and over again on the favorable situation of our right wing. All the possibilities of continuing the fighting up to the final victory were examined. But when it was made

Above: LVG B.I attached to a Bavarian unit. This early production aircraft has the skid and small wheel attached to the undercarriage to reduce the chance of nosing over, a feature omitted from later production aircraft. The sturdy and reliable LVG B.I was the most numerous German reconnaissance biplane in the early months of the war. (Aeronaut)

clear that the decision of Second Army had been reached that morning, that its troops were already in full retreat in the afternoon and that the decision could not be recalled, the army command had to submit. Even a victory over Maunoury could not have prevented us from being encircled by superior forces on our left wing and being cut off from the rest of the German Armies. The First Army would have been isolated.

I appeared before the commander-in-chief to make my report. With a heavy heart, General von Kluck was obliged to accept the order (and initiate the army's retreat). Upon my return Oberstleutnant Hentsch departed, after being informed of the decision arrived at.[210]

In his account of the meeting, Bergmann accurately depicted how little Hentsch had known of First Army's situation and therefore how foolish it had been to order Second Army's retreat that morning before visiting Kluck.

I was present from the beginning to end of the discussion which took place on September 9th with Oberstleutnant Hentsch in the office of General von Kuhl. Oberstleutnant Hentsch recounted what he regarded as the rather unfavorable turn which events had taken in the western theater and insisted particularly on the situation of the neighboring Second Army. I distinctly recall that he depicted this situation in very gloomy colors and that his description of the morale of Second Army HQ staff impressed me particularly. It was evident to us that Oberstleutnant Hentsch had formed, from the information given to him at Second Army HQ, a completely erroneous picture of the situation of First Army and that the much more favorable appraisal given to him by General von Kuhl surprised him exceedingly. To the serious objections raised by General von Kuhl against the retreat of First Army, whose right was engaged at the time in executing an attack which was progressing favorably,— objections in which I joined, also insisting on the technical difficulties and the condition of the troops– Oberstleutnant Hentsch always replied by saying that the retreat of Second Army behind the Marne, already begun, was an irrevocable necessity. In view of this communication and the further declaration of Oberstleutnant Hentsch to the effect that he had full powers to order a retreat of the army in the name of High Command, nothing remained for General von Kuhl but to propose to the commander-in-chief to retreat, the more so as at that time there was no connection permitting a direct exchange of ideas with the High Command [OHL]. To this end he appeared before General von Kluck to make his report while I remained alone in the room with Oberstleutnant Hentsch until General von Kuhl returned with the approval of the commander-in-chief.[211]

As a result of the Hentsch and Bülow's meeting, First Army was forced to halt its promising attacks and begin a general retreat despite remaining tactically unbeaten as well as holding a sizeable advantage at the sector's most meaningful point—the battlefront's northern flank. As Kuhl and Bergmann's accounts attest, the news of Second Army's retreat is what ultimately compelled First Army's leadership to abandon the offensive and begin a general withdrawal of its own. Despite being enormously difficult considering the circumstances, Kluck and Kuhl's decision was the correct one.

Some historians have argued that Kluck should have refused the order until "formal written orders" from Moltke arrived—thus openly challenging Hentsch's word that he possessed the authority he claimed he did. If Kluck had done this, the argument claims, he would have given his army the time and the chance to win the battle at the eleventh hour. This proposed course of action is totally unrealistic.[212] Although First Army was indeed tactically unbeaten and on the verge of delivering a major victory with its right wing, the truth remained that Second Army had already begun its retreat. If Quast's attack were allowed to continue, First Army's left wing would ultimately have become engaged frontally by the BEF and later surrounded on the left by elements of French Fifth Army—resulting in the encirclement of the army and a major military disaster. Kuhl understood that once Second Army had begun its retreat, no matter the circumstances, First Army was obliged to fall back with it even though that meant the end of the campaign without the prerequisite decisive victory.

First Army's Retreat Begins

In yet another testament to the ability of the First Army HQ staff, highly detailed instructions were quickly drawn up and dispatched to all the necessary commands ordering the cessation of all offensive action and the army's immediate withdrawal—an extremely complex and dangerous maneuver considering the state of Kluck's troops after the forced marches and heavy combat of the previous five days. Fully understanding the risks involved with initiating a general retreat in the midst of an attack, the orders gave the commanders at the front, particularly on the attacking right wing, the freedom to complete any local assaults in order to create sufficient space to safely begin the retreat. Kluck's official army order, dispatched minutes after Hentsch's departure at 2:00pm, read as follows.

First Army

The situation of Second Army has necessitated its withdrawal behind the Marne on both sides of Épernay. By order of the Supreme Command, the First Army is to be withdrawn in the general direction of Soissons, to cover the flank of the armies. A new German Army is being assembled at Saint Quentin. The movement of First Army will begin today. The left wing of the army, under General von Linsingen, including the group of General Lochow, will therefore be withdrawn first behind the line Montigny—Brumetz. The group under General Sixt von Armin will conform to this movement so far as the tactical situation will allow and take up a new line from Antilly to Mareuil. The offensive of the group under General von Quast will not be proceeded with any further than is required for the purpose of breaking away from the enemy so that it will be possible to conform to the movement of the other groups.

(Signed)
von Kluck[213]

Located on the army's extreme northern wing, the 4 KD was informed of the retreat order at 2:00pm and was subsequently directed to "advance to the Aisne River and secure all the bridges and crossings from Soissons to Attichy in order to facilitate the army's withdrawal." Meanwhile, the overworked airman of FFA 12 were ordered to immediately dispatch two flights; the first of which was to fly south in the direction of the Marne to give the Army HQ staff an updated picture of the BEF's progress into the gap while the second was to travel north and confirm that the areas where the army was to retreat were free of enemy troops. The first flight, performed by the proven team of *Oblt.* von dem Hagen and Ltn. Haupt, took off from Mareuil at 3:00pm and returned at 4:45 after a lengthy flight along the route La Ferté Milon—Château-Thierry—Coulommiers—La Ferté-sous-Jouarre—Château-Thierry—La Ferté Milon. Almost two hours after landing, amidst the chaos of the army's retreat, von dem Hagen finally located a First Army HQ staff officer to deliver his report. Hagen had wanted to deliver his report directly to either Bergmann or First Army HQ's Ia officer, *Oberstleutnant* Grautoff, but was unable to locate either of them. After spending an hour slowly riding in the flying section's touring car amongst heavily congested roads full of ammunition and supply columns being turned around for the retreat, Hagen decided to leave his reconnaissance results with a lower level First Army HQ staff officer.[214] The report read as follows:

Airplane #: *Ersatz* A.70   La Ferté-Milon,
                              9 September 1914
Pilot: *Ltn.* Haupt   6:30pm
Observer: *Oblt.* von dem Hagen

To: First Army HQ

1. Strong enemy bivouac observed at 3:15pm west of Bézu-Saint-Germain at Étrépilly and north of Chézy-en-Orxois. Other enemy forces in the area seen moving northwest.
2. At 3:30pm Château-Thierry was free of the enemy.
3. Weaker enemy bivouacs in the areas of Viffort, Montfaucon and north of Fontenelle.
4. A weak enemy column at Villemoyenne marching northwest, stronger forces south of Rozoy-Bellevalle, weaker at Marchais to the north.
5. Train seen stopped north of Hondevilliers. Supply column approaching from the west of Hondevilliers.
6. Weaker bivouacs north of Rebais. Bivouac column and airfield south of Coulommiers.
7. Strong enemy bivouacs observed 3 kilometers southwest of the line Jouarre—Courte Soupe (south of La Ferté-sous-Jouarre).
8. Weaker enemy column at 4:20pm at La Ferté observed crossing the river.
9. Weak enemy positions west of Citry and south of Saâcy as well as north and south of Saint-Cyr-sur-Morin. Only weak forces crossing the Marne at Nanteuil-sur-Marne. East of Caumont enemy artillery in battle firing north.
10. Weak enemy positions south of Charly. Stronger column spotted west of Bassevelle.
11. Bridge at Chézy not destroyed.
12. Strong enemy cavalry observed at 4:30pm east of Montgivrault. (just south of Lucy-le-Bocage)

(Signed)
Oblt. von dem Hagen[215]

Hagen's message proved that the British were unable to make a significant push late in the day beyond the Marne. Thus, with the BEF's failure to move quickly confirmed, the officers of First Army HQ breathed a collective sigh of relief and proceeded to plan the following day's withdrawal without fear of the British pursuit drawing the army back into battle. As it happened, Kuhl personally reviewed Hagen's report later that evening while planning the army's order of the day for September 10th.

FFA 12's second and final flight of the afternoon, conducted by *Hptm.* Detten and *Ltn.* von Kleist, was a brief 30 minute sortie covering the army's projected route of retreat. Taking off at 3:15pm, Detten directed Kleist to pilot their Albatros towards Villers-Cotterêts to determine if there were any French forces in the area that could pose a threat to the retreating German flank. After seeing no hostile forces in the area, the airmen continued west to Crépy-en-Valois and then south to Betz before returning back to the airfield at La Ferté-Milon. Upon landing, Detten quickly penned a note simply stating: "Area: Villers-Cotterêts—Vivières—Crépy completely free of enemy troops as of 3:30pm." Detten's brief message was taken by touring car to First Army HQ where it was handed to a member of the headquarters staff at 4:00pm. With the route of retreat deemed 'safe to travel,' First Army HQ continued with its original plan to retreat on the line Soissons—Fismes to reestablish contact with Second Army. Meanwhile, after dispatching his note to First Army HQ, Detten ordered the enlisted men of FFA 12 to prepare all the aircraft and equipment for road travel in order to join the forward supply columns and find a new suitable airfield for the following day.[216]

Number of Times First Army's Flying Sections Changed Airfields August 28—September 11[217]	
FFA 7	14
FFA 9	11
FFA 11	9
FFA 12	10
FFA 30	15

First Army HQ's retreat orders were quickly dispatched via telephone or automobile to the corps commanders of each of the army's three groups. The first to receive the directive was the leadership of Quast's 'offensive right wing.' After a long and uneventful morning pursuing the retreating forces of French Sixth Army's left wing, Quast's men had finally caught up with their adversaries and were in the process of delivering a series of successful assaults at the time of the message's arrival. Receiving the message with a great deal of shock, Quast's staff "vehemently opposed" the retreat order. Indeed, *Hptm.* Bührmann (First Army HQ's liaison officer to Quast) picked up the telephone receiver and told the First Army HQ staff officer on the other end of the line that a retreat would be a mistake. "The troops have absolute confidence in victory," Bührmann exclaimed, "Their morale is excellent…[and]…the enemy is on the point of yielding everywhere." Undeterred, the officer on the end of the receiver replied to present the order to Quast and to make sure it was executed without delay. A baffled Bührmann responded by demanding to hear the order directly from Kuhl. Seconds later, the army's chief of staff took the phone and calmly described Second Army's situation, further explaining that there was no alternative but a general retreat regardless of how favorable First Army's tactical situation was. With nothing left to do, Bührmann hung up the telephone and then presented the order to Quast. In his official report written several days after the battle, Bührmann explained what happened next.

General von Quast raised a violent protest and urged me to have the order countermanded. I again called General von Kuhl, who once more explained to me in detail the whole situation and the necessity for the retreat. During this time the order had come to the knowledge of the division commands. The Chief of Staff of the 18 ID refused to execute the order, which was quite incomprehensible to him, saying that the enemy was in full retreat upon Paris and that his own troops wished to exploit their victory to the limit and were in the position to do so. He suggested calling the army command once more. His Royal Highness the Grand Duke of Mecklenburg-Schwerin, who was accompanying Quast's headquarters, joined in this demand. First Army HQ once again replied that the decision was irrevocable.[218]

Thus, with a heavy heart Quast and his staff reluctantly accepted their superior's decision and halted their promising attack. In several cases division commanders at the front, seeing the fruits of victory within their grasp, delayed their retreat and proceeded with their pre-planned assaults. These attacks were ultimately successful, resulting in the swift capture of Boissy-Fresnoy and Villers-Saint-Genest. Shortly thereafter, however, a second order explicitly forbidding all further forward movement was dispatched amongst the group bringing the day's combat and the epic five day 'Battle of the Ourcq' to a close. Quast's men ultimately bivouacked in the open on the newly won terrain with orders to begin their retreat the following morning, acting as the army's rear-guard.

Kluck's retreat order similarly reached Sixt von Armin's center group in the midst of a major attack. Around noon, Armin had learned that French infantry in his sector were being shifted north to oppose Quast's advance and had chosen to launch an attack of his own to facilitate the right wing's assault. When Kluck's retreat order arrived, however, Armin quickly obeyed and ordered his group to break off all contact with the enemy. In an excellent example of how badly diminished French Sixth Army's combat value truly was, the exhausted and battered troops of Armin's command were able to disengage and begin a general withdrawal without any interference or pursuit from the French. By sunset, Maunoury's men had lost contact with First Army and would be unable to regain it until the armies met again several days later along the banks of the Aisne River.

Linsingen's southern facing 'defensive left wing' began its northward movement at 4:00pm. These troops were directed to march to the line Montigny-l'Ailler—Brumetz, then push on to the area of Villers-Cotterêts by the end of the following day—September 10th. Once again, the group's withdrawal was completed in good order and was performed without any hindrance from the Franco-British forces in their sector.

That evening, First Army HQ dispatched the army's official order for the following day amongst the troops. Dispatched at 8:15pm, the communiqué read as follows:

To the officers and men of First Army,
The right wing of the army was today advancing victoriously towards Nanteuil-le-Haudouin. On the left wing the 5 ID with HKK 2 attacked the enemy advancing on the front Nanteuil-sur-Marne—Nogent-l'Artaud. By order of OHL, First Army is to be withdrawn towards Soissons and west of it behind the Aisne, so as to cover the flank of the Armies; the Second Army is in retreat to behind the Marne on both sides of Epernay.

I wish to express to the men of First Army my highest admiration for their devotion to duty and for their exceptional achievements during the offensive.

The Army will today continue the movement ordered from the lines already reached, the main body moving up to and north of a line Gondreville—southeast of Crépy-en-Valois—La Ferté Milon—upper half of the Ourcq line. The left wing of the army under General von Linsingen, including the group under General von Lochow, will march east of the Ourcq, below La Ferté Milon, and then with its right flank along the road La Ferté Milon—Villers Cotterêts—cross roads northeast of Villers Cotterêts—Ambleny. Meanwhile, General Sixt von Armin's group will march with its right flank on the Antilly—Vauciennes—Taillefontaine—Attichy road. General Quast's group will keep to the west of it.

HKK 2 with Kraewel's Brigade will cover the left flank of the withdrawal. The 4th KD has been ordered to move ahead [of the main body] to the Aisne to occupy the bridges between Compiegne and Soissons. The Reserve Brigade of von Lepel and the composite 11th Landwehr Brigade of Von der Schulenburg will march by Compiegne to Vic for the same purpose. In addition to the rearguards holding up the enemy, his advance will be further obstructed by breaking up the roads and demolishing the bridges over the upper Ourcq. The 18th Pioneer Regiment will be sent forward by the column it is now with to the Aisne and if possible its transport will accompany it. Preliminary measures for reorganizing corps units

| **First Army *Fliegertruppe*** ||||||
| **Daily Number of Flights: Sept. 5–9, 1914** ||||||
Unit	Sept. 5	Sept. 6	Sept. 7	Sept. 8	Sept. 9
FFA 7	2	3	2	3	1
FFA 9	1	1	3	2	1
FFA 11	0	3	1	2	1
FFA 12	2	4	4	5	4
FFA 30	2	3	2	5	1

will take place tomorrow.

First Army HQ will move tonight to La Ferté Milon. Orders will be issued there at 7 am tomorrow morning.

(Signed)
Kluck[219]

Kluck's order established the route of retreat as agreed upon with Hentsch earlier in the day. According to OHL's wishes, First Army was to withdraw to safety north of the Aisne River in a direction that was to close the breech with Second Army. Furthermore, Kluck's army was to resume its original mission of acting as a flank guard for Germany's western armies. Thus, despite achieving a tactical victory over French Sixth Army and being in position to check the BEF's advance against their rear, First Army was obliged to relinquish its hard-earned victory and the potential of an even greater one in order to retreat to a new position behind the Aisne River from where Kluck's force would ultimately once again be subordinated in a defensive role under Second Army.

With few exceptions, each of First Army's five flying sections ceased their aerial activity late in the day upon receiving news of Kluck's retreat order. Instead, the officers and men of each flying section spent the hours of the late afternoon preparing all of their aircraft and equipment for road column and a hasty relocation to new airfields to the north in the hopes of resuming limited aerial activity the following day. By nightfall all five sections had successfully joined the forward echelons of the army's supply columns—enabling them to establish new airfields further north.

That evening, after settling into their new airfield at Mortefontaine near the army's uncovered western flank, the men of FFA 11 were attacked by a cavalry squadron of the French 16th Dragoon Regiment. The attacking French cavalry were part of a division-sized raid that had been sent into the rear of First Army the night before in the hopes causing confusion and possibly even a general retreat. The endeavor was ultimately a failure, however, after the force was unable to come into action during the critical hours of Quast's assault.[220] Later that evening, as the French cavalrymen were returning to their own lines, a detachment of the 16th Dragoons discovered FFA 11's airfield. Seeing that it was lightly defended, the French cavalrymen elected to attack and cause as much material damage as possible. At 3:00am, the men of FFA 11 were wakened by the commotion of the attack. Within moments the section's officers and enlisted men grabbed their rifles and pistols and entered into a firefight with the dragoons, ultimately driving them back after 15 minutes of hard fighting. This unusual episode cost the section three enlisted men killed; two officers, one *unteroffizer* and six enlisted men wounded and one damaged aircraft. The damage to the aircraft was negligible, however, as it was deemed airworthy the following afternoon after some minor repairs.[221]

The First Army *Fliegertruppe* performed remarkable service during the epic five day battle. When later asked about the importance of German aviation during the engagement, Kuhl answered:

Our airmen's efforts greatly assisted the army's tactical situation during the battle by giving the army leadership outstanding reports that enabled us to make the correct decisions at many different moments of the battle.[222]

A sober look at the *Fliegertruppe's* actions in support of First Army's combat on the Ourcq completely vindicate Kuhl's claim. From the first day of the battle when an aviator of FFA 30 was able to promptly report French Sixth Army's intent to strike First Army in the flank, to the final day of the battle when *Oblt.* Niemoller alerted First Army HQ to the presence of the BEF on the Marne's northern bank, German aviators kept the army and corps leadership consistently informed of the strength, deployment, and actions of the local allied forces. Using this intelligence, First Army's leadership was able to redeploy its troops and smartly fight a defensive battle using economy of force throughout the engagement's first four days. This ultimately enabled Kluck's army to turn over to the offensive on the 9th, an action that had placed First Army on the precipice of victory at the time of Hentsch's arrival at Kluck's headquarters. On the other hand, without the services of their airmen, First Army HQ would have been left blind and unable to hold off French Sixth Army's repeated assaults—resulting in a serious disaster.

Kluck's aviators combined to perform 58 flights over the span of the battle. During that time, the army's five flying sections suffered a material loss of seven aircraft—two lost from enemy ground fire and five from engine failure or crash landings near the airfield. Fortunately, no airmen were lost in any of the crashes.[223] Nevertheless, by the evening of the 9th, First Army's flying sections were in desperate need of replacement aircraft—a luxury that would prove hard to come by. In the meantime, First Army's *Fliegertruppe* would have to make do with what they already possessed while they carried out the monumentally important task of keeping the army leadership informed of the allies' progress as they pursued Kluck's columns to the Aisne River.

Second & Third Armies' Retreat

After a promising start, Second and Third Armies' combined offensive was halted upon the arrival of Bülow's retreat order. Thus, around 4:00pm, Second Army's left wing broke off its attack near the key area of Mondement and began a withdrawal to the north in order to remain in contact with the rest of the army. 30 minutes later, Kirchbach reluctantly initiated the withdrawal of Third Army's right wing under the cover of a strong rear-guard. Kirchbach's troops had spent the morning and early afternoon successfully throwing back the battered remnants of Foch's Ninth Army and securing the villages of Mailly-le-Camp, Euvy, Corroy, and Gourgançon. These gains, combined with a successful defensive engagement fought by the heavily outnumbered forces of Third Army's left wing, had led Hausen and the Third Army HQ staff to view the situation favorably at the time Bülow's order was delivered. Indeed, Hausen had just consented to change the axis of Kirchbach's attack to the west in support of Second Army's left wing at the time of the communiqué's arrival. The arrival of Bülow's message, however, stopped the resumption of the attack before it could begin—effectively ending the 'Battle of The Marne.'[224]

The *Fliegertruppe* of Second and Third Armies spent the afternoon hours preparing themselves for road column and the retreat northwards. Consequently, all aerial activity in Second Army's sector was suspended as of 1:00pm. In Hausen's sector, the airmen of FFAs 22 and 24 each flew one flight during the afternoon.

As Third Army HQ's flying section, FFA 22 was ordered to conduct a reconnaissance sortie along the entirety of the army's front to give the headquarters staff an updated view regarding the enemy's proximity and condition. The section dispatched a flight accordingly at 3:30pm along the line Châlons-sur-Marne—Connantre—Gourgançon—Mailly-le-Camp—Sompuis—Vitry-le-François—Châlons-sur-Marne. After landing at 4:45pm and making the quick drive to Hausen's headquarters, the flight's observer promptly informed the Third Army HQ staff that the army's right wing was lightly opposed. The army's left wing, however, was reported to be in close proximity to the enemy (elements of French Fourth Army), making an immediate withdrawal impossible without heavy casualties.

Meanwhile, FFA 24's flight, conducted exclusively over Third Army's left wing, confirmed FFA 22's findings. The subsequent report, delivered directly to Third Army HQ, stated that there were "strong French infantry and artillery in the area of Sompuis. A hasty withdrawal would endanger the troops of the XIX AK and uncover Fourth Army's flank...."[225] Thus, after considerable debate, Third Army HQ dispatched orders at 4:30pm to begin the immediate withdrawal of the army's right wing while leaving the endangered left wing in place until after nightfall.

The Evening of September 9

Owing to his position far to the rear in Luxembourg, Moltke remained largely uninformed of the situation at the front throughout the morning and afternoon. The German Commander-in-Chief was therefore reduced to waiting by the radio for news from his subordinate commanders. Finally, around 3:00pm, OHL received word of Second Army's retreat, followed shortly thereafter by news of First Army's withdrawal. Later that afternoon Moltke updated the Kaiser on the situation and asked for permission to pull back the entire German battle-line to conform to the right wing's retreat. The Kaiser, backed by several prominent generals, refused to allow a general withdrawal of the center armies until it was clear such a move was absolutely necessary. In his memoirs, Tappen described OHL's concept of operations that evening.

In the evening of September 9th, OHL believed basing its views upon reports received, that there was not yet any occasion to contemplate a general retreat in order to obtain more favorable defensive conditions in positions further to the rear, but on the contrary, that the Fourth and Fifth Armies and if possible the Third as well should take the offensive. It was believed in this very tense situation that victory would belong to the one who held out the longest.[226]

Thus, Moltke's 'September 5th strategy' of achieving a decision with concentric attacks conducted by the German left and center groups remained in effect on the night of the 9th despite the right wing's withdrawal. With this in mind, the following order was dispatched to Third–Fifth Armies:

Third Army will remain south of Châlons-sur-Marne. The offensive is to be taken on September 10th as quickly as possible. Fifth Army will attack during the night of the 9th-10th; Fourth Army will also attack if it has prospects of success. To this end, Fourth Army will establish liaison with Third Army.[227]

Hausen received Moltke's message at 10:30pm with the greatest sense of relief. After three days of hard fighting against an opponent superior in number, Third Army's right wing had defeated Foch's Ninth Army and was in position to turn in concert with Second Army's Guard Korps against the flank of d'Espèrey's greatly diminished Fifth Army. News of Bülow's retreat order however, had forced the Saxons to withdraw and concede the initiative despite holding the tactical advantage—resulting in the hard fought victory escaping from their clutches. Hausen had worried what the retreat would do to his army's morale and furthermore what the army's defensive capabilities would be once the withdrawal concluded. Fortunately, OHL's latest message prescribing Third Army to stop its retreat and turn to the offensive reinvigorated the army and prompted Hausen's men to look forward to the following day.

Further west, Bülow and his staff remained with elements of the army's right wing well to the north of any French pursuers. Despite supposedly being 'burnt to cinders,' Second Army managed to easily detach itself from the enemy and reach the Marne's northern bank by the end of the day. Under the cover of a strong rear-guard left south of the river, all of the army's trains, ammunition columns, and flying sections were across the river with the head of the army's columns. From their new headquarters at Épernay, the Second Army HQ staff dispatched the following message to Moltke at OHL.

In agreement with Oberstleutnant *Hentsch the situation is estimated as follows: The retreat of First Army behind the Aisne was required by the strategic and tactical situation. The Second Army must support the First north of the Marne, if the right wing armies are not to be crowded together and enveloped. The army will reach today (10th) Dormans—Avize with a strong rear-guard south of the Marne in liaison with Third Army. Awaiting further orders.*

Second Army's disingenuous message was clearly crafted to shift the blame for the right wing's general retreat away from Bülow and towards Kluck. Indeed, the communiqué neglected the important fact that Second Army ordered their general withdrawal *before* First Army had started any retrograde movements. Nor did it broach the subject that a retreat had been decided upon at Second Army HQ hours before First Army was required to initiate its retreat. Nevertheless, the fact remained that no matter who was to blame, the right wing was currently retreating and needed to close its sizeable gap before the allied pursuit exposed it.

Meanwhile, Kluck, Kuhl, and the First Army HQ staff moved north ahead of their army's main body to establish their new headquarters at La Ferté-Milon. The army's left and center columns spent the early evening hours pushing further north beyond the reach of the allied pursuers to positions along the line Gondreville—La Ferté-Milon—Boissy-Fresnoy. After reaching their marching objectives, Kluck's exhausted men threw themselves on the ground and bivouacked in the open with orders to begin the retreat before dawn. In the meantime, the army headquarters staff worked late into the night adopting ad-hoc measures to sort out their intermixed unit's trains and supply columns. Amidst the evening's tedious work, Kuhl dispatched the following communiqué to Moltke in far off Luxembourg:

Right wing of the First Army was driving back the enemy in the direction of Nanteuil; the center and left wing were maintaining their positions. The HKK 2 was holding the enemy in check on the Marne at La Ferté-sous-Jouarre and above. The First Army retired in conformity to an order of the High Command, not pressed by the French, on the line Crépy-en-Valois, La Ferté-Milon, Neuilly. The British are progressing beyond the Marne in the region of La Ferté-sous-Jouarre, Château-Thierry. Intention for the 10th, to continue the movement north of the Aisne. Army HQ today: La Ferté-Milon.[228]

Moltke received First Army's message shortly before retiring to bed late that evening. Although he had not yet heard from Hentsch, who was travelling back to Luxembourg, the German General Staff Chief considered himself sufficiently brought up to speed and remained committed to his earlier strategy. As of the evening of the 9th–10th, the First and Second Armies were closing the dangerous breech between them and retreating to safety behind the Aisne and Marne Rivers. Meanwhile, the armies of the left and center

Above: Promoted to army command in late August, Maurice Sarrail remained commander of French Third Army until his dismissal in July, 1915. (Author's Collection)

groups were poised to renew the attack and break the last remnants of French resistance. Unfortunately for Moltke and the Germans, the plan was doomed to failure, as the armies of the left and center were ultimately too weak to break the formidable wall of fortresses in front of them.

The German Left & Center Group's Offensive Stalls

While the armies of the German right wing were engaged in their epic struggle along the Ourcq and the Marne, the center and left groups were busy executing their attacks in accordance with Moltke's September 5th directive. In the center, the Fourth and Fifth Armies struck southeast with the intent to achieve a breakthrough and secure the left wing's passage past the Moselle and the Toul—Épinal fortress line. Once they were beyond the fortresses, according to Moltke's plan, the left wing armies would regain the operational freedom to maneuver and drive the French right wing armies into the arms of the German center group—resulting in a large scale 'modern Cannae.' However, in order for the strategy to be successful, Moltke's center had to quickly defeat the forces in front of them.

Starting at dawn on the 6th, the Duke of Württemberg's Fourth Army attacked French Fourth Army (De Langle) around Vitry-le-François while the Crown Prince's Fifth Army simultaneously struck French Third Army (Sarrail) to the southwest of Verdun. In Fourth Army's sector, the initial attacks aimed at exploiting a gap and enveloping De Langle's left flank faltered, prompting Fourth Army HQ to ask for assistance from Third Army on their right. Although the requested support from Third Army promptly arrived the next day (September 7th), the battle remained a stalemate over the next several days with no substantial progress made by either side. By the end of the 8th, however, a second large gap uncovering the *right* flank of French Fourth Army had been probed and widened by the Germans, resulting in significant uneasiness among the French Fourth Army's forces—who were then operating with two uncovered flanks.[229] Nevertheless, newly arrived reinforcements employed on the French left during the early morning of the 9th ultimately threw back the German attackers and restored the situation for De Langle, effectively ending the battle in a German defeat.[230]

To the east, German Fifth Army focused its energies around the town of Revigny where a gap between French Third and Fourth Armies had opened. After an extremely successful start to the offensive on the 6th, the Crown Prince's forces slowly continued to push forward over the next 72 hours towards their objective at Bar-le-Duc. Unfortunately for the Germans, Fifth Army's advance was unsupported by Fourth Army, who independently decided to concentrate its effort further west against French Fourth Army's open right flank. Consequently, the golden opportunity for the two center armies to act in concert to split the French Third and Fourth Armies was never fully taken advantage of. Instead, Fourth and Fifth Armies fought two separate battles side by side.[231]

Meanwhile, on the other side of Fortress Verdun, Fifth Army's V AK attacked through a string of four small forts with the intent of assaulting French Third Army's rear. The forts—Génicourt, Troyon, Les Paroches, and Camp des Romains—were essentially large redoubts designed to help protect central France from a German sortie out of Metz by serving as a connecting link between the major fortresses of Toul and Verdun. In order to cross the Meuse and strike the French field forces as quickly as possible, Fifth Army HQ ordered V AK (Strantz) to concentrate on reducing Fort Troyon, which covered the river's bridges and the line of communications for any force that would seek to maneuver against the French rear.[232] Thus, shortly after dawn on the 8th, Strantz' guns opened fire on the fort—pounding it incessantly throughout the day and into the evening. The next day, V AK's 10 ID launched several infantry attacks that were beaten back by the French defenders. The attacks having failed, Moltke cancelled the operation and the 10 ID was redeployed to the rear.

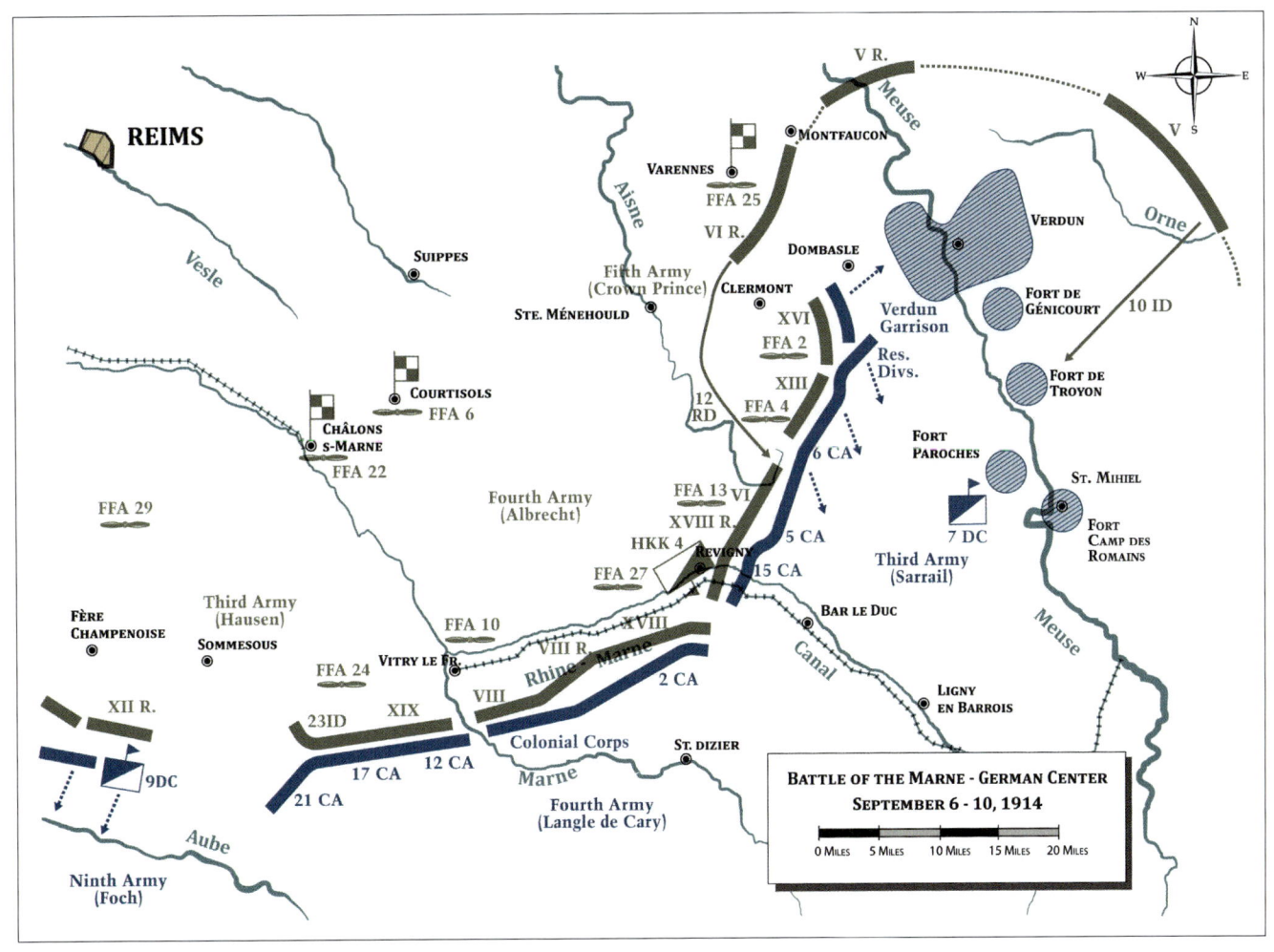

Frustrated by the lack of success, Fifth Army HQ met on the evening of the 9th to evaluate the condition of the army and adopt a new operational plan. In lieu of the right wing armies' recent retreat, it was agreed that quick and dramatic action was necessary to restore the overall situation. It was therefore proposed to launch a pre-dawn assault under the cover of darkness using three corps as a spearhead with elements of three additional corps in support to pierce and roll-up French Third Army's line. If successful, the French right wing armies (First and Second) would likely be compelled to fall back and abandon the Toul—Épinal fortress line—allowing German Sixth Army to resume the offensive in the open and assist the center group in achieving the much sought after major victory. With a direct telephone link to Moltke at OHL, Fifth Army HQ's proposal was quickly transmitted and approved, allowing the attack to proceed in the pre-dawn hours of the 10th.

At 3:00am, after nearly a three hour preparatory bombardment, troops of Fifth Army's VI, XIII, and XVI AKs began their attack which continued unabated through late morning. On the assault's left wing, *General der Infanterie* Bruno von Mudra's XVI AK easily pierced the lines of three French Reserve Divisions, severing French Third Army's contact with Verdun. Indeed, Sarrail's forces were compelled to yield more than six miles of territory across the entirety of their front. By sunset, the Crown Prince's army had made substantial forward progress, turned a flank of the opposing army, and had isolated Fortress Verdun from the mobile troops—marking the operation a large tactical success. Nevertheless, after five days of constant combat, the German attackers were simply too bloodied and exhausted to continue the offensive and drive the French back as the situation required. Instead, the Crown Prince's troops were compelled to spend the afternoon of the 11th resting and consolidating their positions until

Above: French and German troops clash in the Argonne Forest during the Battle of the Marne. (Author's Collection)

the early evening when word of the right wing's worsening circumstances caused OHL to order a general withdrawal for both Fourth and Fifth Armies, bringing their part in the Marne Campaign to a close.[233]

The German center group's *Fliegertruppe* became highly active after the army crossed the Meuse on August 29th. In the days leading up to the battle along the Marne, both Fourth and Fifth Armies' aviators performed long-range reconnaissance flights that kept the army leadership constantly informed of the positions of the French Third and Fourth Armies ahead of them. Several of these flights discovered evidence of the large-scale redeployment Joffre was conducting to create the new French Sixth Army around Paris. As previously discussed, each of these reports were forwarded to OHL, giving Moltke the necessary intelligence to correctly determine that large numbers of troops were being moved to the allied left to eventually strike his right wing in a counter-offensive. For reasons that were never fully explained, Moltke failed to forward the substance of these reports to the right wing army commanders in the days leading up to their battles along the Marne.

The airmen of Fourth and Fifth Army focused on tactical reconnaissance and artillery cooperation sorties once major combat resumed on September 6th. Late in the morning Fourth Army HQ learned from its own FFA 6 that French Fourth Army had halted its retreat and was attempting a large-scale counter-offensive to the north. The airman's report had a tremendous impact on the course of the battle. Prior to receiving the observer's information, Albrecht had remained convinced that the French retreat would continue on the 6th. However, the aerial report changed his mind, prompting Fourth Army HQ to order all available heavy artillery to "deploy for the decisive battle that had finally arrived."[234] A second aviation report dropped at army headquarters several hours later by airmen of VIII AK's FFA 10 confirmed that several large columns of French infantry (correctly estimated to be a division in strength) were seen marching west to reinforce French Fourth Army's left flank. Citing this report, Fourth Army HQ calculated that enveloping the French left would be impossible without the assistance of Third Army on their right. Thus, by mid-afternoon Albrecht had sent a formal request for reinforcements to Hausen which was promptly accepted.[235] Hausen's left wing (comprised mainly of the XIX AK) secured Fourth Army's flank for the remainder of the battle, ultimately preventing a disaster from occurring when two divisions of French reinforcements attacked Albrecht's right wing on September 9th.

On the morning of the 7th, an aircraft of FFA 6 delivered a report to Fourth Army HQ confirming that all of French Fourth Army's reserves appeared to be engaged in the battle-line. Meanwhile, the army's corps flying participated in artillery spotting flights over the next three days to assist with directing heavy artillery fire. East of Vitry-le-François, aircraft of XVIII AK's FFA 27 worked in pairs to coordinate local artillery against the battered French Colonial Corps—causing their line to briefly break during the afternoon of the 8th. Unfortunately, the German infantry in the sector were unable to exploit the situation and the opportunity passed. Elsewhere, the balloon of Fourth Army's FLA 3 was able to ascend on multiple occasions between the 8th and 10th to direct artillery fire against French troops in the area Blesme.[236]

In Fifth Army's sector, the *Fliegertruppe* conducted both long-range and tactical reconnaissance flights aimed at keeping the army's leadership updated on any changes in the French deployment. Accordingly, Fifth Army HQ's FFA 25 performed multiple long-range patrols each day deep behind the front lines. Taking off from the section's airfield near Varennes, these flights would generally follow the route of Varennes—Revigny—Bar-le-Duc—Souilly—Verdun—Varennes. While executing one of these patrols during the afternoon of 7th, an observer reported witnessing "two large columns, estimated strength—one reinforced division—moving towards [the gap in] the area of Revigny."[237] In reality the aviator had identified the entire XV CA marching from the south to plug the troublesome gap that had grown over the course of the battle between French Third and Fourth Armies. Acting quickly, Fifth Army HQ used the aviator's intelligence to strengthen their army's right wing. This successfully foiled the French effort to plug the gap and enabled the German offensive in the area to continue.

Likewise, each of Fifth Army's three corps flying sections conducted detailed reconnaissance sorties throughout the battle on behalf of their commanders. None of these flights, however, managed to significantly impact the battle's outcome, as the combat had by then degenerated into a series of violent frontal assaults against familiar opponents in prepared positions. The lone exception was a sortie performed by XVI AK's FFA 2 the evening before the corps' impressive assault that shattered the French line and isolated Verdun. While reconnoitering the area over Souilly on the evening of the 9th, the observer was able to discern and sketch several weak points in the French line that appeared susceptible to attack. The areas in question were lightly defended by infantry and contained several gaps with little or no artillery. The observer's report and sketches were delivered to XVI AK HQ later that evening whereupon they were a factor in planning the next morning's successful assault—once again demonstrating the *Fliegertruppe's* superb potential to support the ground troops whenever their reports were properly handled and considered.

The Battles on The German Left Wing

After winning a major victory at Morhange-Sarrebourg on August 20th–21st, the armies of the German left wing began a very slow and methodical pursuit of the defeated French First and Second Armies. The German plan, as approved by Moltke the day after the battle, was for the Sixth and Seventh Armies to follow up their victory with an offensive between the fortress cities of Toul and Épinal, culminating in the destruction of the French right wing armies and a subsequent strike beyond the Toul—Épinal fortress line into the French center group's flank/rear. However, by the time contact was regained on August 24th, the armies of the French right wing were reorganized and under the protection of the Grand Couronné de Nancy—a strongly fortified position that defended both Nancy and the 'Trouée de Charmes Gap' between Toul and Épinal.

On the morning of the 24th the French boldly decided to launch a pre-emptive counter-attack into the right flank of the advancing German Sixth Army.[238] Over the next three days the French First and Second Armies pushed the German left wing back, causing Sixth Army HQ to abandon their previous expectations of quickly finishing off the French right wing and then continuing beyond the fortress line into the heart of France. Instead, the two German left wing armies were compelled to adopt a more conservative plan to rest and wait for the German center and right wing groups to complete their wheeling movement to the line Verdun—Paris. Once First–Fifth Armies were in position, the German left wing (reinforced with additional heavy artillery) was to resume the offensive, cross the Moselle, and tie down more French troops in support of the campaign's final decisive battle. Crown Prince Rupprecht's logical reasoning for the change in strategy was simply stated in the German Official History as follows:

Rupprecht was also convinced that the breakthrough (of the Toul—Épinal Fortress Line) could only be accomplished when the Army's wheeling wings advanced in a southwesterly or southerly direction. Therefore, his primary task was to preserve the energy of the armies under his command and to strengthen them for the difficult tasks that lay ahead. As long as the enemy [in front of the German left] had not been forced to withdraw, a premature assault would only lead to serious losses. So, for the 29th Rupprecht again ordered his armies to rest.[239]

Under this plan, the German left wing armies spent August 28th—September 2nd slowly advancing while engaged in a number of smaller battles along the entire front. The situation became confused, however, when a liaison officer from OHL arrived at Sixth Army HQ on August 30th to order an immediate resumption of the offensive. The officer (Major Bauer) told Rupprecht that Moltke expected significant resistance opposite the German right wing. Thus, in order to relieve pressure on the right wing, Bauer ordered the left wing armies to launch an immediate "decisive attack" in the direction of Nancy through the Charmes Gap, culminating in the envelopment of the of the French main body's right flank. For Rupprecht and the left wing, this meant immediately altering their mission from a secondary/support role to a decisive one—a task that would take a great deal of staff work. A larger problem with the visit, however, was that Bauer had not been given the authority to order the offensive. This created further problems, demonstrating once again how chaotic and counterproductive Moltke's command style truly was.[240]

Rupprecht and his staff took Bauer's message to be legitimate and immediately started making plans to reduce Nancy. In order to hasten the city's capitulation, Sixth Army HQ spent the next 48 hours methodically moving large numbers of infantry and recently acquired heavy artillery into smartly deployed attack positions. Meanwhile on August 31st, Sixth Army HQ's liaison officer to OHL, (Major Ritter von Xylander) returned from Luxembourg reporting that, "the French were believed to be withdrawing forces away from the German left and moving them in the direction of Paris to check the right wing armies' advance." "Therefore," Xylander continued, "OHL considered it imperative that the left wing's offensive through the Charmes Gap resume as quickly as possible in order to strike the French main body's flank and, in concert with the right wing's enveloping attack, bring about the decisive campaign winning victory." When asked whether Bauer's prescribed operation against Nancy should proceed, Xylander answered in the negative stating, "A prior or simultaneous attack on Nancy is unnecessary; Sixth Army must only be so strong opposite Nancy as to prevent an enemy breakthrough threatening contact with Metz."[241]

Nevertheless, despite getting an answer Rupprecht remained justifiably confused regarding OHL's true intentions. Was his army group to methodically reduce Nancy and then strike beyond the gap or was he supposed to launch a more limited attack across the entire sector with the objective of merely holding the enemy in place? To answer this question, Rupprecht wisely dispatched his Chief of Staff, Krafft von Dellmensingen, to Luxembourg on the morning of the 2nd to discuss the matter with Moltke in person. After a long morning of talks it became obvious that hastily attacking through the gap was an impossibility. Therefore, left with few options, the two parties agreed to proceed with the reduction of Nancy as a means to fix large numbers of French troops in place until the right and center groups could complete their wheel and engage the allied armies in the campaign's decisive engagement. Before departing, however, Krafft expressed his doubts about the operation's success:

I emphasized that accomplishing our mission—to tie up enemy forces in the approximate strength proportional to the combined strength of 6th and 7th Armies—may be impossible as ordered. With the fortresses in play, no God could prevent the enemy from withdrawing if he desires to do so; we will be confronted with fortifications wherever we turn. The enemy will be able to hold fortified sectors with inferior forces until their main forces are gone.[242]

Moltke and the OHL staff openly agreed with Krafft that the strategy to "tie up the enemy" along the Moselle presented serious difficulties. However, with the rapid redeployment of Rupprecht's forces to another area of the front an impossibility, the attack on Nancy was seen as the only realistic option. Therefore, after several days of indecision, the German left wing's mission was finally established: reduce Nancy and then attack the Moselle line in order to keep as many French troops away from the decisive fighting on the German right wing as possible.

Rupprecht's Sixth Army began their assault on Nancy on the morning of September 4th. The attack started with a heavy preliminary bombardment featuring 272 guns of varying calibers, followed by a costly infantry advance into the French artillery zone. Meanwhile to the south, Heeringen's Seventh Army attempted unsuccessfully to push through the heavily wooded eastern foothills of the Vosges towards Rambervillers. This was significant as the success of the entire operation depended upon Seventh Army advancing to the Moselle and covering Sixth Army's flank. The following day, however, OHL was compelled to order a new Seventh Army be formed in the north on the German right wing. Accordingly,

Above: Used to quickly reduce enemy fortifications, the 210mm Howitzer was a formidable weapon. Ammunition shortages, however, greatly diminished their effectiveness in the latter stages of the Marne Campaign. (Author's Collection)

the Seventh Army HQ staff (and attached formations) immediately withdrew themselves and one corps (XV AK) from the action on the German left in order to carry out the redeployment as quickly as possible.

The loss of Seventh Army HQ and two divisions from the front doomed the already faulty German strategy in Lorraine to certain failure. Upon learning of the redeployment, Krafft said: "The whole matter is very bad indeed. (If such a redeployment occurs) we will not be able to avoid passivity."[243] Nonetheless, despite the redeployment, Moltke ordered the left wing to continue with the plan. In his general order of September 5th, the German General Staff Chief ordered what remained of Rupprecht's forces to "advance as soon as possible to attack the line of the Moselle [position] between Toul and Épinal, covering themselves toward these fortresses."[244]

As ordered, the German left wing armies continued their drive against Nancy for three more days. Due to shell shortages, all of Rupprecht's artillery batteries were issued restrictions on ammunition expenditure. Most notably, the firing of heavy howitzer rounds in support of infantry operations was expressly forbidden. These guns, such as the excellent 210mm howitzer, were extremely effective in counter-battery work and were considered to be an indispensable component for the attempted reduction of Nancy. Nevertheless, despite several protests, the restrictions remained in effect—leaving Rupprecht's unsupported infantry to suffer terrible casualties in their attempts to take the heights surrounding the city. By mid-afternoon of the 8th the futility of the operation was fully recognized and the offensive was suspended. The next morning Rupprecht ordered his troops to disengage and prepare for a withdrawal to the rear-area to be entrained and redeployed to the north in support of the right wing. The subsequent retreat began on the morning of the 10th and continued without interference from the French until the 14th when Rupprecht's troops were all within the Metz fortress zone.[245]

Despite the offensive's failure and high casualty rate, the German left wing came closer to success than is generally suspected. French Second Army experienced a crisis on the offensive's second day (September 5th) when their army's left wing, made up of a group of heavily damaged reserve divisions, showed serious signs of cracking under the stress of Rupprecht's heavy artillery. Upon learning of his left wing's predicament, de Castelnau formally asked Joffre for permission to abandon Nancy and retire behind the Moselle. Joffre denied the request with an order to stand firm in support of the allied counter-offensive around Paris. After a temporary stabilization, the situation deteriorated once again for de Castelnau when a German attack on the morning of the 7th caused elements of the 59th Reserve Division to abandon the position of Ste. Geneviève—a dominating series of heights overlooking the Grand Couronné northwest of Nancy. If captured, the Germans could have used the position as an artillery platform to bombard Nancy as well as the flank and rear of Second Army. News of the division's retreat prompted de Castelnau to prepare an order for the immediate retreat of First and Second Armies to prepared positions behind the Moselle. Standing firm, Joffre ordered de Castelnau to wait 24 hours to see if the situation improved. Fortunately for the French, the unsupported German infantry failed to advance and take the important position, leaving it vacant until French troops reclaimed it shortly before dark. The next morning, a message arrived from OHL ordering Sixth Army to return the majority of the heavy artillery shells in its possession back under the former's control. Thus, without any tangible support from their artillery, Rupprecht's corps commanders knew the offensive was hopeless and ordered a halt to the advance accordingly. The following day, the left wing's retreat began—bringing their part of the campaign to a close.[246]

The *Fliegertruppe* of the German left wing played a significant role in the attempted breakthrough around Nancy. In his history of the German Air Force, Ernst von Hoeppner summarized the task and importance of the left wing aviators during the period as follows:

The German aviators on the left flank of our army were ordered on missions that were no less important and equally dangerous [as the aviators of the right wing armies.] Here too, there was no longer any question about their usefulness. The Army Cavalry [HKK 3] encountered opposition that was too great and the Bavarian Cavalry Division lost 1,400 horses and their best patrol leaders during the first days of August, after having penetrated no more than 15 kilometers into the hostile territory. For this reason, information concerning the enemy was most important in this sector where we were fighting in the dark. The [left wing's] leaders came to base their decisions almost entirely on the Fliegertruppe's intelligence, which gave them a clear and correct picture on the activities of the enemy. Acting on the basis of these air reports, Sixth Army avoided the powerful hostile offensive in Lorraine until the flyers reported only a slow advance on the part of the enemy. Air intelligence played an important part in the conclusion of the battle in Lorraine. Thus, in this sector of the front the Fliegertruppe *became from the very beginning of the war the only carrier of operational intelligence.*[247]

Since reliable intelligence regarding the French dispositions in Lorraine was at a premium from the onset of the campaign due to the failure of the army cavalry to penetrate the French reconnaissance screen, both Sixth and Seventh Army HQs had become fully dependent upon their aviators for information since the time of their victory at Morhange-Sarrebourg on August 20th. In the days following the battle, German left wing aviators were ordered to keep visual contact with the retreating French forces ahead of them while the ground forces reorganized for the pursuit.

The left wing's advance began on August 23rd when Seventh Army was ordered to resume the march and reestablish contact with Sixth Army's flank along the Meurthe. Scouting far ahead of the army that morning, airmen of Seventh Army HQ's FFA 26 discovered large numbers of French troops marching through Baccarat and Saint Dié towards Épinal. Believing the area directly in front of the army to be weakly held, Seventh Army HQ immediately ordered the I Bavarian AK (temporarily under their control) to hasten their advance. Around noon, however, the Bavarians made unexpected contact with a strong rear-guard in the area of Montigny—Migneville. After five hours of heavy fighting, the French were defeated and thrown back—allowing the Bavarians to press forward to the outskirts of Baccarat by midnight.

The pursuit continued on August 24th with Sixth and Seventh Armies moving westward into the jaws of the French preemptive offensive. During the morning's combat, aviators from FFA 3(b) discovered an entire French corps advancing against the right flank of Bavarian II AK. Using the information, Bavarian III AK's commander, *General* Ludwig Gebsattel, ordered his troops to march to his neighbor's aid. Later that afternoon, FFA 3(b) attempted to resume flights in support of their corps' advance but was unable to keep their aircraft over the French lines due to increasingly aggressive tactics being employed by the local French aviators. In the official Bavarian history of the campaign, the flying section's older Otto B-Type pushers were mentioned to be "too heavy to evade the French machines which continually chased and harassed them."[248]

The battles across Sixth Army's front increased in severity over the next several days. On the 25th, airmen of FFA 2(b) performed several tactical reconnaissance sorties that kept the corps and division leadership informed of the general situation at the front and behind French lines. At 3:30pm, the section attempted to resume artillery cooperation work and launched an aircraft to the front for that purpose. Unfortunately, the flight was in vain. The observer's note documenting the position of several French heavy batteries was dropped too far to the rear to be used by German batteries before dark. Nevertheless, the left wing's aviators continued to fly useful tactical and long-range reconnaissance flights in support of the army during the final days of the French attack.

To better facilitate his army's advance, Rupprecht ordered the reduction of Fort Manonviller on August 25th. Indeed, the fort's guns overlooked the German supply lines, constituting a serious threat to Sixth Army's future operations. Thus, recognizing the need to quickly destroy the fort, Rupprecht sent a large detachment of heavy artillery batteries and siege guns to the area with orders to commence firing as quickly as possible. Sixth Army HQ likewise attached FLA 1(b) and FEA Germersheim to the local commander in support of the operation. By the morning of the 25th these *Fliegertruppe* units, along with the aforementioned siege train and a brigade of infantry, were in position to begin the attack. At 8:00am, FLA 1(b)'s balloon was sent aloft in an area east of Chazelles where the operation's senior artillery commander had established his command post. Working in concert with forward spotters on the ground, the observers of both aviation sections directed efficient artillery fire that resulted in the fort's destruction and surrender two days later. Over the course of the two day operation, FLA 1(b)'s balloon was moved forward several times to enhance its effect. Likewise, FEA Germersheim's three available aircraft were employed in useful artillery spotting duties by taking turns circling directly over the fort throughout the cannonade while periodically returning to the German lines to drop handwritten accuracy correction notes at the site of the artillery commander's command post. Thanks to actions like these, the fort's guns were quickly silenced, prompting the 779 man garrison to surrender despite only suffering minimal casualties.[249]

Manonviller's capitulation immediately opened up several routes of re-supply for Rupprecht's Sixth Army. The increase in supplies and reinforcements brought forward on these routes strengthened the German resolve and helped Sixth Army decisively defeat and end the French offensive by nightfall on August 28th. Over the course of the next week the armies of the German left reorganized while their leadership debated the next move. During that time, Sixth and Seventh Armies' aviators were ordered to conduct long-range reconnaissance flights to determine the enemy's strength and intentions. According to the subsequent reports, French First and Second Armies remained in strong positions east of the Moselle River along the Toul—Nancy—Épinal Fortress Line. This intelligence caused Rupprecht and his staff to openly question any future attempt to breakthrough the enemy lines. The situation changed however, on the evening of 31st when Major Bauer erroneously ordered Sixth Army to begin preparations for the attack on Nancy. Moreover, reconnaissance reports produced by Sixth and Seventh Army flying sections on September 2nd, 3rd, and 4th all showed signs of significant French forces being redeployed away from the area and moved towards the west—evidently to help in the defense of Paris. Thus, with a seemingly weaker opponent in front of them and a major redeployment out of the question, it was agreed to proceed with the attack on Nancy.

Rupprecht's flying sections attempted unsuccessfully to conduct artillery cooperation flights during the three day battle around Nancy. Despite making the necessary arrangements with battery commanders, the airmen were simply unable to safely circle above the heavily fortified areas of Nancy for long enough to make an impact. In several reports, observers from FFAs 2(b) and 3(b) reported heavy anti-aircraft and ground fire, resulting in their pilots flying at altitudes too high for effective artillery correction.[250] Furthermore, the artillery's ammunition restrictions prompted local battery commanders to neglect their airmen in favor of more reliable ground observers linked by telephone. However, the loss of these forward observers, combined with damaged equipment, forced many German artillery commanders to ultimately resort to inaccurate "unobserved area fire," which suffered from a lack of accurate maps and surveyed positions. The lone bright spot during the battle for the Sixth Army *Fliegertruppe* was the contributions of FLA 1(b). The section's balloon was brought forward on the 6th and used successfully in directing heavy artillery fire against key French positions located on hilltops and reverse slopes.

Heavy rains grounded all flying activity during the first three days of the left wing's retreat. Consequently, the German leadership was compelled to rely upon their cavalry for all information. Once it was confirmed that the French left wing were not pursuing, Sixth Army HQ ordered their aviators to remain grounded and move ahead to assigned airfields along the Metz defensive line. From these positions, the left wing's airmen remained grounded until the start of the next campaign on September 16th.

The Center's Withdrawal from the Marne

On the afternoon of the 10th, with the German right wing in retreat and the center in the midst of its attack, *Oberstleutnant* Hentsch finally returned to Luxembourg to make his report to Moltke and the OHL staff. Upon the conclusion of Hentch's presentation, Moltke issued a new set of orders to all western armies effective immediately:

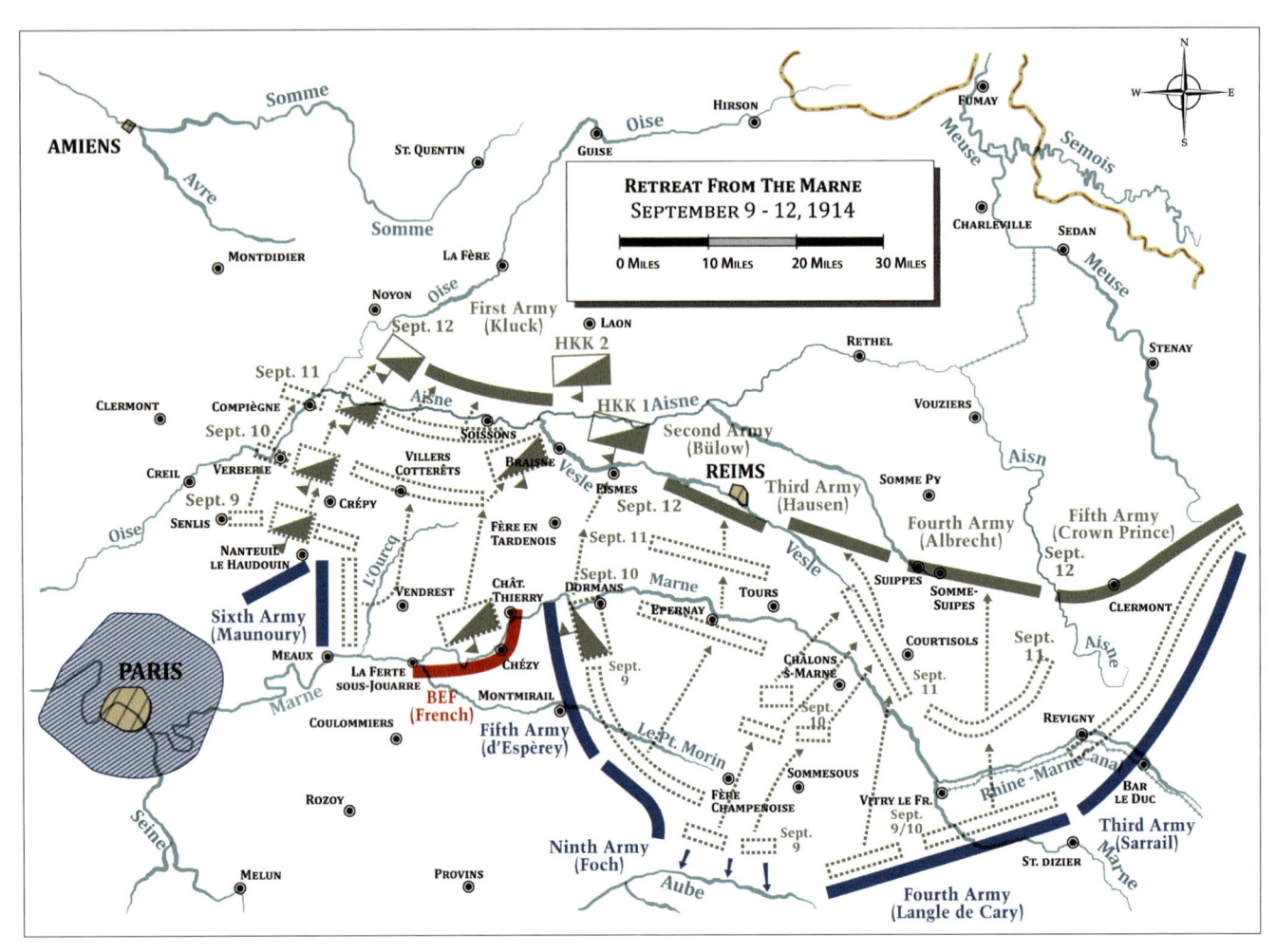

His Majesty Orders:
The Second Army will retire behind the Vesle, left wing Thuizy. The First Army will receive its instructions from the Second Army. The Third Army, maintaining contact with the Second, will hold the line Mourmelon-le-Petit—Francheville-sur-Moivre. The Fourth Army, in liaison with the Third, will hold positions north of the Rhine-Marne Canal as far as Revigny. The Fifth Army will remain on the positions it has attained. The V AK and the General Reserve from Metz will attack the forts of Troyon, Les Paroches and Roman Camp. The positions reached by the respective armies will be organized and held.

The first elements of the Seventh Army—XV AK and VII RK—will reach approximately the region of St. Quentin—Sizy by noon of September 12th. Liaison will be established at the latter point by Second Army.[251]

Moltke's order signaled the denouement of the Marne Campaign. According to the directive, the right wing armies were to withdraw behind the Aisne—Vesle and remain on the defensive until Seventh Army was fully redeployed. In the meantime, Fourth Army was to initiate a short withdrawal in order to maintain contact with Third Army's flank. Finally, Fifth Army was instructed to stop its attacks, and dig-in on the ground it had won. In other words, Moltke believed the situation warranted the withdrawal of First—Fourth Armies and the postponement of all further offensive action until the arrival of the new Seventh Army on the right wing could stabilize the crisis. Later that afternoon, however, Moltke learned that the gap between First and Second Armies remained open and susceptible to attack. Furthermore, Seventh Army HQ reported that they would be unable to deploy their forces on schedule—leaving the breech exposed for an additional 24 hours. Thus, with the situation rapidly deteriorating and his armies seemingly on the brink of a disaster, Moltke finally decided to personally go to the front to see if any further measures could be taken to ameliorate the crisis.[252]

On the morning of the 11th, Moltke, Tappen, and

Dommes left Luxembourg to make their tour of the front. Over the course of his visits to Fifth, Fourth, and Third Armies Moltke learned of a potentially dangerous threat which menaced the right and center of Third Army. If the French acted aggressively, it was possible that Fourth and Fifth Armies' rear would be fully exposed to an attack—possibly creating the conditions for their encirclement. Moltke therefore considered the retreat of his center armies to have become a necessity. Thus, at 2:30pm, he issued a general directive to withdraw the entire center group back to the line Suippes—Sainte-Menehould with the objective to reestablish a united front and buy time for Seventh Army's deployment on the right wing. Next, Moltke motored to Reims to meet with Bülow and the Second Army HQ staff and discuss strategy for the next phase of operations. At the conclusion of the meeting, Moltke subordinated First and Seventh Armies under Bülow's command to ensure a unified strategy in the sector in the days to come. That evening, having driven through the pouring rain, a visibly anxious Moltke and his entourage returned to Luxembourg to begin preparing for the inevitable allied offensive.[253]

On the 10th, Bülow's Second Army crossed the Marne, leaving only a rear-guard south of the river. Later that day, Kluck's First Army safely reached Villers-Cotterêts ahead of the slowly pursuing allies. The next day, First Army reached relative safety across the Aisne between the towns of Soissons and Attichy. Kluck and Kuhl smartly used the opportunity to reorganize their intermixed units back to their assigned corps and restore order. Consequently, by the morning of the 12th, the highly confident First Army was prepared once again to accept battle. To the east, Second Army had moved into prepared positions along the Vesle during the afternoon of the 12th. Then, early in the afternoon of the 13th, the forward units of Heeringen's newly arrived Seventh Army deployed on the Chemin des Dammes Ridge located between First and Second Armies—closing the gap between First and Second Armies just hours before the first large-scale allied attacks in the sector began. Elsewhere, the Third, Fourth, and Fifth Armies successfully completed their withdrawals to positions along the line: Reims—east of Verdun. Thus, by the afternoon of September 13th Germany's western armies were well established in strong defensive positions, ready to repulse any frontal attacks attempted by the allied forces. Undeterred, the Franco-British leadership decided on a series of assaults against the entrenched German right wing armies on the morning of the 14th—ushering in a new campaign and a form of warfare that would ultimately characterize combat on the Western Front for the next four years.[254]

Chapter 3 Endnotes

1 BA:MA RH 61 1223, *Tätigkeit der Fliegerverbände der Deutschen 1. Armee im Zussammenhange mit den Operationen vor und während der Marneschlacht: 2nd–9th September 1914*, p.68.
2 Kuhl, *The Marne Campaign 1914*, pp.116–117.
3 Ferko–UTD, Box 20, Folder 5, "Marne 1914 papers."
4 BA:MA RH 61 1223, *Die Fliegerverbände der 1. Armee während der Marneoperationen von Ende August bis zum 9. September 1914*, pp.6–7.
5 Kuhl, *The Marne Campaign 1914*, pp.118–119.
6 BA:MA RH 61 1223, *Die Fliegerverbände der 1. Armee während der Marneoperationen von Ende August bis zum 9. September 1914*, pp.8–9.
7 BA:MA RH 61 1223, *Die Fliegerverbände der 1. Armee während der Marneoperationen von Ende August bis zum 9. September 1914*, pp.9–10.
8 Kuhl, *The Marne Campaign 1914*, pp.120–121.
9 Alexander von Kluck, *The March on Paris*, (East Sussex, 2004) p.75.
10 Ferko–UTD, Box 20, Folder 5, "Marne 1914 papers."
11 Sächsisches Hauptstaatsarchiv—Dresden (hereafter: SHStA), 11355 Armeeoberkommando der 3. Armee, Nr. 0179.
12 Hausen, *Memoirs of the Marne Campaign*, p.201.
13 Hausen, *Memoirs of the Marne Campaign*, pp.210–212.
14 Personal Letters of Josef Suwelack–FFA 24.
15 Personal Letters of Josef Suwelack–FFA 24.
16 BA:MA RH 61 1223, *Die Fliegerverbände der 1. Armee während der Marneoperationen von Ende August bis zum 9. September 1914*, pp.13–14.
17 Humphries and Maker (ed.), *Germany's Western Front 1914*, Part 1, pp.407–408.
18 Kuhl, *The Marne Campaign 1914*, pp.126–128.
19 Kluck, *The March on Paris*, pp. 86–88; Kuhl, *The Marne Campaign 1914*, p.128.
20 BA:MA RH 61 1223, *Die Fliegerverbände der 1. Armee während der Marneoperationen von Ende August bis zum 9. September 1914*, p.15.
21 Ferko–UTD, Box 20, Folder 5, "Marne 1914 papers."
22 Humphries and Maker (ed.), *Germany's Western Front 1914*, Part 1, p.416.
23 *Der Weltkrieg*, III, pp.202–203.
24 Kuhl, *The Marne Campaign 1914*, p.130.
25 Humphries and Maker (ed.), *Germany's Western Front 1914*, Part 1, pp.422–423.
26 BA:MA RH 61 1223, *Die Fliegerverbände der 1. Armee während der Marneoperationen von Ende August bis zum 9. September 1914*, pp.15–16.
27 BA:MA RH 61 1223, *Die Fliegerverbände der 1. Armee während der Marneoperationen von Ende

August bis zum 9. September 1914, p.15.
28 Humphries and Maker (ed.), *Germany's Western Front 1914*, Part 1, p.423.
29 Kuhl, *The Marne Campaign 1914*, pp.133–136.
30 Kuhl, *The Marne Campaign 1914*, p.136.
31 Humphries and Maker (ed.), *Germany's Western Front 1914*, Part 1, pp.425–427.
32 OHL's single Morse-type telegraph transmitter had a range of 186 miles, which meant Moltke could only reach First and Second Armies by way of relay stations. First Army HQ could only be reached through the stations at Péronne, Noyon, Compiègne, and Villers-Cotterêts. Second Army HQ was only accessible through relay stations at Marle, Laon, and Soissons. Experience quickly proved that the delays caused by relaying messages between OHL and the right wing armies could render the reports/orders useless by the time they were finally received. Furthermore, the radio equipment was often subject to failure, jamming, or interference from a multitude of factors such as weather or the Eiffel Tower. A definitive study on German radio communications during the Marne Campaign can be found in: Carlsward, *The Strategic Signal Communications with the German Right Wing*, pp.9–14, pp.128–139; Herwig, *The Battle of the Marne*, pp.171–172.
33 Indeed, Tage Carlsward's history of the campaign stated that: "the picture of the situation [on the evening of September 2nd] which the Supreme Command visualized was built upon assumptions rather than facts." Carlsward, *The Strategic Signal Communications with The German Right Wing*, p.138.
34 Kuhl, *The Marne Campaign 1914*, p.138.
35 First Army HQ's decision to act as it did on the evening of the 2nd is one of the most controversial command decisions of the entire Marne Campaign. Many historians have placed the blame for the outcome of the subsequent Battle of the Marne squarely on Kluck and Kuhl's shoulders for making the choice to continue their advance on the 3rd. However, the full context behind First Army HQ's decision is rarely given. It is an unavoidable fact that Moltke's ignorance of the developments on the right wing played a major role in the extremely poor wording of the order. Furthermore, by repeatedly directing his right wing to, if possible, push the allied forces away from Paris and then restating it in the first sentence of the new order, Moltke confirmed that it remained OHL's top priority to drive the French main body away from the capital. Thus, given the position of Second Army, Kluck and Kuhl felt obligated to support IX AK, who was already in close proximity to Lanrezac's flank. To act otherwise would be allowing the last great opportunity to turn the allied flank before it was safely anchored against Fortress Paris. See: Hermann von Kuhl, *The Marne Campaign 1914*, pp.138–139. Robert C. Schuette, Effects of Decentralized Execution on the German Army During the Marne Campaign of 1914, Master of Military Art and Science Thesis Paper (US Army Command and General Staff College, 2014), pp. 53–55.
36 Humphries and Maker (ed.), *Germany's Western Front 1914*, Part 1, pp.449–452.
37 Ferko–UTD, Box 20, Folder 5, "Marne 1914 papers."
38 Ferko–UTD, Box 20, Folder 5, "Marne 1914 papers."
39 Humphries and Maker (ed.), *Germany's Western Front 1914*, Part 1, pp.452–454.
40 Kuhl, *The Marne Campaign 1914*, pp.144–145.
41 Kuhl, *The Marne Campaign 1914*, pp.140–141.
42 Kluck, *The March on Paris*, p.93.
43 BA:MA RH 61 1223, *Tätigkeit der Fliegerverbände der Deutschen 1. Armee im Zussammenhange mit den Operationen vor und während der Marneschlacht: 2nd–9th September 1914*, pp.14–15.
44 BA:MA RH 61 1223, *Tätigkeit der Fliegerverbände der Deutschen 1. Armee im Zussammenhange mit den Operationen vor und während der Marneschlacht: 2nd–9th September 1914*, p.14.
45 BA:MA RH 61 1223, *Tätigkeit der Fliegerverbände der Deutschen 1. Armee im Zussammenhange mit den Operationen vor und während der Marneschlacht: 2nd–9th September 1914*, pp.14–15.
46 Cuneo, *The Air Weapon 1914–1916*, pp.63–64.
47 BA:MA RH 61 1223, *Die Fliegerverbände der 1. Armee während der Marneoperationen von Ende August bis zum 9. September 1914*, p.25.
48 Bucholz, Joe Robinson, Janet Robinson, *The Great War Dawning*, p.131.
49 On September 3rd, Mengelbier was instructed not to send the important aerial reports in a separate message, but include them instead in the body of the corps' daily report. In a post-war interview with the Kriegsarchiv, II AK HQ's Chief of staff, *Oberst* Hans Freiherr von Hammerstein, stated that he: "believed First Army HQ had lost the orders after II AK HQ dispatched them." Hammerstein also stressed that although the aerial reconnaissance reports of September 3rd indicated the army's dangerous situation and were of great importance that "often times important aerial reconnaissance reports and important details were not immediately sent but instead sent in the corps' evening report." As it happened, any report containing intelligence of great importance should have been sent at once

in a separate message. The fact that it wasn't is only a reflection of poor staff work on the part of II AK HQ.
50 BA:MA RH 61 1223, *Die Fliegerverbände der 1. Armee während der Marneoperationen von Ende August bis zum 9. September 1914*, pp.23–25.
51 As "Alfred von Schlieffen's favorite student," it is hard to believe that Kuhl would have continued First Army's advance after being supplied with his aviators' critical intelligence.
52 Although it is pure speculation to say what would have occurred after First Army's halt and redeployment north of the Marne, one can confidently say the German position would have been improved by its execution. The movement of forces to First Army's right wing would have enabled Kluck's forces to better resist an allied attack against their right flank or continue an advance with the flank well defended. Kuhl later described in his memoirs what he believed would have occurred had First Army stopped north of the Marne on the 3rd: "After obtaining knowledge of the events, we shall come to the conclusion that the German Army ought to have halted on the Marne in order to regroup and strengthen its right wing at the expense of the left. That would, to be sure, have required time; the enemy might have displaced forces more rapidly than we. The continuation of the operations would not have been easy, and a quick decision definitely out of the question. But even if our offensive had been stopped, our situation would have been incomparably better than the one that actually came about. If we had succeeded in holding out on the Marne facing Paris and on the lower Seine, our position would have been much nearer and much more menacing for the enemy than the line on which we remained until the year 1918, a line starting from Soissons, passing in front of Lille and extending all the way to the coast near Nieuport. Above all, the Channel ports would have been in our hands." Hermann von Kuhl, *The Marne Campaign 1914*, p.139.
53 Peter Kilduff, *Iron Man*, p.35.
54 Kuhl, *The Marne Campaign 1914*, p.142.
55 Hausen, *Memoirs of The Marne Campaign*, pp. 219–222. Humphries and Maker (ed.), *Germany's Western Front 1914*, Part 1, p.458.
56 Humphries and Maker (ed.), *Germany's Western Front 1914*, Part 1, pp.458–466.
57 Hausen, *Memoirs of The Marne Campaign*, pp.222–224. Humphries and Maker (ed.), *Germany's Western Front 1914*, Part 1, pp.462–464.
58 Ferko–UTD, Box 20, Folder 5, "Marne 1914 papers."
59 BA:MA RH 61 1223, *Tätigkeit der Fliegerverbände der Deutschen 1. Armee im Zussammenhange mit den Operationen vor und während der Marneschlacht: 2nd–9th September 1914*, pp.17–18.
60 Along with reports from their cavalry, II AK HQ could confidently report that at least three British divisions had retreated south across the line west of Coulommiers—Rebais. Humphries and Maker (ed.), *Germany's Western Front 1914*, Part 1, p. 461.
61 BA:MA RH 61 1223, *Tätigkeit der Fliegerverbände der Deutschen 1. Armee im Zussammenhange mit den Operationen vor und während der Marneschlacht: 2nd–9th September 1914*, p.18.
62 BA:MA RH 61 1223, *Tätigkeit der Fliegerverbände der Deutschen 1. Armee im Zussammenhange mit den Operationen vor und während der Marneschlacht: 2nd–9th September 1914*, pp.18–19; *Der Weltkrieg*, III, p.250.
63 Kuhl, *The Marne Campaign 1914*, pp.146–149; Humphries and Maker (ed.), *Germany's Western Front 1914*, Part 1, pp. 461–462. Baumgarten-Crusius, *German Generalship in the Marne Campaign*, pp.86–88.
64 Humphries and Maker (ed.), *Germany's Western Front 1914*, Part 1, pp.478–480.
65 Cuneo, *The Air Weapon 1914–1916*, pp.65–66.
66 Wilhelm Groener, *Commander Against His Will*, Translated by Martin F. Schmitt, Unpublished manuscript held at USAHEC (US Army War College, 1943), pp. 275–295. Humphries and Maker (ed.), *Germany's Western Front 1914*, Part 1, pp. 479–481. Tyng, *The Campaign of the Marne*, pp. 204–209.
67 Translation of the order provided in: Tyng, *The Campaign of the Marne*, pp.381–383.
68 Kuhl, *The Marne Campaign 1914*, pp.150–151.
69 Kuhl, *The Marne Campaign 1914*, p.154.
70 First Army HQ knew from their aviators, cavalry, and lead columns that French Fifth Army as well as the British had continued their retreat on the 5th. Therefore, it was considered highly unlikely that the allies would be able to turn around and launch an offensive the following day; as Kuhl wrote regarding September 5th: "The danger to our right flank appeared to be increasing, though it did not yet appear to be imminent." Kuhl, *The Marne Campaign 1914*, pp.148–155.
71 Tyng, *The Campaign of the Marne*, p.219.
72 Tyng, *The Campaign of the Marne*, pp.211–220.
73 Senior, *Home Before the Leaves Fall*, pp.192–202.
74 Reinhold Dahlmann, *Die Schlacht vor Paris: Das Marnedrama 4. Teil. Bd. 26. Schlachten des Weltkrieges, im Auftrage des Reichsarchivs* (Berlin, 1925), pp.12–49.
75 Both sections patrolled over 30 miles behind French

Fifth Army's rearguards. BA:MA RH 61 1223, *Tätigkeit der Fliegerverbände der Deutschen 1. Armee im Zussammenhange mit den Operationen vor und während der Marneschlacht: 2nd–9th September 1914*, pp.19–21.

76 Ferko–UTD, Box 20, Folder 5, "Marne 1914 papers."

77 BA:MA RH 61 1223, *Tätigkeit der Fliegerverbände der Deutschen 1. Armee im Zussammenhange mit den Operationen vor und während der Marneschlacht: 2nd–9th September 1914*, pp.24.

78 Kuhl, *The Marne Campaign 1914*, p.209.

79 Dahlmann, *Die Schlacht vor Paris: Das Marnedrama 1914 4. Teil*, pp.43–49.

80 Tyng, *The Campaign of the Marne*, p.224.

81 The Allies numbered 980,000 French and 100,000 British with 3,000 guns between Verdun and Paris. This force was opposed by just 750,000 German troops equipped with 3,300 guns. Herwig, *The Marne, 1914*, p.244.

82 Kuhl, *The Marne Campaign 1914*, p.218.

83 Tyng, *The Campaign of the Marne*, p.257.

84 Ian Senior, *Home Before the Leaves Fall*, p.208

85 "More Marne Through German Spectacles: The Actions of the Guard Korps and of the Right Wing of the Third Army from the 5th to the 8th of September, 1914," *The Army Quarterly*, Edited by Major-General G. P. Dawnay, Vol. XVIII, No.1 (1929), pp.38–39.

86 BA:MA RH 61 1223, *Tätigkeit der Fliegerverbände der Deutschen 1. Armee im Zussammenhange mit den Operationen vor und während der Marneschlacht: 2nd–9th September 1914*, pp.29–30.

87 Thilo von Bose, *Das Marnedrama 1914. I Teil. Bd. 22. Schlachten des Weltkrieges, im Auftrage des Reichsarchivs* (Berlin, 1928), pp.16–26, pp.53–103; Tyng, *The Campaign of the Marne*, pp.245–246.

88 BA:MA RH 61 1223, *Tätigkeit der Fliegerverbände der Deutschen 1. Armee im Zussammenhange mit den Operationen vor und während der Marneschlacht: 2nd–9th September 1914*, p.30.

89 With extended supply lines, the German artillerymen were forced to conserve their ammunition, resulting in the eventual success of the French attack.

90 BA:MA RH 61 1223, *Tätigkeit der Fliegerverbände der Deutschen 1. Armee im Zussammenhange mit den Operationen vor und während der Marneschlacht: 2nd–9th September 1914*, p.31.

91 BA:MA RH 61 1223, *Tätigkeit der Fliegerverbände der Deutschen 1. Armee im Zussammenhange mit den Operationen vor und während der Marneschlacht: 2nd–9th September 1914*, pp.33–34.

92 Ferko–UTD, Box 20, Folder 5, "Marne 1914 papers."

93 For a good narrative of the battles in First Army's sector on September 6th, see: Dahlmann, *Die Schlacht vor Paris: Das Marnedrama 1914 4. Teil*, pp.50–104; Kuhl, *The Marne Campaign*, pp.211–217; Tyng, *The Campaign of the Marne*, pp.231–234; Senior, *Home Before the Leaves Fall*, pp.224–233.

94 Kuhl, *The Marne Campaign*, p.217.

95 Tyng, *The Marne Campaign*, p.249.

96 Hausen, *Memoirs of the Marne Campaign*, pp.245–251.

97 Kuhl, *The Marne Campaign*, p.227.

98 BA:MA RH 61 1223, *Die Fliegerverbände der 1. Armee während der Marneoperationen von Ende August bis zum 9. September 1914*, pp.55–59.

99 Ferko–UTD, Box 20, Folder 5, "Marne 1914 papers."

100 BA:MA RH 61 1223, *Tätigkeit der Fliegerverbände der Deutschen 1. Armee im Zussammenhange mit den Operationen vor und während der Marneschlacht: 2nd–9th September 1914*, pp.39.

101 Kuhl, *The Marne Campaign*, p.231.

102 The finest account of the two cavalry corps' actions during the Battle of the Marne can be found here: Louis J. Fortier, *The Defense of the Kluck–Bülow Gap by the Marwitz and Richthofen Cavalry Corps (September 6–noon September 8 1914)*, Unpublished Paper held at USAHEC (1934).

103 There can be no doubt that *Ltn.* Fink's aviation report showing the BEF's reluctance to advance in the face of weak opposition further bolstered this assumption.

104 Kuhl, *The Marne Campaign*, pp.229–230.

105 Kuhl, *The Marne Campaign*, p.240.

106 IX AK covered the greatest distance. In 48 hours, from the morning of the 7th until the morning of the 9th, Quast's men marched 75 miles to its attack positions on the far right of the German line. At Austerlitz, Davout's men travelled 70 miles in 48 hours. This heroic performance should be compared to the BEF, whose *cavalry* advanced an average of just 9 miles per day from September 6th–9th! Scott Bowden, *Napoleon and Austerlitz* (Chicago, 1997), p. 310; Louis N. Davout, *Opérations du 3e Corps, 1806–1807. Rapport du maréchal Davout, duc d'Auerstadt*, Edited by Général Léopold Davout (Paris, 1896).

107 13 ID had only one brigade present on the Petit Morin. The other brigade was left behind by Bülow at Maubeuge.

108 "The French Official Account of the Marne, 1914. Franchet d'Espèrey's Army." *The Army Quarterly*. Vol. XXVII, No. 1 (1933), pp.33–34.

109 As shown in: "More Marne Through German Spectacles," *The Army Quarterly*, Edited by

Major-General G. P. Dawnay, Vol. XVII, No.2. (1929), p.241.
110 Ferko–UTD, Box 20, Folder 5, "Marne 1914 papers."
111 Hausen, *Memoirs of the Marne*, pp.247–248.
112 "More Marne Through German Spectacles: The Actions of the Guard Korps and of the Right Wing of the Third Army from the 5th to the 8th of September, 1914," pp.38–39.
113 Senior, *Home Before the Leaves Fall*, pp.234–240.
114 Ferko–UTD, Box 20, Folder 5, "Marne 1914 papers."
115 Hausen, pp.248–250.
116 Herwig, *The Marne, 1914*, pp.256–258; Tyng, *The Campaign of the Marne*, pp.258–263.
117 Kuhl, *The Marne Campaign*, pp.227–235; Senior, *Home Before the Leaves Fall*, pp.245–254.
118 Ferko–UTD, Box 20, Folder 5, "Marne 1914 papers."; BA:MA RH 61 1223, *Tätigkeit der Fliegerverbände der Deutschen 1. Armee im Zussammenhange mit den Operationen vor und während der Marneschlacht: 2nd–9th September 1914*, p.39.
119 Later in the afternoon, Roos and his pilot, *Ltn.* Rohdewald, flew another sortie along the same route to determine if any new changes in the French deployment had occurred. Unfortunately, the second flight yielded nothing new to report. BA:MA RH 61 1223, *Tätigkeit der Fliegerverbände der Deutschen 1. Armee im Zussammenhange mit den Operationen vor und während der Marneschlacht: 2–9 September 1914*, pp.39, A1 p.4.
120 Ferko–UTD, Box 20, Folder 5, "Marne 1914 papers."
121 Tyng, *The Campaign of The Marne*, pp.234–236.
122 Ferko–UTD, Box 20, Folder 5, "Marne 1914 papers."
123 BA:MA RH 61 1223, *Tätigkeit der Fliegerverbände der Deutschen 1. Armee im Zussammenhange mit den Operationen vor und während der Marneschlacht: 2nd–9th September 1914*, pp.42–43.
124 Kuhl, *The Marne Campaign 1914*, p.232.
125 Kuhl, *The Marne Campaign 1914*, pp.232–233.
126 Perhaps the biggest factor in the fighting along the Ourcq Front was the French superiority in artillery shells. French field batteries could afford to fire more often than their German counterparts on account of First Army's distance from its closest railhead. It should be noted, however, that despite the great distances involved that First Army remained adequately supplied throughout the campaign. Kuhl and Bergmann, *Movements and Supply of the German First Army During August and September, 1914*, pp.114–124. Martin van Crevald, *Supplying War* (New York, 2004), pp.109–141.
127 *Der Weltkrieg*, IV, p.58.
128 An excellent overview of the communications (or lack thereof) between the German army commands during the Battle of the Marne can be found: Carlswärd, *Strategic Signal Communications with the German Right Wing* (trans.), pp.160–177.
129 Herwig, *The Marne, 1914*, pp.257–258.
130 Hausen, *Memoirs of the Marne Campaign*, pp.251–256; Thilo von Bose, *Das Marnedrama 1914*, 1 Abschnitt des III Teil, *Der Ausgang der Schlacht*, Bd. 24, *Schlachten des Weltkrieges, im Auftrage des Reichsarchivs* (Berlin, 1928), pp.138–266.
131 This river is not to be confused with the major river of the same name in northern France that empties into the English Channel. The Somme River mentioned in the order is a tributary of the Somme-Soude River, which empties into the Marne in an area northwest of Châlons-sur-Marne.
132 Ferko–UTD, Box 20, Folder 5, "Marne 1914 papers."
133 Ferko–UTD, Box 20, Folder 5, "Marne 1914 papers."
134 Hausen, *Memoirs of the Marne Campaign*, pp.251–252.
135 Ferdinand Foch, *The Memoirs of Marshal Foch*, Translated by Colonel T. Bentley Mott (New York, 1931), pp.67–78. Elizabeth Greenhalgh, *Foch in Command* (New York, 2011), pp.32–34.
136 Hausen, *Memoirs of the Marne Campaign*, pp.255–258.
137 Schell, *Battle Leadership*, pp.32–37.
138 Ferko–UTD, Box 20, Folder 5, "Marne 1914 papers."
139 Ferko–UTD, Box 20, Folder 5, "Marne 1914 papers."
140 "More Marne Through German Spectacles," pp.242–244.
141 "More Marne Through German Spectacles," p.243.
142 Ferko–UTD, Box 20, Folder 5, "Marne 1914 papers."
143 Senior, *Home Before the Leaves Fall*, p.265.
144 Bülow, Report on the Battle of the Marne, pp.85–86; Tyng, *The Campaign of the Marne*, pp.273–277.
145 Tyng, *The Campaign of the Marne*, p.273.
146 *Der Weltkrieg*, IV, p.186.
147 *Der Weltkrieg*, IV, p.233.
148 For the definitive work on Hentsch's mission see: Wilhelm Müller-Loebnitz, *The Mission of Lieutenant-Colonel Hentsch on Sept. 8–10, 1914 (trans.)*, Unpublished manuscript held at USAHEC, pp.5–6.

Above: Euler B.413/14 was one of the motley collection on unarmed two-seaters used early in the war. (Aeronaut)

149 John E. Dahlquist "Study of the Mission of Lieutenant-Colonel Hentsch, German General Staff, on 8th–10th September 1914 During the First Battle of The Marne," Unpublished manuscript held at USCGSC-CARL (Fort Leavenworth, 1931), p.3.
150 Dahlquist "Study of The Mission of Lieutenant-Colonel Hentsch," pp.3–5.
151 J.E. Edmonds, "The Scapegoat of the Battle of the Marne, 1914. Lieutenant-Colonel Hentsch, and the Order for the German Retreat," *Field Artillery Journal*, Vol. XII (Washington DC, 1922), pp.49–50.
152 Both vehicles had enlisted men as drivers in addition to the officers mentioned. *Hptm.* König was a member of Hentsch's own Intelligence Section. *Hptm.* Koeppen was a member of Tappen's Operations Section. Interestingly, Koeppen participated before the war in the famed "Great Auto Race of 1908," serving as the German team's captain. The race was from New York to Paris, with the participants crossing the Pacific Ocean from Seattle by ship; ultimately resuming the race in Russia and driving overland through Czar Nicholas' Empire into Europe and finally onto Paris. Koeppen, with his driver and mechanic onboard, reached Paris first but was penalized 15 days over a technicality regarding the use of a railcar. Despite the penalty, the German team still finished in second place. For more on Koeppen and the 'Great Auto Race of 1908,' see Koeppen's own account of the race titled: *"Im Auto Um Die Welt"* (Berlin, 1909). An excellent contemporary account of the event in English is Julie M. Fenster's *"Race of The Century: The Heroic True Story of the 1908 New York To Paris Auto Race,"* (New York, 2005).
153 Müller-Loebnitz, *The Mission of Lieutenant-Colonel Hentsch*, pp.9–10.
154 *Der Weltkrieg*, IV, p.232.
155 Senior, *Home Before the Leaves Fall*, p.271.
156 Herwig, *The Marne 1914*, pp.273–274.
157 Herwig, *The Marne 1914*, pp.273–276.
158 BA:MA RH 61 1223, *Tätigkeit der Fliegerverbände der Deutschen 1. Armee im Zussammenhange mit den Operationen vor und während der Marneschlacht: 2nd–9th September 1914*, p.49.
159 Ferko–UTD, Box 20, Folder 5, "Marne 1914 papers."
160 Everyone present at the meal later remarked how serious and depressing the mood at the table was. Regarding Hentsch, Prince August Wilhelm of Prussia, wrote, "He made a very serious, almost gloomy impression which undoubtedly had a strong influence on the usual cheerful character of the army commander." Matthes claimed that Hentsch was "suffering very badly from the task which he had been given." Senior, *Home Before the Leaves the Fall*, p.309.
161 BA:MA RH 61 1223, *Tätigkeit der Fliegerverbände der Deutschen 1. Armee im Zussammenhange mit den Operationen vor und während der Marneschlacht: 2nd–9th September 1914*, pp.44–45.
162 Kuhl, *The Marne Campaign 1914*, pp.232–233.
163 Ferko–UTD, Box 20, Folder 5, "Marne 1914 papers."

164 BA:MA RH 61 1223, *Die Fliegerverbände der 1. Armee während der Marneoperationen von Ende August bis zum 9. September 1914*, pp.52–53.

165 Linsingen's request for reinforcements shows how desperate he believed the situation to be. The phrase "Help urgently needed through Lizy" was underlined twice. Senior, *Home Before the Leaves Fall*, p.280.

166 Senior, *Home Before the Leaves Fall*, p. 80.

167 Dahlmann, *Die Schlacht vor Paris*, pp.187–206.

168 BA:MA RH 61 1223, *Tätigkeit der Fliegerverbände der Deutschen 1. Armee im Zussammenhange mit den Operationen vor und während der Marneschlacht: 2nd–9th September 1914*, p.47.

169 BA:MA RH 61 1223, *Tätigkeit der Fliegerverbände der Deutschen 1. Armee im Zussammenhange mit den Operationen vor und während der Marneschlacht: 2nd–9th September 1914*, pp.47–48.

170 BA:MA RH 61 1223, *Tätigkeit der Fliegerverbände der Deutschen 1. Armee im Zussammenhange mit den Operationen vor und während der Marneschlacht: 2nd–9th September 1914*, p.55.

171 Senior, *Home Before the Leaves Fall*, p.282–284.

172 In his history of the role aviation during the Ourcq Battles, French historian Paul Langevin cited FFA 30's excellent artillery direction work in admitting that German aviators were superior to their French counterparts in artillery cooperation at the start of the war. (Langevin, "Une etude allemande sur les reconnaissances aériennes dans l'armée von Kluck à la bataille de la Marne de 1914," *Revue des Forces Aériennes*, vol. 2, (Paris, 1931) pp.1383–1411.

173 To make good on these losses, the army reserve of one infantry regiment and one artillery group was to be placed at the disposal of IX AK upon their arrival to the Ourcq. Nevertheless, Kluck's decision (necessary as it was) further diluted the force that was to launch First Army's grand enveloping attack at dawn on the 9th. Kuhl, *The Marne Campaign 1914*, p.238.

174 It is a legitimate question as to why the right wing army commanders neglected to take advantage of their flying sections for communications between the various headquarters. Prior to the war, army commanders had used their flying sections during field exercises to relay messages. For whatever reason, however, no army commander thought to employ their aviators in this manner during the Marne Campaign.

175 Ferko–UTD, Box 20, Folder 5, "Marne 1914 papers."

176 BA:MA RH 61 1223, *Tätigkeit der Fliegerverbände der Deutschen 1. Armee im Zussammenhange mit den Operationen vor und während der Marneschlacht: 2nd–9th September 1914*, pp.55–56.

177 Hermann von Kuhl, *The Marne Campaign 1914*, pp.240–242.

178 Hermann von Kuhl, *The Marne Campaign 1914*, p.242.

179 Indeed, it has been argued that Germany lost the war in the west on the Marne on September 9th.

180 Sewell Tyng, *The Campaign of the Marne*, pp.276–277.

181 Sewell Tyng, *The Campaign of the Marne*, pp.287–293.

182 Ferko–UTD, Box 11, Folder 1, "FFA 1 papers."

183 Ferko–UTD, Box 20, Folder 5, "Marne 1914 papers."

184 Ferko–UTD, Box 11, Folder 1, "FFA 1 papers."

185 Ferko–UTD, Box 20, Folder 5, "Marne 1914 papers."

186 BA:MA RH 61 1223, *Tätigkeit der Fliegerverbände der Deutschen 1. Armee im Zussammenhange mit den Operationen vor und während der Marneschlacht: 2nd–9th September 1914*, p.62.

187 Tyng, *The Campaign of the Marne*, pp.276–277.

188 Bülow himself was the first to receive it. It should be remembered that the medal was highly coveted and rarely awarded during the first year of the war. Kilduff, *Iron Man*, p.36.

189 At first glance it is curious why Bülow communicated that his aviator observed four columns when he was in fact presented with evidence that there were five. Upon closer inspection, it is clear that after discussing the matter in person with Berthold, it was agreed that column #1 marching to Chézy-sur-Marne was suspected to merely be a covering force for the river crossing and not an independent column for offensive use north of the river and was therefore considered part of column #2 crossing at Nogent l'Artaud. NOTE: Upon its transmission, the village Daméry was written in error. The communiqué intended to list Dormans as the location where the army's right wing was withdrawing. Louis Marie Koeltz, *Le G.Q.G. allemand*, (Paris, 1931) p.384.

190 The division's reinforcements were two battalions of infantry and two batteries of heavy artillery.

191 While it is hard to believe that 13 ID could have held its position on the 9th *and* 10th, it is interesting to note that the commanders at the front charged with defending the gap were all in agreement that they could have held their position for at least another 12 hours. Kuhl, *The Marne Campaign 1914*, pp.265–266.

192 Herwig, *The Marne, 1914*, p.303.

193 Even in the worst case scenario of First Army's right wing attack failing, First, Second and Third Armies could have all disengaged and withdrawn during the night of the 9th under the cover of darkness.

Although the attack had failed in this hypothetical scenario, the offensive action would doubtlessly have spoiled any offensive action on the part of French Sixth Army that would result in interfering with a general withdrawal. In other words, First Army would have been no worse off if their attack had been allowed to continue until sunset.

194 Kuhl, *The Marne Campaign 1914*, p.281.
195 A unique look at how the Germans won the battle of the Marne at a tactical and operational levels can be found in: John Mosier, *The Myth of the Great War: How the Germans Won the Battles and How the Americans Saved the Allies* (New York, 2002), pp.83–99.
196 Throughout the campaign Bülow selfishly viewed operations as they affected his own army rather than looking at the bigger picture of the right wing. As a result, the army commander repeatedly ordered both First and Third Armies to assist his army in ways that only would help matters tactically. In contrast, First and Third Army HQs conducted operations with a view of gaining a major strategic victory.
197 Ferko–UTD, Box 20, Folder 5, "Marne 1914 papers.
198 Kuhl recognized that the lack of aircraft was seriously impacting his army's ability to win the battle on the ground. Indeed, in a postwar interview Kuhl stated that "The number of the airmen was much too small. The lack of available aircraft was especially felt during the battle." BA:MA RH 61 1223, *Die Fliegerverbände der 1. Armee während der Marneoperationen von Ende August bis zum 9. September 1914*, p. 6; Hermann von Kuhl, *The Marne Campaign 1914*, pp.241–242.
199 Ferko–UTD, Box 20, Folder 5, "Marne 1914 papers."
200 BA:MA RH 61 1223, *Tätigkeit der Fliegerverbände der Deutschen 1. Armee im Zussammenhange mit den Operationen vor und während der Marneschlacht: 2nd–9th September 1914*, pp.58–59.
201 Hermann von Kuhl, *The Marne Campaign 1914*, pp.248–249.
202 BA:MA RH 61 1223, *Tätigkeit der Fliegerverbände der Deutschen 1. Armee im Zussammenhange mit den Operationen vor und während der Marneschlacht: 2nd–9th September 1914*, pp.60–61.
203 Fortier, *The Defense of the Kluck–Bülow Gap by the Marwitz and Richthofen Cavalry Corps*, p.155; Tyng, *The Campaign of the Marne*, p.281.
204 Kuhl, *The Marne Campaign 1914*, p.249.
205 *Der Weltkrieg*, IV, p.215.
206 *Der Weltkrieg*, IV, p 260.
207 Müller-Loebnitz, *The Mission of Lieutenant-Colonel Hentsch on Sept. 8–10, 1914*, pp. 19–23; Kuhl, *The Marne Campaign 1914*, p. 252.
208 The logistics of the maneuver should also be considered. The army's ammunition and supply wagons were heavily concentrated in the rear-area, which congested the roads and left the army exposed to encirclement by French cavalry (which outnumbered German cavalry on the Ourcq Front's northern wing).
209 *Der Weltkrieg*, IV, p.264.
210 Kuhl, *The Marne Campaign 1914*, p.253.
211 Kuhl, *The Marne Campaign 1914*, p.253–254.
212 Besides, Hentsch would have never misrepresented the authority he carried in Moltke's name. For any historian to suggest that Kluck could have or should have demanded a retreat order from Moltke in writing reveals a complete misunderstanding of military etiquette as it existed in 1914. Herwig, *The Marne 1914*, p. 283, pp.310–312.
213 Kluck, *The March on Paris*, p.141.
214 Ferko–UTD, Box 20, Folder 5, "1914 papers."
215 BA:MA RH 61 1223, *Tätigkeit der Fliegerverbände der Deutschen 1. Armee im Zussammenhange mit den Operationen vor und während der Marneschlacht: 2nd–9th September 1914*, p.65.
216 Ferko–UTD, Box 20, Folder 5, "Marne 1914 papers."
217 BA:MA RH 61 1223, *Die Fliegerverbände der 1. Armee während der Marneoperationen von Ende August bis zum 9. September 1914*, p.74.
218 Kuhl, *The Marne Campaign 1914*, p.256.
219 Kluck, *The March on Paris*, pp.141–143.
220 Hermann von Kuhl, "French Cavalry Raid in The Battle of The Marne," The Cavalry Journal (Washington DC, October 1922), Vol. XXXI, No. 129, pp.356–360.
221 Hans-Edouard von Heemskerck, *Das Gefecht der FFA 11 bei Mortefontaine 1914*, Ünsere Luftstreitkräfte 1914–18 (Berlin, 1930), pp.435–438.
222 BA:MA RH 61 1223, *Tätigkeit der Fliegerverbände der Deutschen 1. Armee im Zussammenhange mit den Operationen vor und während der Marneschlacht: 2nd–9th September 1914*, p.69.
223 BA:MA RH 61 1223, *Tätigkeit der Fliegerverbände der Deutschen 1. Armee im Zussammenhange mit den Operationen vor und während der Marneschlacht: 2nd–9th September 1914*, p.67.
224 Bülow's message was first sent to Kirchbach, directly ordering Third Army's left wing to retreat at 1:00pm. Kirchbach decided to wait until 4:30pm to follow the instructions. Incredibly, Third Army HQ did not learn of Bülow's decision to order the retreat until 3:00pm. The order read as follows: "First Army is falling back; Second Army has thus commenced retreating to Dormans—Tours; order

Aviatik (P14) B.268/13

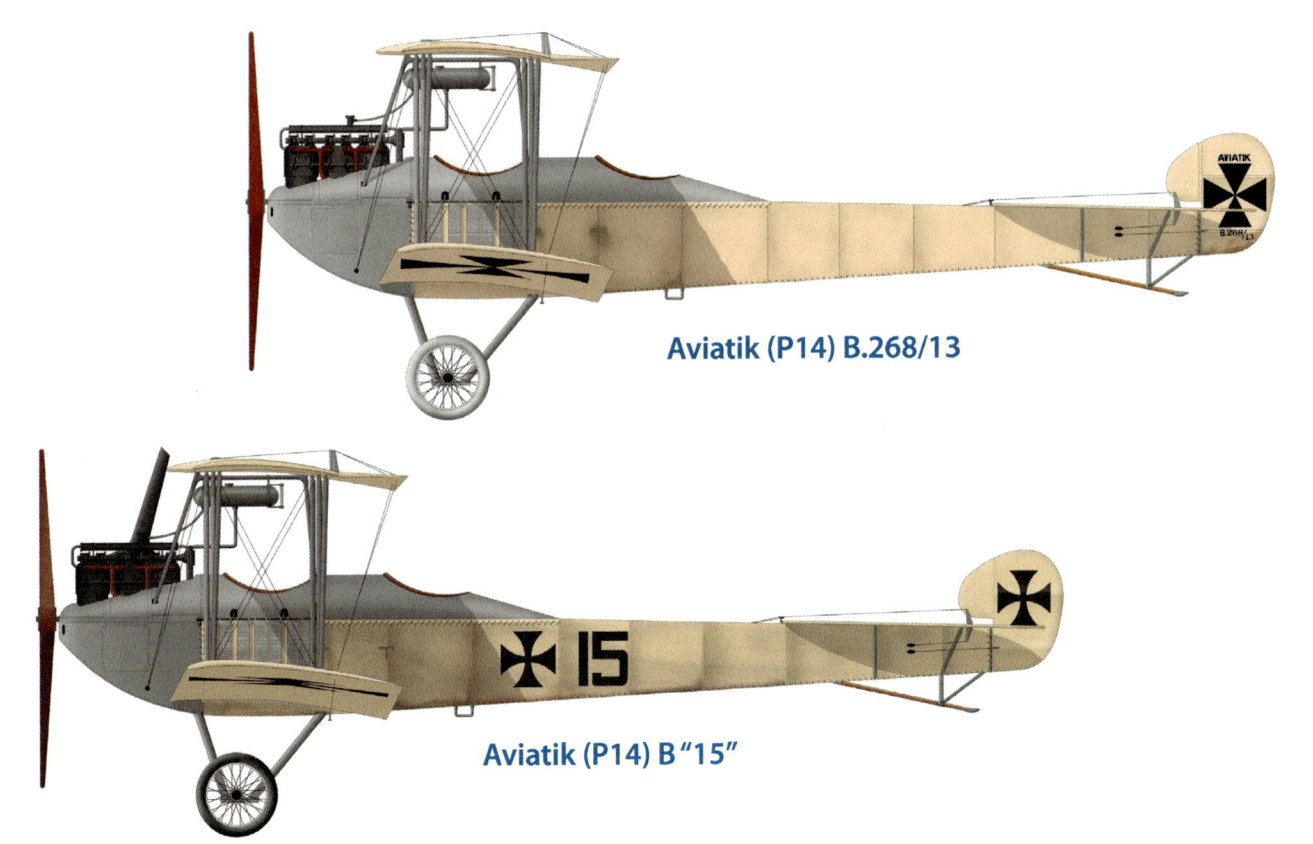

Aviatik (P14) B "15"

to retreat has been given to Kirchbach." Hausen, *Memoirs of the Marne Campaign*, p.257.
225 Ferko–UTD, Box 20, Folder 5, "1914 papers."
226 Gerhard Tappen, *Movements, transportation and supply of the German right wing in August and September, 1914*, 3 vols. Translated by W.C. Koenig, Unpublished manuscript held at USAHEC, vol. 1 pp.8–14.
227 Tyng, *The Campaign of the Marne*, p.323.
228 Tyng, *The Campaign of the Marne*, p.324.
229 General Cordonnier, *L'Obéissance aux Armées*, (Paris, 1924).
230 Tyng, *The Campaign of the Marne*, pp.295–299.
231 Mosier, *The Myth of the Great War*, pp.92–96.
232 Donnell, *Breaking the Fortress Line*, pp.165–167.
233 Tyng, *The Campaign of the Marne*, pp.308–309.
234 Ferko–UTD, Box 20, Folder 5, "1914 papers."
235 Hausen, *Memoirs of the Marne Campaign*, p.241.
236 Ferko–UTD, Box 20, Folder 5, "1914 papers."
237 Ferko–UTD, Box 20, Folder 5, "1914 papers."
238 Storz, "This Trench and Fortress Warfare is Horrible!" pp.156–157.
239 Humphries and Maker (ed.), *Germany's Western Front 1914*, Part 1, p.395.
240 Storz, "This Trench and Fortress Warfare is Horrible!" pp.160–162.
241 *Der Weltkrieg*, III, pp.286–287.
242 Humphries and Maker (ed.), *Germany's Western Front 1914*, Part 1, p.433.
243 Storz, "This Trench and Fortress Warfare is Horrible!" p.166.
244 Tyng, *The Campaign of the Marne*, p.382.
245 A translation of the best and most detailed history of these battles can be found in: Deuringer, *The First Battle of the First World War*, pp.167–371.
246 Tyng, *The Campaign of the Marne*, pp.311–319.
247 Hoeppner, *Germany's War in The Air*, (Nashville, 1994) pp.13–14.
248 Deuringer, *The First Battle of the First World War*, p.175.
249 Deuringer, *The First Battle of the First World War*, pp.225–227.
250 Ferko–UTD, Box 20, Folder 5, "1914 papers."
251 Tyng, *The Campaign of the Marne*, p.325.
252 Tyng, *The Campaign of the Marne*, p.326. Annika Mombauer, *Helmuth von Moltke and the Origins of the First World War* (New York, 2005), pp.256–257.
253 Tyng, *The Campaign of the Marne*, pp.327–328.
254 Although trench warfare had been featured in the battles on the German left wing armies throughout early September, the Battle of the Aisne is considered by most to be the true "dawn of trench warfare on the Western Front." See: Jerry Murland, *Battle on the Aisne 1914–The BEF and the Birth of the Western Front*, (South Yorkshire, 2012).

4
The 'Race to the Sea'
September–October 1914

"The lack of any staff officers for aviation at the various commands under which the flying sections operated without doubt contributed much to the failure of air reconnaissance to achieve greater results in 1914... The defect was obvious... To Oberst Eberhardt, the solution was a new organization to stamp out the rampant individualism which was destroying any semblance of order."

John R. Cuneo[1]
World War One Aviation Historian

After his inspection at the front was completed, Moltke returned to Luxembourg at 2:00am on the morning of September 12th. Upon his arrival, the German Chief of Staff drew up orders to relieve the severely ill Max von Hausen of command of Third Army. Hausen had come down with a case of Typhoid and was too sick to continue his duties. Thus, Hausen was officially relieved and replaced by VII AK's commander, *General der Kavallerie* Karl von Einem. Next, after a brief nap, the visibly nervous Moltke appeared before the Kaiser to report upon his tour of the front. According to all accounts, the interview was a disaster. The stress from the defeat along the Marne had clearly affected Moltke, who was judged by many at OHL to be on the verge of a mental breakdown. While giving his report, the General Staff Chief reportedly mentioned the possibility of continuing the retreat. The Kaiser, enraged, slammed his fist on the table and explicitly forbade any further mention of the move.

By the end of their meeting, the Emperor's confidence in Moltke had been irreparably broken. Thus, after a brief discussion with his military cabinet, the Kaiser decided to relieve the unreliable General Staff Chief of his duties on the evening of the 14th. In truth, the move was made not a moment too soon. Moltke's ineffectual command style, combined with several key errors made during the course of the campaign, had ultimately cost the Germans a chance at a quick victory in the West. The Kaiser, recognizing this fact, saw no other option but to make an immediate change in the hopes of restoring the army's morale and confidence before the next major battle.[2]

In a somewhat controversial decision, the Kaiser named 53-year-old Prussian War Minister Erich von Falkenhayn as Moltke's successor. As mentioned in Chapter 1, Falkenhayn was a far-sighted officer with a seemingly unlimited work ethic. More importantly, he was a strong-willed leader who vehemently opposed delegating OHL's power unduly. Thus, after the debacles of the Marne Campaign, Falkenhayn was seen by the Kaiser's advisors and the Emperor himself as the perfect man for the job.[3]

Falkenhayn had witnessed Moltke's mismanagement of the army's operations first hand throughout the Marne Campaign as a member of the Kaiser's Imperial Headquarters (GHQ). From this position, the Prussian War Minister made a series of well-timed criticisms which ultimately helped propel his name to the top of the list of potential replacements for Moltke. Indeed, in the days after the victory on the Frontiers, as Moltke continued to make mistake after mistake, Falkenhayn took it upon himself to voice what many other excellent officers were thinking. For example, when Moltke prematurely made the decision on August 26th–27th to declare the French decisively defeated and to therefore withdraw two corps from the right wing for operations in the East, Falkenhayn stated the obvious: "This is not a battle won; it is a planned withdrawal. There is no evidence of a major victory. Show me the trophies or prisoners of war."[4] Nevertheless, the redeployments proceeded, leaving the right wing greatly weakened as their advance into France continued. In the days that followed, the Prussian War Minister privately criticized, among other things, Moltke's decision to keep his headquarters in the rear in Luxembourg instead of operating closer to the critically important right wing. Consequently, when the Battle of the Marne began and the situation deteriorated on the right wing due to lack of communication between OHL and the right wing armies, Falkenhayn's assessments were instantly validated throughout the high command.

These sound criticisms were brought to the Kaiser's

Above: Starting the war as commander of VII Armee Korps, Karl von Einem was quickly named commander of Third Army after the Battle of the Marne—a post he held until the end of the war. (Author's Collection)

Above: As the new Chief of the General Staff, Erich von Falkenhayn was determined to establish his own command style and quickly reverse many of his predecessor's biggest mistakes. (Author's Collection)

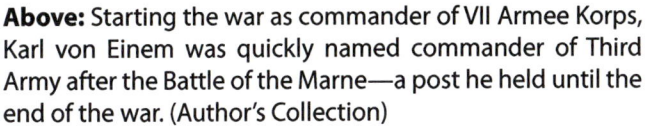

attention during the campaign through Falkenhayn's friend and confidant, *Generaloberst* Moriz von Lyncker, who served at GHQ as Chief of the Kaiser's Military Cabinet. As a result, the Emperor's already positive opinion of the War Minister's capabilities increased exponentially as news from the front continued to vindicate his claims. Then, after the disastrous defeat at the Marne and Moltke's subsequent mental breakdown during the retreat, it became clear to everyone in Luxembourg that Moltke needed to be relieved of his duties. Thus, after holding another meeting on the morning of the 14th, the Kaiser formally decided to fire Moltke and replace him with Falkenhayn.[5]

The First Battle of the Aisne: Falkenhayn's First Decisions

Falkenhayn ascended to command at an extremely dangerous time. The defeat on the Marne and the collapse of Moltke's revised operational strategy had left the Kaiser's western armies exposed to disaster as the pursuing allies sought to reengage the vulnerable German right wing along the Aisne River with the intention of delivering a massive, crippling blow that would prematurely expel the Germans from French soil. Fortunately, the retreat and redeployment of the German right and center groups had already been completed by the time the first large-scale allied attacks materialized on the afternoon of the 13th. Consequently, each German field army was well entrenched and in position to repulse the frontal assaults launched by the Entente's forces. The weak point, however, remained the German's uncovered right flank. Joffre and the allied leadership had recognized this but chose to attack the gap that remained between First and Second Armies at the start of the battle. The arrival of Seventh Army, however, had filled the breech before any allied attacks could be delivered. Thus, the subsequent fighting on the 14th and 15th featured nothing but failed allied frontal assaults delivered uphill over open ground against formidable German defenses on the *Chemin des Dames* ridgeline.[6]

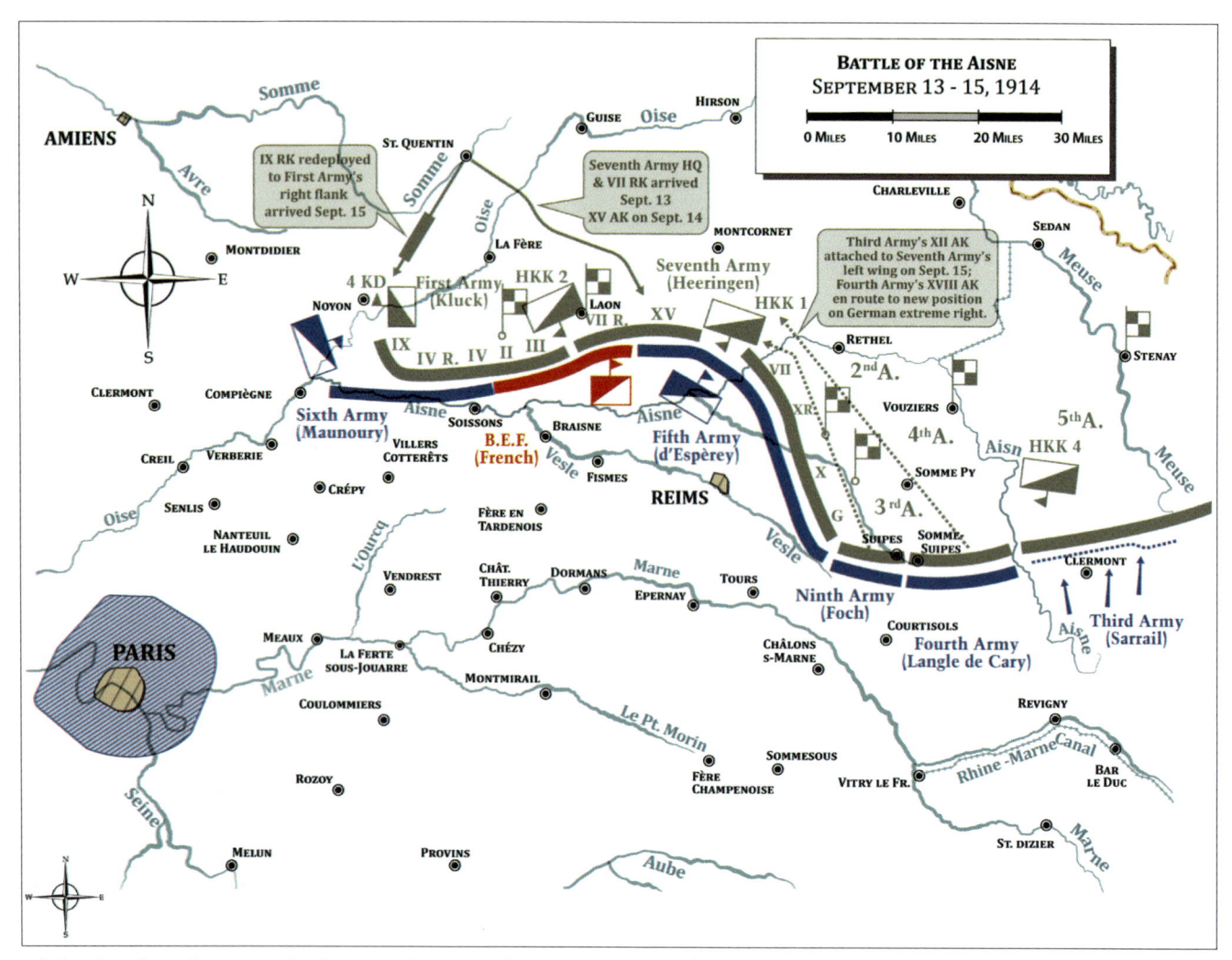

The German retreat from the Marne had effectively concluded by the afternoon of the 12th when First Army's rear guard joined the rest of the army in their positions along the heights north of the river Aisne. As the reader will recall, Moltke had transferred Heeringen's Seventh Army to the right wing, hoping to use it on the battle line's extreme right for offensive action. During the evening of the 12th, however, it was decided that there was no choice but to rush forward Heeringen's force to permanently close the breech between First and Second Armies before any Franco-British troops attempted a breakthrough the next day. Thus, throughout the evening of the 12th–13th, troops of Seventh Army completed a frantic forced march to grab the strong defensive positions north of the Aisne. This ad-hoc deployment was ultimately completed by mid-day on the 13th, just hours before the first British troops arrived in the area. By nightfall, the German right wing was entrenched along the heights of the northern bank, fully prepared to resist the allied attacks that were expected the following day.

Due to weather, the *Fliegertruppe* were unable to play any substantial role during the retreat from the Marne. Thick fog, strong winds, and various precipitation had combined to ground the right wing armies' airmen throughout the 10th and 11th as each flying section retreated northwards in road column with the army's supply troops. The resulting loss of aerial intelligence significantly impacted the mood at First Army HQ, who had grown accustomed to acting on the basis of their aviators' excellent reporting. Without any aerial reports describing the activity opposite their uncovered right flank, Kluck and Kuhl remained greatly concerned that their vulnerable command would be attacked before reaching safety across the Aisne. Fortunately, the absence of intelligence did not result in a catastrophe as the pursuing Franco-British troops moved too slowly to take advantage of the situation.

Above: German troops take up firing positions behind the Aisne River. (Author's Collection)

The weather cleared on the morning of the 12th, allowing aerial activity to briefly resume across the German right wing until storms returned that evening. Reports from First and Second Army flyers were delivered to their respective army commanders, detailing the location of the allied armies' main bodies. To the relief of both First and Second Army HQs, the reports confirmed that the Franco-British armies had timidly pursued the Germans from the Marne and were not yet in position to begin an attack. Specifically, the British were reported to be advancing in the direction of the gap between First and Second Armies while French Fifth and Sixth Armies maintained contact with their flanks. The airmen's important reports were forwarded directly to Bülow (in command of the entire right wing—First, Second, and Seventh Armies) that evening. Reading the reports, Bülow quickly recognized the grave danger his forces would be in if the allies were to enter the breech and therefore ordered Seventh Army to immediately begin a forced march to get there first.[7] Twelve hours later, in the early afternoon of the 13th, Heeringen's Seventh Army moved into position between First and Second Armies along the *Chemin des Dammes*—plugging the gap and ending the Entente's chance of a quick follow-up victory in the aftermath of the Marne. The next day, the "Battle of the Aisne" began as the Franco-British forces launched a series of bloody yet futile frontal assaults against the German defensive positions located on the ridges north of the river.

After two days of bitter fighting, the allied armies remained unable to breakthrough the German right wing. German artillery and machine guns, deployed skillfully on the crests of slopes throughout the entire sector, inflicted horrific casualties against the attacking Franco-British forces, who were not expecting such strong resistance from the Germans after their setback along the Marne. The situation therefore quickly devolved into a stalemate as the allies and Germans were each compelled to entrench in order to hold their current positions while limiting the power of the other side's artillery. Indeed, by the evening of the 14th both sides had given up on the idea of a breakthrough and had begun to focus on outflanking their opponent's uncovered northern flank.

Falkenhayn arrived at OHL in his new role as Chief of the General Staff on the afternoon of the 14th, immediately getting to work on a new strategy to restart the stalled western offensive. First, the new commander-in-chief sought to reverse many of Moltke's errors by establishing a new and more efficient command model. Unlike his predecessor, Falkenhayn believed that OHL should exercise a much stricter control over the army's operations. To that end, OHL was moved forward some 63 miles from Luxembourg to Charleville—Mézières behind the critically important right wing armies. From there, Falkenhayn and the Supreme Headquarters staff could better communicate with each of the field armies and ensure that the significant mistakes of the Marne Campaign were not repeated.[8] Moreover, Falkenhayn resolved to fundamentally change *how* the German Supreme Command operated. Whereas Moltke had believed in a totally decentralized command style that failed to actively coordinate the actions of the different armies, Falkenhayn desired a system where communication and synchronization at the army level were prerequisites for any future operation.[9] Thus, upon immediately taking command, Falkenhayn resolved to play a much greater role in operations and, on a strategic level, to "assume sole responsibility for Germany's conduct of the war."[10]

True to his beliefs, Falkenhayn first reviewed the overall situation on the evening of the 14th alone in his office, outside the influence of OHL's various section heads. After looking over all the available intelligence, the new General Staff Chief held steadfast to his belief that a speedy decision in the West was still possible through a properly planned and executed envelopment attack against the allies' uncovered left flank. Although Moltke had already transferred part of Heeringen's Seventh Army to the area, Falkenhayn recognized that Rupprecht's Sixth Army was also needed on the extreme right wing for any offensive against the allied left to succeed. Thus, according to Falkenhayn's plan, Sixth Army was to be sent north to the area of Maubeuge to deliver the main assault. In the meantime, in order to secure his own exposed flank and to buy time for Rupprecht's redeployment (which would not be completed until the 21st), Falkenhayn proposed withdrawing Kluck's First Army to the line Artemps—La Fère—Nouvion-et-Catillon. By pulling his right wing back, Falkenhayn would temporarily secure his own flank while inviting the French to advance and further expose themselves to Sixth Army's eventual attack. According to the ambitious plan, the renewed offensive would start on September 18th with Fifth Army and would gradually continue *en échelon* up the entire German battle line—culminating with Sixth Army's knockout blow on the 22nd.[11]

Falkenhayn's bold plan was fully in accordance with the well-established German military tradition of destroying the enemy through bold operational-level maneuvers.[12] Unfortunately, the new General Staff Chief was ultimately dissuaded of implementing his plan by Tappen, a believer in the breakthrough, who returned from the front during the afternoon of the 15th with information suggesting that the plan required additional refinement. First, Tappen believed from what he had observed at the front that the French were exhausted and on the verge of collapse. Next, he argued that any further withdrawal of the troops would seriously diminish the army's morale. Finally, Falkenhayn was informed that Bülow had already ordered the bulk of the First, Seventh, and Second Armies to renew their attacks west of Reims with the added strength of three corps transferred from Third, Fourth, and Fifth Armies.[13] Thus, despite disagreeing with elements of Tappen's argument, Falkenhayn ultimately decided to defer to a trusted subordinate and not to order First Army and the rest of the right wing to conduct a strategic withdrawal to better facilitate Sixth Army's offensive. Instead, Bülow's renewed offensive at Reims was allowed to proceed while the remainder of the right wing was to hold its current positions and await the arrival of Sixth Army to begin its attack.[14]

The Race to the Sea: An Overview

Bülow's attempted breakthrough predictably failed to achieve the desired result. For four consecutive days from the 15th through the 19th, troops from the First, Seventh, and Second Armies vigorously attacked the allied positions west of Reims with little success. Shortages in artillery ammunition as well as losses amongst the junior officer corps combined to deeply impair the attacking forces, which therefore lacked the firepower and discipline to overcome the allies' fortified positions.[15] Nonetheless, the attacks were successful in disrupting Joffre's attempt to quickly envelop and crush the vulnerable German right wing. Thus, after multiple offensives both sides remained locked in a stalemate, unable to turn their opponents' flank. Falkenhayn, who still believed a major success in the West was possible, recognized the need to move as many troops as he could to his right wing. He therefore ordered Sixth Army to hasten their redeployment and to "achieve the decision of battle on the right wing of the army as soon as possible."[16] To support this endeavor, the new German commander ordered the reduction of Antwerp (which had contained the entire Belgian Field Army since the fall of Liège) as well as a series of diversionary attacks south of Verdun at St. Mihiel—thus taking action to secure his line of communications, free up some additional forces, and to further weaken the French and prevent them from redeploying any more divisions to the north to oppose the German right wing.

Rupprecht's Sixth Army began their redeployment from Metz to the right wing on the morning of the 17th. However, the rail lines in German occupied France and Belgium remained in disarray—severely delaying the important redeployment. Consequently, Sixth Army was unable to become even partially

Above: German infantry and engineers storm Fort 'Camp des Romains,' located southeast of Verdun. (Author's Collection)

concentrated until September 21st. This gave Joffre time to use the allied advantage in working railways to quickly transfer forces from Lorraine in the south to his northern wing. Thus, in order to protect his flank and buy time for Rupprecht's arrival, Falkenhayn ultimately moved two corps (II and XVIII AKs) to the extreme right of the line. Fortunately for the Germans, these two corps successfully halted the French attack, allowing Sixth Army's first two corps to arrive later on the 24th to begin their attack.

Upon their arrival, Rupprecht's vanguard (XXI AK and I Bavarian AK) engaged elements of the newly constituted French Second Army in the area of Chaulnes—Roye with the objective of turning and rolling up the allied battle line. The offensive quickly picked up momentum and pierced the French left wing at multiple points, creating temporary panic at French Second Army HQ (still under the command of Édouard de Castelnau). The French rail superiority, however, enabled Joffre to quickly redeploy additional forces from quieter areas of the front, ending the crisis in its second day. Conversely, the Germans' lack of working rail lines in the occupied areas of Belgium and France had forced Sixth Army HQ to feed its formations into the battle piecemeal and not together as a larger united force. Consequently, Rupprecht's army never possessed the strength necessary to achieve the 'decisive victory' envisioned by Falkenhayn. Instead, the new German Commander was obliged to use Sixth Army as a means to forestall a renewed French attack against the German right flank, which became increasingly more likely as Joffre continued to redeploy more troops to the area.[17]

The final days of September saw the opposing sides continue to transfer forces to the northern end of the battle line with the intent to envelop their opponent's uncovered flank. Using the Entente's rail advantage, Joffre quickly moved three corps, plus another three infantry and two cavalry divisions, to the area of Arras—Bapaume north of the Somme River, directing them to attack as quickly as possible.[18] All of these troops were in position on October 1st, making Castelnau's Second Army a formidable collection of eight corps deployed along a front stretching 60+ miles from the Oise River in the south to the environs of Lens in the north. The army's offensive never materialized, however, as Rupprecht's Sixth Army successfully protected the German flank through a series of spoiling attacks launched on September 26th, 27th, and October 1st as the army's various corps completed their deployments and became available for action. In his memoirs, French General Joseph

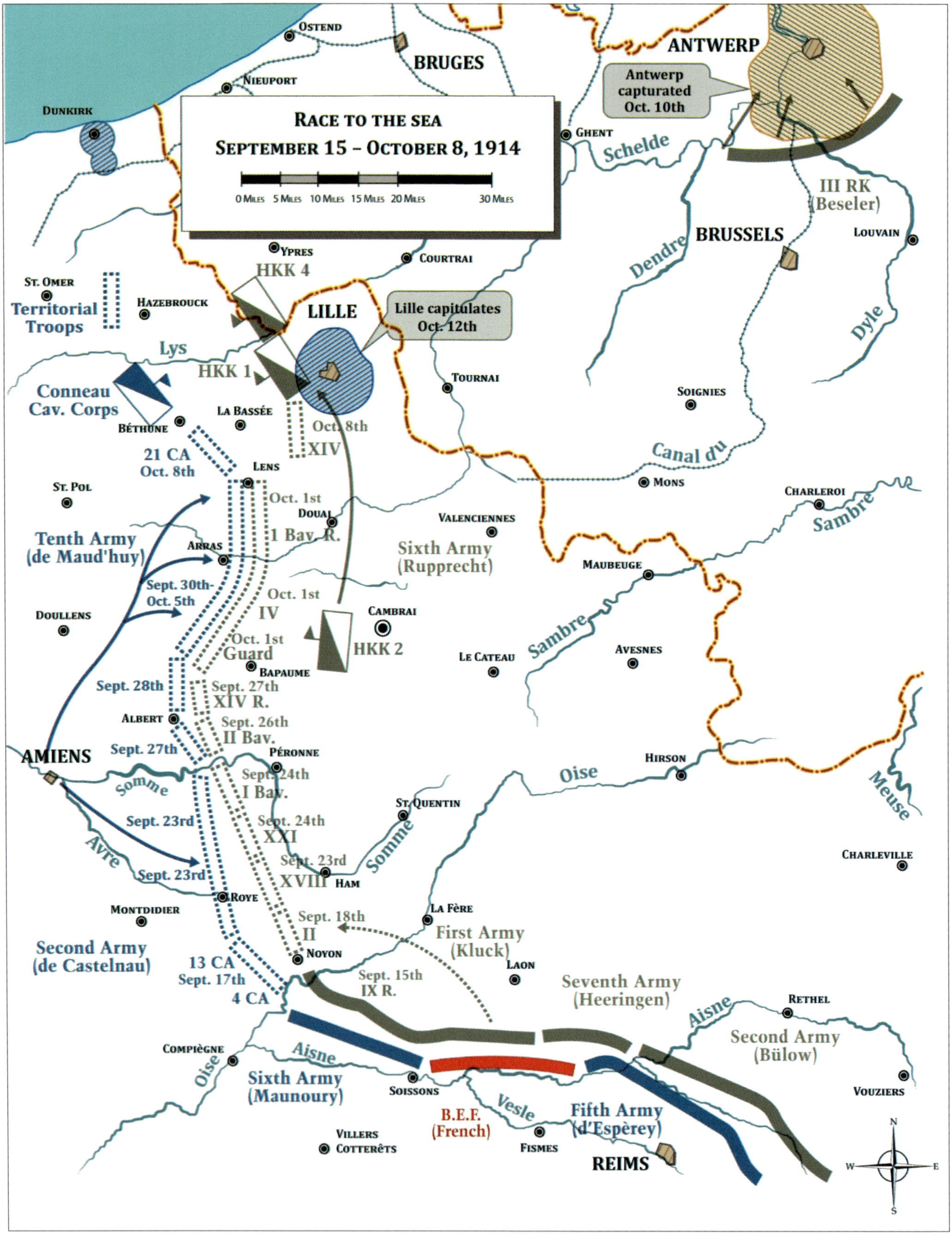
Race to the sea
September 15 – October 8, 1914
0 Miles 5 Miles 10 Miles 15 Miles 20 Miles 30 Miles
Ostend
Bruges
Nieuport
Dunkirk
Antwerp
Antwerp capturated Oct. 10th
Ghent
Schelde
III RK (Beseler)
Brussels
Louvain
Dendre
Ypres
Courtrai
HKK 4
St. Omer
Hazebrouck
Territorial Troops
Lille
Lille capitulates Oct. 12th
Dyle
Lys
HKK 1
Conneau Cav. Corps
Tournai
Soignies
Béthune
La Bassée
Oct. 8th
XIV
21 CA
Oct. 8th
Canal du
Lens
Mons
St. Pol
Oct. 1st
Douai
1 Bav. R.
Valenciennes
Charleroi
Sambre
Tenth Army (de Maud'huy)
Arras
Sixth Army (Rupprecht)
Maubeuge
Sept. 30th- Oct. 5th
Oct. 1st
IV
Cambrai
Doullens
Oct. 1st
Guard
HKK 2
Le Cateau
Sambre
Avesnes
Bapaume
Sept. 28th
Sept. 27th
XIV R.
Albert
Sept. 26th
II Bav.
Amiens
Sept. 27th
Péronne
Sept. 24th
I Bav.
Hirson
Oise
Somme
Meuse
St. Quentin
Sept. 23rd
Sept. 24th
XXI
Somme
Avre
Sept. 23rd
Sept. 23rd
XVIII
Ham
Charleville
Roye
Montdidier
La Fère
Sept. 18th
II
First Army (Kluck)
Second Army (de Castelnau)
Noyon
13 CA
Sept. 17th
4 CA
Sept. 15th
IX R.
Laon
Seventh Army (Heeringen)
Rethel
Aisne
Second Army (Bülow)
Compiègne
Oise
Aisne
Sixth Army (Maunoury)
Soissons
Vesle
B.E.F. (French)
Fifth Army (d'Espèrey)
Vouziers
Villers Cotterêts
Fismes
Reims
N
W
E
S

Galliéni correctly summed up the episode (and the entire 'Race to the Sea' campaign) as follows: "The French and the British were always twenty-four hours and an army corps behind the Germans."[19] By the evening of October 4th, Rupprecht's forces had tenaciously fought their way through superior numbers to threaten three of Castelnau's northernmost corps (collectively under the command of *Général* Louis de Maud'huy) with encirclement. Joffre, sensing a disaster, quickly acted to save the situation. First, Castelnau was ordered to hold his ground at all costs while Maud'huy's force was organized into an independent command dubbed the "French Tenth Army." Next, in order to guarantee firmer control, these two northern armies were collectively placed under the command of Ferdinand Foch, who had recently been appointed as Joffre's deputy.[20] These measures, in concert with a substantial resupply of artillery ammunition, collectively secured the allied front by October 6th—prompting Falkenhayn to suspend all further offensive action until a new plan could be determined.[21]

The *Fliegertruppe's* Actions During the 'Race to the Sea'

Foul weather continued to inhibit air operations as Bülow's attack along the Aisne commenced. Constant rain, fog, and mist on September 13th, 14th, and 15th diminished the number of flights and the quality of information delivered into the German leaderships' hands. In a message sent to OHL at 11:55pm on September 15th, Kluck voiced his frustration regarding the lack of quality intelligence First Army HQ had received over the previous five days:

If the absence of any clear report by First Army HQ regarding the danger threatening the right flank of the armies has given rise to dissatisfaction, it must be stated that, unfortunately, nothing in the nature of a clear report could be obtained. Both the cavalry and Fliegertruppe *have failed in the past few days.*[22]

With the opposing sides now entrenched in strong defensive positions, cavalry was no longer a viable option for reconnaissance. Indeed, with the onset of trench warfare across the front, aviation became the only method to collect meaningful intelligence. Fortunately, the weather finally broke late in the afternoon of the 15th, allowing consistent long-range aerial reconnaissance flights to be conducted for the first time since the retreat from the Marne. That evening, an observer from FFA 12 delivered a report to Kluck and Kuhl which clarified the situation. According to the airmen's message, a strong column of all arms, estimated strength to be a reinforced division, was observed approaching Ribécourt from the direction of Compiègne. Three more columns, consisting of cavalry and artillery, were also witnessed moving east from Beuvraignes toward Noyon with small cavalry detachments pushed as far forward as Lassigny. Finally, all areas to the north of this force was found to be free of enemy troops.[23]

FFA 12's timely report had unknowingly discovered the IV and XIII CAs, supported by cavalry, moving into position north of the Oise on French Sixth Army's extreme left wing with orders directly from Joffre to quickly attack and envelop Kluck's uncovered northern flank. Sunny skies on the morning of September 16th enabled German airmen to be airborne throughout the day. That afternoon First Army HQ was flooded with reports from FFAs 9, 11, and 12 describing newly arrived French forces moving into position on French Sixth Army's left

Above: As Joffre's deputy, Ferdinand Foch skillfully managed the battles on the French left wing during the latter stages of the 'Race to the Sea.' (Author's Collection)

northeast of Compiègne. Thus, Joffre's intention to strike First Army's open right flank, as suggested by FFA 12's report from the previous evening, was confirmed. Citing this intelligence, First Army HQ immediately pulled back Max von Boehn's recently arrived IX RK, which was located on the army's extreme right wing, to stronger defensive positions north of the Oise. This redeployment of the IX RK temporarily removed any threat of envelopment from the French offensive. Nevertheless, the IV and XIII CAs attempted unsuccessfully throughout the 17th to turn IX RK's position north of the Oise at Noyon. Alarmed by these developments, First Army HQ hastily redeployed the II AK into positions north of Noyon on IX RK's right. The arrival of II AK on the 18th brought the French advance to a halt, thus ending the German Army's greatest crisis during the 'Race to the Sea' period thanks largely to the work of First Army's flyers.[24]

German airmen in the other armies' sectors had likewise made significant contributions during this period. In Seventh Army's sector, the airmen of FFAs 3 and 26 completed their redeployment from Lorraine on September 17th, allowing flying operations to commence the following day. The subsequent reports produced by Seventh Army's *Fliegertruppe* illustrated the Franco-British inability to resume the attack along the Aisne. According to these reports, Bülow's counter-attack had successfully forced the BEF to throw in their last reserves, making further offensive action from that direction highly unlikely. Further east, Second Army's airmen, still extremely handicapped from a lack of serviceable aircraft, delivered useful reports which confirmed Seventh Army *Fliegertruppe's* findings. Elsewhere, airmen from the other western armies carried out reconnaissance flights which confirmed the redeployment of forces from quieter sectors around Lorraine and Verdun to the French left wing.[25]

Citing these reports, Falkenhayn deduced that Joffre was seeking to strike the uncovered German right flank.[26] To counter this, OHL ordered Rupprecht's Sixth Army on the 18th to deploy their forces piecemeal as soon as they became available instead of together as a united force as originally desired. Although this would reduce the striking power of the army, Falkenhayn recognized that it was necessary in order to guarantee the safety of the German position in France. From the 18th–22nd, German aerial reports describing additional troop movements to the allied left wing continued to be forwarded to OHL. A concerned Falkenhayn ordered additional reinforcements to be immediately sent to the right wing to assist Sixth Army (who's leading units had yet to arrive). Accordingly, General Heinrich Schenk's XVIII AK was transferred from Second Army to the battle line's extreme right wing on the 23rd. This redeployment, which secured the German right flank, was a direct result of aerial intelligence. Multiple aerial reports had alerted OHL of new French formations opposite the German right wing along with local reconnaissance in Second Army's sector that illustrated the removal of Schenk's corps would not endanger Second Army's position. Consequently, when French Second Army's lead formations arrived in their assigned sectors opposite the German right wing, they were unable to effectively deliver Joffre's flank attack due to the presence of Schenk's corps.[27]

The subsequent battles of the 23rd and 24th confirmed the arrival of even more French troops on the allied left flank. These freshly redeployed formations comprised Castelnau's new Second Army, which had been moved up from Lorraine by Joffre to attack and envelop the German battle line. Fortunately for the Germans, two corps from Rupprecht's Sixth Army had entered the line on the 24th—giving the Germans a total of four corps between the Somme and the Oise to stop Castelnau's advance.[28]

Joffre's plan to quickly envelop the German line had failed. Nonetheless, additional intelligence regarding Castelnau's movements and strength was necessary. Therefore, recognizing the value of aviation, Falkenhayn dispatched a message to First and Sixth Army HQs ordering new aerial reconnaissance flights over the sectors opposite the German right wing:

Information on the army's right flank is urgently needed to gain clarity. Aviation and automobile reconnaissance over the line Lille—Arras—Amiens—Beauvais, preferably to the coast, is hereby ordered. Sixth Army is assigned the line Lille—Arras—Amiens and First Army Amiens—Beauvais.[29]

In accordance with the above directive, all five of Sixth Army's flying sections were ordered airborne on the 25th to reconnoiter the area assigned to them by the Supreme Command. By late afternoon, Rupprecht had gained a complete picture of the French deployment in his army's sector. According to the reports, three corps were between Amiens and Montdidier marching northeast in preparation to strike the German right flank. Further north, a reinforced cavalry division was observed crossing the Somme in the environs of Bray. Significant military traffic was also witnessed in the French rear-area around Amiens. Most importantly, significant rail traffic was noted moving north through the city to the allied left flank—a clear indication of Joffre's desire to extend his line further north to strike Rupprecht's uncovered right.[30]

Sixth Army's aviation reports left no doubt as to the

French designs to turn the German northern flank. Nonetheless, Falkenhayn and Rupprecht remained committed to using Sixth Army offensively. After some discussion, Rupprecht's troops were ordered to launch a series of 'spoiling attacks' against French Second Army's positions in order to disrupt Joffre's planned offensive and stop the attack before the troops were deployed. The first of these attacks occurred on the 26th, as the I and II Bavarian AKs struck the French left flank on either side of the Somme in the area of Bray.

Two months of constant campaigning had taken its toll on the Bavarian flying sections—reducing each section's strength to below 50% with no replacements available at Sixth Army's Air Park. Consequently, Rupprecht's flying sections were compelled to forego tactical flights and concentrate only on long range reconnaissance of the French rear-area.[31] FFA 1b spent the morning of the 26th reconnoitering south of the Somme, discovering French reinforcements moving east from Amiens. Meanwhile, FFA 2b alerted Rupprecht of new French formations on the Somme's northern bank. According to the reports there were heavy troop concentrations in the area Albert—Bapaume as well as a sizeable column marching toward Thiepval. Together, these aerial reports proved that French Second Army had once again extended their line northward, making envelopment of the French line impossible. A further message from Sixth Army HQ's FFA 5 confirmed the redeployment of a new French corps (XI CA) to Amiens—thus reinforcing the French numerical superiority in the sector. Undeterred, Rupprecht used his own newly arrived XIV RK (in concert with Bavarian I and II AKs) to launch a second spoiling attack on the 27th against the French left west of Bapaume. Although the attack failed to envelop the French line, it successfully disrupted Castelnau's plans of launching his own assault.

Constant fears of French formations north of Sixth Army's open flank caused Rupprecht to order his own FFA 5 to patrol the area on the 29th. The subsequent report confirmed northward rail traffic but found no troops in the area Lille—Arras—Douai or Roubaix—Hazebrouck.[32] On October 1st, Rupprecht launched his third and final 'spoiling attack,' using the newly arrived forces of IV AK, Guard Korps, and I Bavarian RK to strike the extreme French left (a group of three corps under the overall command of Maud'huy) in the north while the XVIII and XXI AKs attacked a weak point in the French line to the south near the village of Roye. Over the next 72 hours Sixth Army's offensive successfully pierced the French line to the north and south of Arras, threatening Maud'huy's three corps with encirclement. During the attack, aircraft from

Above: Louis de Maud'huy, commander of French Tenth Army. (Author's Collection)

FFAs 1b and 8 reported three strong columns marching from Amiens to reinforce Maud'huy's wavering defenses. Additional messages, delivered on the 5th, confirmed large numbers of fresh reinforcements detraining at St. Pol and marching to Maud'huy's aid at Arras. To the south, airmen from XVIII AK's FFA 4 similarly reported the strengthening of the French positions around Roye—ending any possibility of a breakthrough in that sector. Upon receiving these reports, Rupprecht and Falkenhayn wisely agreed to pull back their troops, thus inconclusively ending the offensive.[33]

Throughout the offensive, observers from FFAs 5 and 2b continued to monitor the situation north of Sixth Army's right flank, looking for new threats that could potentially roll up the German line. The subsequent reconnaissance reports promptly informed Rupprecht that the sector Lille—Orchies—Douai—Carvin was free of troops and rail traffic.[34] Consequently, by the time Sixth Army's offensive had petered out on the 6th, Falkenhayn felt confident that

Above: French and German cavalry patrols clash near La Bassée. (Author's Collection)

his right flank was secure from envelopment. In order to maintain the initiative, however, OHL ordered the three cavalry corps (HKKs 1, 2 and 4) covering Sixth Army's right flank to advance through Flanders and strike the Allied left flank and rear. HKKs 1 and 2 were eventually checked by Franco-British cavalry around La Bassée over the following 48 hours. Meanwhile, the three divisions of HKK 4 pushed through Ypres and moved west to the suburbs of Hazebrouck.[35] However, by this time (October 10), substantial allied forces had been moved into the sector, forcing HKK 4 to withdraw to the area of Armentières.[36] Back at OHL, Falkenhayn realized that cavalry alone would be insufficient to achieve a decision against the allied left. Thus, the new commander began formulating a plan to bring up large numbers of fresh troops from the German interior for an offensive in Flanders on the extreme right of the battle line. This offensive would ultimately become known as the 'First Battle of Ypres.'

A Growing Crisis: Aircraft Production & Delivery

The heavy losses sustained during the Marne Campaign significantly reduced the *Fliegertruppe's* ability to support the army as the war progressed into its second and third month. Indeed, aircraft losses during the month of August alone were estimated to be 40% of the *Fliegertruppe's* original strength.[37] This "wastage" was a result of a multitude of factors including engine/mechanical failures, pilot error, and ground fire (among others). In theory, replacement aircraft were to be provided through the various army air parks, which were charged with keeping the flying sections at the front operational. However, with each *AFP* equipped with a meager two replacement machines, the strength of the front line flying sections was rapidly reduced. Moreover, with the overwhelming majority of Germany's 254 military pilots and 271 observers having been deployed to front line units upon mobilization, the *Fliegertruppe's* replacement units simply had an insufficient number of trained reserves available to send to the front as the losses mounted. Consequently, by mid-September many flying sections in the west were operating at or below half strength.

The disruption in the supply of aircraft to the front was a direct result of the German government's reluctance to develop a plan for the wartime mobilization of the aircraft industry. In the days leading up to war, the Prussian War Ministry had attempted to compensate for this by ordering an

Quantity of Monthly Aircraft Deliveries Agreed Upon at the August Parliamentary Commission by Firm*								
LVG	Albatros	DFW	Rumpler	Aviatik	Jeannin	AEG	LFG
40	48	25	25	20 min.	20	16 min.	4

* John H. Morrow Jr., *German Air Power 1914–18*, p.20.

additional 220 aircraft to be delivered as speedily as possible. On August 1st, the Ministry went further by sending a memorandum to each of the manufacturers ordering maximum production and an increase in productive capacity.[38] As German aviation historian John Morrow points out, these measures ended up doing more harm than good:

Yet such an order was dysfunctional, as firms seized the opportunity to hoard materials, and thus exacerbated shortages and price increases, for the production of aircraft that had yet to prove their battle worthiness.[39]

Representatives of the aviation industry met with the War Ministry and the Inspectorate of Flying Troops (*Idflieg*) on August 4th in order to establish some measures to maximize aircraft production in accordance with the Ministry's aforementioned emergency order. The aviation firms quickly discovered, however, that the army would do very little to assist them. The aircraft manufacturers had requested long term contracts guaranteeing purchases for the next three months, aid in material procurement, and assurances on employee exemptions from front line service. Furthermore, many firms requested immediate price increases on all delivered aircraft to cover the rising wartime costs of raw materials, parts, and labor. In response, the army's representatives offered to assist the industry only with procurement and exemptions—stating that price increases and long term contracts would lead to the industry's "unjustified enrichment."[40]

On the following day, seven members of the *Reichstag* decided to further address the problem of aircraft production by forming a parliamentary commission that would place political pressure on the army. First, the commission met with officers from the Prussian War Ministry and the Naval Office to obtain the military's official opinion. Next, representatives from eight aircraft firms (AEG, Albatros, Aviatik, DFW, Jeannin, LFG, LVG, and Rumpler), three aircraft motor manufacturers (Mercedes Daimler, NAG [Neue Automobil Gesekkschaft], and Argus), and five accessories factories were called in front of the commission to testify. After two full days of discussion, a compromise was reached between the two sides. First, the military agreed to pay the manufacturers one third of the cost for "proven aircraft" in advance.[41] This measure greatly assisted the manufacturers, who were crippled by the wartime cost of raw materials and skilled labor. Second, the army announced their intent to reestablish and subsidize military flight training at the factories' flying schools. In exchange, the industry accepted the continuation of pre-war aircraft/engine prices until the War Ministry, in good faith, declared that the labor and material costs warranted a price increase.[42] The truth was, however, that rising costs in production had already made a price increase necessary. The War Ministry, for whatever reason, failed to grasp this fact. As a result, the manufacturers were collectively never able to produce and deliver the monthly quota of 200 aircraft and 170 engines that was agreed upon during the meetings—to the detriment of the *Fliegertruppe* and the army.

Meanwhile, despite obtaining several promises from the army, the aviation industry quickly discovered that the exemptions for their employees were not being honored. The exemption system was significantly important to the firms, who needed to retain their skilled workers to build their products. Unfortunately, the army's mobilization orders, written in peacetime, failed to include a provision that would keep the firms' employees in the factories. Consequently, the mobilization orders did not distinguish a difference between the aviation industry's workers and anyone else. Thus, many highly skilled laborers were obligated to leave the factories and join the army in the field, leaving the firms understaffed at a time when production was to increase. Clearly, the army's needs took precedent over the industry, as the various Deputy General Commanders (*Stellvertretender KG*) in Germany's 24 corps districts were ordered to grant exemptions to "absolutely indispensable men" only. However, with the Deputy Commanders themselves making the determination regarding who was "absolutely indispensable," many skilled workers remained at the front in spite of the aviation firms' repeated protests.

The lack of laborers at the various firms deeply contributed to the industry's inability produce the 200 aircraft per month agreed upon during the Parliamentary Commission meetings in early August.

Material Price Increases—Fall 1914*				
Raw Rubber	**Steel**	**Aluminum & Steel Tubing**	**Copper & Tin**
100%	15%	20%	30%
* John H. Morrow Jr., *German Air Power 1914–18*, p.23.			

Finally, on October 5th, after two months of poor production, General Franz von Wandel (head of the War Department) took action to address the problem. After investigating the matter, Wandel identified the army's exemption policy as the prime cause for the industry's inability to keep up with demand. The general therefore sent a memorandum to each of the Deputy Commands declaring that the army must exempt all personnel that the *manufacturers* believed were indispensable. Thus, with the lack of aircraft at the front severely affecting the army's operations, the Deputy Commanders had no choice but to acquiesce to Wandel's proposal.[43]

While the army's exemption policy was undoubtedly problematic, the bigger issue affecting aircraft production remained the War Ministry's refusal to increase their purchasing price for aircraft in spite of sharp wartime rises in the cost of parts and labor necessary for production. Although the manufacturers had agreed to prewar delivery prices in August as part of the aforementioned Parliamentary Commission compromise, they had done so only under the strict understanding that the War Ministry would raise prices on delivered aircraft when "labor and material costs warranted a price increase." Thus, when prices of raw materials jumped 10–30% by September (with raw rubber spiking an incredible 100%), the industry petitioned the War Ministry for an increase. In addition to the high cost of materials, the factories cited rising interest rates, freight charges, and wages as further justification for a price increase. Most notably of these was the wages, which had risen due to the army's demand for shorter delivery schedules. Indeed, in order to deliver the aircraft on time, many factories had to remain open at night, resulting in temporary wage increases of up to 50%.[44]

Incredibly, despite the jump in costs for parts, raw material, wages, and interest rates, the War Ministry refused to increase their delivery prices. Speaking at a reconvened session of the parliamentary commission, a representative from the War Ministry stated that it would grant a price increase only after it was determined that the aircraft firms were cooperating to keep prices low.[45] This mistrust of the aircraft firms was entirely unjustified, as the costs of raw materials in other industries proved. Nevertheless, the Ministry stuck to its guns and kept paying prewar delivery prices throughout the calendar year—keeping Germany's various aircraft firms down in the process.

The War Ministry did, however, make several attempts to ameliorate the problem. First, all prewar debts that were owed to each of the firms were immediately paid. Next, substantial advances were given for ordered aircraft and engines. These measures gave the manufacturers much needed liquidity at a time when interest rates were soaring yet failed to fix the underlying problem of price inflation.

The army's error in refusing to increase prices for delivered aircraft was further compounded by its failure to check the rising costs for parts and materials necessary for aircraft production.[46] Despite rising prices, the War Ministry refused to set price ceilings on key items such as aluminum, oak, steel, tin, and copper (among others). As a result, by October prices had skyrocketed up to 50% for most of these items, leaving the manufacturers dangerously exposed financially. Although the War Ministry had established a "raw materials" department to monitor the market for unproductive inflation, the department was unscrupulously directed by the same business interests who stood to profit from the inflated prices—resulting in no meaningful change.

The remainder of 1914 continued to see delivery prices frozen despite continued rises in the cost for production. Under these circumstances, the *Kaiserreich's* aviation manufacturers strained to produce the number of aircraft necessary for the war's successful prosecution. Instead of producing 200 aircraft each month (about 47 per week) as promised, the industry could only muster a total of 462 new aircraft between August 3rd and November 29th—an average of only 28 per week. The meager addition of 28 new machines each week alarmingly translated to the delivery of only one replacement aircraft per flying section every two weeks.[47] Consequently, each of the *Fliegertruppe's* various flying sections dwindled in strength as the war progressed, resulting in dramatically less support for the ground troops at the front. Moreover, the army was unable to create new flying sections, which were sorely needed to support the newly raised reserve corps, consisting of men who had volunteered for service at the outbreak of the conflict.[48] This problem was temporarily solved, however, by forming new FFAs out of disbanded fortress flying sections (Fest.FA). Nonetheless, the overall situation remained unsatisfactory as the understrength *Fliegertruppe* routinely failed to provide adequate support to the army during the final 2–3 months of the year.[49]

Circumstances were no different in Bavaria where local leaders resisted Prussian assistance to the detriment of the troops and overall war effort. Under pre-war agreements, Bavaria's two domestic aircraft firms, Otto and Pfalz, had been expected to produce the Bavarian *Fliegertruppe's* aircraft in wartime. It was quickly determined, however, that the two firms were unable to meet wartime demands. Nevertheless, the Bavarian War Ministry declined an offer from Albatros to deliver aircraft to Bavarian Flying sections—illustrating the Kingdom's desire to retain internal autonomy, even at the risk of severe shortages. It should be noted that LVG was allowed to build aircraft for Bavaria due to a prewar agreement.[50] However, the Prussian Transportation Inspectorate only allowed Bavaria two LVG B-Types and two Mercedes engines per week.[51] These numbers illustrated the Bavarian War Ministry's model to be grossly insufficient.

As with the Prussian units, shortages in aircraft had become a major issue for the Bavarians. The Bavarian *Fliegertruppe's* three flying sections experienced heavy losses in the war's first six weeks. By September 1st all replacement aircraft had been sent forward, leaving the replacement units and flight schools without serviceable aircraft. Although the Bavarians' refusal to accept additional aircraft from Albatros was ill-advised, the bigger problem facing Bavarian production was engines. Because the aviation industry's three engine manufacturers were located in Prussia, *Idflieg* controlled all engine allocation. Thus, feeling the shortages within their own units, the Prussian Inspectorate decided to give their Bavarian counterparts only two engines per week. It was therefore irrelevant how many aircraft Bavarian companies produced each week as long as they only had two engines to equip them.

Bavarian aircraft production remained stifled throughout the rest of the year due to many of these poor policies.[52] When pressed again regarding Albatros' wish to open a plant in Bavaria, the War Ministry demurred; absurdly stating that it couldn't support three aviation factories with contracts *after* the war's conclusion. Aviation historian John Morrow later summarized this remarkably foolish logic: "Two months after the setback on the Marne, the Bavarian War Ministry was more preoccupied with the postwar period and its independence from Prussia than with winning the war."[53] By the end of 1914 Otto had produced 50 aircraft while the tiny Pfalz factory delivered 14. These poor numbers were a direct result of Bavaria's desire to retain internal autonomy—a

Frontline Inventory—1914

Type	8/31	10/31	12/31
Class A Unarmed Monoplanes			
Albatros A/13	5	2	7
Albatros A.I	2	—	—
Fokker A	—	5	9
Gotha A/13	13	3	6
Gotha A/14	3	5	14
Halberstadt A.II	—	—	2
Hirth A	1	—	—
Jeannin A	9	9	7
Kondor A	—	1	4
LVG A	1	—	—
Otto A	—	2	—
Rumpler A/13	2	—	—
Rumpler A/14	8	3	—
Subtotal Class A	**44**	**30**	**49**
Class B Unarmed Biplanes			
AEG B.I	3	2	3
Albatros B.I	25	32	65
Albatros B.II	5	9	7
Aviatik B Misc	35	24	13
Aviatik B.I	12	43	51
Brandenburg B.I	—	—	2
DFW B.I	2	17	12
Euler B/13	2	—	—
Euler B.I	3	—	—
Fokker B/13	1	—	—
Gotha B.I	—	1	—
LVG B.I	84	73	95
LVG B.II	—	—	13
Otto B.I	—	1	—
Rumpler B.I	1	6	21
Subtotal Class B	**173**	**208**	**281**
Miscellaneous			
Brandenburg	—	3	—
DFW Mars*	1	2	—
Farman	—	1	—
Parasol	—	—	1
Total Inventory	**218**	**244**	**331**

* Perhaps a Mars monoplane since some Mars biplanes were given B-type serial numbers.

policy that kept Bavarian troops inadequately supported throughout the war's first five months.

The army's shortsighted attitude toward the aviation industry can be directly linked to Germany's short war strategy. In accordance with their war plan, many senior leaders believed that rapidly concluding the war in the West was Germany's best and only option. Indeed, Schlieffen and Moltke both believed that the Central Powers could not survive a prolonged two-front war.[54] Therefore, the army's leadership staked everything upon winning the war in the West as rapidly as possible. Consequently, the army considered it unnecessary to assist the aviation industry mobilize for war and build itself up for long term production. Instead, they allowed raw material prices to skyrocket, kept skilled workers at the front, and refused to increase prices for delivered aircraft—all in the hopes that the war would be over before the negative effects of these policies were felt. As it happened, this strategy stifled the industry's growth and kept aircraft production low, effectively crippling the *Fliegertruppe* by the onset of winter.

The *Fliegertruppe's* Organizational Deficiencies Revisited

Although the *Fliegertruppe's* contribution during the Marne Campaign was unquestionably valuable, close inspection reveals that Germany's airmen could have performed even better if it wasn't for a grossly inefficient organizational model. Prior to the war and upon mobilization all Prussian aviation units were subordinated to *Idflieg*, which was comprised of the army's highest ranking and most knowledgeable aviation officers. Under the deficient system of 1914 however, *Idflieg* lacked the authority to control their units in the field during wartime. Similarly, all of Bavaria's flying sections were controlled by an external agency, the Bavarian War Ministry, which lacked the power to directly influence events at the front. In other words, each of the army's 33 flying sections was inexplicably forced to operate independently throughout the first year of the war without any coordination or direction from the organization to whom they were subordinated in peacetime. As a result, the potential level of support given to the army by the Germany's airmen was severely diminished.

Perhaps the single most damaging component of the 1914 organization model was the lack of an aviation staff officer at each army HQ. The primary role of an aviation staff officer, as exhibited by the French during the Marne Campaign and later by the Germans after the reorganization of 1915, was to function as an intermediary between the flying troops in the field and the army's leadership. Each day the officer was to use his expertise to ensure that the flying sections under his jurisdiction were operating efficiently and in accordance with the army commander's concept of operations. For example, if the commander desired reconnaissance in a particular direction, the officer would issue orders to make sure the action was executed. At the end of the day he would gather and interpret all the reconnaissance reports in his jurisdiction in order to accurately brief the commander on the day's results. Other responsibilities would include assigning approximate locations for airfields when the army was on the march, ordering replacement aircraft to be transferred from the army airpark to a flying section at the front, and maintaining constant liaison with various aviation commands in the rear. In short, an aviation staff officer was to manage all the *Fliegertruppe* units under his command so that they could provide as much support to the ground forces as possible.

The absence of aviation staff officers at the various army commands significantly reduced the *Fliegertruppe's* level of support throughout the war's first six months. In his book *The Air Weapon 1914–1916*, aviation historian John Cuneo succinctly described the effects of the German Air Service's poor organizational model:

The lack of any staff officers for aviation at the various commands under which the flying sections operated without doubt contributed much to the failure of air reconnaissance to achieve greater results in 1914. No particular officer had the responsibility of assuring the regular issuance of orders and receipt of reports from airmen. There were no specially trained men to collect interpret and integrate the air intelligence to assure its dissemination to all units affected by the information. No one at headquarters understood the capacity and limitations of aircraft. No one curbed the individualism of the leaders of the various aviation units which sometimes did more harm than good. There was no intelligent cooperation between the commands and the aerial observers. This could have been attained if the former had seen to it that the airmen were familiar with the general scheme of maneuver taken by the supported forces and with the details of operations.[55]

Six weeks of experience during the Marne Campaign validated Cuneo's thesis. First, the absence of an aviation officer during mobilization affected how quickly the army's flying sections were "campaign ready." Airparks were often approached during the period of concentration by representatives from flying sections commandeering supplies—resulting in the unequal distribution of matériel and further delays in equipping all flying sections within the air park's zone. The role of an aviation staff officer during mobilization, if the position existed, would have been to coordinate the supply and deployment of each of his flying sections under his command to improve efficiency and speed. Secondly, the lack of an aviation specialist on staff reduced a commander's ability to properly use his aviation assets. With military aviation still in its infancy, many senior army commanders had little faith in their airmen's abilities to make an impact on operations. Consequently, many flying sections mobilized for war without any substantial contact with their army headquarters. If an aviation staff officer were present on each army commander's staff, one of his duties during the period of mobilization would have been to build a rapport with the commander and communicate the airplane's uses and limitations so that the commander could properly use them when the period of active operations commenced. As it happened, many commanders went into the field without any real knowledge of how their flying sections operated or what they could contribute once the campaign began.

Although many problems existed during mobilization, the true effects of the *Fliegertruppe's* organizational flaws weren't felt until after the campaign began and the various armies became engaged with the Franco-British forces. Numerous instances of lost reconnaissance reports, mishandled orders from army headquarters to a flying section, misinterpreted aerial reports, and poor liaison procedures between commanders and their airmen all contributed to the army's inability to achieve a swift and decisive victory as prescribed by the German War Plan. Despite there being numerous examples throughout the war's first six weeks, the following four cases are perhaps the most egregious and influential upon the outcome of the campaign and deserve further attention.

The first noteworthy example occurred on August 22nd as Kluck's First Army desperately searched for the location of the BEF. The reader will remember that the army's leadership believed the English would be to the west in the vicinity of Lille and had sent the majority of their aircraft and cavalry in that direction. However, IX AK's FFA 11 was sent south to search around the area of Mons and had discovered the BEF's presence there late in the afternoon. Unfortunately for the Germans, FFA 11's report was never forwarded to First Army HQ, resulting in further uncertainty regarding the BEF's whereabouts. If an aviation staff officer had been present to diligently collect all aviation reports, First Army HQ could have been informed of the British force's location in time to direct adequate forces to the area to take part in the battle the next day—resulting in a larger German victory.

Two days later, Kluck's airmen provided First Army HQ with no less than five reports confirming the BEF's direction of retreat to the southwest towards Le Cateau. In spite of this, Kluck and Kuhl decided to switch the army's direction of pursuit to the *southeast* in the direction of Maubeuge based upon a single aviation report furnished by III AK's FFA 7. Once again, if an aviation specialist were on Kluck's staff, he would have reviewed all the available reports and came to the correct conclusion that the BEF was headed towards Le Cateau and would have strongly advised First Army continue in that direction. While it is always possible that Kluck would have continued toward Maubeuge, it seems highly unlikely if all the aviation reports had been gathered by a staff officer and smartly presented together rather than being reviewed piecemeal as they were delivered at headquarters. Indeed, it appears that Kluck and Kuhl made their decision because FFA 7's report was the last to arrive. This was a critical mistake. Had First Army continued the pursuit southwest, they would have enjoyed a decisive numerical superiority at Le Cateau on the morning of the battle, likely resulting in the complete destruction of the British II Corps.

Missed opportunities also occurred in Second Army's sector where Bülow's troops had pursued Lanrezac's French Fifth Army to the Oise River in the area of Guise—St. Quentin. True to his cautious nature, Bülow was hesitant to order his troops across the Oise until he gained confirmation that the river line was not heavily defended. After a thunderstorm passed, an aircraft of FFA 21 reconnoitered the area and discovered that only weak rear guards occupied the sector. After landing, the observer's report was quickly dispatched by motorcycle to Second Army HQ. The courier inexplicably did not reach Second Army HQ until the following morning, delaying Second Army's advance across the river and allowing the French to redeploy in preparation for their counter-attack at Guise-St. Quentin the following day. If the message had been received and had Bülow ordered his troops forward, Lanrezac's troops wouldn't have possessed the freedom of maneuver to prepare for their successful counter-stroke, giving the rest of campaign a dramatically different character.

Perhaps the most conspicuous example for the need for an aviation staff officer was First Army HQ's miserable liaison record with II AK's FFA 30 in the days leading up to the Battle of the Marne. It will be recalled that on September 3rd, FFA 30 discovered newly redeployed forces in the vicinity of Paris near First Army's vulnerable right flank. Regrettably, the overburdened II AK HQ staff failed to recognize the report's importance and chose not to forward it to

First Army HQ. The following day, First Army HQ explicitly ordered FFA 30 to perform flights in the area of Paris to verify that the army's flank was secure. Once again, the II AK HQ staff failed to forward the orders to the flying section. Thus, unaware of the order, FFA 30 failed to send a single aircraft in the direction of the French capital. Consequently, First Army's leadership remained unaware of the French buildup on their right flank until after the army had crossed the Marne and it was too late to react. The addition of an aviation staff officer on staff at First Army HQ would have undoubtedly helped facilitate communication between the flying section and the army leadership—ensuring that the airmen's reports reached army headquarters. Moreover, he would have confirmed that the army commander's September 4th orders for "aerial reconnaissance to be conducted in the area of Paris" were properly executed. In a post-war interview with the *Reichsarchiv*, Kuhl stated that if First Army HQ would have received the reconnaissance reports, the army would have halted the advance north of the Marne and taken up defensive positions to resist an attack from Paris.[56] As a result, Joffre's counter-offensive would not have been delivered against a weak and unsuspecting German right wing, but a highly concentrated and well-prepared one.

As the head of *Idflieg*, *Oberst* Walther von Eberhardt tried to point out the insanity of the German organizational model. In 1912 and 1913, Eberhardt and Moltke unsuccessfully attempted to reorganize the flying troops into a semi-independent force where all flying sections would be under the direction of a 'Chief of the Air Forces in the Field.' Furthermore, they recommended attaching aviation staff officers on the staffs of each army headquarters. The War Ministry, headed by Heeringen at the time, ultimately rejected these proposals—forcing the *Fliegertruppe* to operate for the first year of the war under its poor organizational model.

Without any control over his airmen in the field, Eberhardt remained in Berlin during the first two weeks of the war, writing in his diary on August 2nd that the *Fliegertruppe's* organizational model would lead to trouble.[57] Although he was powerless to control his units in the field, Eberhardt fulfilled his role as Inspector of Flying Troops by overseeing the production and transportation of aircraft to the front. Once again, however, the army's substandard organizational model created problems. In his diary on August 12th, Eberhardt wrote: "without aviation staff officers at the various commands, the flying troops lack the central agencies in the field to collect and evaluate frontline experience for technical and operational use." Nevertheless, Eberhardt and his staff attempted to collect the information themselves during the war's first two weeks, but were unable to establish any meaningful line of communication with the airmen at the front, who were continuously on the move. Thus, on August 15th, Eberhardt took the initiative to leave Berlin and visit the front to conduct a personal investigation.[58]

The primary cause of the trip was multiple reports concerning a disruption in the supply of aircraft to the front line flying sections. After a brief investigation Eberhardt discovered that several aviation parks were not following their mandate to supply the flying sections with replacement aircraft or supplies. For example, in Second Army's sector, the commander of AFP 2 had ordered his aircraft to be used directly for frontline service. Upon hearing this Eberhardt personally visited the airpark at Liège and scolded *Hptm*. Spangenberg, the park's commander, thus quickly restoring order. Similar instances in Sixth and Seventh Armies' sectors were also addressed.[59] Although these measures improved matters, Eberhardt quickly learned that the problem was much greater than several maverick airpark commanders.

With losses quickly mounting and each airpark possessing only two replacement aircraft, many aviators found themselves at the front without an aircraft. Desperate for replacements, the airmen absented themselves from the front and drove the long journey back to Germany to pick up newly produced aircraft directly from the manufacturers. The resulting confusion and disorganization made matters worse as prearranged aircraft deliveries were constantly disrupted by individual airmen commandeering aircraft from the factories that were destined to be delivered elsewhere.[60] Moreover, these trips reduced *Idflieg's* ability to closely monitor losses and allocate replacements to high priority sectors.[61] Thus, Eberhardt concluded that the prime cause in the disruption of the supply of aircraft was the *Fliegertruppe's* haphazard organizational model. Without any centralized command to control the flying sections in the field, the airmen were forced to act independently, and the result was chaos.

Throughout his trip, Eberhardt quizzed many airmen regarding their experiences during flight operations and their liaison with the army's leadership. It quickly became apparent that many of the army/corps commanders were not using their flying sections correctly. He learned that many flying sections were often receiving orders requesting reconnaissance over distances that far exceeded their aircraft's capabilities. Orders would also frequently neglect to mention important information such as the position and march objectives of local friendly ground troops or the estimated position and strength of the enemy.

Above: The Albatros B.I was powered by a variety of engines using drag-producing side radiators. Larger vertical tail surfaces improved the aircraft's in-flight stability and were adopted for most production B.I aircraft.

In some cases, commanders would completely forget to dispatch orders to their flying section—leaving the airmen to perform ad-hoc flights throughout the day without any knowledge of the commander's concept of operations.[62] To Eberhardt, it was clear that the various Army/Corps HQ staffs lacked the knowledge and experience necessary to properly utilize their flying sections in wartime.[63] Indeed, Eberhardt's following diary entry demonstrates his concern:

Methods which would have assisted the discharge of the missions were unknown. The system of ordering, receiving, interpreting, and disseminating reports was haphazard to the extreme. No conception existed of the necessity for defining and limiting missions, for outlining the scheme of maneuver in order to obtain an intelligent performance of the air missions or for specifying the desired information. There was no acknowledgment of the fact that the ground commands bore any responsibility for the successful discharge of air reconnaissance missions.[64]

Eberhardt considered the situation extremely dire. The poor organization of the army's flying troops had resulted in delays in the supply of aircraft as well as gross mismanagement of the flying sections at the front. The only remedy, Eberhardt opined, was a massive reorganization immediately placing all *Fliegertruppe* units under a centralized command. On August 23rd, while still on his trip at the front, Eberhardt telephoned the War Ministry to inform them of his findings and to make his proposal. While most officers at the War Ministry reluctantly agreed with Eberhardt's conclusions, representatives from the Inspectorate of Military Air and Motor Transport and the General Inspectorate of Military Transport quickly moved to block the reorganization. Eberhardt's case was persuasive enough, however, for the War Ministry to order him back to Berlin to make a presentation in person.

Two days later Eberhardt formally proposed his plan at the War Ministry's morning conference. According to the proposal, each army command was to have an aviation staff officer in charge of all aviation matters within his army's sector. All of these staff officers were to be responsible to a "Chief of the Air Forces in the Field" (*Feldflugchef*) who would therefore control all aviation units at the front. Under the proposed system, the *Feldflugchef* would establish a unified doctrine for all the flying sections as well as a standardized system of liaison based on the experiences of the airmen at the front.[65]

Eberhardt's presentation was ultimately persuasive enough to sway the former naysayers and gain the War Ministry's authorization, "subject to the approval of OHL." The plan quickly found renewed opposition however, upon Eberhardt's arrival in Luxembourg. Falkenhayn, who was serving as War Minister at the time, opposed making such a dramatic reorganization in the midst of such an important campaign.[66] The War Minister believed that immediately instituting a major reorganization as the western armies advanced into France would create unnecessary confusion and lower the *Fliegertruppe's* ability to support the army in the short term. Falkenhayn's argument, made on August 27th as the Kaiser's forces appeared to be on the precipice of victory, was yet another product of the General Staff's 'short war' mentality. Like most senior officers on the General Staff, Falkenhayn believed Germany's best opportunity to win the war was to quickly knock France out in the initial campaign.[67] Thus, the War Minister was simply unwilling to consent to a reorganization that might temporarily reduce the *Fliegertruppe's* capacity, no matter how slight, to support ground operations. Consequently, Eberhardt's proposal was formally blocked.[68]

With the reorganization thwarted, matters for the *Fliegertruppe* continued to deteriorate. Liaison between the airmen and the army's leadership remained problematic. Losses at the front as well as anemic aircraft production at home combined to reduce the strength of all the front line flying sections. As a result, the number of flights and quality of support given to the army was severely diminished. Nevertheless, despite all these problems, Eberhardt's *Fliegertruppe* would ultimately be called upon to play an important role in the army's operations as Falkenhayn launched 1914's final major offensive on the Western Front.

Chapter 4 Endnotes

1 Cuneo, *The Air Weapon 1914–1916*, p.144–146.
2 The story of Moltke's dismissal can be found in: Mombauer, *Helmuth von Moltke and the Origins of the First World War*, pp.265–269; Herwig, *The Marne 1914*, pp.301–302.
3 At his age, Falkenhayn was younger than all the army commanders. This 'meteoric' rise soon became the source of envy amongst some senior members of the German officer corps, which would ultimately have a negative impact on the course of operations in 1915 and 1916. Foley, *German Strategy and the Path to Verdun*, pp.87–93.
4 Holger Afflerbach, *Falkenhayn* (Munich, 1996), pp.181–182.
5 The definitive biographical work on Falkenhayn's life, which treats its subject strictly but fair, discusses at length the circumstances surrounding Falkenhayn's promotion. See: Afflerbach, *Falkenhayn*, pp.186–192. A brief summary of Moltke's replacement is also provided in: Hew Strachan, *The First World War: Volume I To Arms* (New York, 2003), pp.262–263.
6 This engagement would ultimately become known as 'The First Battle of The Aisne."
7 Bülow, *Report on the Battle of the Marne*, p.98.
8 Although steps were immediately enacted to move OHL forward, it took several days to make all the necessary preparations. Falkenhayn's OHL therefore completed the move on September 26th, with the Kaiser's GHQ arriving 2 days later.
9 Throughout the Marne Campaign, many senior Generals, including Falkenhayn, had criticized Moltke's passivity—claiming that the lack of guidance from OHL created an environment where the various army commanders determined the best course of action and acted accordingly without synchronization or knowledge of the broader strategic situation. For more on Falkenhayn's culture change at Supreme Headquarters, see: Foley, *German Strategy*, pp.92–94.
10 Erich von Falkenhayn, *General Headquarters, 1914–1916, and its Critical Decisions*, (Nashville, 2000) p.1. For an excellent overview of Falkenhayn's command style, see: Foley, *German Strategy and the Path to Verdun*, pp.92–98.
11 Jehuda Wallach, *The Dogma of The Battle of Annihilation* (Westport, 1986), pp.128–130.
12 Robert M. Citino, *The German Way of War* (Lawrence, 2005).
13 As noted WWI historian Hew Strachan suggests, Falkenhayn's approval of Bülow's attack was against his better judgement and was a major mistake. It is probable that Falkenhayn, in his first full day in command, was reluctant to overrule Tappen, who had just returned from the front. It should also be noted that in addition to the military concerns, Tappen dissuaded Falkenhayn from employing his initial plan by bringing up the political ramifications of a withdrawal—reminding him of Italy and Romania's possible response to a further German withdrawal. Strachan, *The First World War: Volume I To Arms*, pp.264–266; Regarding Bülow's plan and subsequent attack, see: Bülow, *Marne Campaign*, pp.100–110.
14 Falkenhayn's decision to abandon his original plan of September 15th has been identified by some historians as the real turning point of the First World War. While this is debatable, one can easily argue in hindsight that the decision was the Germans' final chance to return to "a war of movement" until Ludendorff's Offensives in 1918. The clarity of this fact, combined with the lessons learned by his greatest mistake a month later at 1st Ypres, is what prompted Falkenhayn to trust no judgment other than his own in the pursuit of a suitable peace for Germany. For a full description of Falkenhayn's original plan see: Wallach, *The Dogma of The Battle of Annihilation*, pp.128–130.
15 *Der Weltkrieg*, V, pp.34–55; GLA 456 F 1 Nr. 209.
16 As quoted in: Foley, *German Strategy and the Path to Verdun*, p.101.

17 An excellent concise history of the 'Race to the Sea' period is featured in: Strachan, *The First World War: Volume I To Arms*, pp.262–274.
18 Doughty, *Pyrrhic Victory*, pp.99–101.
19 Joseph Galliéni, *Mémoires de Général Galliéni* (Paris, 1920), p.197.
20 Foch's 'group of the north' consisted of the French Second and Tenth Armies as well as all previously independent territorial units operating in the area. Strachan, *The First World War: Volume I To Arms*, pp.267–270.
21 Within 72 hours of his arrival, Foch had single handedly restored the situation on the Allied left flank and saved the important city of Arras from falling into German hands. For Foch's role during the crisis, see: Greenhalgh, *Foch In Command*, pp.45–51. For more in depth coverage of the battles, see: FOH, Tome 1, Volume 4, pp.177–235.
22 Kluck, *The March on Paris*, p.166.
23 Cuneo, *The Air Weapon 1914–1916*, p.97.
24 Cuneo, *The Air Weapon 1914–1916*, p.97.
25 Cuneo, *The Air Weapon 1914–1916*, p.99; GLA 456 F 1 Nr. 209; *Der Weltkrieg*, V, p.52.
26 *Der Weltkrieg*, V, p.61.
27 Strachan, *The First World War: Volume I To Arms*, pp.264–269.
28 In addition to his own two corps (XXI AK and I Bavarian AK), Rupprecht was given control of II AK and XVIII AK, which had entered the line on the 18th and 23rd respectively. *Der Weltkrieg*, V, pp.73–75.
29 *Der Weltkrieg*, V, p.75.
30 *Der Weltkrieg*, V, pp.90–91.
31 Ferko–UTD, Box 11, Folder 5, "Bavarian *Fliegertruppe*."
32 Cuneo, *The Air Weapon 1914–1916*, p.102.
33 Frustrated by his troops' inability to make better progress, Falkenhayn visited Sixth Army HQ on October 5th. When the German commander questioned Rupprecht as to why the offensive had stalled, Rupprecht responded by showing the aviation reports illustrating the most recent French redeployments. After being led to believe that the French were nearly tapped out in terms of manpower, Falkenhayn responded that he was shocked to find such large numbers of new formations being transferred to the allied left wing. Nonetheless, the *Fliegertruppe's* reconnaissance clearly proved the allied numerical superiority was too vast for the Germans to secure victory. Thus, OHL instructed Sixth Army to pull back their troops and abandon any further offensive action until further notice.
34 Armengaud, *La Renseignement Aérien Sauvegarde des Armées* (Paris 1934), pp.14–15.
35 Several aircraft were attached to HKK 4, but were of little use due to the corps' high mobility and the commander's unfamiliarity with using aerial reconnaissance. Cuneo, *The Air Weapon 1914–1916*, p.401.
36 Jack Sheldon, *The German Army at Ypres 1914* (South Yorkshire, 2010), pp.12–25.
37 Cuneo, *The Air Weapon 1914–1916*, p.147.
38 Ferko–UTD, Box 58, Folder 1, "Idflieg no. 643."
39 Morrow, *German Air Power 1914–18*, p.18.
40 Morrow, *German Air Power 1914–18*, p.19.
41 This measure, wisely limited to proven aircraft, rewarded the firms that produced the *Fliegertruppe's* best performing aircraft. Examples included aircraft built by Albatros, Aviatik, LVG, and DFW. Meanwhile, the smaller firms with unproven models (Euler, Jeannin, etc.) were forced to prove their aircraft's value.
42 *Transcript of the Commission of Parliamentary Deputies 7 August 1914 Bundesarchiv* DE N 1103.
43 Morrow, *German Air Power 1914–18*, pp.22–23.
44 Morrow, *German Air Power 1914–18*, p.23.
45 Morrow, *German Air Power 1914–18*, p.23.
46 Roland Eisenlohr, *Flugwesen und Flugzeugindustrie der Kriegführenden Staaten* (Berlin, 1915), pp.31–36.
47 Cuneo, *The Air Weapon 1914–1916*, p.147.
48 On August 15, the army had ordered four new flying sections to be established. However, due to the industry's poor production numbers, the first newly formed unit wasn't dispatched for the Eastern Front until September 28. Cuneo, *The Air Weapon 1914–1916*, pp.146–47.
49 For example, many FFAs were forced to temporarily suspend tactical and artillery support flights in October due to a lack of available aircraft.
50 Hubert von Hösslin, *Die Organisation der K.B. Fliegertruppe, 1912–1919* (Munich, 1924), pp.11–17, p.27; Morrow, *German Air Power 1914–18*, p.29.
51 In October, under pressures to deliver as many aircraft as possible to their own units, the Prussian Inspectorate reneged on the deal—keeping the two LVG aircraft for themselves. Only after the Bavarian War Ministry's 'vehement protests' was LVG's aircraft released back to the Bavarian *Fliegertruppe*. Morrow, *German Air Power 1914–18*, p.30.
52 Harald Potempa, *Die Königlich-Bayerische Fliegertruppe als Teil der Deutschen Luftstreitkräfte bis 1918* (Published in Blätter zur Geschichte der Deutschen Luft- und Raumfahrt XIX)
53 Morrow, *German Air Power 1914–18*, p.30.
54 For the German leadership's rationale in developing the 'short war strategy' see: Foley, *German*

Above: Gotha LE3 A.301/14 *Taube*; the military serial and national insignia confirm it was in operational service. (Aeronaut)

Strategy, pp.56–81.
55 Cuneo, *The Air Weapon 1914–1916*, p.144.
56 BA:MA RH 61 1223, *Die Fliegerverbande der 1. Armee während der Verfolgung bis zur Marne*, p.21.
57 Cuneo, *The Air Weapon 1914–1916*, p.144.
58 Walter von Eberhardt, "Wie wir wurden: Tagebuchblätter aus dem Anfang des Weltkrieges," *In der Luft unbesiegt: Erlebnisse im Weltkrieg erzählt von Luftkämpfern* (Munich, 1923).
59 Neumann, *Die Deutschen Luftstreitkräfte im Weltkrieg*, (Munich 1923) pp.12–15.
60 Perhaps more alarming, Eberhardt discovered a message from a group of aviation officers to the Aviatik Factory requesting several aircraft to form a new 'wildcat' flying section that was being established without the knowledge or authorization from OHL or *Idflieg*. Cuneo p.144.
61 Morrow, *German Air Power 1914–18*, pp.17–18.
62 The most well-known example of this occurred with FFA 30 and the II AK HQ staff in the day's leading up to the Battle of the Marne. II AK HQ's failure to forward First Army HQ's message to their own flying section on September 4th left the section to perform flights in the wrong direction, which almost resulted in the destruction of First Army.
63 Indeed, many of the army's senior officers mobilized for war still viewing the airplane with great suspicion. Although the younger officers were more likely to believe in the airplane's value, most remained completely unaware of its true military potential. See: Brose, *The Kaiser's Army*, pp.161–165.
64 Ferko–UTD, Box 58, Folder 1, "Ferko's Eberhardt notes."
65 Without a centralized command, each flying section would operate differently, creating confusion and inefficiency. For example, the quality and style of aerial reconnaissance reports was dramatically different amongst the various flying sections. During the Marne Campaign, an army command with four flying sections was likely to receive four unique reports; each focusing on different subjects, delivered by different means, and written in different styles. Eberhardt's proposal for a standardized doctrine and system of liaison would have undoubtedly improved matters and would have greatly assisted older army/corps commanders with understanding their aviators' reports.
66 Cuneo, *The Air Weapon 1914–1916*, p.146.
67 Afflerbach, *Falkenhayn*, pp.147–171, pp.179–184.
68 After the meetings at OHL, Eberhardt returned to Berlin to resume his duties as the Inspector of Flying Troops. For much of the remainder of 1914, Eberhardt sought to improve matters with the *Fliegertruppe* as best he could, working with the aircraft firms on production and pushing the army to honor their prewar exemption policy.

5
The First Battle of Ypres
October–November 1914

"The British are very tough here. Our troops have to fight for every meter of earth with his fists. The German artillery shoots poorly and there are always appalling night fights. However, it appears that our side has made some gains of late. Some English formations have withdrawn this morning. In general, they do not deviate from their mission and remain in their trenches to die."

Josef Suwelack[1]

Pilot—FFA 24

From the moment he first ascended to command, Falkenhayn had sought to achieve a decisive victory by striking the allies' vulnerable left flank. First, he consented to Bülow's proposed offensive using the First, Second, and Seventh Armies along the Aisne. Next, he attempted to use the recently redeployed Sixth Army to envelop and roll-up the French left flank around the town of Albert. Deteriorating circumstances on his own northern flank ultimately forced the new German General Staff Chief to send Sixth Army's divisions into battle piecemeal rather than *en masse* as desired. Consequently, the two French armies of the Allied left wing were able to hold their ground and barely avert disaster after 12 days of intense combat.

Although the redeployment of Sixth and Seventh Armies had been unsuccessful in crushing the allied left, they had at least fulfilled their original purpose of securing the uncovered German right wing in the aftermath of the retreat from the Marne. Their arrival had successfully stopped Joffre's attempts to strike the German right flank—enabling the Kaiser's forces to maintain the initiative throughout the latter stages of 'The Race to the Sea.' Thus, on October 6th, with his flank secure and the initiative still firmly in his possession, Falkenhayn sought to resume offensive operations against the allied left. He recognized, however, that a renewed attack against Maud'huy or de Castelnau was unfeasible due to German shell shortages and the French army's construction of field works in the area. Launching a new offensive further north was therefore the only option.

Falkenhayn recognized the possibility that the war in the west might continue indefinitely. He therefore believed in the necessity of extending the German line to the coast and capturing the strategically important Channel ports as evidenced in the following passage from his memoirs:

It still seemed possible, if the present German front held, to bring the northern coast of France, and therefore the control of the English Channel, into German hands. It was all the more inadvisable to abandon this possibility, since the Chief of the General Staff clung to the object which was the root of the original plan in the West; at any rate to keep the forces employed in the East to a minimum, so long as the front in the West was not securely established. There is no need to prove that this condition had not [yet] been realized...[2]

In addition to gaining control of the ports, Falkenhayn saw an offensive in the north, specifically in Flanders, as his last opportunity to deliver a decisive blow to the allied left.[3] German military doctrine since the time of Scharnhorst and Gneisenau had stressed the value of envelopment. Under Schlieffen, flank attacks and encirclements were taught as the best possible method of attack to quickly and thoroughly destroy an enemy.[4] Falkenhayn embraced this philosophy and believed that the only chance to end the war was to envelop and roll up the uncovered allied battle line from the north, thereby creating the conditions for a decisive victory. Thus, with that goal in mind, the new General Staff Chief and the entire OHL staff immediately began planning the offensive they hoped would end the war in the West.

Antwerp

Before proceeding with an attack in Flanders, Falkenhayn first had to destroy a major pocket of resistance at Antwerp. Indeed, after two months of

Above: Hans von Beseler, commander of III Reserve Korps. (Author's Collection)

active campaigning, Fortress Antwerp (known as Belgium's National Redoubt) had remained a thorn in the German's side. The Belgian Field Army, under the command of King Albert, was still operating within the fortress' defenses, constituting a major threat to the German right wing's line of communications. Indeed, during the Marne Campaign Albert's forces launched two sorties from the city that gravely threatened the German supply lines— successfully drawing large numbers of German troops away from the crucially important battles around Paris. By late September the number of troops covering the fortress numbered 125,000 officers and men.[5] With a major offensive in Flanders pending, Falkenhayn could ill afford to waste employing large numbers of troops in a mere covering operation. Moreover, the fragile state of the German supply lines in Belgium mandated that all major threats in the rear-areas be neutralized. Thus, Antwerp had to be reduced before any operation in Flanders could commence.

Forcing Antwerp's capitulation would be no simple task. With 35 forts and 12 redoubts spaced in regular intervals around a 59-mile perimeter, Antwerp was considered by many to be an impregnable defensive position. The city's defenses consisted of moats, ditches, open air batteries, and concrete forts equipped with steel revolving turrets housing 15cm, 12cm, or 7.5cm guns. Defensive dykes had also been established to create flood zones in the outlying portions of the perimeter.[6] Altogether, these were formidable obstacles for any potential attacker. Fortunately for Falkenhayn, he had a trump card—the German heavy siege artillery.

Back on August 20th, Hans von Beseler's III RK had been detached from First Army to observe and contain the Belgian Field Army inside Antwerp. Two days later, Boehn's IX RK was ordered to relocate from Schleswig-Holstein to Antwerp to assist III RK with the operation. Together, these two corps held the Belgian forces in check throughout the final four weeks of the Marne Campaign. By mid-September it had become clear that the fortress needed to be reduced, releasing the covering force for service at the front. To that end Falkenhayn placed the German Western Armies' entire complement of siege artillery under Beseler's command. On September 28th, Falkenhayn ordered the reinforced III RK, consisting of five divisions supported by 173 heavy guns, to initiate a siege of the city.[7] Fortunately for the Germans, Antwerp's outer defenses were not yet fully completed. Many of the trenches were shallow and poorly placed, giving the Belgian defenders minimal cover. More importantly, several of the forts' concrete defenses were not yet finished—forcing the garrisons to use sand bags instead.[8] Beseler was informed of the fortress' shortcomings while planning the attack with his staff. Thus, after some discussion, it was agreed to destroy a sector of the enemy's vulnerable defenses with the heavy guns, then send an infantry attack into the breach to secure a foothold within the defensive perimeter. Once the infantry was installed inside the enemy defenses, a second heavy artillery bombardment would be ordered to assist the attacking forces widen the gap before ultimately taking the city.

Beseler's heavy batteries began their initial bombardment on September 28th. Playing a major role in the operation were the heavy siege batteries, which collectively consisted of four massive Austrian 30.5cm Skodas and four giant 42cm Krupp 'Big Bertha' Howitzers. Also participating were 40 10cm and 13cm guns, 72 15cm howitzers, 48 21cm mortars and five 30cm coastal mortars.[9] Firing at the rate of 10 shells per minute, Beseler's heavy ordnance quickly destroyed key parts of Antwerp's outer defenses, driving the

Above: German infantry storm the city of Antwerp. (Author's Collection)

Belgian Army's 3rd and 6th Infantry Divisions from their trenches during just the second day of the battle. On October 1st the German bombardment intensified, destroying several key forts and redoubts. Following each bombardment, German infantry would move forward and capture important intervening ground between the forts, allowing Beseler's heavy guns to relocate closer to the city. Over the course of the next few days, the German heavy artillery systematically tightened the noose around the city by pummeling the outlying forts and redoubts to rubble.

Seeing the bombardment first hand, King Albert recognized the futility of keeping his army within the fortress to await certain annihilation.[10] Thus, the Belgian Field Army was ordered to prepare for evacuation to the west to join with the allied armies' left wing. Meanwhile, Antwerp's garrison of 80,000 (with the assistance of 10,000 British naval troops) was ordered to defend the city and delay the Germans as long as possible.[11] Starting on October 2nd, the evacuation of the Belgian Army was conducted by rail through Ghent to Ostend. By the 7th, Albert's army was successfully installed in new positions on the Allied left flank in the area of Dixmude—Ostend. Two days later Antwerp officially capitulated.

Planning the Year's Final Offensive

With Antwerp's fall and his line of communications finally secured, Falkenhayn turned his attention to Flanders and striking the allied left wing. The German commander recognized that this was his final opportunity to deal a decisive blow to the western allied armies before positional warfare was established across the entirety of the Western Front. He therefore considered it imperative to attack the vulnerable allied left as quickly and forcefully as he could. To that end, Falkenhayn initiated a significant reorganization of the Western Armies—creating a new Fourth Army comprised of Beseler's III RK as well as four newly formed reserve corps to deliver the attack from the extreme right wing of the German line.

More than a million volunteers had entered German barracks at the outbreak of the war to offer their services in the defense of the Fatherland. On August 16th, Falkenhayn (then serving as Prussian War Minister) ordered these volunteers to be quickly trained and formed into five reserve corps to be numbered XXII–XXVI.[12] Elsewhere, volunteers from Bavaria formed the 6th Bavarian RD while Saxony and Württemberg combined their numbers to create the XXVII RK. Roughly 75% of these new formations consisted of volunteers, the remainder being reservists, *Landwehr*, and *Landstrum* who had been called back to the colors in September.[13] Although these troops were inexperienced, Falkenhayn had faith that, if properly led and supported, they could be successful in accomplishing the tasks assigned to them in his October 9th order to Fourth Army HQ:

Fourth Army… will comprise XXII, XXIII, XXVI and XXVII Reserve Korps. Once Antwerp has fallen, the III RK and 4th Ersatz Division will also come under its command. The leading elements of the XXII, XXIII, XXVI and XXVII RKs will begin arriving west and southwest of Brussels throughout the morning of October 13th and, with detachments pushed forward within each Korps' sector, are to assemble along the line Lokeren—Lessines.

The Army is tasked, as soon as the Korps are ready to move, to advance north of Lille, clear the enemy out of Belgian territory, then pivot south and continue the advance with its right flank west of St. Omer. The

Commander of Fourth Army, HRH Duke Albrecht of Württemberg, together with his staff officers, is to proceed to the army's new headquarters in Brussels via Mezières to report to His Majesty the Kaiser and to be briefed by the staff of Supreme Army Headquarters.[14]

As instructed, Duke Albrecht and the Fourth Army HQ staff met with Falkenhayn in Mezières to discuss the forthcoming operation. During the course of the meeting, Albrecht was instructed to "advance without regard of casualties, with the army's right wing on the coast." Elaborating further, Falkenhayn said that the "fortified ports of Dunkirk and Calais are to be isolated—then captured later—and then, leaving St. Omer to the left, to swing down towards the south."[15] These instructions clearly illustrated Falkenhayn's intention to aggressively use the newly created Fourth Army to strike and roll up the Allied left flank in a final attempt to achieve a major victory before 'maneuver warfare' ended in the West.

However, Albrecht's Fourth Army would not be available for operations until October 17th. In the meantime, Falkenhayn instructed all available forces in the area (namely the XIII and XIX AKs on Sixth Army's right wing as well as the cavalry corps) to engage the Allied left in order to prevent their forces from reorganizing and establishing a fortified position. It was hoped that these attacks would fix the French forces in place long enough for Fourth Army to arrive and deliver the main attack. Unfortunately for the Germans, Falkenhayn was quickly forced to abandon his initial plan for Sixth Army after a series of aerial reconnaissance reports alerted him to the arrival of large allied formations in Flanders opposite the German right flank—a clear indication that the Allies were planning to soon launch an offensive of their own.

Aerial reconnaissance within Fourth Army's sector was initially handled by III RK's FFA 38, under the command of *Hptm.* Bruno Volkmann. Officially formed on August 31st from the men and machines of Fest.FA Cologne as well as a few newly trained replacements from the interior, FFA 38 initially operated ten Taube monoplanes from an airfield just outside Brussels. On the morning of October 11th, FFA 38 was ordered to conduct several sorties along the front Ostend—Ypres to locate the Belgian Army and determine whether the III RK could be safely pushed forward ahead of the rest of Fourth Army (who was not yet ready to begin their advance to the front). Five aircraft, each assigned its own narrow sector, were ordered airborne during the early morning hours. All five of the subsequent reports reached a consensus: the Belgians were in the midst of a rapid withdrawal by rail and were showing no signs of preparing to halt. Only a small token force of one or two artillery battalions with some supporting infantry had been spotted covering the withdrawal.[16] The bulk of the remaining Belgian forces were observed on the rail lines moving through the town of Ghent to the west or southwest; particularly along the route Ghent—Bruges—Ostend as indicated by the following report delivered to III RK HQ around midday:

Three trains were observed in Ghent at 8am under full steam. Four trains were seen moving west between 8:05 and 9:45am on the line through Bruges; two additional trains were observed [on the same line] moving east. In Ostend at 8:50am there were no fewer than eight trains under steam, two additional trains also noted moving back to the east. Heavy traffic observed on the double tracked main transport line from Selzaete through Bassevelde and Ecloo then through Bruges to Ghent. Trains were seen on both tracks moving southwest. All roads radiating from Ghent in a westerly and southwesterly direction are free of enemy columns.[17]

FFA 38's reconnaissance confirmed the Belgian's post-Antwerp strategy. By detailing the high level of rail activity around Ghent while also reporting the area's roads mostly free of traffic, Beseler's airmen had discovered the Belgian leadership's intention to immediately link with the French main body's left wing—resulting in a large portion of Belgian territory being voluntarily surrendered in the process. As a result of these messages, III RK was quickly pushed forward to seize the uncontested ground.

In the meantime, another aviation report arrived which generated some controversy. Delivered at 10:30am to the III RK HQ staff, the airmen's message described considerable naval activity at Ostend. Indeed, ten large ships had been observed in the harbor at 9am, suggesting disembarkations had recently occurred. It remained unknown, however, whether the allies had landed new forces or were using the ships to evacuate allied troops from Antwerp. Not knowing which, Beseler resolved to send up more flights the following morning before reporting definitively to OHL.[18]

> **Three Main Rail Lines Used During the Belgian Withdrawal**
> 1: Rail line Ghent—Aeltre—Bruges—Ostend
> 2: Rail line Ghent—Eecloo—Bruges—Blankenerghe
> 3: Rail line Ghent—Deynze—Lichtervelde—Dixmude
>
> (The interconnecting line Bruges—Torhout was also used)

The aviators of FFA 38 were joined by Fourth Army HQ's FFA 6 on the morning of the 12th in conducting aerial reconnaissance over the western Belgium. FFA 6's six various B-Types had been flown from their previous airfield at Vouziers to Brussels on the 11th while the section's mechanics, trucks and equipment redeployed by rail with the other units attached to Fourth Army HQ. Despite the foggy weather, which had forced several early flights to be aborted, the afternoon's sorties proved to be extremely effective. The continued rail withdrawals of Belgian troops through Ghent's three main rail lines were continually spotted throughout the day. Indeed, long trains moving west in quarter hour intervals plainly illustrated the impressive scale and desperation of the Belgian retreat.

III RK HQ
At 9:00am large quantities of rolling stock were observed at Brugge [Bruges] and Ghent as well as around Deynze and Lichtervelde. Lighter traffic also seen at the smaller intermediate stations. Between 10:10am and 10:55am, four trains were seen moving in rapid succession along the rail line Ghent—Deynze—Dixmude. At 1:30pm, several transport trains moving west and southwest at full speed through Ghent and Brugge were observed using the leading rail line as well as the line through Thourout [Torhout] and Cortemarck.[19]

At this time, the creation of the new Fourth Army was still unknown to the Entente powers. To help keep it that way, III RK's airmen were instructed to identify the Belgian positions between Bruges and the coast. With the Belgians' whereabouts known, the westward march and deployment of the III RK could be smartly conducted in order to conceal Fourth Army's arrival in Flanders. To this end, several flights were performed to locate the Belgian troops that had been seen at Ostend the previous day. At 2:30pm, while over Dixmude, an airman of FFA 38 found the Belgian's main body: "several long columns observed marching west through the village [Dixmude] towards Dunkirk." Another report delivered shortly thereafter echoed the sentiments of the first: "all observed enemy columns moving either west or southwest."[20]

With the confirmation of the Belgians' continued withdrawal in hand, Beseler ordered his three divisions to immediately advance along a broad front—placing his right flank at Torhout and the left at Courtrai in order to "take possession of the coast and to throw the enemy back against the sea."[21] Fourth Army HQ, having reviewed the day's aviation reports, approved Beseler's decision—notifying Falkenhayn at Mezières via the following communiqué:

Multiple reports indicate the withdrawal of strong enemy forces from Ghent and Bruges towards Ostend and the southwest. Corps Beseler has been ordered to advance to Courtrai on a broad front to screen the arrival of Fourth Army.[22]

The situation at Ostend, however, remained unclear. While flying over the area at 10:10am, an observer from FFA 6 noted "five cruisers, two transport steamers, and between 16 and 20 large, apparently empty, vessels in the harbor." The airman also reported newly erected troop bivouacs southeast of the city.[23] Another flight, performed later that afternoon by FFA 38, found two battalions of infantry and a large number of transport vehicles near the exit of the city.[24] These reports collectively indicated either the landing of new troops or an evacuation of the allied forces from the coast. III RK HQ leaned toward the latter view, reporting to Fourth Army HQ at 7pm that: "Three cruisers and other observed naval activity indicate enemy troop redeployments from Ostend." However, after reviewing the aviation reports for themselves, the Fourth Army HQ staff came to the opposite conclusion—believing that significant landings had just occurred.

In truth, it was not the landing of new troops nor was it a naval evacuation that had been witnessed at Ostend. The German aviators had instead found empty vessels that had been used for earlier landings. At the beginning of October, the English IV Corps, consisting of the 7th Infantry Division and 3rd Cavalry Division, had been sent to Flanders via Ostend to assist the Belgian troops at Antwerp. The French Marine Brigade had also been dispatched through the same port on the evening of October 8th for the same purpose. Together, these forces remained in the area of Bruges through the 11th to cover the Belgian army's retreat. By the 12th, the Franco-British formations were in retreat along the line Torhout—Roulers, well to the south of Ostend. At the same time, the retreating Belgian field army was passing through Ostend en route to their new positions at Nieuport.[25] Thus, it was exhausted Belgians, not newly landed British, that had been spotted by German airmen near the city on the 12th. Falkenhayn nonetheless, interpreted the aerial reports to believe fresh troops had landed on the continent as part of an impending allied offensive directed against the German right wing.

In the south, Sixth Army's aviators (acting in concert with the cavalry corps) found definitive evidence of a major allied troop redeployment in Flanders. It should be remembered that Sixth Army's northern wing had spent the first week of October battling Maud'huy's Tenth Army around Arras in an attempt to turn the French flank. Foch's arrival, however, ultimately stiffened the French troops' resolve, compelling Falkenhayn to break off the offensive until the arrival of the new Fourth Army in Flanders. In the days that followed, OHL ordered HKKs 1, 2, and 4 to swing around the vulnerable allied left flank to attack their rear. Although these formations were ultimately unable to decisively strike the allied line, they were successful in screening the army's movements as well as locating signs of newly arrived enemy forces. Intrigued by their cavalry's reports, Sixth Army HQ ordered their flyers to perform a large-scale search of the area on the 11th in order to gauge the scale of the allied redeployment.

> **Specified Boundaries for Sixth Army's Flying Sections—October 11 and 12**
>
> XIX AK—FFA 24
> On the line Seclin—Armentières—Belgian Border (Specifically: La Bassée—Merville—Hazebrouck
>
> XIV AK—FFA 20
> Up to and including the line Lens—Noeux-les-Mines—Marles—Ferfay—Coyecque (9 miles south of St. Omer)
>
> VII AK—FFA 18
> Up to line Neuville-St. Vaast—Acq
> Including railline from St. Pol—Blangy
>
> IV AK—FFA 9
> Up to Roussent—Auxi-le-Château
>
> Guard Korps—FFA 1
> Up to line Serre—Condé Folie

Sixth Army's *Fliegertruppe* had quietly grown into an impressively large force by October 11th. In addition to Army HQ's own FFA 5, they had six flying sections attached to its various corps. As one of these six corps, the I Bavarian RK had recently become one of the first reserve formations to be equipped with its own flying section—the newly created FFA 4b formed from the men and machines of Fest.FA Germersheim. Lastly, HKK 4 was temporarily given four aircraft to assist the corps locate and attack the French rear. This temporary unit was unofficially dubbed 'Flying Section St. Amand' after the town where it was officially created. In all, the number of airworthy aircraft available within Sixth Army's sector totaled 48. Although only half this number was used on any given day, Sixth Army's *Fliegertruppe* quickly proved themselves up to the task of covering their army's zone of operations—which had a width of more than 50 miles and a depth of 30 miles into the allied rear (approximately up to the line Condé-Folie—Hesdin—St. Omer).[26]

Unfavorable weather disallowed all long-range reconnaissance on the 11th. Consequently, only tactical flights of negligible importance could be performed within the army's sector. The weather cleared during the night of the 11th–12th, however, allowing each of the flying sections to conduct flights over their previously assigned sectors. First, the Guard Korps' FFA 1 reported several large columns, estimated to be a division in strength, marching north towards Flanders.[27] Two more reports delivered around noon from FFAs 9 and 20 stated that there were numerous columns consisting mostly of vehicles on the roads from Le Boisle to Hesdin, from Brias to Dieval, from Villers-au-Bois to Hersin, and from Aix-Noulette to Béthune. These columns represented newly arrived British troops moving to the north and northeast to reinforce the Allied left. Later that afternoon, messages arrived from FFAs 4 and 24 confirming that strong forces were concentrating in the area of St. Omer and Hazebrouck on the Allied left wing. Moreover, an observer of FFA 20 found large numbers of troops flowing eastward from St. Omer through the major railroad nexus at Hazebrouck towards Ypres. The observer noted that, "all roads leading west out of Hazebrouck were covered with eastward moving columns." Finally, around dusk, 'Flying Section St. Amand' delivered a message describing a "mixed infantry brigade with very strong cavalry at Hazebrouck." Together, these reports strongly suggested that the Allied forces were now redeploying a large number of their forces toward the channel coast in order to achieve a decision against the German right.

Rupprecht's airmen also noted substantial rail traffic between the coast and Hazebrouck where more reinforcements were suspected of arriving. The accumulation of locomotives and rolling stock throughout this area appeared to confirm that the Allies were attempting to quickly move large numbers of troops to Flanders to extend and strengthen their left wing in preparation for an offensive. Elsewhere, throughout western Belgium, airmen from several flying sections reported heavy traffic on the tracks at Iseghem, Deynze, and Oudenaarde from the direction of Courtrai—proving just how hastily the forces from Antwerp were being withdrawn. Several lengthy trains were also observed moving north and south along the route Dunkirk—Hazebrouck at 3:30pm by 'Flying Section St. Amand.' Finally, perhaps most alarming, airmen from FFA 20 reported six long trains steaming along the line St. Omer—Calais, confirming that fresh British troops were being introduced into the theater.[28] Thus, all of the *Fliegertruppe* messages

Above: As commander of Sixth Army, Rupprecht played a major role in the Battle of Ypres. (Author's Collection)

received throughout the day suggested that the allies were reinforcing their left wing in preparation for an offensive action against the German right.

Late in the evening of the 12th, Sixth Army HQ forwarded several of the day's most convincing reconnaissance reports along with an intelligence summary of their own to OHL for Falkenhayn's consideration. Falkenhayn had originally planned for the XIX and XIII AKs, located on Sixth Army's right wing, to kick off the Flanders offensive by attacking and seizing the high ground on the line Lillers—Béthune before Fourth Army's deployment and assault would commence.[29] However, the October 12th *Fliegertruppe* reports caused Falkenhayn and the OHL staff to reluctantly abandon the idea. Falkenhayn believed that the naval activity at Ostend, combined with heavy rail traffic behind the allied lines at St. Omer, Hazebrouck, Dunkirk, and Calais, suggested that an allied offensive was fast approaching.[30] Thus, upon carefully reviewing his airmen's reconnaissance,

Falkenhayn concluded that both the XIX and XIII AKs would open themselves up to a potentially fatal flank attack if they were to indeed take the heights at Lillers—Béthune. Therefore, after reviewing his options, Falkenhayn decided to use the coming allied offensive to his own advantage. According to the new plan, the two corps were to immediately stop their westward advance and pivot to face north along the line Menin—Armentières—La Bassée (using the cavalry corps to screen their movements). From this position, both corps would be able to strike the attacking allied force's flank and rear as they advanced.

Sixth Army HQ was first notified of Falkenhayn's decision by an OHL directive sent on the 13th:

It is desirable that the two corps, XIX and XIII AKs, not be deployed towards the west. Instead, their front should face north to counter the enemy attack and then, in concert with III RK, attack the enemy.[31]

In order to better convey his concept of operations, Falkenhayn ordered Sixth Army HQ's Chief Operations Officer, Major von Mertz, to Mézières to discuss the matter in person.[32] After a lengthy interview, Mertz returned to Rupprecht's headquarters to give his report. Rupprecht's diary entry from the following day illustrated Sixth Army HQ's disproval of the plan:

Major von Mertz, who had been summoned to OHL at Mézières, brought us a mission that did not please me much. While III RK of Fourth Army is being dispatched in an offensive role to Brugge, the remaining corps of Fourth Army are to form the base of a sack into which, it is hoped, the enemy will thrust in an attempt to surround Sixth Army. To this end, Sixth Army is to act purely defensively in a line stretching from Menen [Menin] via Armentières and La Cassee to the right flank of Second Army. The task we have been given is analogous to that we fulfilled right at the start of the campaign and it is equally as questionable as it was then if the enemy will really enter the open sack. In acting this way, we are yielding the initiative completely to the enemy. They will be able either to attempt to breakthrough somewhere along the overextended length of Sixth Army's frontage, whose sole reserve is the 14 ID, probably near to its right flank, or—and this is most likely of all—will dig in, in a strong position approximately on the line Poperinghe—Bailleul—Béthune, there to await calmly the arrival of reinforcements and to extend their line to the sea. That will mean an end to the war of movement and it will be very difficult to force a decision in the foreseeable future. As before, I am in favor of cutting the Gordian knot; that is to say to launch an attack with the fresh troops of the 14

ID, XIX AK, and XIII AK on my right flank. I do not believe that there are overwhelmingly strong forces opposing our current northern flank.[33]

Although everything stated in Rupprecht's diary entry was factual, the author's words proved how blind to the overall strategic situation he truly was. It should be remembered that upon the outbreak of war, German strategy was to achieve a decisive victory in the West in order to rapidly end the war on that front and subsequently shift the entirety of their forces to the East. From the moment he first ascended to command, Falkenhayn understood and followed this strategy wholeheartedly, believing it to be Germany's only logical path to an honorable peace. By mid-October, Falkenhayn was aware that Germany could no longer afford winning 'ordinary victories' such as Charleroi, Mons, Le Cateau, and Guise–St. Quentin. The young German commander therefore felt justified taking great risks in order to achieve the type of victory he knew that the circumstances required. Thus, despite the likelihood of a minor success if Sixth Army's right wing were permitted to attack, Falkenhayn decided instead to pull the XIX and XIII AKs back in the hopes of luring the allied northern armies into becoming destroyed in a decisive victory.

Aerial reconnaissance flights conducted on the 13th further clarified the Allies' latest movements and intentions in Flanders, causing Falkenhayn to make several revisions to his new plan. In Fourth Army's sector, airmen from III RK's FFA 38 confirmed that the Allied retreat from Antwerp was still continuing westward. Heavy rail traffic headed from Ghent and Bruges towards Ostend was noted throughout the day. "Substantial infantry formations" were also observed moving westbound on the nearby roads toward Bruges and Ostend. Meanwhile, all vessels inside the channel ports appeared to either be empty or departing at full steam.[34] Thus, upon reviewing these messages, Beseler promptly abandoned his previous plan of advancing in the direction of Courtrai. Instead, III RK's three divisions were sent to the area Thielt—Bruges with the objective to harass and disrupt the allied plans of fortifying the area. Falkenhayn approved Beseler's decision after reading FFA 38's reports for himself later that evening—ordering III RK to push ahead of Fourth Army in order "to begin forming the sack along with the withdrawn flank of Sixth Army."[35]

Sixth Army's airmen were also able to produce several key reports on the 13th despite experiencing persistent low-lying clouds and intermittent precipitation throughout the day. Indeed, all rail lines leading out of Ghent were reported to still be heavily congested with traffic. Between 10:00 and 10:20am, an observer from FFA 4 reported large quantities of rolling

Above: As General Staff Chief, Falkenhayn boldly decided to commit his fresh reserves to Flanders in search of a decisive victory. (Author's Collection)

stock and empty trains clogging traffic along the rail line from Courtrai to Roulers. The same report also stated that, "all traffic between Izegem and Roulers was completely blocked due to seven accumulated trains that were stopped one behind the other."[36] Yet another patrol performed by FFA 4 over the rail line Furnes—Dixmude discovered a long transport train pulling into the Dixmude station at 10:00am. This flight also found "lively" railroad and naval traffic at Nieuport, proving the Allied leadership's intent to strengthen their extreme left using forces that were previously deployed to the south.

While several successful sorties were performed on the 13th, the dismal weather proved to once again significantly affect the *Fliegertruppe's* work. Thick cloud cover between Roulers and Dixmude continually restricted Sixth Army's airmen from inspecting the area despite several failed attempts. Most importantly, the cloud cover concealed several large allied columns from the German aviators that patrolled the area.[37]

Citing his airmen's reconnaissance, Falkenhayn was able to successfully determine the allied strategy in Flanders. The latest reports all suggested that the allied left had been considerably strengthened. Large columns of troops and heavy rail traffic observed in the area Ypres—Dixmude—Nieuport, as well as increased activity in the Franco-British rear-areas, proved to Falkenhayn that the allies had successfully redeployed new formations to their left wing before Fourth Army's flank attack could commence. This meant that Fourth Army's initial thrust would now have to be a frontal assault. Falkenhayn recognized this and immediately forwarded an updated directive to Albrecht at Fourth Army HQ.

Based on all reports to date, it must be assumed that the enemy intended to assemble strong forces southeast of Dunkirk with the intention of enveloping our right wing. Whether this is still the aim is doubtful. In any event, direction has been given that Sixth Army is to withdraw its right flank behind the Lys, roughly along the line Menen—Armentières—La Bassée, and to go into defense. If the enemy threatens it, it will be the task of Fourth Army, once the fresh forces are completely deployed, to thrust against the flank and rear of the enemy. To this end, by the 17th of this month, the new corps are not to be located along the verbally directed line Eekloo—Anzegem. Instead, after III RK has today moved to Bruges, the line Ursel—Anzegem is to be occupied [by the same date]. The maximum possible number of advance guards are to be in place and ready to move. It is then planned to launch the army, with Corps Beseler [III RK] well to the fore and moving offensively along the coast, in an advance towards Merville, with its left flank moving via Menen.

Corps Beseler, complete with its allocated 4th Ersatz Division and HKK 4 (which is currently operating in conjunction with Sixth Army southeast of Bailleul and which, therefore, will temporarily remain subject to its instructions) will be subordinated to the [Fourth] Army from this evening. Sixth Army HQ has been directed, in the event that HKK 4 comes under enemy pressure, to withdraw in the General direction of Courtrai. In the most probable situation; viz. that the enemy renounces their original intentions as soon as they detect the arrival of the fresh troops, it is improbable that there will have to be any change in the above mentioned direction of advance. Regardless of whether the enemy choose to stand on the line Dunkirk—Poperinghe—Hills east of Bailleul or on the heights southwest of the line Ardres—St. Omer—Merville, the Army will be obliged to attack them frontally; whilst it would remain the task of Sixth Army, after a breakthrough from the general direction of Arras, to operate against the flank and rear of the enemy.

In this event, I should like to draw the attention of Your Royal Highness to the requirement to make the headquarters of the new corps aware of the need to exploit the heavy artillery effectively. There is also a pressing need, because of the necessity to make best use of the strictly limited ammunition, to lay down in advance exactly where the fire is to be placed, so as to ensure that it has a decisive effect. Finally, I wish the marching performance demanded of the fresh troops to match their actual capabilities.[38]

Although Falkenhayn believed an allied envelopment of his right flank was highly improbable, he had to consider the possibility of a breakthrough occurring before Albrecht's new Fourth Army could be fully deployed. Thus, he considered it imperative that the German forces already in the theater would engage the allies and deny them freedom of action. To that end, Falkenhayn ordered Sixth Army HQ (which was still advocating an immediate all-out offensive) to push forward their right wing beyond the Lys River. Specifically, the XIX AK was directed to move north of La Bassée and proceed in the direction of Bethune while the XIII AK was to advance via Lille to the line Menin—Comines—Warneton—east of Armentières.[39] These movements were to be screened by the cavalry corps operating on Sixth Army's right flank, who were already engaged with strong Franco-British forces found to be *advancing* in the direction of Lille. Thus, Falkenhayn's concern of a preemptive allied offensive proved to be entirely justified. Fortunately for the Germans, however, Sixth Army's redeployment was ordered and completed in time to sufficiently counter the threat.

The Entente's Plan in Flanders: A Converging Attack Around Lille

On September 29th Sir John French sent a letter to Joffre requesting the BEF's three corps be relocated from their existing positions on the Aisne to a location near the Channel coast where they would be closer to their bases of supply. Joffre ultimately consented to the request under the condition that the move was made in increments, thus allowing him to slowly plug the gap left by the departing British without overburdening France's rail lines. The BEF's II and III Corps were the first to be redeployed, detraining in the area of Doullens—Arras—St. Pol from October 8th–13th under the protection of Maud'huy's Tenth

Above: German troops storm the citadel at Lille. (Author's Collection)

Army. From there, the British troops were to march northeast and take up positions on Maud'huy's left.

While traveling north to their new headquarters, Sir John and his staff met with Foch to try and establish a unified strategy for the coming phase of operations. After some discussion, it was agreed to launch a combined Franco-British offensive to save the important city of Lille, which at the time was still held by a small French garrison.[40] Foch's proposal called for the newly arriving British forces to advance north of the city while Maud'huy's left wing pushed around to the south. Next, the two groups were to rejoin in the area between Courtrai and Tournai, where the Lys and Scheldt Rivers would act as protective barriers against a German counter-attack.[41] The attack would then resume against the German flank and rear once adequate reinforcements were brought up from south.

Foch's plan was entirely dependent upon speed. Coming from the Aisne, the British II and III Corps were required to quickly detrain, take up positions on the French left, and begin the assault before the Germans could bring up reinforcements. By the morning of the 12th both of the British corps had arrived and were ready to advance. Unfortunately for the allies, their movements were preempted by the German capture of Lille the night before.[42] Thus, German Sixth Army's right wing was now extended further north in positions capable of resisting an attack such as the one the allies were planning. Foch nonetheless remained optimistic, writing to Joffre later that evening stating that, "the situation is considered to be good."[43] The French commander's confidence was based on the false assumption that the German troops in the area were comprised of third-rate *Landwehr* who could easily be defeated. Consequently, he maintained faith in his plan and ordered the offensive to proceed.

The Allies' offensive formally began on the morning of the 13th. The initial objectives for the advance were the towns of Bailleul—Armentières in the north and Givenchy—La Bassée in the south. Participating in the attack from north to south were the British III and II Corps and the French XXI CA.[44] Opposing them was a thick defensive screen composed of three German cavalry corps.[45] Fighting dismounted with the assistance of some advance units from Sixth Army's right wing, the German cavalrymen conducted a fierce delaying action in order to slow the allied advance until Falkenhayn's newly formed Fourth Army could arrive.[46] In the south, the British II Corps quickly became bogged down in heavy fighting, sustaining heavy casualties while making no gains. Meanwhile to the north, the BEF's III Corps took full advantage of their superiority in field artillery to launch an attack along a five mile front resulting in the successful capture of the village of Meteren (west of Bailleul).[47]

The battles of the 13th convinced the German leadership that the Allies were launching a major offensive in Flanders. Thus, with Fourth Army's arrival still days away, Falkenhayn decided to strengthen his northern flank in order to avert a potentially disastrous allied breakthrough. OHL therefore ordered Sixth Army's right wing to push forward to the aforementioned line: Menin—Comines—Warneton—east of Armentières while the three badly exhausted cavalry corps, with no room left for maneuver, were withdrawn south of Lille and transformed into an army reserve. From this position, Rupprecht's right wing would be able to delay the allied advance until Fourth Army's attack could be delivered.

Heavy fighting continued over the next two days as the Allies attempted in vain to push eastward through the stout German defenses. Nevertheless, despite their lethargic advance, the Allies managed to make gains in several locations, particularly in the north where a brief numerical superiority enabled the two Franco-British cavalry corps, as well as elements of the British III Corps, to capture Bailleul and the bridges over the Lys at Sailly, Erquinghem, and Bac St. Maur. By nightfall of the 15th, the Allied armies had reached the line: Pont Fixe—Estaires—Pont de Nieppe—west of Commines—Ypres—Dixmude (with the Belgian Army extending the line to the sea).[48]

Encouraged by the latest reports of German withdrawals throughout his sector and believing there to only be Beseler's III RK moving against Ypres, Sir John French ordered a general advance to immediately resume on the morning of the 16th. To support this endeavor, Foch ordered the Belgian and French forces deployed north of the BEF to advance as well, covering the British left.[49] Moreover, with the BEF's IV Corps now participating in the offensive, the Allies fully expected to sweep away all German resistance in Flanders—then turn and finally strike the German army's right flank and rear.

Little progress was ultimately made on the 16th due to the British corps commanders' decision to first consolidate their previous gains before continuing the advance. Nonetheless, by the end of the day the British II, III, and IV Corps had all established contact with one another's flanks and were ready to resume the offensive the next day. Meanwhile, the three German Cavalry Corps were finally being withdrawn from the front line and replaced by Sixth Army's right wing corps. Thus, with reorganizations occurring on both sides, the 16th saw little changes to the front lines.

The Allies renewed their offensive across the entire Flanders front on the morning of the 17th. In the south, Smith-Dorrien's II Corps, having just recaptured Givenchy the day before, established itself on Aubers Ridge after a late bayonet charge finally drove the Germans back. Meanwhile to the north, Sir William Pulteney's III Corps hastily captured the important town of Armentières before Rupprecht's infantry could arrive. Owing to their location on Pulteney's left, Sir Henry Rawlinson's IV Corps remained in their positions east of Ypres while waiting for the II and III Corps' advance to conclude. Near the coast, the Franco-Belgian troops were pinned down by local counter-attacks performed by Beseler's III RK, who was still screening the arrival of Fourth Army's new reserve corps.

Sir John French ordered the offensive to resume at 6:30am the following morning, insisting that all attacks should be delivered "vigorously against the enemy wherever met."[50] The latest intelligence had suggested that the Germans were entrenched on a formidable series of heights near Pérenchies, four miles east of Armentières. Sir John nevertheless remained confident in his men's capability of dislodging the Germans from the impressive position and directed the III Corps to attack accordingly. The corps' two divisions were divided for the attack in order to gain passage across the Lys River in multiple locations. The left division (4th), under heavy fire, was unable to make any noteworthy progress and ended the day well short of their initial objective. The 6th Division on the right, however, smartly employed both of their artillery brigades to enable the division's infantry to secure a foothold on the Pérenchies Ridge at multiple points.[51] Meanwhile, the IV Corps cautiously advanced without incident in the direction of Menin with the intention of delivering an attack the following day. In the south, the exhausted men of II Corps struggled to make minor gains in their drive to capture La Bassée. This incidentally later became known as the "high water mark" of the offensive, as the BEF's II Corps was closer to the city on the 18th than any British troops would be over the next four years.[52] The following day the BEF made contact with leading elements of German Fourth Army as it made its advance, marking the end of Sir John's offensive and the start of the First Battle of Ypres.

Fliegertruppe Operations During the Allied Advance

After their successful performance on the 13th, the right wing armies' flying sections were each directed to immediately increase the number of flights performed each day. The German leadership had issued the directive in order to first obtain more information before making a final decision regarding the implementation of Fourth Army's coming offensive. Unfortunately, the onset of Belgium's rainy season on the 14th greatly restricted aerial activity as the Allied advance began to gain momentum. Rain, low lying clouds, and dense fog from the sea prevented important long-range flights from being carried throughout the day. Several brief local flights were successfully performed however, in Fourth Army's sector when circumstances permitted. In the north, III RK's FFA 38 conducted three of these flights along the line Zeebrugge—Dunkirk—Lille. Although one of the aircraft was shot down by British ground fire, the two others returned with excellent reports.[53] Flying amidst stormy weather, the airmen found the town of Ostend to be "abandoned and clear of enemy troops." Next, in stark contrast to previous reports, there was no traffic observed on the rail lines between Ghent and the Dutch border. Thus, the Allies were confirmed to be moving all their forces to the west and southwest *away* from Ostend.

FFA 38's reports were promptly sent to III RK HQ where the corps staff inexplicably dismissed them as false. Beseler believed the retreat of the Belgian Army from Ostend and the subsequent evacuation of the city to be highly unlikely. Instead, he assumed the Belgians would choose to make a stand there in an attempt to further delay the German advance until allied reinforcements could arrive. As a result, Beseler and his staff were needlessly busy throughout the 14th preparing for the bombardment and reduction of the city despite already possessing reports from both the *Fliegertruppe* and German agents on the ground that concluded it was abandoned.[54]

III RK HQ messaged the Fourth Army HQ staff at 7pm informing them of their plans to bombard Ostend the following day. To avoid any chance of the operation being cancelled, III RK HQ chose *not* to forward FFA 38's two reports. Consequently, Albrecht believed there had been no flights performed that day and remained ignorant of the true situation in his army's sector.[55] Later that evening, Fourth Army HQ received a communiqué from OHL describing large concentrations of allied troops seen southeast of Dunkirk the day before by Sixth Army's airmen. Falkenhayn, believing there to also be strong troops in Ostend, did not want III RK to risk prematurely starting the battle before the allies advanced into his trap. Therefore, OHL's message ordered Beseler's troops to halt their advance and not exceed the line Ostend—Torhout. As it happened, the Belgian Army was at that moment already in full retreat back to the Yser River.[56] It will forever remain an open question whether Fourth Army HQ would have pushed III RK forward if they had possessed FFA 38's reports and knew Ostend to be evacuated. Nevertheless, if they had indeed chosen to do so then the 4th Ersatz Division would have been available to seize control of the harbors of Zeebrugge, Blankenberge, and Ostend while Beseler's remaining divisions secured a position up to the line Ostend—Roulers (all without risking a major engagement as Falkenhayn's plan required).[57] As we shall see, this would have greatly improved Fourth Army's circumstances once the battle to cross the Yser began several days later.

Persistent rainstorms and heavy fog on the 15th, once again restricted aerial operations across Flanders. However, with the beginning of Falkenhayn's offensive fast approaching, the German leadership remained anxious for additional intelligence from their airmen. In Sixth Army's sector, each flying section had been given specific instructions issued by Rupprecht's headquarters during the night of the 14th–15th establishing boundaries for long-range reconnaissance flights.

Order to All Sixth Army Aviators

Corps HQ Flying Sections' Assigned Boundaries:

XIII AK [FFA 4]: Up to and including: Warneton—Ypres—Furnes. The main railway from Ostend to Menin is to be observed.

XIX AK [FFA 24]: To the line Don—Estaires—Hazebrouck, Cassel—Dunkirk included.

XIV AK [FFA 20]: Up to and including Béthune—Aire—St. Omer, Calais excluded.

VII AK [FFA 18]: Up to and including the rail line Acq—St. Pol—Anvin—Rumilly, line Desvres—Boulogne excluded.

IV AK [FFA 9]: To the line Ransart—Auxi le Château, over Authie.

Guard Korps [FFA 1]: To Gerres—Condé-Folie, over the Somme.

Army HQ's FFA 5

FFA 5 is to reconnoiter up the sea, especially the north sector; Troop deployments and rail traffic are especially important to note.

The *Fliegertruppe's* instructions made it clear that Rupprecht's headquarters prioritized intelligence related to the allied rear-area and rail lines over tactical front line reports. Under Falkenhayn's plan, it was crucially important to determine whether the allies were advancing into "the sack" and if so, how they were deployed. With this information known, OHL could devise Fourth Army's counter-attack to maximum effect. Thus, only two flying sections—FFA 4b and HKK 4's 'Flying Section St. Amand'—were needed to conduct short-range flights on the 15th while the remaining seven were ordered deep behind allied lines.

A break in the weather enabled FFA 5 to opportunistically send an aircraft airborne in the direction of the coast. After a lengthy flight, the airmen returned to report that the area of Dunkirk was bustling with activity. Large trains carrying troops and significant quantities of matériel were observed on the railway moving south from the city, which was now reported to be heavily fortified on its eastern perimeter. Further east, newly arrived French troops were found moving into well-prepared positions alongside the Belgians in the dunes at Nieuport. Finally, while returning home, the observer noted "lively rail traffic" along the line Hazebrouck—Poperinghe—Ypres, signaling the arrival of more reinforcements behind the British left wing in the area of Ypres.[58]

Located on the extreme right flank of Sixth Army, XIII AK's FFA 4 had a very important assigned search area. The section was ultimately able to complete two successful flights during the afternoon, bringing back intelligence regarding how the British troops previously seen at St. Omer were to be employed. The first flight, conducted shortly after noon, discovered strong mixed arms formations marching from Cassel to Bailleul. Closer to the front, a large column consisting of one infantry brigade with artillery attached was seen around Ploegsteert supporting British forces near Armentières. Several smaller infantry columns were also noted marching in the direction of Armentières—confirming the allies' intention to quickly seize the city. The second flight, performed at 3:00pm, discovered several columns marching through Poperinghe, continuing to Langemarck—Zillebeke (southern suburb of Ypres) and Messines. Finally, HKK 4's 'Flying Section St. Amand' reported "strong forces, estimated to be about one division" at Poperinghe.[59]

Meanwhile, shortly before sunset, a crew from FFA 20 landed at Sixth Army HQ to inform Rupprecht of new activity deep in the Allied rear-area. According to this observer's report, heavy traffic had been seen between 3:30 and 4:00pm on the rail lines St. Omer—Hazebrouck—Béthune as well as the road St. Pol—Béthune. In addition to the unloading of matériel seen at the rail stations, the aviator noted "numerous newly struck tents and bivouacs—each between 8 and 10 of the round English types—in all localities in the area St. Omer—Aire—Béthune."[60] These forces belonged to the British I Corps' 2nd Division, who had left the Aisne sector on the 14th and were awaiting the remainder of the corps to arrive before they began their march to the front.[61]

Unlike other army headquarters throughout 1914, Sixth Army HQ maintained excellent liaison with their airmen. Indeed, Rupprecht and his staff became accustomed to consulting their aviators' reports every night before drawing up orders for the following day. On the evening of the 15th, Sixth Army HQ reviewed all available aviation reports delivered during the day. However, despite the quality of the reports, Rupprecht admittedly remained unable to definitively predict the Allies' intentions in his sector. Whether the British were aggressively advancing into Falkenhayn's trap in an attempt to envelop the Sixth Army's right wing or were merely avoiding an encirclement of their own forces by shifting troops toward Dunkirk remained uncertain.

Late in the afternoon of the following day (16th), Falkenhayn arrived at Sixth Army HQ in Douai to personally discuss the coming offensive.[62] Low clouds, strong winds, and heavy rain throughout Flanders had prevented almost all flights in Fourth and Sixth Army's sectors from being carried out that day. Only a single aircraft of 'Flying Section St. Amand' managed to bring any insight by reporting to Sixth Army HQ about the advance of strong British forces from Ypres to Menin, whose lead units were seen halted at Gheluvelt.[63] Thus, with little information on hand, Falkenhayn and Rupprecht agreed that the British intentions remained unclear.[64] Therefore, the two leaders decided to draft an updated plan for Sixth Army with three contingencies based upon the actions of the British.[65] First, *if* the British aggressively advanced against Rupprecht's right, Sixth Army was to launch a counter-attack and assist Fourth Army as it delivered the main thrust against the allies' flank and rear. In this instance, "the XIII, XIX, parts of VII, and perhaps the XIV AKs, as well as the army cavalry, were to contribute to Fourth Army's offensive against the British flank." Meanwhile, the left wing of Sixth Army was to attack and 'fix' the French units in their sectors to prevent them from being redeployed to the north to resist Fourth Army's offensive. *If*, however, the British were to fall back defensively towards Dunkirk, elements of Fourth Army would be used to pursue and act as a covering force while the remainder of the army would continue with the offensive to the southwest to roll up the allied line. Finally, *if* the Allies withdrew their forces to the strong defensive positions along the heights of St. Omer—Béthune, then Sixth Army was to immediately "attempt a decisive breakthrough to the west on high ground Aix-Noulette—Mont-St.-Éloi, center around Houdain."[66] This attack would have been delivered south of the Allied defensive line, creating the conditions (along with Fourth Army's attack) for the possible double envelopment and destruction of the Allied armies in Flanders.

Aerial activity in Sixth Army's sector was grounded throughout the 17th and 18th due to persistent rain and thick clouds, which pilots had experienced almost incessantly since the morning of the 14th. Sixth Army HQ nonetheless managed to obtain additional information regarding the British based upon their ground troops' forward reconnaissance. Reports of the BEF's vigorous push to take Aubers Ridge and Armentières as well as their advance in the direction of Menin suggested that the Allies intended to act aggressively and attack directly into Falkenhayn's trap. Therefore, in their final meeting before the battle was to commence, Falkenhayn informed Rupprecht to prepare Sixth Army to turn to the offensive and assist Fourth Army with delivering the "annihilating blow" against the allied flank rear.[67]

Fourth Army's Arrival

The four newly created reserve corps that comprised the bulk of Albrecht's Fourth Army began their journey to Flanders on October 10th, just two weeks after completing their training and receiving their final inspection. By the 13th, the detrainment of these formations was underway in the region west and southwest of Brussels.[68] Feeling the need to convey the true seriousness of the situation to his troops, Albrecht circulated the following order:

It is with pleasure that I have assumed the command of Fourth Army with which I have been entrusted. I am completely confident that the corps which have been assembled will bring about the final decision in this theater of war; that with good old German courage they will faithfully do their duty to their last breath and that each officer and each soldier is ready to shed the last drop of blood for the righteous and blessed cause of our Fatherland![69]

On the evening of the 14th all four of the reserve corps began their westward march to the line Ecloo—Deynze—area west of Audenarde.[70] Unfortunately, none of the new reserve corps were equipped with a flying section, thus depriving Fourth Army of badly needed intelligence. Indeed, with only half of a flying section's aircraft airworthy on any given day at this point of the war, Fourth Army's two sections had only six aircraft available each day to cover a 30+ mile front. In an attempt to improve the problem, both FFA 6 and 38 were placed under Fourth Army HQ's control on the 16th. Both of these flying sections were ordered to perform long-range flights to assist Albrecht prepare for the coming offensive.[71] In the meantime, Albrecht issued the following orders to each of his subordinate corps commanders:

The III RK is to march to the line Coxyde—Furnes—Oeren, west of the Yser.

The XXII RK to the line Aertrycke—Torhout.

The XXIII RK to the line Lichtervelde—Ardoyle.

The XXVI RK to the area Emelghem—Iseghem.

On the left wing, the XXVII RK to the line Lendelede—Courtrai.[72]

Poor weather grounded Albrecht's airmen on the 17th while the ground troops approached the front. In order to reduce the distance between themselves and the rest of the army, Beseler's III RK was directed to temporarily hold their positions on the Yser's east bank and not to exceed the line Nieuport—Stuyvekenskerke. Two flights performed the next day from FFA 38's forward airfield near Ostend successfully navigated their way around heavy clouds and fog to conduct reconnaissance along the Yser between Nieuport and Dixmude. Both reports illustrated in detail the strength of the allied defensive positions on the river's west bank. According to the observers' messages, allied troops were entrenched in a series of fortified positions running from Dixmude to the coast. Large groups of artillery as well as isolated batteries were also discovered behind the allied trenches. The main strength of the allies' forces was concentrated just north of Dixmude at Kaaskerke, while weaker forces held the area near Nieuport. This group was apparently supported by the presence of seven armed vessels operating nearby in the channel. Finally, the second aviation report delivered to III RK HQ at 1:40pm confirmed that there was "no westward movement beyond the Yser Canals." It was therefore concluded that the Belgian Army had stopped their retreat and had taken up defensive positions on the Yser in order to resist any further German advance along the coast.[73]

FFA 38's two reports were of significant importance. There was now no doubt whether the allies were going to defend the Yser line in strength, something Albrecht and Falkenhayn had both feared.[74] Beseler's decision on the 14th to disregard his airmen's messages regarding the Belgian withdrawal from Ostend, choosing instead to halt his forces for the bombardment and storming of the city, was now acknowledged as a mistake. If Beseler had trusted the reports and actively pursued the retreating Belgian forces, the allies would not have had the opportunity to freely construct the fortified positions along the Yser as they had. Instead, the III RK would have pushed forward to the Yser much sooner and would have harassed any attempt to construct field works.[75] As it happened, the Franco-Belgian troops in the area were now firmly installed in fortified positions well supported by artillery. Consequently, III RK would ultimately be compelled to make a dangerous frontal assault beginning on the 19th to dislodge the allied troops from their positions—allowing the rest of Fourth Army to deliver its attack as planned.[76]

By nightfall of the 18th all of Fourth Army had arrived in their assigned destinations along the line Nieuport—Torhout—Courtrai. Consulting all the available intelligence, Falkenhayn remained unsure whether the allies were advancing into his trap. Nonetheless, with no evidence suggesting an enemy withdrawal, Falkenhayn decided to proceed with the attack as planned. First, he recognized the need to obtain a decision on the two wings, in the north at Nieuport—Dixmude and in the south at La Bassée—Armentières, where the allies appeared to be strongest. Therefore, the formations in Flanders were ordered to continue their advance on the morning of the 19th in a "staggered deep arc, with both wings pushed sharply forward for the attack."[77] These two forward wings, particularly the northern component made up of Beseler's III RK, were expected to breakthrough the Allies' strongest positions, enabling the remainder of Fourth Army to deliver 'the annihilating blow' that Falkenhayn had envisioned.

The Battle's Initial Stages: October 19–23

The efforts of the German cavalry and the III RK to screen Fourth Army's deployment were ultimately successful. As a result, Sir John French had awakened on the morning of the 19th completely unaware of the fact that there were eight fresh German divisions positioned between Menin and Dixmude. The British commander assumed that his BEF was only opposed by determined pockets of cavalry supported by some badly damaged infantry formations and was therefore optimistic that his men's advance would successfully continue across the entire front.[78] Thus, the morning of the 19th began with the opposing sides' forces advancing headlong towards one another with the allies unaware of the forces opposite them.

Albrecht's Fourth Army was deployed from north to south as follows: III RK, XXII RK, XXIII RK, XXVI RK, and XXVII RK. In the north, Eugen von Falkenhayn's XXII RK had established contact with III RK's flank and was directed to push towards Dixmude.[79] Meanwhile, Georg von Kleist's XXIII RK spent the entire day heavily engaged in close combat. By late evening the 45 RD took the towns of Handzaeme and Gits while the 46 RD reached Staden. To the south, Otto von Hügel's XXVI RK seized the town of Roulers from the French after a lengthy afternoon of house-by-house city fighting. Finally, located on the army's extreme left, Adolph von Carlowitz' XXVII RK pushed back the British 3rd Cavalry Division to reach their objective west of Rolleghem—Cappelle.[80]

Fourth Army's attack had completely taken the Allies by surprise. On the BEF's left, Sir John had ordered Rawlinson's IV Corps to mount a vigorous attack that morning to seize Menin and begin placing pressure on Rupprecht's flank (which was believed to be uncovered). Rawlinson, having been wrongfully informed there were little or no German troops in the area of Courtrai on his left flank, ordered the advance to quickly proceed without a flank guard. Thus, at 6:30am, all three of 7th Division's brigades were deployed in the front line to assault the city. By 10:30am, Rawlinson received word from his own aviators describing strong German columns approaching Menin and Roulers. British and French cavalry were reporting strong German troops to the north making 'vigorous attacks' across a broad front. Thus, facing superior numbers and with his left flank severely threatened, Rawlinson immediately recalled the 7th Division, whose leading units had already become heavily engaged. Fortunately for the Allies, Rawlinson's troops and artillery were able to safely extract themselves and withdraw to safety along the line Zandvoorde—Zonnebeke, some three and a half miles *behind* the positions they had started from that morning.[81]

Later that evening, Sir John held a meeting at BEF HQ to determine the following day's plan. Incredibly, despite the day's actions and the testimonies of German prisoners, French still held the view that there were only weak German forces north of the Lys. Consequently, he ordered Haig's I Corps, who had just completed their redeployment from the Aisne, to move into line north of Ypres on the BEF's extreme left. Believing the area to be lightly contested, French ordered Haig's troops to advance through Torhout to Bruges and Ghent. Meanwhile, II and III Corps were to continue their advances while Rawlinson's IV Corps was directed to cover Haig's arrival and deployment. The realists at BEF HQ, however, were not impressed with the plan as Sir Henry Wilson noted in his diary: "Bruges for all practical purposes is as far as Berlin."[82]

Meanwhile, the German leadership was pleased

First Battle Of Ypres
October 20 - 26, 1914

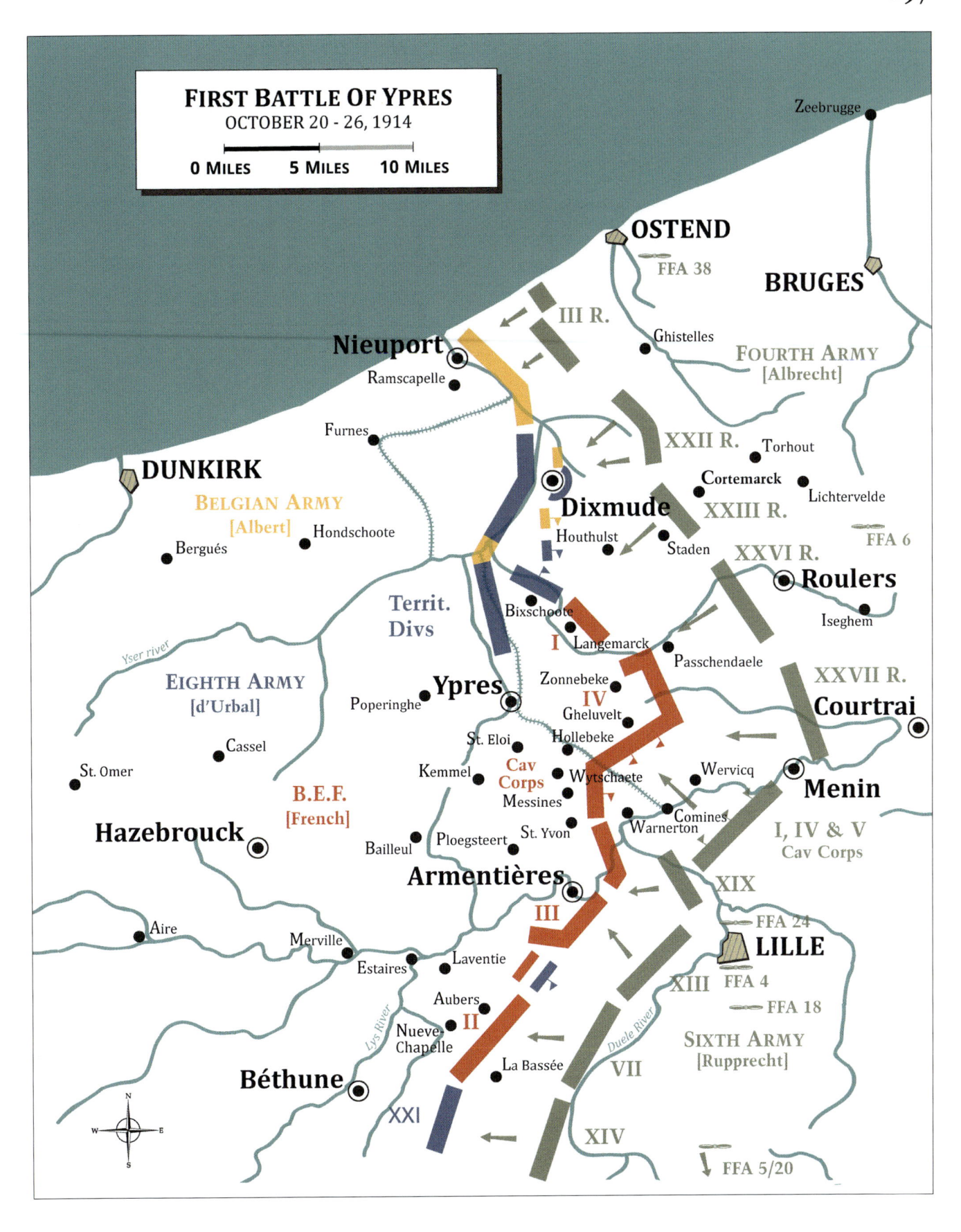

with their men's progress. Each of Fourth Army's four 'volunteer corps' had advanced between six to nine miles and had successfully pushed back all mentionable allied resistance while maintaining contact with one another's flanks. By nightfall, the entire army had moved into line next to Beseler's III RK on the Yser. Nonetheless, Albrecht and his Chief of Staff, *Generalmajor* Emil Ilse, still remained cautious. With mostly inexperienced troops, Fourth Army was obligated to achieve a breakthrough in a sector defended by strong numbers of allied troops utilizing field fortifications, trenches, and the area's natural terrain features to their advantage. Thus, to assist Fourth Army, Falkenhayn ordered Sixth Army to immediately launch a "ruthless offensive" using all available troops.[83] The XII AK was therefore moved from its defensive footing along the line Menin—Warneton into attack positions closer to the allied line. Meanwhile, the men of HKK 4 were ordered forward to man the corps' former positions.

Poor weather on the 19th once again grounded the majority of aerial operations throughout Flanders Front. In Sixth Army's sector, the only successful flights were performed late in the afternoon around Arras. The two flights, conducted by the Guard Corps' FFA 1, discovered that the French positions in the area had been strengthened considerably over the previous 72 hours. Moreover, two brigades of infantry and several supply trains were found west of Arras marching in the direction of the front line. Southwest of the city, dismounted cavalry formations were observed dug in along the Doullens—Arras rail line facing north. This proved to the German leadership that the allies were taking measures to protect their flank and rear in the event of a German breakthrough.

The strengthening of field fortifications around Arras coupled with the arrival of strong reserves directly behind the front line prompted Sixth Army HQ to abandon any plan of concentrating their renewed offensive in that area. Instead, Rupprecht and his staff decided to launch a massive assault with the army's right wing along both banks of the Lys. In the north, Laffert's XIX AK was to advance along the river and strike the allied line. Meanwhile, the cavalry of HKK 4, HKK 1, and 9 KD were to cross the river and strike the flank and rear of the allied forces engaged with XIX AK. On Laffert's left, the XIII, VII, and XIV AKs were to each advance and attempt to break through the allied positions. Finally, the I Bavarian RK, comprising the attacking wing's extreme left, was ordered to attack the allied positions at Ecurie—St. Laurent and then proceed north of Arras to cover the attacking group's left flank for subsequent actions.

Aerial reconnaissance in the north over the British and Belgian sectors was entirely unsuccessful on the 19th. Reports from the troops on the ground, however, confirmed that the day's combat had merely pushed back forward elements of the Allies' forces. It was therefore expected that the battle would resume the following morning as Fourth Army's advance continued. Albrecht and his staff issued their attack orders accordingly on the evening of the 19th fully aware that the following day's battle would take the form of a hazardous front assault.[84]

The long awaited German offensive began on the morning of the 20th along a 60 mile front from Arras in the south to Nieuport and the channel coast in the north. Thunderstorms, wind, and heavy rains grounded aerial operations throughout the day, forcing the troops on the ground to begin the offensive without any aerial support. Near the coast, the III RK concentrated almost its entire allotment of heavy artillery to support the advance of its own 5 RD at Schoorbakke, who made only negligible progress.[85] Positioned on III RK's left, the XXII RK threw back a weak Franco-Belgian screen, crossed the Handzame Canal and took possession of the town of Esen, less than two miles southeast from the corps' main objective at Dixmude. The town of Dixmude was of extraordinary significance for the allies. It served as a nexus for all of the area's lateral rail lines and roads. The allied position, as constituted on the morning of the 20th, would have therefore quickly become untenable if Dixmude fell into German hands. However, the allies had skillfully prepared the area for defense, causing the German attack to breakdown on the town's outskirts.[86] Nevertheless, the troops of XXII RK were reorganized that evening and made ready to storm the city the following morning.

To the south, the XXIII RK successfully advanced through heavy street fighting at Staden to the line of Clercken—Houthulst. Deployed on their left, the XXVI RK initiated their drive toward Langemarck and Ypres. After a chaotic and bloody morning, the corps' two divisions managed to reach a line less than two miles to the east of Langemarck with expectations to renew the advance the following morning.[87] Finally, on Fourth Army's extreme left, the XXVII RK had a miserably unsuccessful day. The corps had been expected to make great gains in the direction Ypres. However, the inexperienced units quickly became bogged down and disorganized under heavy British rifle and artillery fire. Progress was finally resumed late in the day upon the arrival of the corps reserve and some additional artillery, allowing for the capture of Beselare that evening.[88]

Meanwhile, the heavily seasoned troops in Sixth Army's sector were eager to resume the attack after reluctantly remaining on the defensive for nearly a week. Unfortunately, the heavy fighting of the past

Above: German cavalry seizes control of a village. (Author's Collection)

month had significantly reduced the army's combat value. As a result, Rupprecht's planned offensive along both banks of the Lys failed to achieve the desired result of a breakthrough. Nonetheless, Sixth Army's vigorous attacks, launched on consecutive days between the 20th and 29th, placed great pressure on the Franco-British forces in their sector. This caused the allied leadership to commit men and material away from the critically important battles in Fourth Army's sector where the Germans were attempting to envelop the allied left and seize control of the channel ports.[89]

The German offensive continued as anticipated on the morning of the 21st. On Fourth Army's right wing, the III, XXII, and XXIII RKs made minimal progress, pushing back French and Belgian troops to gain possession of the Yser's western bank. To the south, Sir Douglas Haig's newly arrived I Corps delivered an attack against the XXVI RK north of Ypres. Haig had been given orders from Sir John French (who still somehow believed that there was only one German corps in the area[90]) to advance in the direction of Torhout to envelop the German line. Haig's men quickly discovered the numerical strength of Fourth Army and ultimately halted their advance around midday. With this, Sir John was finally forced to admit the futility of continuing his offensive and ordered the entire BEF to entrench where they stood.

Further south, Albrecht's left wing launched a series of determined attacks against the British III and IV Corps. Outnumbered in many places six to one, the British defenders valiantly stood their ground, denying the Germans any mentionable gains. In Sixth Army's sector, Rupprecht's right wing successfully pushed back British II Corps and the French formations defending its flanks. Outside of these minor gains, however, Sixth Army failed to secure any new territory.[91]

Although the storms of the previous day had passed, thick low-lying clouds lingered over Flanders throughout the 21st, causing significant problems once again for flyers. FFA 38 was able, however, to perform two flights around midday in the area of Dixmude. The subsequent reports, delivered directly to III RK HQ and later forwarded to Fourth Army HQ, confirmed that the allies had successfully constructed strong field fortifications between Dixmude and Nieuport—the area of III RK's main thrust. Trenches and strengthened artillery positions were also observed west of Dixmude, thus suggesting the construction of a second defensive line directly behind the first.[92] These reports validated

Above: German infantry attacking along the coast. (Author's Collection)

Falkenhayn's previous concern that Fourth Army's offensive would degenerate into a series of drawn out frontal assaults that would consume the army's artillery ammunition before the advance against the allied flank and rear could occur. However, there were no reserves available, nor was there anything else Falkenhayn could provide to immediately better Fourth Army's chances for success.

In Sixth Army's sector, FFAs 4 and 24 conducted several successful flights over Armentières on the army's right wing. FFA 24's report briefly described the fighting around the Lys—specifically XIX AK's assault on the Ploegsteert Wood and HKK 1 and 4's thrust to seize the Messines Ridge. While cloud cover prevented a detailed report, the observer managed to note the general British deployment opposite XIX AK as well as the existence of a large gap in the allied line east of Wytschaete at Kortewilde. Meanwhile, FFA 4's report (produced late in the day after FFA 24's report had been delivered) dealt with the movement of reserves and matériel in the British rear-area. The message, delivered in person to XIII AK HQ, read: "Lively train traffic on the routes Estaires–Armentières and Hazebrouck—Armentières. North of Armentières there are two large troop encampments." Perhaps unwilling to risk having his offensive once again suspended, Rupprecht failed to share FFA 4's report with either Fourth Army HQ or OHL. Nonetheless, it could safely be assumed that Sixth Army HQ was aware that the Allied defenses were strengthening by the day and that time was running out to achieve a major victory.[93]

The morning of the 22nd brought a welcomed break from the dismal weather of the previous week. Indeed, calm winds and a sunny sky enabled aerial operations to be conducted without interruption throughout the day. The pleasant weather also gave the attacking German ground forces an opportunity to better coordinate their attacks. Recognizing the necessity to restore mobility to the front as quickly as possible, Falkenhayn directed Fourth Army on the morning of the 22nd to hasten their breakthrough of the Yser Canal at all costs. Sixth Army's right wing was similarly directed to make a concerted push to split the allied line in the area of Armentières and La Bassée. Consequently, the 22nd turned into an extremely bloody day across both armies' sectors as the intensity of the German attacks increased.

After two days of heavy fighting, lead elements of Beseler's III RK had successfully fought its way across the area's 12 feet wide and deep drainage ditches to establish a foothold on the Yser's western bank. A significant number of troops from the corps' main body had also pushed forward to the river's eastern bank, giving the engineers an opportunity to begin constructing ad-hoc bridges for the rest of the corps to cross the following day. Reports from the troops that had crossed over confirmed the previous day's aviation report describing strong field fortifications in the area, as the Belgian trenches were deep and in close proximity to the river. Shortly behind the front trenches was a 'second position' protected by a maze of drainage ditches, obstacles, watercourses, and hedgerows.[94] Despite these formidable defenses, the German troops felt confident that their qualitative superiority over their Belgian counterparts would result in a large-scale breakthrough.

To the south, the XXII and XXIII RKs conducted a series of weak, disorganized assaults against the Yser Canal near Dixmude. Before the attacks began, the southern division of the XXIII RK (46 RD) was ordered southwest against the line Bixschoote—Langemarck in order to assist the XXVI RK's attack north of Ypres.[95] The subsequent battle around Langemarck fought by the 46 RD and the XXVI RK quickly became one of the most famous battles in German military history. The attacking infantry, many of whom were young patriotic volunteers that dutifully answered the call to serve in August, assaulted the British I Corps' positions *en masse* in tightly closed ranks—demonstrating the troops' incredible courage as well as their grossly inadequate training. XXVI RK's repeated attacks at Langemarck were ultimately unsuccessful, resulting in nothing but heavy casualties for the Germans. On XXVI RK's right, the 46 RD successfully pierced the enemy's defenses, forcing the British to hastily withdraw. By 6:30pm, Bixschoote was in German hands. Later that evening, however, the half-trained troops of 46 RD misinterpreted an order and withdrew from the town before the relieving force arrived. By morning, the British had recognized the error and reclaimed their previous positions.[96]

Meanwhile on Fourth Army's extreme left, the XXVII RK continued its drive to take Ypres directly from the east. Here, the British 7th Division stubbornly stood their ground against the German attackers, whose half-trained troops foolishly launched their assaults in dense platoon columns over open ground. As a result, the attacks predictably failed with heavy losses. That evening, the Germans dug a new trench line a mere 400–500 yards from the British positions with plans to make a renewed assault the following day.[97]

While the ground troops remained unable to achieve a major breakthrough, Fourth Army's aviators performed long-range reconnaissance flights deep behind allied lines for the first time in nearly a week. First, an aircraft from FFA 5 was sent at 9:30am along the coast to Dunkirk to gain an updated estimate of the allies' overall strength in the coastal sector. Deployed along the coast, Beseler's III RK was the only formation in Albrecht's Fourth Army that consisted of fully trained troops and large quantities of heavy artillery. Thus, Beseler's command served as Fourth Army's "battering ram."[98] If the III RK could successfully breakthrough the Belgian positions along the Yser, the allied forces to the south would be compelled to withdraw, thereby radically changing the strategic situation. Thus, FFA 5's flight was conducted in order to assist Fourth Army HQ determine the proper strategy for III RK's stalled offensive.

According to the airmen's report, the Belgians were showing signs of beginning to crack. Shortly after passing over the front lines, the observer found the floodplains to the east of Nieuport and south of the Yser canal severely inundated—thus confirming the German leadership's suspicions that that Allies would attempt to flood their own territory in order to cover their flank and delay III RK's advance. To the west, a system of newly dug trenches anchored to several strong points was observed around the city of Bergués. At Dunkirk, heavy rail and road traffic was observed leaving the harbor. Despite the high level of activity, the observer reported there to be "no new troops present; only vehicles and war matériel."[99]

Delivered to Fourth Army HQ shortly after noon, FFA 5's report illustrated the allies' resolve to defend the major port at Dunkirk. As serious students of military history, the German General Staff Officers at Fourth Army HQ had anticipated the flooding of the terrain around Nieuport.[100] The seawater served to protect the allied flank and the town of Nieuport from encirclement—forcing III RK to shift its forces further south and deliver all subsequent attacks on a narrower frontage.[101] Nevertheless, the German leadership took the news of the flooding and the construction of new trenches in the rear-area as positive news that a breakthrough was imminent. News from ground troops confirming the completion of a bridgehead across the Yser on the Belgian front reinforced this optimistic viewpoint amongst the Fourth Army HQ staff.

FFA 5 conducted a second flight later in the afternoon to better determine the strength of the Belgian reserves in the area. During the flight, the airmen found the roads leading south and east out of Dunkirk to be "completely free of enemy troops." Meanwhile, immediately behind the Belgian lines, the only infantry observed to be unengaged was a small pocket of infantry at Adinkerke (just west of Furnes). In other words, the Belgians appeared to have already committed the overwhelming majority of their reserves to the front lines to resist III RK's breakthrough efforts.

Elsewhere, the airmen of FFA 38 performed several flights in support of the battles to the south around Dixmude and Ypres. After two unsuccessful flights during the morning hours, it was decided by the section's commander, *Hptm.* Volkmann, to cease all further attempts of tactical reconnaissance flights until further notice.[102] Volkmann ordered his men instead to resume longer range flights over the allied communications zone until a better system of liaison with the ground troops could be enacted. Thus, later that afternoon an aircraft was sent along the route Dixmude—Furnes—Poperinghe—Ypres. The observer's report found the areas in the north to be free of reserves while the town of Poperinghe (west of Ypres, deep behind allied lines) was described as "heavily occupied by enemy troops." Additional reinforcements, as well as large numbers of vehicles and supplies, were seen around Ypres as well.

FFA 38's report instantly doused all remaining feelings of optimism amongst the Fourth Army HQ staff. According to the report's accompanying sketch, the strength of the allied reinforcements at Poperinghe was at least one full infantry division. It was therefore assumed that the allied resistance would continue to strengthen in the days to come. Consequently, Fourth Army HQ decided to liberally employ the army's heavy artillery in support of all future infantry attacks. Ammunition shortages however, were already being felt across the Western Front—resulting in strict orders from OHL to conserve. Therefore, in addition to several other changes, Fourth Army HQ ordered their airmen to attempt artillery cooperation flights beginning on the 23rd to help the infantry achieve a breakthrough.

Meanwhile, reconnaissance flights conducted on the 22nd in Sixth Army's sector by FFA 9 showed no reinforcements in the French rear-area west of Arras. The observer's report stated:

To: IV AK HQ
All streets, roads, and railways in the sector west of Arras were, between 10 and 10:30am, free of any movements. It is currently estimated that no reinforcements are currently on the march, at least nearby.[103]

FFA 9's report was regrettably never forwarded to Sixth Army HQ. As a result, the army leadership remained unaware of what was really occurring behind the allied front lines. Sixth Army HQ was instead compelled to act upon the only information available to them—a faulty intelligence report based on misleading testimonies given by several recently captured French soldiers. According to these POWs, three *Corps d'Armée* were to arrive around Arras on October 25th.[104] However, these statements were merely a ruse to mislead the German leadership and disrupt the ongoing offensive. Thus, Rupprecht and his staff wrongly believed a successful breakthrough around Arras was impossible when chances were in reality favorable. Nonetheless, with Fourth Army's failure to independently achieve a breakthrough, Rupprecht understood that decisive action *somewhere* in Sixth Army's sector was urgently needed. Thus, under the false impression that Arras was to be heavily reinforced, Rupprecht ordered his army's offensive to continue in the north, specifically against the line Lens—Béthune.[105] This attack used all the remaining reserves available to Sixth Army, including the 6th Bavarian RD and one division of the XXIV RK that was temporarily assigned to Sixth Army from OHL's reserve. Despite the employment of these fresh divisions, the offensive quickly stalled with no meaningful gains.[106]

The fighting of the 23rd brought little success across the entire front. Throughout the day all five of Fourth Army's reserve corps continued the previous day's attacks with anticipation of a breakthrough. It was quickly discovered however, that the allied defenses had been strengthened during the night. Thus, each of the attacks failed to accomplish a breakthrough. In the area of Langemarck, it was determined that the allied troops identified the previous afternoon at Poperinghe by FFA 38 was an entire infantry corps, the French IX CA, who took over the British line from Bixschoote to Zonnebeke.[107] Further north, the 42nd DI deployed south of Nieuport to reinforce the deteriorating Belgian defenses near the coast. Finally, the British forces in the Bailleul–Béthune sector were strengthened by the arrival of the lead elements of the Indian Corps, who had landed at Marseilles on September 26th and had spent much of October travelling to northeast France and preparing for battle.[108]

The Allied reinforcements' arrival radically changed the operational situation. After two consecutive days on the offensive, Albrecht's Fourth Army found themselves on the morning of the 23rd to be on the verge of gaining the upper hand at several unique locations across the Flanders Front. In the Belgian sector, troops of III RK had secured a bridgehead across the Yser and were finally poised to push beyond the river in the direction of the channel ports. North of Ypres, the British I Corps' left flank was weakened and threatened after the German offensive at Bixschoote—Langemarck nearly succeeded the previous afternoon.[109] In the south, the

Above: German troops launch an attack on the coastal village of Nieuport. (Author's Collection)

British II Corps had been compelled to withdraw to a reserve line during the evening of 22nd–23rd due to the heavy casualties sustained over the previous 72 hours. The Germans had plans to attack each of these points on the 23rd to create the conditions for an allied withdrawal and a return to open warfare. However, the arrival of the IX CA, 42 DI, and the Indian Corps enabled the allies to shore up their defenses in these threatened areas and prevent the Germans from achieving breakthrough.

After the failures of the 21st and 22nd, Fourth Army HQ decided to further support their ground forces by employing all available aircraft on short-range tactical reconnaissance and artillery direction sorties. To that end, FFA 6 and 38's aircraft were redistributed amongst each of the army's five corps commands. Beginning on the 23rd, FFA 38 was to dedicate its five remaining airworthy aircraft as follows: three to the III RK and two for the XXII RK. Meanwhile, FFA 6 was to commit two aircraft to both the XXVI and XXVII RKs and one aircraft to the XXIII RK. Thus, Albrecht kept no machines back at Fourth Army HQ for long-range flights. Indeed, the commander's earlier experiences during the Marne Campaign had convinced him of the airplane's greater value in a tactical role once the army had become heavily engaged.

For Fourth Army's 'volunteer corps', the *Fliegertruppe's* presence was a welcome sight. Without any organic aviation assets, these corps had struggled during the first days of the offensive to locate many troublesome allied batteries that had inflicted terrific damage on their advancing infantry.[110] By midday Fourth Army's airmen had spotted several of these batteries and had initiated artillery cooperation with artillery command posts on the ground. In the north, FFA 38 first dispatched two flights to identify the location of troublesome allied batteries. After the aircraft landed, reports were made and dispatched to corps headquarters. Then, after the enemy's position was discussed in detail, flights were sent airborne to assist with artillery direction. Meanwhile, FFA 6 divided its crews up at the start the day—some were assigned tactical reconnaissance flights while others were to cooperate with the artillery.

While Albrecht's airmen succeeded in pinpointing a large number of previously unknown allied batteries, the artillery direction sorties (attempted exclusively within III and XXVI RK's sectors) were a failure. For varying reasons, the observers from FFAs 6 and 38 were entirely inexperienced with artillery cooperation work.[111] As a result, the *Fliegertruppe's* attempts to direct the artillery, whether by dropping notes or by landing and giving coordinates verbally, were all equally unsuccessful, as the following report written by a XXVI RK artillery officer suggests:

Our aviators located enemy artillery by messages and sketches, but the accuracy of their positions within 500 meters was uncommon. Thus, an effective bombardment to support the infantry's advance was often not possible.[112]

Sixth Army's *Fliegertruppe* were very active on the 23rd performing flights over all of the Flanders Front. Contrary to Albrecht's decision to employ all of his aircraft in a tactical role, Sixth Army HQ decided to order a number of their aircraft deep behind British lines to Hazebrouck, St. Omer, and Dunkirk. These flights were to reconnoiter the allied rear-areas and bring back details regarding the allied reserves.

Taking off from Lille at 8:35am, the first flight of the morning was performed by FFA 4. The crew, pilot *Leutnant* Zeumer and observer *Oberleutnant* Schinzing, were ordered to explore along the line: Armentières—Ypres—Poperinghe—Hazebrouck—Estaires opposite Sixth Army's right wing.

FFA 4   Lille, October 23 1914
Pilot: *Ltn.* Zeumer   Time of flight: 8:35—10:15am
Observer: *Oblt.* Schinzing
                To: XIII AK HQ

Itinerary: Lille—Armentières—Ypres—Poperinghe—Steenvoorde—Hazebrouck—Aire—Merville—Estaires—Armentières—Lille

Observations:
1) In Ypres-two trains ready for departure heading south.
2) In Ypres-large quantities of lorries and motorcars.
3) Radio station seen to the south of Ypres.
4) A lot of troops and convoys in the villages south of Ypres.
5) Roads south and southeast from Ypres free of traffic.
6) No enemy sighted in Poperinghe.
7) Long convoys on the Steenvoorde—Hazebrouck road.
8) A lot of munitions trains on the Hazebrouck—Bailleul road.
9) Heavy traffic at the Hazebrouck Rail Station. Three trains departed, one after the other in a south-easterly direction.
10) An artillery unit heading for Morbecque on the Aire—Hazebrouck road. The head of the convoy near the southern houses of Hazebrouck.
11) Dropped bombs on the railway station at Hazebrouck.

                                  (Signed)
                                 Oblt. Schinzing

Upon landing, Schinzing's report was forwarded by courier to Sixth Army HQ. Attached to the report was a message from General Max von Fabeck (commander of XIII AK) complaining of the "heavy traffic around Ypres." Fabeck pessimistically ended his note stating, "the heavy groupings of infantry and vehicle traffic seen around Merville—Estaires—Laventie by our airmen strongly suggest that the enemy's defenses in XIII AK's sector have been strengthened considerably and will soon prove unassailable with the forces currently at my disposal."[113]

Later that morning, a Rumpler *Taube* from FFA 9 was dispatched to the coast to gather intelligence regarding the channel ports and the nearby railways. Around 12:45pm, the aircraft reached its first objective—Dunkirk. After circling the harbor, the observer identified "15 large vessels, significant quantities of offloaded matériel and heavy automobile traffic departing the city." The observer also noted roughly 100 merchant ships anchored at Calais, whose streets were described as "bustling with military activity." While on their journey home, the airmen noticed rail activity at Hazebrouck and St. Omer. Although no trains were seen moving in any direction, the observer took it upon himself to report that: "several apparently empty trains were seen stopped in short intervals along the route from St. Omer back to Boulogne."[114]

Meanwhile, FFA 5 investigated the rail lines and roads leading out of Hazebrouck. On the Bailleul—Armentières portion of the line, the tracks in both directions were described as "lightly used" at 9am. A second flight however, performed later that afternoon, found the same rail line bristling with heavy traffic moving in both directions.[115] In both reports the observers found large bivouacs near the rail stations, suggesting that additional allied infantry was about to reach the front lines.

Sixth Army's long-range aviation reports proved that allied reinforcements would substantially strengthen their positions in Flanders—possibly even allowing a counter-offensive to soon take place. The Sixth Army HQ staff believed that the recent strengthening of the allied line around Ypres was impossible without a proportionate weakening of the allied positions to the south, opposite the army's left wing. Thus, after receiving the morning's aviation reports concerning the allied reinforcements opposite their right wing, Sixth Army HQ ordered the flying sections of the army's left wing, (FFAs 1, 9, and 4b) to each dispatch aircraft to locate precisely where the withdrawals (if any) had occurred.[116] Each of the subsequent reports, however, produced no evidence to suggest that any recent large-scale withdrawals in their respective corps' sector had taken place. Sixth Army's leadership

Above: The Gotha LD1 was powered by a 100 hp Mercedes D.I. The single example built was used at Gotha for training, but was commandeered by officers of *Feld-Flieger-Abteilung* 4 on 31 August 1914. It was subsequently purchased by the *Fliegertruppe* and assigned the serial number B.458/14. (Aeronaut)

was therefore forced to make a difficult decision. Should Sixth Army extend its offensive further south where the formidable allied trenches were known to be held by large numbers of troops, or should the offensive continue on the army's right wing despite the arrival of large numbers of reinforcements in the sector?

After some discussion, Rupprecht decided to continue the offensive with his right wing, where he believed success was imminent in spite of the allied reinforcements. However, the commander remained fearful that OHL would once again suspend his army's offensive if they learned of the allied reinforcements' arrival. OHL had practically no means of gathering intelligence independently and had grown to be dependent throughout the campaign upon Sixth Army's aviators for the majority of their information. Rupprecht's staff understood this and decided to only share information that would allow them to continue the offensive with the army's right wing.

All of the day's most important reports, such as *Oblt.* Schinzing's highly detailed account, were carefully absented from the communiqués sent to OHL. As a result, Falkenhayn remained unaware of the new allied formations in the Ypres sector. Fortunately, OHL received some meaningful information that evening from First Army HQ. According to the message, First Army's aviators had reported large numbers of allied troops withdrawn from their sector, with several large columns marching north in the rear-area towards Flanders. First Army HQ's message concluded that, "only a group of inferior and disorganized troops now stand opposite the III AK."[117] For Falkenhayn, there was no longer any doubt whether the allies were reinforcing themselves in Flanders. The question was: when would these new formations arrive?

Unaware of the allies' true strength in the Ypres sector, Falkenhayn was faced with a momentous decision regarding the continuation of his offensive. The failures of the 21st–23rd had proved that reinforcements were needed for his desired breakthrough. Therefore, upon reviewing all the information available to him, Falkenhayn resolved to send the XV AK, II Bavarian AK, 48 RD, and 6 Bavarian RD to reinforce Sixth Army's right wing. Furthermore, OHL ordered Sixth Army HQ to cease all offensive action around Arras in order to conserve the preciously scarce artillery ammunition for the more important action on the army's right wing where the battle was to be decided upon the arrival of reinforcements.

The evening of October 23rd was likewise a crucially important moment in the battle for the Allies. The Franco-British leadership had by then become fully aware of the true strength of German Fourth and Sixth Armies and had gathered enough intelligence to determine their opponent's objectives. With this information in hand, the Franco-British commands felt confident that the arrival of their latest reinforcements would allow themselves to launch a successful counter-offensive, which began on the 23rd. According to their plan, the Belgians were to remain on the defensive along the Yser while the French and British troops immediately to the south were to attack in concert toward Roulers—eventually continuing past the city and pivoting on the Lys to strike the right wing of Sixth Army.[118] The Allies' main thrust was to therefore be concentrated against the largely undertrained troops of Fourth Army's 'volunteer corps.'

The Battle Continues: October 24–28

Three days of heavy combat had left the opposing battle lines relatively unchanged. In the Belgian sector, Beseler's III RK was unable to breakout beyond their recently established bridgehead on the Yser's west bank. At Dixmude, the French had stoically held their ground in the face of repeated assaults delivered by Fourth Army's XXII and XXIII RKs. Meanwhile, the British semi-circle defense around Ypres remained intact after the arrival of the French IX CA enabled the Allies to launch a successful counter-offensive—resulting in the requisition of all territory lost over the previous 48 hours. In Sixth Army's sector, the Germans had pushed back the British II Corps, but did not possess sufficient strength to achieve a meaningful breakthrough. Thus, the morning of the 24th dawned with both sides turning to the offensive in a desperate attempt to break out of the deadly stalemate.

Fourth Army's failure to breakthrough the allied defenses around Ypres prompted Albrecht to halt all further attempts in that direction. Instead, beginning on the 24th, Fourth Army shifted its focus to the north where the III and XXII RKs' efforts still offered promise. Accordingly, a major assault was ordered against the Franco-Belgian defenses at Dixmude. Supported by the army's 350mm and 420mm siege guns, the attacks seriously weakened the allied resistance, prompting the local French commander to write Foch that evening warning that the city would fall into German hands in the next 24 hours.[119]

Further north, troops of the III RK successfully forced their way across the Yser along a three mile front.[120] However, the corps' right wing was unable to make any progress in the direction of Nieuport due to highly effective gunfire from allied vessels in the channel.[121] Nonetheless, III RK's success in crossing the Yser to the south caused the local allied forces to abandon their positions and withdraw to a new defensive line along the Nieuport—Dixmude Railway.

With their forces beaten and disorganized, the Allies sought to permanently protect their flank with a complete inundation of the Yser plain, which covered the area from the Yser to the allies' new position on the rail embankment.[122] The flooding, which was ordered on the evening of the 24th, was accomplished by opening the floodgates of the sluices at Nieuport at high tide and directing the seawater towards the Yser Plain with the assistance of siphons and hastily constructed embankments.[123] After the preparations were made, the floodgates were opened on the evening of the 27th–28th. The water levels steadily rose over the following 48 hours, forcing III RK HQ to order a withdrawal despite having not been bested by force of arms.

Under direct orders from Fourth Army HQ, the XXIII, XXVI, and XXVII RKs were instructed to "maintain and strengthen their positions and take every opportunity of seizing important points on their immediate front."[124] Albrecht therefore understood from the previous days' engagement that his volunteer corps' combined combat value was insufficient to warrant a continuing of the offensive in the direction of Ypres. Instead, these formations were to hold their ground, resist any allied assaults, and launch local counter-attacks with limited tactical goals when or if the situation presented itself. As it happened, these three corps became heavily engaged resisting the main thrust of the allied counter-offensive aimed at breaking through their positions and rolling up Sixth Army's right flank. For three days from the 25th–27th, the troops of the XXVI and XXVII RKs were subjected to 'exceptionally heavy' attacks directed against their inner flanks. Albrecht, recognizing the gravity of the moment, dispatched Fourth Army's last reserves, the 37th *Landwehr* and 2nd *Ersatz* Brigades, into the firing line to assist his beleaguered left wing. With the assistance of these reinforcements as well as several smartly placed field batteries, the XXVI and XXVII RKs launched a local counter-attack that successfully repulsed the allied attackers, enabling elements from the XXVII RK to capture and hold the town of Reutel (south of Zonnebeke).[125]

The Franco-British attacks against Fourth Army's left wing continued unabated through the 28th. The heavy combat of the prior seven days had left the opposing sides utterly exhausted. Consequently, the 27th and 28th saw little meaningful change as the allied commanders attempted in vain to continue their offensive and breakthrough the German defenses.[126] The operational situation radically changed, however, on the evening of the 28th when Belgian engineers finally succeeded in breaking open the Nieuport sluice gates, allowing the seawater to begin flooding the Yser plain. For the Allies, the flooded plain secured their vulnerable left flank and allowed valuable forces to be shifted further south to assist in the battles around Ypres.[127] For the Germans, the inundations left the soldiers of Fourth Army's right wing no choice but to abandon their offensive and retire to the east bank of the Yser. Thus, with the important territory near the coast flooded, the most important battles of the campaign were thenceforth fought to the south around Ypres and Zandvoorde.

Located between the Fourth and Sixth Armies, HKKs 1, 3, and 5 were ordered on the 22nd to attack up the Menin Road in the direction of Ypres in support of Fourth Army's stalled left wing. The cavalrymen's initial attacks, however, were easily beaten back by dug-in British infantry. On the morning of the 25th, OHL ordered the three cavalry corps to redouble their efforts in order to take pressure off Fourth Army's left, which was bearing the brunt of the allied counter-offensive. Thus, after a 36-hour preliminary bombardment, troops from HKK 5 stormed and captured the village of Kruiseecke, throwing back the British 7th Division in the process.

The fighting in Sixth Army's sector between the 25–28th was similarly inconclusive as Rupprecht and his staff anxiously awaited the arrival of the aforementioned reinforcements sent by Falkenhayn for offensive use on the army's right wing. Under orders from OHL, Sixth Army's left and center suspended all unnecessary offensive operations. Therefore, the fighting around Arras and to the south remained relatively quiet as the opposing sides remained in their trenches. Elsewhere, in order to maintain pressure on the British (who had been pushed back to a second position on the 22nd) and prevent them from strengthening their positions, Rupprecht ordered his right wing to remain active and conduct limited attacks whenever possible. The timely appearance of the Indian Corps, however, considerably strengthened the British resolve—allowing them to drive off the German attacks on the 24th. The following day brought little action other than a prolonged artillery duel that saw German aviators accurately direct heavy fire on the British trenches, temporarily driving the infantry out only to be reoccupied later that evening under the cover of darkness.[128]

Sixth Army's right wing launched a series of vigorous assaults throughout the 26th–27th that succeeded in capturing the town of Neuve-Chapelle from British II Corps. Seemingly enraged by the loss, Sir John French ordered an immediate counter-attack to recapture the town. Using Indian troops and French cavalry to support the assault, the British troops pierced the German lines and entered the town around midday on the 28th. Soon thereafter, the outnumbered attackers were once again driven out by a determined German counter-attack, resulting in both sides eventually returning to the same trenches they had started the day in.[129]

With only two flying sections to support the entire army, Fourth Army's airmen were highly active throughout the first ten days of the battle. In accordance with the army order to divide the flying sections amongst each corps, all flights from the 23rd–28th were either tactical or artillery direction sorties. Thus, despite the previous day's issues, FFA 38 performed two artillery cooperation flights late in the afternoon in the area of Dixmude directing fire for a preliminary bombardment. Unlike the section's earlier sorties, these flights were able to successfully direct artillery fire by employing an experimental communication method using colored flares. Using this method, a representative from the flying section would first meet with a local artillery commander prior to the flight to discuss the tactical situation and confirm what each flare color represented. Once airborne, the observer would identify a target for friendly batteries by firing a flare directly over the enemy position. Then, according to how the shells land, the observer would fire flares to improve accuracy. If the observer ran out of flares, the pilot often resorted to tipping the aircraft's wings or other predetermined maneuvers to communicate with the ground troops. Although FFA 38 was not the first to use the flare method, their success prompted several other flying sections to temporarily adopt the method during the final weeks of the Ypres Campaign.

Another aircraft of FFA 38, flying a tactical reconnaissance sortie on the 24th, submitted a report to III RK HQ describing the significant impact the allied ships' gun fire was having on the advance of the 4th *Ersatz* Division near Nieuport. As a result, Beseler decided to deploy his right wing division *en échelon* along the coast, directing his main thrust further south where a bridgehead on the Yser's western bank had already been established. Lastly, an aircraft of FFA 38 flying in support of XXII RK delivered an outline sketch of the allied deployments around and to the south of Dixmude.

In contrast to FFA 38, the airmen of FFA 6 decided to suspend artillery cooperation flights in favor of performing tactical flights. Unfortunately, the nature of the local terrain with its hedges, canals, dykes, and tree-lines made reliable tactical support nearly impossible. Thus, the majority of the flights performed in the area of Ypres from the 24th–27th produced very little meaningful information. Fortunately, Albrecht's Fourth Army was given an additional flying section on the evening of the 26th, FFA 40, to relieve the enormous pressure placed upon the army's two original *Fliegertruppe* units.

Formed on September 13th, FFA 40 had an assortment of four different A and B-Type aircraft piloted by reserve pilots. Fourth Army HQ took advantage of FFA 40's arrival by reorganizing their flying sections on the evening of the 27th. According to the order, the entirety of FFA 38 was placed back under the control of III RK to support the corps' attacks against the Franco-Belgian positions along the Dixmude-Nieuport rail line. Meanwhile, FFA 40 was assigned to the XXII and XXIII RKs while FFA 6 was given the dual role of supporting the army's left wing corps as well as resuming long-range reconnaissance duties on behalf of army headquarters.[130]

Meanwhile, with the ground troops locked in a stalemate, Sixth Army's *Fliegertruppe* were directed, beginning on the 24th, to purely conduct tactical and artillery direction sorties until further notice. Particular emphasis was to be placed upon "noting the location, condition, and strength of the allied front lines" so that the army leadership could determine where to properly use OHL's reinforcements, which were expected to be in action on the 30th. Thus, in accordance with Sixth Army HQ's request, each flying section performed as many flights as possible on the 24th, making the day the busiest of the campaign.

After a relatively uneventful morning, Rupprecht's airmen spotted important troop movements late in the afternoon all across their front. First, four large transport trains were discovered unloading troops at Ypres. Several additional reports described large numbers of rolling stock and stopped trains on both major lines leading out of Hazebrouck.[131] To the south, four short trains were seen halted just west of Armentières by an aircrew of FFA 4. A later flight conducted by the same flying section discovered newly arrived allied troops and supply columns immediately behind the front in the area between Laventie and Armentières—confirming the German leadership's suspicions of allied plans to reinforce themselves in the region between Ypres and Armentières opposite Sixth and Fourth Army's inner wings.

Elsewhere, while on the afternoon's sole long-range reconnaissance flight, an observer of FFA 5 reported a large number of transport ships in the harbor at Dunkirk and a number of warships in the channel nearby. Two trains were also seen leaving the harbor in the direction of Hazebrouck—presumably to deliver reinforcements and/or supplies to the British troops at Ypres—Armentières. The report further described the expansion of the inundations originally seen on the 15th east of Dunkirk. According to the observer's findings, "the inundations had swelled to the area south of the Canals de la Bassée, reaching all the way to the east of Hondschoote." Thus, upon reviewing the day's aviation reports, Rupprecht had no alternative but to admit that the allies would take advantage of the floods to shift their forces south and strengthen their defenses opposite Sixth Army's right wing. Later that evening, Sixth Army HQ dispatched all of the valuable intelligence to OHL for Falkenhayn's consideration.

Because ground troops used the cover of darkness each night to deepen and strengthen their positions, it became increasingly important for the *Fliegertruppe* to reconnoiter the allied lines each morning and quickly report any changes to army leadership. Sixth Army's airmen were therefore airborne on the morning of the 25th, whereupon they discovered important troop movements behind the allied front. On the rail line St. Omer—Hazebrouck—Armentières, an observer from FFA 4 reported eight long trains, including two troop transports steaming east at 9:30am. These movements represented the arrival of the Indian Corps opposite Sixth Army's right wing. Rupprecht, recognizing the report's importance, immediately forwarded its contents to OHL and Fourth Army HQ.[132]

By noon, Sixth Army HQ was practically flooded with aviation reports concerning heavy allied rail traffic across Flanders. FFAs 4, 5, 9, 18, and 24 each submitted at least one message regarding rail movements in the allied rear-area. FFA 24's flight reported seven empty trains departing Armentières and moving back towards Hazebrouck—an obvious sign that reinforcements had recently detrained. FFA 5 similarly reported three westbound trains leaving the rail station at Poperinghe. Other reports described rail traffic moving east towards the front. Most notable among these was FFA 4's message, which declared that: "three lengthy trains were recognized this morning in various areas moving to the front in the direction of Poperinghe and Armentières." This confirmed that the allies had sufficient numbers of men and transports to reinforce both the northern and southern halves of the Flanders sector.

After delivering their initial reports, the flying sections of Sixth Army's five right wing corps spent the remainder of the day performing tactical and artillery cooperation sorties. Four aircraft of VII AK's FFA 18, working in pairs throughout the day, brilliantly directed heavy artillery fire against British 5th Division's positions between Givenchy and Neuve-Chapelle. The British Official History later briefly acknowledged the German aviators' success:

The II Corps on the 25th had a quiet day except for artillery fire, the enemy's being more than usually accurate as the weather permitted aeroplane observation... During the morning the bombardment became so heavy that some battalions were withdrawn from their trenches, to reoccupy them at night.[133]

FFAs 4 and 24 similarly conducted artillery direction flights in their respective sectors but were not as effective due to differing methods. This simply reconfirmed the organizational need for an aviation command in the field that would have the ability and authority to, among other things, establish a unified doctrine for all of the army's *Fliegertruppe* units in order to maximize efficiency and productivity.

Allied aviators were also heavily active over Sixth Army's right wing throughout the battle. Thus, aerial combat became increasingly common as the establishment of trench warfare compelled the opposing airmen to consistently work in close proximity to one another for the first time. These early air battles were usually concluded without result due to a combination of the primitive weapons used (carbines and pistols) and the airmen's inexperience in using them. Nonetheless, German pilots remained greatly impacted by the allied aircraft, which were frequently equipped with machine guns and could thus afford to be more aggressive than their German counterparts, who carried pistols and carbines. For example, FFA 18's *Oblt.* Fritz von Zagen and *Vzfw.* Wilhelm Schlichting were shot down on October 5th by a machine gun equipped French Voisin III, making them the victims of the first successful aerial combat in world history.[134] Although Zagen and Schlichting's case was the exception rather than the rule, allied aircraft were still able to make life difficult for Sixth Army's *Fliegertruppe* by routinely attacking and driving off any German airmen attempting to operate over the battle-area. Thus, with no means of protecting themselves, the German aircraft would often sustain some damage as they attempted to flee, resulting in a large number of forced landings on the German side of the line throughout October and November.[135] Even if they escaped the aerial battle unscathed, the

Above: Armed with self-loading carbines, German aviators earned their first aerial victory during the First Battle of Ypres. (Author's Collection)

Albatros B.I B.30/14 flown by Dietze and Rosenmüller of *Fled-Flieger-Abteiling* 24

Above: Piloting a Jeannin *Stahltaube*, Ltn. Karl Caspar became the first pilot to cross the channel and drop bombs on England. (Bundesarchiv, Bild 183-2012-0702-506)

German flyers would still return to their airfield, likely leaving the ground troops without any aerial support for the remainder of the day.[136]

With the ground troops in desperate need of support, Rupprecht's flying sections collectively resolved to continue tactical flights over the frontlines and not be driven off without a fight. To help curb losses, all tactical flights were to be flown at higher altitudes and for shorter durations. Furthermore, German observers were ordered to actively defend their aircraft and open fire on any allied machines they were to encounter. These measures, implemented as a result of an internal memorandum passed amongst Sixth Army's flying sections on October 16th, also mandated that each aircraft be equipped with a weapon for every flight. Whether it was the preferred '*Mondragon* self-loading carbine' or multiple preloaded bolt-action carbines was subject to availability and circumstance.[137] Even with carbine fire it was believed (and later proven) that a vigorous defense would act as a deterrent to an allied attacker, regardless whether or not they were equipped with a superior weapon such as the machine gun.

Experience quickly proved the wisdom of the various changes implemented by Rupprecht's flying sections. Stories from all over Flanders, spread amongst the *Fliegertruppe*, illustrated the defensive value of performing flights at higher altitudes, improving the airmen's morale. Consequently, the number of tactical flights performed in Sixth Army's sector steadily grew as the battle progressed. More importantly, the German aviators learned to successfully use their weapons to stave off attackers. In many cases Rupprecht's airmen would aggressively fire on nearby allied aircraft to clear the airspace for their assigned tactical work. On the other hand, if an enemy aircraft was the attacker, the observer was to keep firing and defend the aircraft until his pilot was able to pick up airspeed and safely return to German airspace. In the former instance, it was found that allied airmen were likely to turn and flee after experiencing the sensation of being suddenly attacked and fired on. In the latter case, German observers reported that defensive fire often flustered the attacking enemy—giving themselves enough time to reach safety.

Although Sixth Army's aviators had successfully dispersed allied aircraft on several occasions, none of the initial air battles had ended in a German victory. That changed, however, on the morning of the 25th when pilot *Oblt*. Dietze and observer *Oblt*. Rosenmüller of FFA 24 forced down what they reported was a British Sopwith east of Bailleul. Dietze and Rosenmüller had spotted the British aircraft while conducting a tactical flight in their Albatros B.I over St. Yvon.[138] Using their altitude advantage, the German airmen quickly descended into an advantageous position and began firing. After only eight shots from Dietze's self-loading carbine, the British machine began to emit smoke and rapidly lose altitude, quickly disappearing from Dietze and Rosenmüller's view. Having secured the local airspace, the German airmen continued with their tactical work before returning to FFA 24's airfield at Lille to make their report.[139]

A second noteworthy aviation feat was accomplished on the 25th when a Jeannin *Stahltaube* of FFA 9 crossed the channel and bombed Britain for the first time. FFA 9's commander, *Hptm*. Musset, selected his best men for the sortie—pilot *Ltn*. Caspar and observer *Oblt*. Roos. Flying aircraft #A 174, Caspar and Roos passed over the city of Dover and dropped two bombs on what appeared to be a factory. Strong winds, however, pushed the projectiles away from the intended target, resulting in their explosion in a garden. On their return home, Caspar and Roos ran into a bank of clouds, causing Caspar to lose his bearings and inadvertently fly east. After flying off course for more than an hour, the airmen realized their error and turned southwest. Almost out of fuel, Caspar managed to locate and safely land at Sixth Army's Air Park at Mons, thus ending the impressive 4 hour and 20 minute flight. As it happened, Musset ordered a second aircraft airborne just two hours after Caspar and Roos had taken off to cross the channel and bomb Dover. The aircraft was ultimately forced to turn back, however, due to a large body of heavy clouds found southeast of Calais. Once the aircraft was turned around, the observer directed the pilot to fly over the important allied rail hub at Hazebrouck where he ultimately dropped his two bombs to no effect.[140]

Carrying bombs became commonplace for the airmen of all flying sections participating in the Ypres Campaign. Indeed, the German aviators were required to take bombs with them on every flight, regardless of the purpose of the mission. Although the projectiles were light and their damage trivial, the men of the *Fliegertruppe* were delighted to see their impact on the enemy's morale, as illustrated by the following

Above: Max von Fabeck, commander of XIII Armee Korps and Armee Group Fabeck during the battle. (Author's Collection)

letter written by FFA 24's Josef Suwelack:

Yesterday and the day before I went up; on both occasions, I threw ten bombs at the Englishmen. We are currently operating nearby from Armentières, somewhat northwest of Lille. You would not believe what a terrible moral impression a bomb makes!

Just imagine—You sit quietly in a garden and see in the distance an aircraft coming at great height. A cellar (to hide) is not for a soldier. The airplane is now exactly vertically above you. Now the moment has come when the bomb drops. Looking around, you see nothing. Meanwhile, the airman begins to circle overhead. Suddenly a sinister hiss comes. Then, a crash and a house explodes—or the bomb falls into a column of troops. All is random in war. The horses are then spooked. Then a second bomb comes, then a third, etc. Now you can see the airplane is bombarded by artillery. 10, 20, 30 explosive points, all appear very close from below. Individual carbines are now out, all starting a murderous gunfire. Then, the airman above, turns around for a second time and drops five more bombs.

We conduct this experiment every day. In our army's sector alone, the English receive about 20 bombs daily. We enjoy giving these friendly greetings to our dear cousins. When we see the clouds of smoke rising among them, we always come back home satisfied knowing we have done our day's work.

Even on the reconnaissance flights, we always take some bombs to throw on the enemy troops. Today, Leutnant Dietze had an air battle with an English pilot at 2000m. He downed the guy after just the 8th shot! In general, the enemy airmen immediately go down, when you fly toward them. I myself have not taken part in any successful aerial battles, only a forced landing.

The British are very tough here. Our troops have to fight for every meter of earth with his fists. The German artillery shoots poorly and there are numerous appalling night fights. However, it appears that our side has made some big gains of late. Some English formations have withdrawn this morning. In general, they do not deviate from their mission and remain in their trenches to die. Thank goodness that we are deployed at Lille, otherwise we would have a hard nut to crack![141]

Unfavorable weather grounded all flights on Sixth Army's right wing on the 26th. However, several tactical reconnaissance flights were performed by the army's left wing flying sections over the Arras sector. The subsequent reports produced as a result of these flights confirmed that no strengthening of the allied positions around Arras had occurred.[142] As the reader will recall, Sixth Army HQ had expected the region to be heavily reinforced by three corps based upon the false testimonies of several captured French soldiers and had formulated a plan on the 22nd to continue the army's offensive further north around Lens—Béthune. This strategy was promptly amended however, once Rupprecht and his staff learned the reality of the situation.

Rupprecht messaged Falkenhayn at OHL on the evening of the 26th to inform him of the army's progress as well as his *Fliegertruppe's* latest intelligence. Intrigued by what he heard, Falkenhayn visited the Bavarian prince at Sixth Army HQ the following day, October 27th, to review the situation.[143] After a lengthy meeting, the two commanders came to the following conclusions: 1) Fourth and Sixth Army's combined offensive had been decisively checked well short of any of its assigned objectives; 2) the two armies, as they were currently constituted, no longer possessed the necessary strength to successfully continue the offensive on their own; and 3) the campaign's success now depended upon the resumption of the offensive and a breakthrough with fresh forces brought from other areas of the front. Thus, with Rupprecht's aviators confirming there to be no threat from the area around Arras and to the south, Falkenhayn felt confident in redeploying troops away from that sector and using them as part of a new 'Army Group' that would be used to achieve the battle's decisive breakthrough—a plan that the German Supreme Commander had been entertaining since the 25th.[144]

Falkenhayn's newly created 'Army Group' was to initially consist of the II Bavarian AK (pulled from Sixth Army's left wing south of Arras), XV AK (from Seventh Army), XIII AK's 26 ID, and HKK 2. The 6th Bavarian RD was to also be used if it wasn't already engaged inside Fourth Army's sector. The group was to be subordinated to Sixth Army HQ and be under command of General Fabeck and his staff. Thus, Fabeck and all principal subordinate commanders of his group were called into Rupprecht and Falkenhayn's meeting to develop an operational strategy for the coming offensive. It was ultimately decided to deploy the group along a 5½ mile front along the line Wervicq—Warneton between Fourth and Sixth Armies and attack northwest with the objective to split the allied line south of Ypres. To support the endeavor, all of Sixth Army's heavy artillery was to be placed under Fabeck's command.

After bidding his subordinates farewell, Falkenhayn immediately returned to his headquarters at Mézières and wrote the following General Order.

The situation on the army's right wing requires a speedy decision. Therefore, in addition to Fourth and Sixth Armies' attacks, a breakthrough between the two armies must be carried out. To this end, the XV, II Bavarian Armeekorps, *26* Infanterie-Division, *and 6 Bavarian* Reserve-Division *(the latter if not yet being employed by Fourth Army) are to launch a combined attack under General Fabeck, who is subordinated to Sixth Army HQ. The attack is to commence on 30 October from the general line Wervicq—Deûlémont in a northwest direction. The army corps under the command of Sixth Army HQ are to remain ready for combat until the evening of October 29. All of Sixth Army's available heavy artillery shall be used to support the breakthrough... A preparatory bombardment with the heaviest caliber guns against*

Messines and the forest of Ploegsteert to support Sixth Army's right wing should be considered at Sixth Army HQ's discretion. Cooperation of the Fourth and Sixth Armies is necessary for the operation's success. HKKs 1, 2 and 4 are to remain under the command of Sixth Army HQ.

Thus, the Germans' hopes were pinned upon Group Fabeck's ability to pierce the allied line and restore mobility to the German right wing. Reports from the front, however, received during the evening of the 27th, informed Falkenhayn that Fourth Army's right wing had been stopped just beyond the canals. Equally pessimistic reports from Sixth Army HQ caused Falkenhayn to suspect that the Allies would reinforce and strengthen their defenses before Fabeck's attack could begin. Therefore, in order to increase the operation's likelihood of success, OHL ordered First Army's 3 ID to redeploy to Lille to further reinforce Fabeck's forces as the offensive began.

While remaining confident in victory, Sixth Army's leadership continued to be fearful of the allied potential to reinforce and strengthen their defenses before Fabeck's attack could be delivered. Consequently, a great deal of pressure was placed upon the *Fliegertruppe* in the days leading up to the offensive. The airmen were directed to identify new troop movements in the allied rear-areas as well as to deliver an updated appraisal of the Franco-British deployments in the sector south of Ypres where the offensive would be delivered. As it happened, poor weather greatly restricted the number of flights that were performed on the 28th. Fortunately, a break in the gloomy weather enabled several sorties to be conducted late in the afternoon. The most important of these was a reconnaissance of the Hazebrouck rail line carried out by airmen from FFA 5 that found "lively rail traffic" between Bergués and Hazebrouck running in both directions. Specifically, five trains were spotted between 4:00–4:30pm moving to and from the coast while three additional trains were also reported under steam at the Hazebrouck rail station. While these discoveries suggested that the allies were shifting troops south to bolster their defense around Ypres, it remained an open question whether this was indeed the case.[145]

Operational and tactical reconnaissance flights carried out throughout the morning and afternoon of the 29th successfully clarified the situation for the German leadership. Between 10am and noon, aviators from FFAs 3 and 4 found "very strong concentrations" of reinforcements deploying around Ypres. According to both reports the overall strength of the newly arrived formations was "at least one army corps."[146] Additional troop and support columns were also discovered marching south through Ypres in the direction of Messines, proving that the allies were taking active steps to reinforce their front in the area of Fabeck's planned assault.

Meanwhile, Sixth Army HQ's FFA 5 surveyed the highly active allied rail lines along the route: Ypres—Poperinghe—Hazebrouck—Dunkirk, reporting heavy traffic in the direction of Ypres. Most importantly, 11 troop transports and large quantities of rolling stock were discovered at the western suburb of Vlamertinghe along with three additional supply trains headed in their direction from Hazebrouck. Several smaller trains were also noted between Hazebrouck and Ypres moving in both directions. The origin of all the activity was confirmed to be at Dunkirk, where 20 transport ships were spotted anchored in the harbor.[147] Furthermore, the Bergués—Poperinghe Road leading southwest from the harbor was found to be bristling with motor supply columns.

Sixth Army HQ received all of the day's aerial reports with great interest. In their totality, the airmen's messages painted a picture of a large-scale resupply/reinforcement of the Ypres sector. Recognizing the importance of this information, Sixth Army HQ promptly sent the following message to OHL at 7:30pm:

Our aviators' messages confirm that very strong forces are now concentrated around Ypres. Rail traffic and road columns from Hazebrouck and Dunkirk are also currently headed in this direction.[148]

A similar message was also dispatched to the Fourth Army HQ staff, who had already learned of the allied movements through their own FFA 6. Indeed, late in the afternoon on the 28th, Fourth Army HQ had ordered a reconnaissance of the rail lines between Dunkirk and Hazebrouck. Due to poor weather, the pilot was forced to fly southwest directly to Hazebrouck, whereupon the observer discovered four trains arriving from the south. The next day, the same crew located three trains leaving Dunkirk in the direction of Hazebrouck as well as several motor transport groups headed towards Ypres. Yet another flight conducted by FFA 6 later in the day reported the roads around Furnes to be heavily congested. Thus, it was concluded that the allies were strengthening the Yser front as well as the area around Ypres.[149]

The *Fliegertruppe's* excellent work on the 29th accurately depicted the scale and intent of the latest Franco-British redeployments. While the German leadership had long suspected a shifting of allied forces already in Flanders to strengthen their line at potentially vulnerable points in the line, it remained an open question whether any additional forces would be received from other areas of the front as well. The multitude of aerial reports delivered on the 29th finally answered this question in the affirmative, with estimates of at least one army corps recently arriving in the area of Ypres as well as another division along the Yser. Thus, on the eve of Fabeck's attack, both Fourth and Sixth Army HQs were informed of the general situation and had in their possession fairly accurate intelligence estimates of the Franco-British strength.

In truth, the allies had been quietly moving new formations into Flanders since the 24th. The French 31 DI had arrived by rail from Montdidier to Bailleul on the 25th, continuing the following day by truck to join Dubois' IX CA fighting north of Ypres.[150] Then on the 29th, the 32 DI, together with the staff and corps troops of XVI CA, arrived in the area of Ypres with orders to move south into line at Wytschaete. Next, the 31 DI was detached from IX CA and sent through Ypres to Wytschaete. Although these troops remained unengaged until the 31st, their presence would later prove to be decisive. Elsewhere along the Belgian front, the newly arrived 38 DI joined hands with the 42 DI, 89th Territorial Division, Marine Brigade, and assorted cavalry formations to form the XXXII CA. According to orders directly from Foch, this combined force was to deploy on the Belgian's right flank and launch a counter-offensive with the objective to drive back Fourth Army in the direction of Roulers.[151] Thanks to the timely reporting of their aviators, Albrecht's staff was fully aware of each of these important arrivals and had time to adopt the appropriate counter-measures to stop them.

The Battle Reaches Its Climax: October 29–31

The troops of the recently constituted Army Group Fabeck spent the entirety of the 29th moving into their assigned positions along the line Wervicq—Deûlémont in accordance with Falkenhayn's October 27th General Order. The group's six divisions were to begin the attack at dawn on 30th with the objective to advance northwest along the axis Mt. Kemmel—Ypres and split the allied line. Specifically, Deimling's XV AK was to be placed on the group's right wing near the Menin—Ypres road with orders to advance directly towards Ypres. To their left, the Bavarian II AK was to act as the group's spearhead, advancing along the Comines—Ypres Canal through Hollebeke. Representing Fabeck's extreme left, the 26 ID was to push beyond Messines towards Wulvergem. Finally, the 6th Bavarian RD was initially placed in the second line as the group's reserve in order to be used when and where the opportunity presented itself.[152] To support the operation, both Fourth and Sixth Armies were directed by OHL to launch a general attack on their respective fronts with the goal of pinning down allied front line formations as well as drawing allied reserves away from Fabeck's front.

Thus, the German plan was simply to achieve and exploit a numerical advantage within the sector that the allies were believed to be weakest—the British held line from Ploegsteert Wood to Gheluvelt. Upon Group Fabeck's arrival, the total number of German divisions in Flanders opposite the allies' 11½ divisions rose from 17½ to 23½.[153] Nonetheless, while this appeared to be a sound plan, the operation had two major problems. First, the necessarily hasty redeployment of Fabeck's formations had caused considerable confusion and exhaustion amongst his troops. As indicated previously, most of the group's formations spent the 29th marching to their assigned positions. Therefore, many of the men had to forego any rest during the evening of the 29th–30th in order to push forward into assault positions for the scheduled attack at dawn. Fabeck's troops would consequently be compelled to begin the offensive tired and unfamiliar with the nature of the ground.[154]

Secondly, the army's artillery ammunition supply was nearly exhausted before Fabeck's offensive even began.[155] Under direct orders from OHL, Army Group Fabeck had been assigned all of Sixth Army's heavy artillery and a large number of supplemental field batteries for use throughout the operation. In all, the group was to begin the offensive with 8 batteries of mortars, 20 groups of heavy howitzers (each with three batteries) and a 30.5cm coastal mortar—a total of 262 heavy howitzers to complement the group's 484 field guns.[156] This seemingly remarkable concentration of firepower was deemed critically important to the offensive's success. At dawn on the 30th, prior to the initial assault, the heavy artillery batteries were ordered to initiate a preliminary bombardment against the allied trenches with the intent of destroying the position and inflicting as many casualties as possible on the enemy's forward formations. Next, the infantry were to advance with the support of the field batteries and breakthrough the allies' weakened defenses. After the initial breakthrough had been achieved, Fabeck's heavy artillery was to be employed in reducing allied strongholds. The nature of the terrain throughout the region was well suited for defense and the allies were expected to take full advantage of it. Farm complexes, groups of houses, woods, and canals all served as potential rallying points and redoubts that would need to be reduced by German heavy artillery to keep the offensive going. The severely limited quantities of ammunition, however, severely reduced the heavy batteries' potential impact—causing Fabeck and his chief of staff, Fritz von Loßberg, to privately question the operation's likelihood of success. In his post war

First Battle Of Ypres
October 27 - November 11, 1914

memoirs, Loßberg discussed the mood at Fabeck's HQ prior to the offensive:

We were convinced that the assault of Army Group Fabeck was going to bump up against stiff resistance because the time delay on this previously planned attack to the south of Ypres (the orders for which had only just been given), had provided the enemy time to move up reinforcements. Having strengthened their front along the lower Yser by flooding, they were then able to move the formations thus released south to counter Army Group Fabeck and the left flank of Fourth Army. (In addition) with the exception of 6th Bavarian RD, almost all the units of Army Group Fabeck had been severely reduced in strength as a result of previous battles and, above all, from the very first day of the battle, there was a serious lack of artillery ammunition, which became worse and worse with the onset of offensive operations, thus making the reduction of stubbornly defended woods and strongly-built villages increasingly difficult.[157]

While nothing could be done in the short term about the troops' strength or the quantities of available artillery ammunition, several measures were taken to address these concerns and help improve the offensive's likelihood of success. First, all unnecessary expenditure of artillery ammunition was forbidden. Harassing fire, ranging fire, and long-range counter-battery fire squandered ammunition and was considered too ineffective to waste the army's limited shells. Secondly, orders were issued for division and corps headquarters to be connected by telephone. Next, infantry tactics were altered to rely less on the artillery for tasks that the infantry themselves were capable of accomplishing. The officers and men of Army Group Fabeck were informed of these measures on the eve of the attack via the following directive issued by Sixth Army HQ:

During recent days, several promising opportunities have been wasted because entire corps have allowed themselves to be held up by vastly inferior forces. This situation can be traced back to the fact that attacks are not being pressed home with the utter disregard for danger that each attack which aims at a decisive result demands. Naturally it must be recognized that the high officer losses have seriously damaged the offensive power of the army. However, if as a consequence we frequently conduct half-hearted attacks, the enemy will soon lose respect for us. We shall spare the blood of our infantry if we confine ourselves to attacks which are essential and make sure that they are strongly supported by all available artillery and other technical resources. If, however, a decisive attack is ordered, then every commander, including those at high levels, is <u>personally</u> to ensure, including by example, that he is so engaged that the attack will be conducted with the utmost power. Superior officers are also to appear at the front sufficiently frequent to bring their personalities to bear on the troops.

Valuable advantages are often not being exploited properly because, in the case of formations advancing alongside one another, instead of driving on ruthlessly within their boundaries, troops are pausing to allow their neighbors to catch up, or halting as soon as they come under enfilade fire from neighboring sectors. From receipt of this (directive) there is an explicit ban on delaying or halting if the planned advance is delayed or disrupted on neighboring fronts, or to see what effect the outflanking maneuver might have. Even if neighboring attacks are held up, that is no excuse for failing to act…

An extremely threatening squandering of artillery ammunition, far beyond even the most pessimistic of estimates, has occurred. This has been caused because many units and formations have constantly engaged inadequately reconnoitered enemy positions with wide-ranging, harassing fire; either that or attempts have been made by bombarding enemy positions for days or hours at a time to eliminate them in that manner, instead of the infantry through daring direct action, forcing a weaker enemy to occupy (and attempt to defend) their positions. For the future it is essential that the troops draw the lesson that careful husbandry of artillery ammunition is of decisive importance; that simply bringing fire down will never have an effect and, furthermore, that fire may only be directed against <u>clearly identified</u> targets and only for <u>strictly limited tactical purposes</u>. As a result, the artillery may not meet every request by the infantry for fire support and may only respond to the senior commander. The <u>infantry</u> must always cooperate <u>with the artillery</u> in the attack; the artillery must never try to attack alone…[158]

Sixth Army's aviators were similarly petitioned to alter their tactics in support of Fabeck's offensive. Considering both the importance of heavy artillery as well as the latest restrictions on ammunition expenditure, the German leadership did everything in their power to increase their heavy batteries' effectiveness. Thus, Group Fabeck's flying sections were each ordered to establish liaison with the artillery and focus on fire direction sorties during the first days of the attack. Meanwhile, other FFAs operating nearby were directed to remain highly active and perform tactical reconnaissance flights over the allied positions close to the front.

In order to meet these demands, Sixth Army's *Fliegertruppe* underwent a hasty reorganization. The arrival of the XV and II Bavarian AKs brought two new flying sections, each with a full complement of six aircraft, to the Ypres sector. XV AK's FFA 3 established their airfield on a former parade ground at Lille on the 28th while FFA 2b moved into their new airfield at Lesquin the following day.[159] In addition to these two units, Fabeck retained his own FFA 4 for tactical and artillery direction sorties within his Army Group's sector. Over the previous week, the flying section had divided itself, giving three aircraft to Fourth Army's XXIV RK operating on their right. By the morning of the offensive, however, each of the aircraft had been returned—bringing the section back to full strength.[160] By the morning of the 30th, Rupprecht's Sixth Army possessed ten flying sections with an estimated total strength of 45 serviceable aircraft for operations along the army's 65 mile front.[161] The three sections attached to Group Fabeck were ordered to exclusively conduct artillery direction sorties until further notice. Meanwhile, the three flying sections located on Fabeck's left (XIX AK's FFA 24, VII AK's FFA 18, and XIV AK's FFA 20) were directed to perform tactical and artillery direction work within their own corps' boundaries as well as tactical flights within Fabeck's sector. To the south, the army's remaining flying sections were to operate within their assigned sectors, flying both tactical and long-range sorties in support of the left wing's defensive strategy. Lastly, Sixth Army HQ's FFA 5 was to continue gathering operational intelligence, conducting occasional tactical flights whenever possible upon Rupprecht's request.

With all the necessary tactical preparations having been made, Sixth Army's leadership eagerly awaited Fabeck's offensive to commence. Unfortunately, a thick layer of fog compelled the attack to be delayed for several hours until conditions improved. Finally, at 8:00am, visibility had improved enough to allow the preliminary bombardment to begin. Over the next hour, Fabeck's heavy batteries pummeled the British front line trenches hoping to breakup their defenses before the infantry advanced. Then, at 9:00am, the group's five front line divisions moved out of their starting positions and began their assault.[162] With the regimental bands playing, the troops of XV AK on the group's extreme right punctured the allied line at Zandvoorde and Chapelle d'Epines. To their left, the Bavarian II AK captured the town of Hollebeke and its nearby château, which was fanatically defended until 10:00pm.[163] On the group's extreme left, the 26 ID's assault against the line Wambeke—Messines failed to reach its objective due to extraordinarily heavy fire from Messines and Wytschaete.

Fabeck and his staff were greatly disappointed in the results of the offensive's first day. Despite seizing the towns of Zandvoorde, Hollebeke, and Houthem, any subsequent progress remained limited due to the failure of the group's right and left wings to gain any ground. On the right, the XV AK's 39 ID was unable to safely move forward due to the failure of Fourth Army's XXVII RK to capture Gheluvelt. Only Gheluvelt's capitulation could allow the group's right wing to continue its advance without any fear of a British thrust into their uncovered flank. Meanwhile on the left, the 26 ID remained locked in a fierce battle with British forces at Messines and Wytschaete. In both cases, the wings' failures were caused by the heavy artillery's inability to breakup the British defenses like it had at the other areas of the front. Thus, on the evening of the 30th Fabeck's staff arranged for the group's heavy batteries to be moved forward under cover of darkness into positions to support an assault on Messines the following morning.

The heavy artillery's inability to properly support the attacks was largely due to the cloudy and foggy weather, which had completely ruled out any tactical and artillery direction flights that day. In fact, all artillery direction on the 30th was performed via telephone communications by forward ground observers operating within "the foremost infantry lines."[164] While this ultimately sufficed in some areas of the line (particularly II Bavarian AK's sector in the center), the lack of direct aerial support deeply impacted operations on the group's two wings, where circumstances required accurate artillery fire to overcome the tenacious allied defenses. Therefore, realizing the *Fliegertruppe's* growing importance, each of the HQ staffs under Fabeck's command requested additional tactical flights to be conducted on the following day.

While the deplorable weather on the 30th had grounded all tactical flights, several long-range sorties were successfully completed that afternoon. The first was delivered to Fabeck's HQ at 11:30am by FFA 4's *Oblt.* Sperrle. The observer's report described significant concentrations of allied infantry within Ypres as well as large quantities of lorries and other vehicles north of the city moving east in the direction of Fourth Army's left wing.[165] Additional flights conducted by FFAs 3, 5, and 6 later that afternoon confirmed that more reinforcements were arriving around Ypres and the surrounding area. The day's final report, first delivered to Sixth Army HQ by an observer from FFA 5, alerted the German leadership to the large numbers of troops moving into the allied line southeast of Ypres in the area of Army Group

Above: German infantry storm the French trenches. (Author's Collection)

Fabeck's attack.[166] Fabeck and his staff agreed that the *Fliegertruppe's* intelligence signaled a large scale allied redeployment to block their group's offensive. Therefore, with large numbers of allied troops fast approaching, the Army Group's leadership recognized that their window of opportunity for a successful breakthrough was rapidly closing. Thus, it was decided to immediately place the group's reserve, the 6th Bavarian RD, between the II Bavarian AK and 26 ID and redirect the offensive's main focus towards the Messines—Wytschaete Ridge.[167]

Army Group Fabeck resumed the offensive against the Messines Ridge at 2:45am. A few hours later the sun rose to reveal a beautiful fall day perfectly suitable for tactical reconnaissance and artillery direction sorties. Thus, with the group's heavy batteries now deployed close enough to effectively engage the allied troops, expectations for a successful day were high. Ground observers directing the heavy guns' fire, allowed the fortified defenses at Messines to be quickly reduced.[168] Then, at 10:00am, the *Fliegertruppe* entered the fray. Within 30 minutes, aircraft from FFAs 2b, 4, and 5 were each over the Messines—Wytschaete sector directing fire into the heavily fortified villages. Sixth Army HQ's observation balloon from FLA 1b was also airborne making good use of its telephone communications with the local battery commander to direct howitzer fire against Messines. By 11:00am, the British defenses were believed to be sufficiently weakened to assault the town.[169] Next, three regiments of 26 ID (along with a field artillery piece and some engineers) rushed forward and began pushing the British forward troops back house by house. The street fighting developed into a brutal contest that continued throughout the rest of the day with neither side gaining the advantage. By nightfall the Germans had progressed to within 50 yards of the British main line, which ran through the western half of the village.

Located on the 26 ID's immediate right, the II Bavarian AK and the 6th Bavarian RD was unable to attack Wytschaete during daylight hours. As we have seen, the 6th Bavarian RD had spent the previous day in Army Group Fabeck's second line as a reserve. Consequently, after receiving orders to move forward, the men of the division were compelled to spend the morning and early afternoon completing an approach march into their prescribed position between the II Bavarian AK and 26 ID. In the meantime, the II Bavarian AK attempted to assault Wytschaete on their own. Defending the town was the British 2nd Cavalry Division, who had the support of three batteries of horse artillery as well as six batteries of recently arrived French field guns (these French batteries were part of the group that had been observed by the *Fliegertruppe* the day before). Possessing Group Fabeck's largest complement of heavy artillery, II Bavarian AK began a preparatory bombardment at 7:00am. FFA 2b sent two aircraft airborne at 9:30am to help direct the fire. With orders to concentrate against the British infantry, each aircraft spent the next two hours assisting several local batteries in destroying large portions of the opposing trench line. Despite a successful preliminary bombardment, the Bavarians' failed to quickly follow it up with a large-scale assault. By the time the advance was finally ordered, the allied line had been reinforced with a *cuirassier* brigade and four battalions of French infantry. These freshly deployed troops acting in concert with their artillery easily repulsed the German assault.[170] That evening the 6th Bavarian RD launched multiple attacks under the cover of darkness to try and take the town. With minimal experience and no knowledge of the ground, each of the Bavarian volunteers' attacks were easily repulsed, suffering heavy losses in the process.[171]

Meanwhile, troops from XV AK and Fourth Army's XXVII RK jointly attacked the British forces at Gheluvelt in an attempt to free-up Group Fabeck's right flank. Just as it had occurred in the sectors to the south, all available heavy batteries were ordered to carry out a preliminary bombardment as early as possible to prepare the way for the infantry's advance. Unfortunately, XV AK's earlier requests for additional heavy batteries had been denied—leaving the corps with about 25% of the firepower of Bavarian II AK on its left.[172] As a result, the commander of XV AK, General Deimling, was forced to mainly rely upon his infantry to achieve the corps' assigned objective. Accordingly, after a two hour systematic bombardment of the whole British line, troops from both XV AK's 30 ID and XXVII RK's 54 RD launched a converging assault against Gheluvelt at 11:00am. In grand Napoleonic style, both of the division commanders, as well as Deimling, were seen mounted with their staffs directing the attack from near the front lines.[173] With the men under heavy machine gun and rifle fire from all over the field, Deimling's infantry slowly advanced from one position to the next, allowing the field batteries to move forward and provide close support. By 2:00pm, the leading elements of the 54 RD finally entered Gheluvelt, reaching the château and its surrounding park in the process.

The capture of Gheluvelt enabled the Germans to begin raking nearby British trenches with deadly enfilade fire. Almost immediately, the British infantry carried out an orderly withdrawal to a pre-arranged position west of the village. Soon thereafter a British

Above: French infantry fire on a *Taube*. (Author's Collection)

counter-attack carried out by an ad-hoc group of 364 well trained soldiers retook the château from roughly 1,200 German reservists.[174] This small force of British troops later pushed forward to the outskirts of Gheluvelt, where by virtue of their position on 46 RD's uncovered right flank, they were able to stop any further German advance. Nevertheless, by 5:00pm Haig was compelled to order all forward British troops in the sector to withdraw to safety on a reverse slope located about 600 yards west of the village.

Elsewhere, XV AK's left division (39 ID) made almost no progress. After witnessing the morning's preliminary bombardment, the troops had fully expected to build upon the previous day's success and decisively break the weakened British line. Unbeknownst to the Germans however, large numbers of French reinforcements had already arrived by the time of 39 ID's attack.[175] Thus, with their lines strengthened, the British were able to launch a successful, albeit costly, counter-attack that forced back the German advance. Indeed, by nightfall the allies had reclaimed the southern and eastern edges of Shewsbury Forest (west of Gheluvelt).

Aircraft from XV AK's FFA 3 were airborne in large numbers throughout the day in support of their corps' attack. Four artillery direction sorties were conducted between 9am and noon to assist the nearby batteries reduce the various allied strongholds—mainly at Gheluvelt. Instead of dropping written messages over the side of the aircraft, the airmen of FFA 3 chose to communicate with the artillery commanders by establishing a small forward airfield near an artillery command post that allowed the crews to land and verbally relay their instructions and quickly resume the flight over allied lines.[176] Later that afternoon, during the various allied counter-attacks, XV AK HQ ordered two flights be performed over the battle-area to determine the strength and location of the allied line. Unfortunately, with the opposing sides heavily engaged in close combat, the observers' reports were of little value.

To the north, Albrecht's Fourth Army remained on the offensive in order to prevent any allied troops from being shifted south to Fabeck's sector. However, the heavy losses of the offensive's first ten days had severely diminished the army's offensive capabilities—compelling Albrecht and his staff to order their left and center to only carry out limited "harassing attacks" while Beseler's III RK on the right wing carried out the main assault. Unfortunately, the flooding of the Yser plain halted Beseler's men on the evening of the 30th after they had already successfully pierced the Franco-Belgian line earlier in the day. Thus, with a large body of constantly deepening seawater at their backs, Beseler was forced to order a withdrawal on the evening of the 30th–31st and permanently suspend all offensive operations in the Dixmude–Nieuport sector.[177]

With Fourth Army's right wing stopped, the allies were able to direct reinforcements to the Ypres sector. French reinforcements (IX CA) north of Ypres moved into British I Corps' former positions, allowing the British infantry to shift to the south to concentrate against Fabeck's forces. Further north, the XXXII CA arrived in established positions around Bixschoote (south of Dixmude).[178] These fresh troops easily defeated German Fourth Army's diversionary attacks of the 30th and 31st—thereby confirming for the allied leadership that Albrecht's army was no longer a major threat.

As had occurred in Fabeck's sector, the poor weather greatly restricted Fourth Army *Fliegertruppe's* operations on the 30th. Nonetheless, FFA 38 managed to complete two critically important tactical flights during the day. The first, performed between 9:15 and 10am, was flown over the Belgian positions south of Nieuport at the time of III RK's successful infantry assault. While circling over the battlefield, the aircraft's observer noted that the Belgian line had been broken and that a retreat had already begun. However, due to the low lying clouds the airmen had to fly at a low altitude, resulting in heavy ground fire being directed against the crew. Within minutes the flight's observer had been severely wounded, prompting the pilot to immediately return to the airfield. Upon landing, the observer verbally relayed his findings to a III RK HQ staff officer, who promptly took the information to Beseler for his consideration. The observer ultimately died of his wounds an hour later.[179]

Citing the aviator's report, Beseler confidently ordered his men to push forward and keep pressure on the crumbling Belgian defenses. Shortly after 3pm, a gap in the clouds allowed another aircraft to go aloft. Under directions from III RK HQ, the aircraft scouted the Belgian positions around Nieuport. The subsequent report, delivered in person to III RK HQ, described a total victory for Beseler and his men. Near the coast, German infantry were reported to have reached the Yser River and the eastern outskirts of Nieuport itself. Further south, the Belgian line was confirmed to have been shattered with German forces seen advancing beyond the allied "defensive railway embankment line" in the vicinity of Ramscapelle. Finally, located directly in front of III RK's advancing forces, several allied troop columns were spotted moving in the direction of Furnes—a sure sign that a general withdrawal had been ordered.

FFA 38's afternoon message was received by III RK HQ with great excitement. The aviator's observations had confirmed that Beseler's lengthy struggle to break the Franco-Belgian line had finally succeeded. Moreover, despite the allied attempts to check the German advance via naval gun fire and repeated counter-attacks, III RK had successfully captured Ramscapelle in the north and parts of Pervyse to the south.[180] Thus, with victory in his grasps and a clear path to Dunkirk at stake, Beseler drew up plans on the evening of the 30th for the immediate resumption of the offensive in the direction of Furnes the following morning. Later that evening, however, III RK HQ was made aware of the allied inundations and the rapidly rising waters along the Yser, prompting Beseler to reluctantly order a general retreat to safety.

On the 31st, as III RK was carrying out its withdrawal, FFA 38 performed two flights—one in the morning and the other in the late afternoon. Both flights were ordered to reconnoiter the entire allied position between Dixmude and the channel coast in order to determine if there was going to be any counter-attacks launched against III RK as

Above: With their arrival, the Indian Corps helped solidify the allied position in Flanders. (Author's Collection)

it attempted to retreat across the flooded plain. In both cases, the observers delivered detailed reports confirming that the Franco-Belgian forces in the area showed no signs of engaging the German troops as they withdrew. Elsewhere on the 31st, aircraft from FFAs 6 and 40 carried out tactical reconnaissance sorties with minimal success. However, one long-range reconnaissance flight performed by FFA 6 located three large columns of infantry and artillery moving into the allied line near Bixschoote.[181] This report solidified Fourth Army HQ's assumption that the flooding of the Yser plain had enabled allied reinforcements to be shifted south to the Ypres sector. Albrecht and his staff therefore concluded that Fourth Army no longer had any chance of achieving a major breakthrough. Thus, it was resolved to henceforth use Fourth Army in a limited role by launching successive spoiling attacks in order to prevent any Franco-British troops from being withdrawn to Fabeck's sector.

Positioned to the south of Army Group Fabeck, Sixth Army's right wing resumed their offensive on the 30th against the BEF's extreme right. These formations had recently carried out a series of diversionary attacks from the 27th–29th to help cover Group Fabeck's concentration. These attacks, along with Sixth Army's earlier battles, had greatly weakened Rupprecht's right wing by the morning of the attack—resulting in a significant reduction in their offensive potential. Moreover, the vast majority of the right wing's heavy batteries had been transferred under Fabeck's control by the morning of the 30th. Thus, Sixth Army's chances for success were virtually nil. Instead, as with Fourth Army, all that could be hoped for was to launch a series of successful diversionary attacks that improved Army Group Fabeck's chances of success.

With their right flank resting against 26 ID's left, HKK 1 represented the link between Sixth Army's right wing and Army Group Fabeck. On the morning of the 30th, HKK 1's two cavalry divisions (Guard & 4th KDs) launched a dismounted attack against the British line near the Ploegsteert wood.[182] To their left, the XIX AK pushed towards Armentières. Together, these were the most vitally important element of Sixth Army's right wing. Their ability to gain territory and cover Fabeck's flank was paramount to the entire offensive's success. Unfortunately, the lack of heavy artillery severely hampered the attacking troops' ability to secure any of the territory that they initially gained. In both HKK 1 and XIX AK's sectors, the German attackers were able to fight their way into the British trenches, only to have hastily assembled

British counter-attacks throw them out.[183] Thus, the 30th and 31st saw Sixth Army's extreme right wing unable to push the line forward. However, the group's attack was successful in holding down large numbers of British troops that could have otherwise been shifted north into Fabeck's sector.

South of Armentières, the remainder of Rupprecht's right wing (25th RD, 48 RD, VII AK, and XIV AK) carried out a series of limited spoiling attacks aimed at holding the allied forces to that area of the battlefront. By the 30th, the entire Indian Corps had moved into line around La Bassée to relieve the BEF's badly shattered II Corps. Thanks to the work of their flying sections, the local German commanders were aware of this fact and had taken the Indians' arrival into consideration. Consequently, based upon this and several other factors, the German leadership decided to conserve their manpower and harass the allied troops with constant artillery fire, interspersed with frequent small-scale infantry attacks.[184] This plan ultimately continued within this sector until the end of the battle in mid-November.

The majority of Sixth Army's aviators were grounded throughout the 30th due to the poor weather. The few flights that were completed were performed in the south over the army's left wing where the opposing sides remained on the defensive. By contrast, the 31st saw good weather that resulted in each of the right wing flying sections heavily contribute to the battles raging on the ground. At Armentières, FFA 24 conducted four artillery sorties, directing effective counter-battery fire against several well-hidden British heavy batteries that had impacted the German advance.[185] FFA 24 also dispatched an aircraft late in the day with orders to reconnoiter the allied rear-area and determine if any reinforcements were being drawn away from Fabeck's sector. The subsequent report found that while no major troop movements had been seen to the north or west, the British appeared to be shifting troops from south of Armentières to the Ploegsteert wood area to help challenge XIX AKs advance. These movements signaled to the German leadership that Sixth Army's right wing was successfully exerting pressure on the BEF's right wing and preventing allied reinforcements from being moved into Fabeck's sector.

South of the Lys, FFAs 18 and 20 carried out artillery direction flights in support of their respective corps' attacks. Each section also completed a tactical flight in order to provide updated estimates regarding allied troop strengths in their respective sector. These reports were forwarded to Sixth Army HQ and later to OHL for the intelligence departments' consideration. Meanwhile, Sixth Army HQ's FFA 5 continued its long-range flights around Ypres. Completing three flights that day, FFA 5's airmen discovered large numbers of additional French troops moving to reinforce the failing allied line opposite Army Group Fabeck. Multiple large columns, estimated to be a reinforced brigade consisting of infantry, artillery, and supplies were reported south of Ypres near St. Eloi in close proximity to the fighting. German leaders recognized that these troops would undoubtedly be present in the allied line the following day—significantly impacting Fabeck's chances of success. The final flight of the day, however, delivered even more troubling news. The observer located a much larger French force at Hallebast marching east in the direction of Fabeck's sector.[186] Although this force (correctly estimated to be two divisions in strength) was still a day's march from the battle, its presence signaled an end to Fabeck's localized superiority. In other words, if Fabeck's men were to achieve a major breakthrough, they needed to do it before these French reinforcements made their presence felt.

Fabeck's Attack Continues: November 1–5

November 1st was a critically important day to the success or failure of Fabeck's offensive. In the north, Albrecht's Fourth Army was still reeling from the effects of the inundations of the Yser plain and was therefore unable to turn to the attack. Albrecht's troops were therefore ordered to spend the day on the defensive, resting and reorganizing whenever possible. On Fabeck's other flank, Sixth Army's right wing was expected to continue placing pressure on the allied formations in front of them. It can be reasonably assumed, however, that the German leadership did not expect any major victories from that sector. Thus, the entire operation's success rested solely in the hands of Army Group Fabeck itself, a force that had already committed all of the reserves at its disposal.

With no reserves to draw upon and a rapidly dwindling shell supply, Fabeck and his staff had few options other than to vary their tactics and hope that one more big push could break the allied line. In order to minimize casualties, Fabeck's infantry were directed to carry out night assaults against the strong points of the British line. Accordingly, the 6th Bavarian RD attacked Wytschaete at 1:00am. Unfortunately, the division's undertrained troops proved to be incapable of conducting a coordinated attack in the dark and suffered terrible casualties, many of which were a result of friendly fire.[187] By 9:00am, the division was compelled to withdraw back to Wambeke where it spent the remainder of the day. In the north, Deimling's XV AK resumed the offensive against the British positions between the Comines Canal and the Gheluvelt—Ypres road. Fighting through heavily wooded terrain, Deimling's assault was checked at the eastern edge of Shewsbury Forest, resulting in a total gain of less than a thousand yards by the end of the day.

The Bavarian II AK's attack ran straight into the recently deployed French troops that had been observed by the *Fliegertruppe* the previous day. As a result, the corps' two badly damaged divisions struggled to advance throughout the afternoon. Nonetheless, the Bavarians did manage to gain some ground by capturing the small woods west of Oosttaverne (one mile south of St. Eloi).[188] To II Bavarian AK's left, the 6th Bavarian RD returned to the attack around 4pm after spending the morning and early afternoon regrouping after their failed night attack. With the 26 ID in support, the Bavarians fought their way into Wytschaete at 5pm. Unfortunately, the 26 ID (covering the Bavarian's left flank) was unable to seize the heights northwest of Messines until after dark. This allowed the British to subject the Bavarian reservists to murderous enfilade fire during their advance into Wytschaete. What remained of the 6th Bavarian RD was easily driven back out of the town by a French counter-attack.[189]

November 1st was a beautiful fall day perfectly suited for low level flying. By 11:00am each of Fabeck's flying sections had aircraft airborne prepared to begin directing artillery fire. In the north, the airmen of FFA 3 completed two flights before noon, assisting the heavy batteries range in on the new British line after it had been moved back during the night. Once the infantry assault began however, the airmen quickly discovered that they were unable to help any further. Indeed, the heavily wooded terrain that the battle was raging in made it impossible for the observers to discern friend from foe. Thus, with his men unable to direct artillery fire or accurately report on the progress of the battle, FFA 3's commander kept his machines grounded for the rest of the day.[190]

Meanwhile, the Bavarian II AK's FFA 2b performed three flights on the 1st. The initial sortie was an artillery direction flight in support of the corps' right division (4th Bavarian ID) in an area north of the Comines Canal. The section's second and third flights were both tactical missions aimed at determining the strength of the allied force in the corps' sector. Reports of French troops at Wytschaete had confirmed II Bavarian AK's suspicions that the British were being reinforced. Upon receiving news of the French troops' arrival, FFA 2b was ordered to send an aircraft airborne to investigate. The observer ultimately confirmed the presence of several newly placed batteries, as well as several small columns nearing the front. The third flight, conducted late in the day during 6th Bavarian RD's attack, also confirmed the placement of new French batteries but was unable to determine the strength of the allied forces in the area due to the mayhem of the battle.

On the group's extreme left, FFA 4 flew several artillery spotting sorties throughout the morning and afternoon to help locate several British heavy batteries that were bombarding Messines and slowing up 26 ID's progress. The section also dispatched a lengthy tactical flight around midday across the entire breadth of the allied position within Group Fabeck's sector—a front of roughly 11 miles. It was difficult, however, to accurately report on the battle's progress from the air. The observer's testimony was therefore of little value and had no bearing on the offensive's outcome.

After receiving word from Sixth Army HQ regarding FFA 5's discovery of large allied columns marching in their direction, Army Group Fabeck HQ ordered its own FFA 4 to reconnoiter the roads throughout the area. The subsequent report confirmed that several large columns of French troops were indeed nearing the battlefront to further strengthen the allied defenses. According to the observer's estimate, the columns totaled two divisions of infantry, supported by a group of artillery and cavalry. The leading group of columns was seen in the area of Kemmel, marching east on the Wytschaete road. Further west, the second group was recorded in the area of De Klyte with its rear components stretching back to the southwest near Bailleul.[191] This report confirmed without any doubt for Fabeck and his staff that their men's numerical advantage would gradually erode over the next 24–48 hours. Army Group Fabeck HQ henceforth understood that if a breakthrough were to occur, it had to be done *before* the allied line was reinforced. Thus, Fabeck resolved to relax the restrictions on his group's artillery usage and resume the offensive across his entire front the following morning in order to force a decision.

As it happened, Army Group Fabeck's efforts to breakthrough the allied line on November 1st had failed. Indeed, each division was unable to push the German lines forward any further than 1,000 yards. Fortunately, some progress was made after dark when the 26 ID was finally able to move through Messines and occupy the critically important hills on the town's western face. Thus, with the 'Messines Ridge' fully under German control, Fabeck's center was now free to focus against Wytschaete. The best news of the night, however, was the arrival of 3 ID into the battle-area. The division had spent the previous two days en-route from First Army's sector to Flanders to provide Fabeck with an additional reserve. Upon learning of the capture of Messines Ridge, Fabeck ordered 3 ID to move straight into the fighting line and prepare to carry out an attack against Wytschaete the following morning.[192] If successful, the attack would break the allied line, restoring mobility to the sector before the next French reinforcements arrived.

November 2nd was yet another beautiful fall day in the Ypres sector. The capture of Messines Ridge, in combination with the arrival of the elite 3 ID to take the place of the half-trained troops of 6th Bavarian RD, had combined to bolster Group Fabeck HQ's confidence.[193] With knowledge of the pending deployment of more French troops into their sector, Fabeck recognized November 2nd as the group's last chance to decisively breakthrough the allied positions. Therefore, he ordered his troops to resume the offensive, vigorously press home their attacks, and force a decision. To increase the operation's chance for success, the group's artillery (now positioned closer to the front with spotters positioned on Messines Ridge) was permitted to expend more shells on the 2nd than during the first three days of the offensive combined.

For the Allies, November 2nd was also expected to be a decisive day. The major redeployments as a result of the inundations of the Yser Plain had, by the evening of the 1st, given the French a sizeable numerical and qualitative advantage north of Ypres with parity achieved in the south opposite Army Group Fabeck. Foch had directed the IX and XXXII CAs as well as the 2nd French Cavalry Corps to begin a new offensive north of Ypres on the 1st with the objective to split Albrecht's Fourth Army in half.[194] The attacks were quickly halted by well-directed German artillery, alerting Albrecht's reservists who were installed in formidable defensive positions. Completely convinced of German Fourth Army's qualitative inferiority, Foch ordered the attacks to continue on the 2nd. Meanwhile, the deployment of the first French troops in the British sector opposite 'Group Fabeck' had prompted Foch to order an offensive that same day to recapture Messines and all the territory captured thus far during Fabeck's offensive.[195] In other words, November 2nd saw both the French and Germans launch simultaneous offensives against one another within Group Fabeck's sector with both sides attempting to deal a major blow to their opponent.

Fabeck's main effort on the 2nd was directed against the allied stronghold at Wytschaete—now defended by the French. Participating in the assault from left to right were the 11th *Landwehr* Brigade, 26 ID, 3 ID, and the 6th Bavarian RD.[196] Beginning at 7:00am, Fabeck's batteries initiated the German preliminary bombardment. Shortly thereafter, the Franco-British batteries opened up in response, commencing a large artillery duel that continued throughout the morning. At 8:30am, Fabeck's infantry began their advance towards the allied line. Here, newly deployed French batteries (many of which were placed outside the view of forward artillery observers) were able to inflict grievous losses on the advancing German troops before they reached the allied line. Nonetheless, the surviving attackers successfully reached the French position and drove them back through the town in fierce house to house street combat that lasted well into the afternoon. At 3:10pm the 3 ID, which had reached its objective southwest of Wytschaete, was called upon to stamp out the last of the French resistance. By 5:00pm, the majority of French defenders had been expelled from the town—allowing the Germans to establish a defensive line on its western edge.[197] Finally, after three days of bloody fighting, Wytschaete and the important high ground that surrounded it, was in German hands.

To the north, the Bavarian II AK's advance was bluntly stopped by the French troops carrying out Foch's counter-attack. From 11am until dark, the opposing sides clashed along the Comines Canal with neither side making any noteworthy progress. Meanwhile at Messines, the 11th *Landwehr* Brigade and 26 ID successfully repulsed an allied assault late in the day, thereby securing Fabeck's left flank. Located on the group's right wing, the XV AK tried once again to assault the British positions located in Shrewsbury Forest. Despite a simultaneous attack from XXVII RK on their right, Deimling's forces ultimately failed to move their line forward.[198] Thus, with his right wing's failure to achieve a decisive breakthrough, Fabeck was forced to admit that his strategy to push the battle to a conclusion before the last French reinforcements arrived had failed. Nonetheless, his forces were in possession of the heights, both at Messines and Wytschaete and were in a position to continue the offensive the following day towards the Ypres sector's highest point—Mount Kemmel. To help achieve this goal, Fabeck's headquarters consolidated 3 and 26 IDs into an ad-hoc corps under the command of Wilhelm Karl, Duke of Urach (previously commanding 26 ID). This group was to continue the offensive in the direction of Kemmel the following morning.

Fabeck's airmen were once again highly active on the 2nd. Having learned from their experience the day before, XV AK's FFA 3 performed one artillery direction flight in the early morning during the preliminary bombardment as well as one tactical reconnaissance flight in the late afternoon to determine if there were any troop movements into the Shrewsbury Forest. FFA 2b also had an aircraft airborne during the morning's preliminary bombardment. Once Bavarian II AK's troops became heavily engaged in close combat, the airmen recognized they were no longer needed and returned to their airfield at Verlinghem (northwest of Lille). Later that day II Bavarian AK HQ received a report from their ground troops that a gap in the French line had opened up and could possibly be exploited. The corps' leadership promptly ordered a quick tactical flight be made over the battle-area to gather additional details. The observer promptly returned with word that the breech indeed existed but was being closed by a French counter-attack at the time the observer had reconnoitered the area.

FLA 1b's balloon had begun the offensive on October 30th tethered to a position east of the Lys River. Then, the balloon and all of its supporting vehicles and supplies spent the next 48 hours crossing the Lys and moving into a position to resume aerial operations. By the 2nd the airmen were once again airborne in an area north of Messines to perform flash spotting and determine the location of allied batteries.[199] Meanwhile, Fabeck's own FFA 4 worked with heavy batteries attached to 26 ID on the group's extreme left. With the observer's assistance, the heavy batteries engaged and silenced several groups of allied guns operating in support of their counter-offensive to recapture Messines. FFA 4 also dispatched two long-range reconnaissance flights during the afternoon to continue monitoring the progress of the French reinforcement columns. According to the final report, the French column was seen in the area west of Kemmel around 5:00pm. This indicated that the French reinforcements would be present in the allied battle line the following morning.

That evening after reviewing the various reports from his airmen, Fabeck sent a message to Rupprecht at Sixth Army HQ. Fabeck's communiqué informed Rupprecht of the day's lackluster results as well as FFA 4's disturbing discovery. Rupprecht, for his part, had by that time suspected that Army Group Fabeck no longer possessed the necessary strength to achieve a meaningful breakthrough south of Ypres. Thus, the news of the arrival of more French reserves into the battle-area simply confirmed his suspicions. From the very beginning of the campaign, Rupprecht focused on the offensive south of Ypres and continued to believe that the area offered the best chance for success. Consequently, despite Army Group Fabeck's predicament on the evening of the 2nd, Sixth Army HQ refused to support any alternative other than continuing the offensive within the Fabeck salient with additional reinforcements.[200] The German General Staff's official monograph of the battle summed up Rupprecht's logical reasoning as follows:

During the early days of November the commander of Sixth Army came to the conclusion that the offensive of the Army Group Fabeck could lead to no decisive results. The forces available were still too weak to breakthrough the enemy's strongly entrenched positions, particularly as he was continually bringing up reinforcements to the battlefront. If the attempt to breakthrough south of Ypres was not to be entirely abandoned, and a purely defensive war on the Western Front thereby avoided, more troops would have to be brought up for the Ypres battle from other sectors of the front.[201]

A new plan was therefore crafted during the evening of the 2nd to transfer five more divisions (three infantry and two cavalry) to Fabeck's sector as quickly as possible. Additional heavy batteries from Sixth Army's left wing were also to be brought under Fabeck's control. Rupprecht had met with several of the commanders under Fabeck and had learned that the shell restrictions were partially to blame for the offensive's initial failure to secure a breakthrough. Rupprecht therefore gave Fabeck's batteries all the heavy artillery ammunition allotted to Sixth Army. With more troops and heavy guns and substantially more shells to support the attacks, Rupprecht believed Group Fabeck could achieve its mandate.[202] Unfortunately, however, the various reinforcements would take several days to move into the battle-area—leaving Group Fabeck's forces, as they were then constituted, to continue the offensive by themselves.

Despite a continuance of heavy fighting, the battles of the 3rd, 4th, and 5th resulted in very little movement of the front line or change in the general situation.[203] For three days Fabeck's heavily damaged and completely exhausted divisions attempted in vain to breakthrough the recently reinforced allied line. Meanwhile, the allies carried out Foch's planned counter-offensive with similarly unsuccessful results. In short, the opposing offensives had left the opposing lines more or less unchanged by the evening of the 5th.

Fabeck's flying sections continued to be active during the 3rd, which happened to be yet another gorgeous day. On the right, FFA 3 conducted artillery direction sorties during the morning, with each flight dropping bombs on opportune targets before returning to their airfield. FFA 2b similarly performed artillery work throughout the day. One flight accomplished north of the Comines Canal shortly before noon was noted as promptly and accurately directing fire onto a large group of attacking French infantrymen, causing the attack to be aborted. On the right, FFA 4 carried out artillery direction and tactical reconnaissance flights in support of "Corps Urach's" drive towards Kemmel. The allied attacks of the previous day had caused Group Fabeck HQ to suspect another attempt to capture Messines would be forthcoming. Therefore, two reconnaissance flights were ordered (one in the morning and the other in the afternoon) to report on troop movements and activity opposite the group's left flank.[204] As it happened, no attacks came to fruition that day. However, the orders for continual reconnaissance over the important area served as an example of how important aerial reconnaissance had become due to the nature of static warfare—something officers like Fabeck and Loßberg had already showed signs of acknowledging.

November 4th brought thick mist and rain to Flanders, grounding all aerial activity in the region. Likewise, the 4th also saw a significant reduction in the quantity and severity of attacks made by Fabeck's troops. Consequently, the day was little more than an artillery duel, as Fabeck's troops waited for the reinforcements to arrive. Conversely, the allied leadership interpreted the Germans' relative inactivity over the previous several days to mean the campaign had ended.[205] Joffre, for example, ordered Foch on the 4th to abandon any offensive plans, stabilize the front, and reconstitute the reserves—showing the French commander's desire to suspend operations for the winter. Likewise, the British leadership believed the German's attempt at a breakthrough was now over and that any future attacks would be merely a covering action to allow a redeployment of their forces to the east to face Russia.[206]

Favorable weather on the 5th once again allowed Fabeck's aviators to resume flight operations. Keen to conserve their men's strength for the upcoming offensive, Group Fabeck HQ restricted operations to limited attacks only. One of these, carried out by 26 ID in the vicinity of Messines, successfully expelled a French force from Spanbroekmolen—forcing the entire allied line to move back another kilometer towards Kemmel. To the north, the French attacks to recapture Wytschaete failed. Meanwhile in XV and II Bavarian AK's sectors, the German heavy artillery spent the morning and afternoon pounding the allied lines in preparation for the renewal of the offensive. In support of these operations, Group Fabeck's flying sections collectively performed only three flights the entire day. This drop of activity was due to maintenance concerns regarding the remaining aircraft. Indeed, after nearly two weeks of active campaigning, the flying section's remaining airworthy aircraft had been pushed to the limits of their endurance and needed service in preparation for the more important battles to come.[207] Thus, outside FFA 4's two tactical reconnaissance flights and FFA 2b's artillery direction sortie, there was no aerial activity on the 5th, ostensibly so each of the sections could be fully prepared to actively support operations during the resumption of the offensive. Unbeknownst to the airmen, however, the recently excellent flying weather was about to turn—keeping them grounded for nearly the remainder of the battle.

The Battle's Conclusion: November 6th–17th

Army Group Fabeck's failure to breakthrough the allied line south of Ypres was a major setback for Falkenhayn and the German leadership's campaign strategy. It had become clear to Falkenhayn by November 4th that the campaign's original goals of "breaking through the allied line, enveloping their left flank and seizing the channel ports" were not going to be accomplished. Nonetheless, with large numbers of reinforcements (three infantry divisions) currently on their way to Flanders, the German commander was unprepared to suspend offensive operations. Instead, Falkenhayn's OHL issued orders narrowing the offensive's objectives to more attainable "limited goals" that could be accomplished before the onset of winter. Fourth Army, with the assistance of these reinforcements, was directed to capture Ypres, which was correctly identified as "the central point of the enemy's defensive position." Meanwhile, a rested and 'beefed up' Army Group Fabeck was to seize the critically important high ground around Kemmel.[208] Achieving these objectives, while certainly not decisive, would have at least given the Germans something tangible to show for the heavy losses incurred up to that point in the campaign.[209] It would furthermore put the German right wing in an excellent position for resuming its drive to seize the channel ports the following spring, as Falkenhayn had hoped to do.

The city of Ypres was the nerve center of the entire allied position in Flanders. The excellent network of roads and rail lines leading in and out of the city enabled the allied commanders to quickly move troops and supplies laterally across the front. Its loss would therefore greatly inhibit the allies' ability to defend their position in Flanders during future operations. More importantly, the loss of Mount Kemmel would spell certain disaster for the allies. At a height of 400 feet above sea level, the ridgeline at Kemmel held a commanding view over the entire sector. From its summit, it was possible to see Lille, Menin, and even Dixmude.[210] German heavy batteries would therefore be able to use these heights to decimate the allied positions with devastating enfilade fire—forcing the allied armies to abandon their positions and withdraw to safety along a line much closer to Dunkirk.

With knowledge of the new plan and reinforcements on their way, Fourth Army and Group Fabeck HQs spent November 6th–9th actively carrying out small scale 'spoiling attacks' to deny the allied forces any opportunity to rest or reorganize as well as to gain small pieces of tactically important territory whenever possible.[211] German troops took advantage of thick fog throughout the 6th to surprise and throw back a contingent of French troops defending the

Above: Propaganda postcard romantically depicting the combat outside the town of Ypres. (Author's Collection)

allied line north of the Comines Canal. Fabeck's flying sections were grounded throughout the day due to persistent heavy fog. Even without aerial intelligence, the attack pierced the French line in three places, allowing troops of XV AK's 39 ID to seize Zwarteleen (one mile southeast of Zillebeke).[212] Elsewhere, the II Bavarian AK was able to extend its front to within 1,300 yards of St. Eloi—a town that held the critically important road which ran south from Ypres across the entire allied line. By the end of the day, the German advance had pushed forward to within two miles of Ypres itself.

Foggy and misty weather continued to hover over Flanders on the 7th, keeping all aircraft grounded. In an effort to keep pressure on the allied forces, Fabeck's artillery relentlessly pounded their enemy's supply lines. The towns of Ypres and Armentières, as well as all nearby roads, were continuously shelled throughout the day. The absence of the *Fliegertruppe* over the battlefield, however, greatly diminished Fabeck's ability to build upon the gains made the previous day. For example, a severely weakened French force held a large section of the allied line north of the Comines Canal throughout the morning. This group had been hotly engaged and defeated the previous day and was reduced to half their original strength. If heavily attacked, this large and important portion of the allied line would almost assuredly had been shattered, enabling German troops to advance north and envelop the British forces defending the Menin Road. Unfortunately, without any tactical reconnaissance reports, the German leadership remained unaware of how weak the allied north of the canal truly was. Consequently, no serious attacks were ordered against that section of the line. Instead, Group Fabeck's attacks were directed against the British portion of the line further north.[213]

As it happened, the 7th brought little change to the position of the opposing lines. Fighting broke out in five isolated localities. Each of the attacks were separated from one another and unable to lend mutual support, resulting in no significant gains. Shortly after dawn, an allied counter-attack carried out by the BEF's 22nd Brigade made some initial gains in an attempt to reclaim Zwarteleen but was forced to eventually withdraw back to their starting positions under threat of encirclement. Meanwhile, the German attacks carried out north of the Comines Canal near the Menin Road managed to temporarily pierce the allied line and eject several battalions of British troops from their trenches. Consecutive allied counter-attacks, supported by the British 1st Division's artillery,

Above: The German Marine Brigade launches an attack on Lombartzyde during its famous attack on November 10, 1914. (Author's Collection)

successfully reclaimed a majority of the lost territory. When the fighting finally died down that evening, a new allied trench line was established 100 yards short of the old one.[214] Similarly in the south, two regiments of Sixth Army's XIX AK pierced the allied defenses in the Ploegsteert Wood only to be incessantly counter-attacked the remainder of the day. The Saxon infantry were therefore unable to advance beyond the line of their initial gains—a meager 500 yards into the woods between St. Yves and Le Gheer.[215]

November 8th and 9th were relatively uneventful days as preparations for the German renewed offensive continued. Although there was negligible precipitation, flying was yet again ruled out due to the heavy blanket of low lying clouds. While certainly frustrating for the airmen and the army, the four day stretch of inactivity had given the flying sections' ground crews ample time to repair and improve their remaining aircraft, many of which had been grounded due to mechanical issues. Thus, the strength of the German right wing's *Fliegertruppe* significantly increased in the days leading up to the campaign's final offensive. Unfortunately, the winter weather would continue to linger, negating aviation's role in the final phase of the campaign.

Late in the afternoon of the 9th, the long-awaited German reinforcements finally arrived in the battle-area. Deployed directly on Group Fabeck's right within Fourth Army's sector, the 4 ID and General Arnold von Winckler's Guard Division were formed into an ad-hoc corps and combined with the XV AK to form a new Army Group under the command of General Linsingen and the II AK HQ staff. Attached to 'Group Linsingen' were the Guard Corps' FFA 1 and II AK's FFA 30. With five airworthy aircraft (four LVG B.Is and one Albatros B.I), FFA 1 was assigned to the newly arrived ad-hoc corps that was under the overall command of Plettenberg and the Guard Corps HQ staff. Meanwhile, FFA 30 (two Albatros B.Is, one LVG B.I and one Euler B-Type) was attached to Linsingen's HQ and expected to fly close tactical support for either of the group's two corps, or long-range reconnaissance for Linsingen's staff. Additional reinforcements had also arrived in the south where the 25 RD was inserted into Group Fabeck's line near Wytschaete. In addition to six cavalry divisions, the 25 RD was expected to strengthen Fabeck's line and enable the battered group to effectively cover Linsingen's flank and seize their objective at Mount Kemmel.[216]

With all their troops in position, the German's offensive was set to commence across the front on the morning of the 10th. However, the absence of aerial activity caused by the poor weather had left the German leadership with no means of gathering critical information concerning the allied defenses prior to the attack.[217] It was therefore decided to postpone Linsingen and Fabeck's attacks until the 11th in order to give Linsingen's troops the ability to personally reconnoiter the area and make all the necessary preliminary arrangements.

Meanwhile, Fourth Army was ordered to proceed with its attack on the 10th with the objective of tying down allied troops and keeping any additional forces being moved into Linsingen and Fabeck's sector. The day had begun misty and foggy just like the days that had preceded it, grounding the flying sections but allowing the infantry to carry out their advance in relative safety. In the north, the XXII RK's 43 RD attacked the French garrison defending the Yser Canal at Dixmude on three sides. After a lengthy and bloody struggle, the French defenders were driven back across the canal, allowing the Germans to finally gain possession of the important town.[218] On XXII RK's left, the XXIII RK, III RK, and 9 RD launched a series of bloody, yet unsuccessful attacks along the line Hoekske—Bixschoote—Langemarck—St. Julien. While these attacks failed to pierce the French line or make any gains, they were successful in spoiling a large scale allied operation that was to have been made later that morning.

On Fourth Army's extreme left, the XXVI and XXVII RKs jointly attacked the Franco-British troops holding the northern end of the Ypres Salient. Facing the Allies' stout defenses in and near the Polygon Wood, the attacking infantry were unable to press home any of their attacks on account of heavy artillery fire. By mid-afternoon, the local division commands made the decision to call off the attacks and subject the allied defenses to artillery fire in preparation for Linsingen's attack the following morning.

All things considered, the attacks of the 10th improved the overall situation for the Germans on the eve of their final major offensive. The capture of Dixmude and the Yser Canal had secured the attacking German forces' flank. Furthermore, the attacks northeast of Ypres around Langemarck, although unable to pierce the allied line, successfully convinced the allied leadership to send reinforcements to the area. Indeed, both the French and British leadership had suspected a renewed offensive might be possible but did not know when or where it would come. Thus, when the allied leadership learned of the impressive attacks in Fourth Army's sector, they became convinced that the great German offensive

Above: German marines attack Belgian positions near the coast. (Author's Collection)

had already begun and immediately redeployed reserves from Fabeck's sector to Langemarck. This drew badly needed troops away from the very sector that was to be attacked by Fabeck and Linsingen the following day.

By 2:00pm the weather around Ypres had improved enough to safely perform aerial operations. Between 2:00 and 5:30pm, a plethora of eager airmen from both sides crowded the airspace from Dixmude to Messines. In the north, an aircraft of FFA 40 was sent to report on the situation at Dixmude. The same crew ultimately made two flights over the area, giving XXII RK HQ information regarding the actions and strength of allied troops north and west of the canal that were being moved into position for a counter-attack. This valuable intelligence was quickly forwarded to 43 RD HQ, alerting them of the pending attacks which occurred throughout the night without success. In Linsingen's sector, FFAs 1, 3, and

Above: German troops assault the allied field fortifications near Ypres. (Author's Collection)

30 each had aircraft airborne in an attempt to appraise the allied strength in their sectors as well as to map the location of hostile field batteries. To the south, aircraft from FFAs 2b, 4, and 24 performed long-range reconnaissance flights over the allied front line and rear-areas in order to furnish the army's leadership with updated information to assist with last minute changes to the attack plan. Indeed, Linsingen and Fabeck's staffs used several of the *Fliegertruppe's* reports later that evening as they sketched maps of the allied defenses that were to be used during the following day's assault.

Yet another dull and misty morning greeted the Germans on the morning of the 11th as the final preparations for their offensive were being completed. The plan was relatively simple. Deployed on both sides of the Menin—Ypres road, Group Linsingen was to act as the attack's spearhead and "drive back and crush the enemy lying north of the Ypres—Comines Canal" while Group Fabeck attacked south of Ypres in support.[219] Following a highly intense preliminary bombardment, the attacking divisions were directed to pierce the allied line and press forward as quickly and vigorously as possible. It was therefore hoped that the town of Ypres could be seized in the days immediately following the initial breakthrough.

Unfortunately for the Germans, the poor weather prevented all flying activity, which would have almost exclusively consisted of artillery direction sorties in support of the advancing troops. Thus, many of the smartly concealed allied batteries were never located and silenced during the preliminary bombardment—resulting in large numbers of casualties for the attacking German infantry. Nevertheless, Linsingen and Fabeck's bombardments lasted two and a half hours, effectively destroying large portions of the allies' front line trenches.[220] Around 10am the shelling reached its peak, enabling the German infantry to begin their advance. Using both the bombardment and the fog as cover, the attacking infantry rushed to within a few hundred yards of the allied line. Despite clear orders for all units to attack with utmost energy, only the two newly arrived divisions of Plettenberg's corps followed the directive. The other divisions, perhaps too exhausted from nearly two weeks of continuous combat, did not 'press home' their assaults—thus forfeiting any advantages created by the preliminary bombardment.

Attacking as directed, Plettenberg's troops successfully pierced the British line in several places, continuing into their rear. This area of the allied line, however, was well suited for defense, with multiple heavily wooded areas that the Germans were compelled to traverse. Thus, while the Guard Division's two brigades were split up in these woods, a British counter-attack struck the northern brigade in its flank, compelling the division to halt their advance for the remainder of the day.[221] On the Guard's immediate left, the men of the 4 ID were held up by well-placed defensive obstacles and then cut down by British rifle and machine gun fire.[222]

To the south, Deimling's XV AK failed to conduct their attack in the aggressive manner that the Guards had done, resulting in a lengthy and inconclusive firefight.[223] The corps' two divisions immediately became more aggressive however, upon receipt of an order from Linsingen's HQ requesting assistance be given to 4 ID. Soon thereafter, amidst the afternoon's heavy rains, XV AK pushed the French defenders back to Hill 60 near Zwarteleen, which was subjected to heavy fire for the remainder of the day, thus nullifying its use as a useful allied artillery observation platform.[224] Meanwhile, the attacks in Fabeck's sector gained very little ground due to highly effective allied artillery fire originating from the heights on Mount Kemmel. Likewise, all of Fourth Army's renewed attacks failed to even threaten to break the allied defenses.

The attacks of the 11th were a major disappointment for Falkenhayn and the Germans. Despite once again achieving a sizeable numerical advantage, the attacking German troops proved unable to decisively break the allied line. Poor weather certainly played a key role in the failure. Precipitation and fog had prevented both air and ground observers from directing German artillery fire onto problematic targets. More importantly, many of the formations used in the attack were hungry, tired, and dangerously short on junior officers. These factors undoubtedly led to the attacking formations' failure to vigorously press home their attacks in the immediate aftermath of the preliminary bombardment. Nonetheless, regardless of the causes, it was clear to Falkenhayn that the attacks had failed and the offensive was over.

The final five days of the battle (November 12th–17th) were relatively quiet as the exhausted and battered troops on both sides of the line began to dig in, reorganize, and prepare for winter. The weather throughout this five day period saw a dramatic drop in the temperature, accompanied by incessant precipitation. Frigid rain gave way to snow on the morning of the 15th, which continued to fall until the 20th.[225] Needless to say, the *Fliegertruppe* remained grounded throughout this period, which was just as well considering the lack of major combat.

With news of a growing crisis on the Eastern Front, Falkenhayn ordered a halt to all offensive operations in Flanders on November 17th.[226] This order officially ended the German commander's two month long bid to recover German fortunes by enveloping the allied left and winning a decisive victory before the allied front could be fortified.[227] With little alternative, Germany's western armies were all ordered onto the defensive. Thus, "the war of movement" (*Bewegungskrieg*) in the West was now over as the opposing sides dug in and prepared to fight a lengthy attritional war in the trenches.

Chapter 5 Endnotes

1 From the personal letters of Josef Suwelack.
2 Falkenhayn, *General Headquarters, 1914–1916, and its Critical Decisions*, pp.11–12.
3 From his first day in command, Falkenhayn displayed an amazingly farsighted appraisal of the strategic situation. Although, at this point he still believed a quick decision was still possible in the west, it is clear from decisions like these that he believed it imperative to secure the early German territorial gains and protect the resource rich western territory of the Empire in the event of a long drawn out conflict in France, which he understood was a very real possibility. Moreover, he recognized Britain as Germany's most dangerous enemy and knew possession of the coast was essential to eventual German success over the British Empire. For Falkenhayn's own take on altering the original plan see: Erich von Falkenhayn, *Critical Decisions at General Headquarters 1914–16* (Nashville, 2000), p.13.
4 Changes to German tactics during the Napoleonic era are covered masterfully in: Peter Paret, *Yorck And The Era Of Prussian Reform 1807–1815* (Princeton, 1966), pp. 154–190. For more information on the evolution of the theory of envelopment in German Military Tactics, see: Hughes (ed.), *Moltke On the Art Of War*, pp.50–58; Foley (Trans. & ed.), *Alfred von Schlieffen's Military Writings*, pp.183–218; Citino, *The German Way of War*, pp.104–190.
5 Donnell, *Breaking the Fortress Line*, pp.180–183.
6 Donnell, *Breaking the Fortress Line*, pp.181–182.
7 Erich von Tschischwitz, *Antwerpen 1914* (Berlin, 1924), p.34.
8 Donnell, *Breaking the Fortress Line*, p.183.
9 Tschischwitz, *Antwerpen 1914*, p.34.
10 The Franco-British governments had promised the Belgians support. Albert, however, understood that the Allied armies were too far away (125 miles) to raise the siege before the city capitulated. Donnell, *Breaking the Fortress Line*, p. 187. For more on Albert during this period of the war, see: Charles d'Ydewalle, *Albert King of the Belgians*, Translated

by Phyllis Mégroz (London, 1935), pp.125–131.
11 J.E. Edmonds, *Military Operations France and Belgium 1914*, Vol.2 (1925), pp.35–37.
12 Kriegsministerium, MI Nr. 3531/14 A1, 16 August 1914, in USNA, Documents of the Royal Prussian Military Cabinet, M962, Roll 3.
13 Otto Schwink, *Ypres 1914: An Official Account Published by Order of the German General Staff*, Translated by G.C.W (Nashville 1994), pp.4–5.
14 *Der Weltkrieg*, V, p.276.
15 *Der Weltkrieg*, V, p.279.
16 Ferko–UTD, Box 515, Folder 12, "Die Angriffsoperation des Deutschen rechten Heeresflügels in Flandern vom 10. Oktober bis 11. November 1914," pp.4–5.
17 Ferko–UTD, Box 515, Folder 12, "Die Angriffsoperation des Deutschen rechten Heeresflügels in Flandern vom 10. Oktober bis 11. November 1914," p.5.
18 Ferko–UTD, Box 515, Folder 12, "Die Angriffsoperation des Deutschen rechten Heeresflügels in Flandern vom 10. Oktober bis 11. November 1914," p.6.
19 Joint message from FFA 6 and 38. Ferko–UTD, Box 515, Folder 12, "Die Angriffsoperation des Deutschen rechten Heeresflügels in Flandern vom 10. Oktober bis 11. November 1914," p.7.
20 Ferko–UTD, Box 515, Folder 12, "Die Angriffsoperation des Deutschen rechten Heeresflügels in Flandern vom 10. Oktober bis 11. November 1914," p.7.
21 *Der Weltkrieg*, V, p.296.
22 From Fourth Army HQ Ia "Intelligence and reports from 11 through 21 October 1914."
23 Report from FFA 6 to Fourth Army HQ, October 12th. Ferko–UTD, Box 515, Folder 12, "Die Angriffsoperation des Deutschen rechten Heeresflügels in Flandern vom 10. Oktober bis 11. November 1914," p.8.
24 Report from FFA 38 to Fourth Army HQ, October 12th. Ferko–UTD, Box 515, Folder 12, "Die Angriffsoperation des Deutschen rechten Heeresflügels in Flandern vom 10. Oktober bis 11. November 1914," p.8.
25 Edmonds, *Military Operations in France And Belgium 1914*, Volume 2 (1925), pp.65–67.
26 Ferko–UTD, Box 515, Folder 12, "Die Angriffsoperation des Deutschen rechten Heeresflügels in Flandern vom 10. Oktober bis 11. November 1914," p.13.
27 This was the 58th Reserve Division redeploying from the French Second Army to the Tenth Army. See: FOH, Tome 1, Vol. 4, annexes 2653 and 2716.
28 Ferko–UTD, Box 515, Folder 12, "Die Angriffsoperation des Deutschen rechten Heeresflügels in Flandern vom 10. Oktober bis 11. November 1914," pp.15–17.
29 Sheldon, *German Army at Ypres 1914*, p.36.
30 The reports of sizeable French columns marching northward toward Flanders only served to strengthen this assertion. Hauptstaatsarchiv Stuttgart (hereafter: SHStA), M 33/2 Bü 14, *Nachrichten, Meldungen und Skizzen betr. Operationen in den Argonnen, im Gebiet von Valenciennes und in Flandern*.
31 *Der Weltkrieg*, V, p.281.
32 This is an excellent example of the contrast between Moltke and Falkenhayn. Whereas Moltke would give important orders and rarely even send a staff officer to further elaborate his plan, Falkenhayn went to great lengths to personally discuss all aspects of an operation in order ensure all commanders involved fully understood his orders.
33 Kronprinz Rupprecht, *In Treue Fest: Erster Band*, p.206.
34 Ferko–UTD, Box 507, Folder 2, "FFA 38 Reports."
35 Ferko–UTD, Box 515, Folder 12, "Die Angriffsoperation des Deutschen rechten Heeresflügels in Flandern vom 10. Oktober bis 11. November 1914," p.21.
36 Ferko–UTD, Box 515, Folder 12, "FFA 4 Report, October 13 1914."
37 The unrecognized enemy forces were French troops belonging to 'Group Bidon,' sent to reinforce the Allied positions in Flanders. FOH, Tome 1, Vol.4, Annex No. 2772.
38 Sheldon, *German Army at Ypres 1914*, pp.55–56.
39 Sheldon, *German Army at Ypres 1914*, p.53.
40 Lille was an incredibly important asset to the Allied cause. The city was France's fifth largest, producing an amazing 80% of the country's textiles. Greenhalgh, *Foch in Command*, p.53.
41 Greenhalgh, *Foch in Command*, pp.53–54.
42 Troops from the XIX AK captured the city, forcing its garrison to surrender.
43 FOH, Tome 1, Vol. 4, Annex No. 2705.
44 Two allied cavalry corps were also present. The BEF's Cavalry Corps was placed on the extreme left of the Allied line while De Mitry's French Cavalry Corps was deployed between the British II and III Corps.
45 Poseck, *The German Cavalry in Belgium And France 1914*, pp.183–193, pp.201–205.
46 For an excellent description of the cavalry corps' efforts during the opening phase of the battle, complete with numerous first-hand accounts, see: Sheldon, *German Army at Ypres 1914*, pp.12–33.
47 For description of II and III Corps actions on the 13th see: Edmonds, *Military Operations in France*

And Belgium, Vol. 2, pp.80–82, pp.95–97.
48 Peter Hart, *Fire and Movement: The British Expeditionary Force and the Campaign of 1914* (New York, 2015), pp.270–279; Adrian Gilbert, *The Challenge of Battle* (New York, 2014), pp.209–213.
49 Edmonds, *Military Operations in France And Belgium*, vol. 2, pp.103–104. Sir John French, *1914* (Leonaur, 2009) pp.181–190.
50 BEF GHQ Army Operation Order No. 38.
51 Edmonds, *Military Operations in France And Belgium*, vol. 2, pp.110–115.
52 Edmonds, *Military Operations in France And Belgium*, vol. 2, p.85.
53 The crew of the shot down Taube survived the crash and was taken prisoner. For an account of the event from the British perspective see: Sansom, RFC Ypres Report, The National Archives of the UK (hereafter: TNA), CAB 45/142.
54 III RK HQ—orders and messages—October 10–November 28 1914. Ferko–UTD, Box 515, Folder 12, "Die Angriffsoperation des Deutschen rechten Heeresflügels in Flandern vom 10. Oktober bis 11. November 1914," p.25.
55 Albrecht's own FFA 6 had attempted one flight on the 14th but the aircraft was forced to abort the mission and return home due to weather. Thus, Albrecht had good reason to believe that there had been no flights performed that day.
56 FOH Tome 1, Vol. 4, Annex Nos. 2819, 2820; A brief strategic description of the Belgian retreat can be read in: Foch, *The Memoirs of Marshal Foch*, p.129–137; and: Greenhalgh, *Foch in Command*, pp.58–62.
57 Ferko–UTD, Box 515, Folder 12, "Die Angriffsoperation des Deutschen rechten Heeresflügels in Flandern vom 10. Oktober bis 11. November 1914," p.27.
58 Ferko–UTD, Box 515, Folder 12, "Die Angriffsoperation des Deutschen rechten Heeresflügels in Flandern vom 10. Oktober bis 11. November 1914," p.30.
59 Ferko–UTD, Box 515, Folder 12, "Die Angriffsoperation des Deutschen rechten Heeresflügels in Flandern vom 10. Oktober bis 11. November 1914," pp.30–31.
60 Ferko–UTD, Box 515, Folder 12, "Die Angriffsoperation des Deutschen rechten Heeresflügels in Flandern vom 10. Oktober bis 11. November 1914," p.31.
61 Edmonds, *Military Operations in France And Belgium*, Vol. 2, p.102.
62 *Der Weltkrieg*, V, p.287.
63 Ferko–UTD, Box 515, Folder 12, "Die Angriffsoperation des Deutschen rechten Heeresflügels in Flandern vom 10. Oktober bis 11. November 1914," p.34.
64 Indeed, Falkenhayn messaged the Kaiser directly from Sixth Army HQ stating: "There is still no clarity regarding the English movements. One column is marching in from Ypres in the direction of Menin. Have attacked the English unfortunately nowhere."
65 The document created at the meeting was entitled "Sixth Army's Instructions for the Coming Operation."
66 *Der Weltkrieg*, V, p.287.
67 *Der Weltkrieg*, V, p.287.
68 The specific deployment was as follows: XXII RK in Termonde, Alost and Brussels, XXIII RK in Denderleeuw and Brussels, XXVI RK in Grammont and Enghien, and XXVII RK at Lens and Ath.
69 Sheldon, *German Army at Ypres 1914*, p.56.
70 Schwink, *Ypres 1914*, p.6.
71 Based upon their experiences gained during the Marne Campaign, Fourth Army HQ also recognized the importance of tactical reconnaissance. FFA 38 was therefore instructed to place two aircraft at a forward airfield to assist III RK with their screening operation.
72 Schwink, *Ypres 1914*, p.19.
73 Ferko–UTD, Box 515, Folder 12, "Die Angriffsoperation des Deutschen rechten Heeresflügels in Flandern vom 10. Oktober bis 11. November 1914," pp.39–40.
74 *Der Weltkrieg*, V, p.301.
75 Although OHL had issued orders that evening for III RK to *not* continue past the line Ostend–Torhout, the directive was issued based on the assumption there was still strong forces within Ostend itself. If Falkenhayn had been made aware of the abandonment of the city and the Belgian army's withdrawal, it is highly likely that III RK would have been ordered to cautiously pursue.
76 This is yet one more example of the German leadership's failure in 1914 to capitalize upon the *Fliegertruppe's* excellent aerial reconnaissance efforts to better facilitate the army's success.
77 Ferko–UTD, Box 515, Folder 12, "Die Angriffsoperation des Deutschen rechten Heeresflügels in Flandern vom 10. Oktober bis 11. November 1914," p.42.
78 David Lomas, *First Ypres 1914: The Birth of Trench Warfare* (Westport CT, 2004), p.40.
79 Born in 1853, Eugen von Falkenhayn was Erich's older brother.
80 Schwink, *Ypres 1914*, pp.23–24.
81 Edmonds, *Military Operations in France And Belgium*, vol. 2, pp.132–135.
82 Imperial War Museum (henceforth IWM), Wilson Mss, DS/MISC/80, diary, 20 October 1914.

83 *Der Weltkrieg*, V, p.312.
84 Ferko–UTD, Box 515, Folder 12, "Die Angriffsoperation des Deutschen rechten Heeresflügels in Flandern vom 10. Oktober bis 11. November 1914," pp.36–37.
85 Schwink, *Ypres 1914*, pp.26–28.
86 Sheldon, *German Army at Ypres 1914*, pp.180–185.
87 Sheldon, *German Army at Ypres 1914*, pp.93–94.
88 Sheldon, *German Army at Ypres 1914*, pp.134–143.
89 Schwink, *Ypres 1914*, p. 25; Edmonds, *Military Operations in France And Belgium*, vol. 2, pp.85–87, p.171.
90 Beckett, *Ypres 1914: The First Battle* (New York, 2013), p.92.
91 *Der Weltkrieg*, V, p.307.
92 Ferko–UTD, Box 515, Folder 12, "Die Angriffsoperation des Deutschen rechten Heeresflügels in Flandern vom 10. Oktober bis 11. November 1914," pp.44–45.
93 Ferko–UTD, Box 515, Folder 12, "Die Angriffsoperation des Deutschen rechten Heeresflügels in Flandern vom 10. Oktober bis 11. November 1914," p.46.
94 As described in: Sheldon, *Ypres 1914*, p.68.
95 Jack Sheldon and Nigel Cave, *Ypres 1914: Langemarck* (South Yorkshire, 2014), pp.73–92.
96 46 RD's infamous engagement, sometimes referred to as *Kindermord* (murder of the children), is well documented. The following are a few select examples: Karl Unruh, *Langemarck: Legende und Wirklichkeit* Koblenz (Koblenz, 1986); Sheldon, *The German Army at Ypres 1914*, pp.116–130; Beckett, *Ypres 1914: The First Battle*, pp.98–103; Edmonds, *Military Operations in France And Belgium*, vol. 2, pp.178–182.
97 Edmonds, *Military Operations in France And Belgium*, vol. 2, pp.176–178.
98 The term "battering ram" was given to III RK in the German Official Monograph of the battle: Schwink, *Ypres 1914*, p.27.
99 Ferko–UTD, Box 515, Folder 12, "Die Angriffsoperation des Deutschen rechten Heeresflügels in Flandern vom 10. Oktober bis 11. November 1914," p.46.
100 The tactic of inundation as a means of defense around the town of Nieuport had been adopted several times throughout history. The last time it had occurred was in 1814 when Napoleon's troops were defending the area.
101 Guido Demerre & Johan Termote, "The Flooding of the Yser Plain," *De Grote Rede 36: The Great War and the Sea, De Grote Rede: Nieuws over onze Kust en Zee* (2013), pp.47–52.
102 Considering the difficult terrain and the inexperienced/undertrained nature of the troops, it was quickly proven that tactical reconnaissance from the *Fliegertruppe* would be of little use. Only long-range reconnaissance and artillery cooperation flights were henceforth conducted throughout the rest of the battle. Ferko–UTD, Box 515, Folder 12, "FFA 38 Reports, 1914/15."
103 Ferko–UTD, Box 515, Folder 12, "Die Angriffsoperation des Deutschen rechten Heeresflügels in Flandern vom 10. Oktober bis 11. November 1914," p.48.
104 Ferko–UTD, Box 515, Folder 12, "Die Angriffsoperation des Deutschen rechten Heeresflügels in Flandern vom 10. Oktober bis 11. November 1914," p.48.
105 *Der Weltkrieg*, V, p.314.
106 Once again, the liaison failures between the *Fliegertruppe* and the army were costing the Germans dearly. Had Sixth Army HQ been aware of the French strength around Arras, it is probable that Rupprecht would have sent his reserves to participate in an offensive to take that city—an operation that would have had a much higher likelihood of success.
107 David Lomas, *First Ypres 1914*, p.55; Edmonds, *Military Operations in France And Belgium*, vol. 2, p.182.
108 George Morton-Jack, *The Indian Army on The Western Front* (New York, 2014), pp.134–153.
109 Edmonds, *Military Operations in France And Belgium*, vol. 2, p.182.
110 Due to the nature of the terrain, many of these batteries were well concealed from forward ground observers.
111 As an army headquarters section, FFA 6 had spent the first three months of the war exclusively performing long-range reconnaissance flights. FFA 38 on the other hand was formed after the start of the war from the officers and men of the disbanded 'Fortress Flying Section Cologne.' In both cases, the airmen were unfamiliar with performing artillery direction work under fire.
112 Ferko–UTD, Box 515, Folder 12, "Die Angriffsoperation des Deutschen rechten Heeresflügels in Flandern vom 10. Oktober bis 11. November 1914," p.50.
113 Ferko–UTD, Box 515, Folder 12, "Die Angriffsoperation des Deutschen rechten Heeresflügels in Flandern vom 10. Oktober bis 11. November 1914," pp.50–52.
114 This observation strongly suggested that the Allies were also using the port of Boulogne as a base for resupply.
115 Ferko–UTD, Box 515, Folder 12, "Die Angriffsoperation des Deutschen rechten

115 Heeresflügels in Flandern vom 10. Oktober bis 11. November 1914," p.52.
116 Ferko–UTD, Box 515, Folder 12, "FFA 9 notes"
117 Ferko–UTD, Box 515, Folder 12, "Die Angriffsoperation des Deutschen rechten Heeresflügels in Flandern vom 10. Oktober bis 11. November 1914," p.55.
118 FOH, Tome 1, Vol. 4, p. 321, Annex Nos 3191, 3195–3197.
119 Beckett, *Ypres 1914: The First Battle*, p.109.
120 Foch, *The Memoirs of Marshal Foch*, pp.140–143.
121 *Der Weltkrieg*, V, pp.318–319.
122 The allies' new position on embankment was some 3½–4½ feet above the level of the plain to the east. Foch, *The Memoirs of Marshal Foch*, p.144.
123 Demerre and Termote, "The Flooding of the Yser Plain," p.50.
124 Direct quote from the German Official Monograph of the Battle: Schwink, *Ypres 1914*, p.54.
125 Beckett, *Ypres 1914: The First Battle*, pp.117–122.
126 Edmonds, *Military Operations in France And Belgium*, vol. 2, pp.252–257.
127 Foch, *The Memoirs of Marshal Foch*, p.153.
128 Edmonds, *Military Operations in France And Belgium*, vol. 2, pp.205–210.
129 Edmonds, *Military Operations in France And Belgium*, vol. 2, pp.216–220.
130 Ferko–UTD, Box 515, Folder 12, "Die Angriffsoperation des Deutschen rechten Heeresflügels in Flandern vom 10. Oktober bis 11. November 1914," p.59.
131 Hazebrouck—Ypres and Hazebrouck—Armentières.
132 AOK 4. Armee, Tagesachen 25. Oktober 1914.
133 Edmonds, *Military Operations in France And Belgium*, vol. 2, p.210.
134 Eric and Jane Lawson, *The First Air Campaign: August 1914–November 1918* (Cambridge, 1996), pp.43–44.
135 Although these incidents often ended in the total loss of the aircraft, the German aircrews were rarely injured and were available to fly again the next day (given an aircraft was available).
136 There is no documented case during the Ypres Campaign where a flying section dispatched another aircraft to the scene of an earlier aerial combat on the same day of the event. Several flying sections did, however, dispatch aircraft later in the day to a different area of their sector for another purpose.
137 The common carbine used by the *Fliegertruppe* and all other army support branches was the Gewehr Kar.98.a. These weapons were introduced into service in January of 1908 after deficiencies with the army's existing carbines had warranted the creation of a new weapon. Weighing 8 lbs. and with a total length of 43.3 inches, the Kar.98.a was designed and created to be a shorter/lighter version of the standard infantry rifle, the Gewehr 98, for use in "non-infantry formations." The carbine had a five round magazine for its 7.92mm "S Bore" cartridges and was found to have excellent accuracy with very little recoil. Due to its more compact nature, the Kar.98.a ultimately became a better weapon than its standard Gewehr 98 cousin once trench warfare became established throughout the Western Front. Consequently, later in the war the German General Staff decided to equip their "storm troop" formations with the carbines for use in close-quarters combat.
Regarding the *Fliegertruppe*, the Kar.98.a was given to each flying section in sufficient quantities to give the airmen and ground crew the means to protect themselves if their airfield was ever attacked. As the reader will recall, several flying sections were indeed compelled to use their carbines during the Marne Campaign when allied troops penetrated into the German rear-area and threatened the airfield. In each case the *Fliegertruppe* successfully protected their aircraft and supplies from sabotage— even capturing a large number of French prisoners on one occasion. For air combat, several Sixth Army observers opted to carry two or three preloaded Kar.98.a in order to give the observer more shots without having to reload. The more logical and successful alternative to the Kar.98.a, however, was the "model 1909 Mauser Self-loading carbine." As a self-loader, the Mauser allowed the observer to fire repeatedly without having to cycle the next round into the chamber with the bolt. Thus, with a larger magazine and a significantly faster rate of fire, the Mauser quickly proved to be the *Fliegertruppe's* best weapon for aerial combat in 1914.
138 Dietze and Rosenmüller's Albatros B.I was aircraft # B 30/14.
139 Ferko–UTD, Box 515, Folder 12, "Die Angriffsoperation des Deutschen rechten Heeresflügels in Flandern vom 10. Oktober bis 11. November 1914," p.60.
140 Ferko–UTD, Box 515, Folder 12, "Die Angriffsoperation des Deutschen rechten Heeresflügels in Flandern vom 10. Oktober bis 11. November 1914," p.62, "FFA 9 notes."
141 Personal letters of Josef Suwelack, letter of Oct 25 1914.
142 The flights were conducted by FFAs 1, 8, and 1b.
143 *Der Weltkrieg*, V, pp.330–331.
144 *Der Weltkrieg*, V, pp.328–329.
145 Sixth Army HQ's message to OHL that evening

read: "Enemy has reinforced at several points. Rail movements behind the front were observed on several occasions." Ferko–UTD, Box 515, Folder 12, "Die Angriffsoperation des Deutschen rechten Heeresflügels in Flandern vom 10. Oktober bis 11. November 1914," p.63.
146 The location of these formations were the suburbs of: Boesinghe (N. Ypres), Elverdinghe (N. Ypres), Brielen (NW. Ypres), Vlamertinghe (W. Ypres), as well as the towns of Crombeke and Westvleteren (6 miles northwest of Ypres).
147 The observer's report went on to describe Dunkirk's recently improved defenses (previously seen on the 15th and 25th), stating that the original semi-circle defenses running from the town of Leefrickoucke through Coudekerque had now been expanded to the southern and eastern edges at Hondschoote and Bergués. See: Ferko–UTD, Box 515, Folder 12, "Die Angriffsoperation des Deutschen rechten Heeresflügels in Flandern vom 10. Oktober bis 11. November 1914," pp.64–65.
148 Ferko–UTD, Box 515, Folder 12, "Die Angriffsoperation des Deutschen rechten Heeresflügels in Flandern vom 10. Oktober bis 11. November 1914," p.66.
149 Ferko–UTD, Box 515, Folder 12, "Die Angriffsoperation des Deutschen rechten Heeresflügels in Flandern vom 10. Oktober bis 11. November 1914," pp.66–67.
150 FOH, Tome 1, vol. 4, Annex Nos: 3301, 3339, 3340, 3355, 3405, 3410.
151 In addition to these reinforcements, the allies had used their rail superiority to move in large quantities of guns, ammunition, and supplies to keep their formations well supplied throughout the battle. By comparison, many German formations had to continually operate with inadequate provisions. This ultimately turned out to be a pivotal factor in the outcome of the battle.
152 *Der Weltkrieg*, V, pp.330–331.
153 Beckett, *Ypres 1914: The First Battle*, p.123.
154 Sheldon, *The German Army at Ypres 1914*, pp.225–226.
155 Shell shortages were already being felt as early as the first week of September during the Battle of the Marne. By early October the problem had grown to the point where OHL was forced to put severe restrictions on the number of shells fired per day. The first week of Falkenhayn's large-scale offensive in Flanders only compounded the problem. In an attempt to free up extra ammunition for Fabeck's attack around Ypres, an order was dispatched to all artillery units fighting south of Arras, allowing only 2 or 3 shells to be fired per gun each day. These measures, however, did very little to change the overall situation in the final two weeks of the battle. Consequently, the German advantage in quantity of heavy artillery was essentially nullified throughout the battle. Sheldon, *The German Army at Ypres 1914*, p. 224. For a history on the army's internal struggle to increase the number of artillery munitions before the war began, see: Brose, *The Kaiser's Army: The Politics Of Military Technology In Germany During The Machine Age 1870-1918*.
156 Schwink, *Ypres 1914*, p.63; Beckett, *Ypres 1914: The First Battle*, p.123.
157 Loßberg, Mein Tätigkeit im Weltkriege (Berlin, 1939), p.94.
158 Excerpt of a directive issued by Sixth Army Chief of Staff Konrad Krafft von Dellmensingen: Sheldon, *German Army at Ypres 1914*, pp.228–230.
159 The Lesquin airfield was located on the southern edge of Lille almost exactly where the modern *Aéroport de Lille* stands today.
160 Meanwhile, Sixth Army Airpark loaned two rare replacement aircraft to XXIV RK in order to keep the important corps properly supported.
161 Ferko–UTD, Box 515, Folder 12, "Die Angriffsoperation des Deutschen rechten Heeresflügels in Flandern vom 10. Oktober bis 11. November 1914," p.70.
162 Sheldon, *German Army at Ypres 1914*, p.230.
163 Sheldon, *German Army at Ypres 1914*, pp.235–236.
164 Schwink, *Ypres 1914*, pp.66–67.
165 Deneckere, *Above Ypres: The German Air Force in Flanders 1914–1918* (Brighton, 2013), pp.24–25.
166 These troops were the vanguard of the previously spotted French relief force that would ultimately take up their position in the allied line along the Comines Canal in the area of St. Eloi—Zandvoorde. The initial strength of this force was five battalions and three batteries. Edmonds, *Military Operations in France And Belgium*, vol. 2, p.297.
167 The capture of the Messines Ridge was of "decisive importance" to Fabeck, who recognized its value as an artillery platform capable of enfilading the allied line. The lessons of the offensive's first day had illustrated the value of heavy artillery support. Without the capture of Messines Ridge, 'Army Group Fabeck' would be unable to resume the advance or even defend their initial gains out of a lack adequate artillery support. However, if several heavy batteries were established on the ridge, the group's offensive power would be increased exponentially.
168 Edmonds, *Military Operations in France And Belgium*, vol. 2, pp.303–306.
169 Shortly before 11am, about the time the

Fliegertruppe's aerial direction had begun in earnest, the British garrison manning the trenches on the eastern side of the town were compelled to withdraw to its western outskirts. This, in concert with a growing gap in the defenses on the town's southeastern side, undoubtedly persuaded the local German leadership to order a general assault. Edmonds, *Military Operations in France And Belgium*, vol. 2, p.306.

170 Edmonds, *Military Operations in France And Belgium*, vol. 2, pp.310–311.
171 Sheldon, *German Army at Ypres*, pp.247–252.
172 German artillery expert *Oberst* Max Bauer argues in his book, *Der grosse Krieg in Feld und Heimat*, that the heavy batteries were too diluted and should have been fully concentrated at a single pre-chosen breakthrough point. If this had been done it is very possible that a decisive breakthrough could have occurred. Nonetheless, Deimling, for his part, was simply expressing his desire to have some additional guns to help him achieve his objective. Max Bauer, *Der grosse Krieg in Feld und Heimat*, (Tübingen, 1921) p. 65; Sheldon, *German Army at Ypres 1914*, p.237.
173 Soon after arriving on the field, Deimling was personally wounded by a shell fragment. Schwink, *Ypres 1914*, p.73.
174 Lomas, *First Ypres 1914*, pp.75–77; Edmonds, *Military Operations in France And Belgium*, vol. 2, pp.320–321, pp.328–329.
175 In this sector, three battalions of French infantry and three batteries of artillery were initially available to strengthen the damaged British defenses. Edmonds, *Military Operations in France And Belgium*, vol. 2, pp.332–333.
176 Ferko–UTD, Box 515, Folder 12, "Ypres 1914."
177 Schwink, *Ypres 1914*, pp.49–54.
178 Foch, *The Memoirs of Marshal Foch*, p.154.
179 This story was found in the 1914 *Fliegertruppe* file within the Ferko Collection. It is also briefly mentioned in the German General Staff's official monograph of the battle. Unfortunately, the observer's name doesn't appear in either source. Ferko–UTD, Box 20, Folder 5, "1914 papers"; Schwink, *Ypres 1914*, p.49.
180 Schwink, *Ypres 1914*, p.50.
181 Ferko–UTD, Box 515, Folder 12, "Ypres 1914."
182 Poseck, *The German Cavalry in France & Belgium 1914*, pp.214–215.
183 *Der Weltkrieg*, V, p.337.
184 Edmonds, *Military Operations in France And Belgium*, vol. 2, pp.222–223.
185 FFA 24 was continuing to operate from an airfield in Lille at this time.
186 Ferko–UTD, Box 11, Folder 4, "FFA 5 notes."
187 Sheldon, *German Army at Ypres 1914*, pp.249–252.
188 Schwink, *Ypres 1914*, p.84.
189 The French forces participating in this counter-attack were from the 32 DI and the 9 DC. Supported by large quantities of field guns, this force easily pushed the half trained German reservists out of the town for the second time in less than 24 hours. This group of French troops, under the overall command of General Taverna [commander of XVI CA] was the same group that had been observed by German airmen advancing towards the battle-area over the previous 72 hours. Edmonds, *Military Operations in France And Belgium*, vol. 2, pp.352–353.
190 FFA 3 performed four sorties on the 1st. The first two, done before noon, consisted of artillery direction and tactical reconnaissance. The final two flights flew over the battle-area only to discover that the opposing sides were locked in combat on the eastern edge of the wood. Ferko–UTD, Box 515, Folder 12, "Ypres 1914."
191 Ferko–UTD, Box 515, Folder 12, "Ypres 1914."
192 Schwink, *Ypres 1914*, pp.85–86.
193 In addition to the 3 ID, Fabeck's group also received the 11th *Landwehr* Brigade as further reinforcements. This brigade was sent to Messines to be placed on 26 ID's left—making it the new extreme left of Fabeck's command. Positioned on the *Landwehr* Brigade's left was HKK 1, acting as the link between Fabeck and Sixth Army's XIX AK. Der Weltkrieg, V, p. 336; Edmonds, *Military Operations in France And Belgium*, vol. 2, p.364.
194 Foch, *The Memoirs of Marshal Foch*, pp.158–161.
195 Edmonds, *Military Operations in France And Belgium*, vol. 2, pp.362–363.
196 Opposing this force was two French infantry divisions [32 and 39 DIs] and two cavalry divisions [1st and 9th DCs]. Edmonds, *Military Operations in France And Belgium*, vol. 2, p.364.
197 Schwink, *Ypres 1914*, pp.87–88.
198 In reality, XV AK simply did not possess sufficient strength to break the British positions with the quantity of heavy guns at their disposal.
199 Balloon observers primarily performed flash spotting missions at this time in the war. Flash spotting is the act of looking for the flash of enemy artillery to determine their location so a local battery can silence them with successful counter-battery fire.
200 This was most likely the correct decision when considering that inaction would guarantee static trench warfare across the entirety of the western front—the consequences of which was something the German leadership fully understood and were

deathly afraid of. The potential gains outweighed the risks.
201 Schwink, *Ypres 1914*, pp.91–92.
202 Falkenhayn was also a contributor in crafting this strategy.
203 Edmonds, *Military Operations in France And Belgium*, vol. 2, p.375.
204 Fortunately for Fabeck and the Germans, the group's left flank was secured earlier that morning when elements of HKK 1 seized the Douve Farm from local British troops. Poseck, *The German Cavalry in Belgium And France 1914*, p.218.
205 Beckett, *Ypres 1914 The First Battle*, pp.196–198.
206 Edmonds, *Military Operations in France And Belgium*, vol. 2, p.378.
207 While Army Group Fabeck's flying sections did not have any of their own aircraft shot down behind allied lines, they were forced to "write a number of their aircraft off" due to damage sustained by enemy ground fire. Furthermore, mechanical issues had caused several of the aircraft to become grounded during the first week of Fabeck's offensive.
208 Foley, *German Strategy and the Path to Verdun*, p.103.
209 German casualties for the Ypres Campaign have always been difficult for historians to determine. The casualty reports for Fourth Army's reserve corps were filed irregularly by their HQ staffs. Nonetheless, the best estimate for German losses during the entire Ypres Campaign, that is from October 13–November 24, as 123,910 killed, wounded or missing. British losses during the same period were roughly 88,000 while French estimates range between 50,000–80,000. For their part, the Belgians suffered 21,562. Thus, although the German losses were high, they won the casualty exchange despite being the attacker. Indeed, the estimated losses were 189,562:123,910, or 1.52:1 in the Germans' favor.
210 Beckett, *Ypres 1914: The First Battle*, p.55.
211 The period between the 6th and the 10th is one of the most often overlooked turning points in the battle. The Germans' questionable strategy of expending large quantities of precious artillery ammunition in support of separate, disconnected attacks ultimately did more harm than good. The idea behind the separated assaults was so Group Fabeck could "attack all along the line and prevent the British from recovering from their fatigue or strengthening their positions." Under this strategy, however, the attacking German infantry suffered a large number of casualties and gained very little ground. Had Army Group Fabeck HQ chosen to concentrate against a single area of the line, it is highly likely that more gains would have been made while the goal of "preventing the British from strengthening their positions" could still have been accomplished in the areas directly related to the upcoming renewed offensive.
212 Edmonds, *Military Operations in France And Belgium*, vol. 2, p.395.
213 Edmonds, *Military Operations in France And Belgium*, vol. 2, pp.398–399.
214 Edmonds, *Military Operations in France And Belgium*, vol. 2, pp.400–403.
215 Andrew Lucas and Jurgen Schmieschek, *Fighting the Kaiser's War: The Saxons in Flanders 1914–18* (South Yorkshire, 2015), pp.38–39.
216 The six cavalry divisions temporarily placed under Fabeck's control were: HKKs 1 [4th and Guard Divisions] and 2 [3rd and 7th Cavalry Divisions], as well as the 2nd and Bavarian Cavalry Divisions.
217 Beckett, *Ypres 1914: The First Battle*, p.212.
218 Control of Dixmude had been hotly contested since Fourth Army's initial attack on the town on October 21. See: Charles le Goffic, *Dixmude: French Marines in the Great War 1914–1918* (Leonaur, 2012).
219 Quote taken from the OHL's General order. See: Schwink, *Ypres 1914*, p.103.
220 The British Official History stated that the November 11th bombardment was "the most terrific fire the British had yet experienced; and it increased in intensity as 9am [10am German time] approached...For the infantry in the front line and the fighting staffs there was nothing to do but lie at the bottom of the trenches and in the holes in the ground..." Edmonds, *Military Operations in France And Belgium*, vol. 2, pp.420–421.
221 Schwink, *Ypres 1914*, pp.113–119.
222 Schwink, *Ypres 1914*, p.119; Edmonds, *Military Operations in France And Belgium*, vol. 2, pp.427–428.
223 XV AK's lackluster advance is puzzling. The corps had a sizeable numerical superiority and had the support of an excellent preliminary artillery bombardment. The most likely explanation is the exhaustion from the previous 13 days of incessant combat.
224 Hill 60 was entirely under German control by the end of the month. Possession of the hill would prove useful the following year when multiple British attacks were launched against the nearby German positions.
225 Edmonds, *Military Operations in France And Belgium*, vol. 2, p.448.
226 After the setback at Tannenberg, the Russians had re-focused their efforts against the Austrians with great effect. By early November, the Russians

Right: Unusually, this Mercedes-powered Albatros B.I carries its national insignia far forward on the fuselage. The rudder insignia is much smaller than that on the aircraft in the background.

had won several victories and were on the verge of knocking the Austrians out of the war entirely. As overall commander of German forces in the east, Paul von Hindenburg requested additional troops be sent from the Western Front to take part in a relief offensive to help take pressure off the Austrians. Once it had become clear that the efforts at Ypres had failed, Falkenhayn made the decision to order the Western Armies onto the defensive and send reinforcements to the East to assist with Hindenburg's operation. (See: Norman Stone, *The Eastern Front 1914–1917*, pp.97–108; Robert B. Asprey, *The German High Command at War* (New York, 1991), pp.114–116, pp.126–128.

227 Falkenhayn's decision to launch the offensive at Ypres has always been highly criticized by historians. A sober look at the facts, however, shows that the decision to attack in Flanders, although ultimately unsuccessful, was likely the right one. First, it should be remembered that Falkenhayn's offensive coincided with multiple allied offensives that were launched in Flanders at the same time. It is very possible that the allied armies would have enveloped the German line—forcing a general retreat towards the Franco-German frontiers had the newly created Fourth Army not been present in Flanders to check the allied advance. At a minimum, Foch's armies would have been able to reclaim a substantial part of Belgian territory, which would have had disastrous consequences, particularly for the German Navy. Secondly, many have argued that the newly raised reserve corps should have been used in the east to assist Hindenburg and Ludendorff with their offensives in Poland. Falkenhayn dismissed the suggestion, claiming that the area's poor roads would not have been able to sustain large numbers of troops during the rainy season. The General Staff Chief was ultimately proved correct when the troops that were used during those offensives became bogged down due to "knee high mud on the roads." Moreover, had the reserve corps been sent east, it should be remembered that they would not have possessed the vast quantities of heavy artillery as they had enjoyed in Flanders. Lastly, even if the reserve corps were successfully deployed against the Russians, one must always consider the Russian's Fabian tactic of merely withdrawing into their country to save themselves from annihilation. Had the Russians done this, the strategic situation would have looked very similar to how it did anyway. One must also consider how Foch's offensives would have concluded under this scenario. In short, the risks taken and losses incurred by deploying the reserve corps at Ypres were worth it because Flanders was the only location where a possibility of achieving a decisive victory still existed. Falkenhayn was well aware of the long odds Germany faced if she was forced into a prolonged two front war against enemies with superior numbers and manufacturing capabilities. Thus, the German commander considered it worth the risk to commit the reserve corps in one last effort to gain victory before static warfare became a certainty. As Falkenhayn himself later said: "the prize to be won was worth the stake." See: Falkenhayn, *General Headquarters 1914–1916*, p.28.

6

General Situation at the End of the Year
November/December 1914

"The worst enemy of the German airmen [during the winter of 1914/15] was the rapidly dwindling confidence they received from the leadership and the troops, who did not possess an understanding of the real causes of the airmen's inferiority."

Hilmer Freiherr von Bülow[1]
Observer Officer: 1914–18; Aviation Historian; *Luftwaffe* General: 1939–43

The end of Falkenhayn's offensive in Flanders heralded an entirely new chapter in military operations on the Western Front. Despite the Germans' best efforts, the 'war of movement' was over and had given way to *Stellungskrieg* (positional warfare). Indeed, from the Channel Coast in the north to the Swiss Frontier in the south, the opposing sides were now entrenched in a series of rudimentary, yet interconnected trench systems that were covered by well concealed machine guns and artillery. After closely following the offensive in Flanders, Falkenhayn recognized that the existence of these field works meant winning a decisive military victory in accordance with pre-war doctrine was no longer possible.[2] Instead, the German commander believed a new formula for victory, based partly on diplomacy, would have to be refined and then implemented for Germany and the Central Powers to reach an honorable peace.[3] In the short-term, however, the Kaiser's Western Armies were to dig in and hold every inch of conquered territory until the situation in the East could be restored.[4]

The Allied December Offensives

Although they had failed to drive the German armies back to the frontier as planned, the allied leadership was nonetheless pleased with the general situation in mid-November. Together, the Franco-British armies had thwarted each of the major German offensives in Flanders and had successfully avoided losing any more valuable territory. Thus, with their position secured and the German offensives suspended, the allies now held the initiative.

The Franco-British leadership was aware that Falkenhayn's failed push had left the German troops exhausted, disorganized, and vulnerable to attack. Moreover, Joffre had detailed knowledge of significant German troop redeployments to the Eastern Front in order to take part in an upcoming offensive against the Russians. It was therefore anticipated that the German positions in the West would become greatly weakened—prompting the French leadership to begin planning a renewed offensive. In the meantime, the Russians had suffered another major defeat and had begun sending appeals to the French and British Governments requesting a relief offensive be immediately launched in the West.[5] Joffre ultimately obliged and, together with Foch, launched an over-ambitious two-pronged attack against the German positions in the Flanders and Champagne sectors.[6]

The northern half of the offensive began in Flanders on December 14th. Over the next five days the French Eighth Army, with the BEF's support, launched a series of unsuccessful attacks against the German troops occupying the Ypres Salient. Across the front, the German defenders were discovered to be in a much better condition than the allied leaders had originally suspected. Deep trenches, barbed wire, machine gun posts, and other field fortifications had been rapidly erected in anticipation of an allied attack before Christmas. Furthermore, German field batteries were pre-ranged onto the allied front lines to pummel their infantry as they formed for an attack. Thus, as a result of these measures, each allied attack was easily repulsed with minor losses for the German defenders.[7]

To the south around Arras, French Tenth Army's offensive was just as unimpressive. The attack's objective was the capture of Vimy Ridge—a dominating terrain feature that gave its occupier a commanding view of the entire Arras—Douai sector.[8] On the morning of December 17th, with Foch personally commanding the operation from Tenth Army HQ, three corps assaulted the German positions holding the 8½ mile-long ridge. These attacks, however, were poorly supported with insufficient quantities of artillery and quickly broke down under the pressure of well-executed local German counter-assaults.[9] Foch therefore reluctantly suspended the offensive on the 19th with the intention of resuming the attack on a much smaller scale once all of Tenth Army's artillery and reserves were fully concentrated and in position to support the endeavor. Poor weather ultimately disallowed this second phase of the attack from every being delivered, leaving the opposing front lines more or less unchanged at the end of the year.[10]

Meanwhile in Champagne, de Langle's Fourth Army launched the largest of Joffre's December offensives against Einem's Third Army. Deploying four corps in the first line with a fifth in reserve, de Langle's forces were expected to achieve a major breakthrough. After considerable debate, it was decided to charge the army's two left wing corps (XII and XVII CAs) with launching the main assault, with the corps on their right (Colonial Corps) in support and the army's right wing corps (II CA) remaining purely on the defensive. The army reserve (I CA) was deployed directly behind the two attacking left wing corps to be used wherever the opportunity presented itself. For artillery support, Fourth Army was assigned 742 guns, including 80 heavy caliber pieces.[11] Recognizing the nature of trench warfare, De Langle's staff planned for the advance to be systematically accomplished in successive waves, relying on their artillery to soften up German defenses before each wave made its attack. The army's reserves were to be fed into the battle-line wherever progress had stalled, thereby allowing the advance to continue without losing momentum.[12]

Unfortunately for the French, de Langle's army was woefully under-equipped in artillery pieces and shells. As a result, the offensive's preparatory bombardment failed to damage the German defenses, particularly the barbed wire that had been laid in front of the trenches. Thus, the attacking French infantry were unable to make any gains during the first two days of the battle (December 20th–21st). Likewise, the introduction of I CA into the battle-line over the course of the following week also failed to push the front forward.[13] Joffre nevertheless remained committed to applying pressure on the Germans and ordered de Langle to continue the offensive on the morning of the 30th.[14] The attacks were preempted however, by a well-timed German counteroffensive launched by XVI AK in the Argonne Forest against French Fourth Army's defensive right wing.[15] These attacks, in concert with a temporary lift on artillery shell rationing in Third Army's sector, allowed the

Above: French troops launch their unsuccessful attack against the German fortified positions in the Champagne. (Author's Collection)

Germans to easily repulse de Langle's third attempt at a breakthrough, thus bringing an end to the final battle of 1914.[16]

The *Fliegertruppe*: December 1914

German aerial activity on the Western Front dropped considerably during the final six weeks of 1914. Simply put, the strain of four months of constant campaigning in support of the army's various offensives had taken its toll on the men and the machines of the *Fliegertruppe*. As a result, with many flying sections at or below half their original strength, Germany's airmen collectively welcomed the suspension of the offensives in the West as an opportunity to refit and reorganize.[17]

The sharp decline in the number of flights performed each week significantly impacted the quality of support offered to the ground troops during the

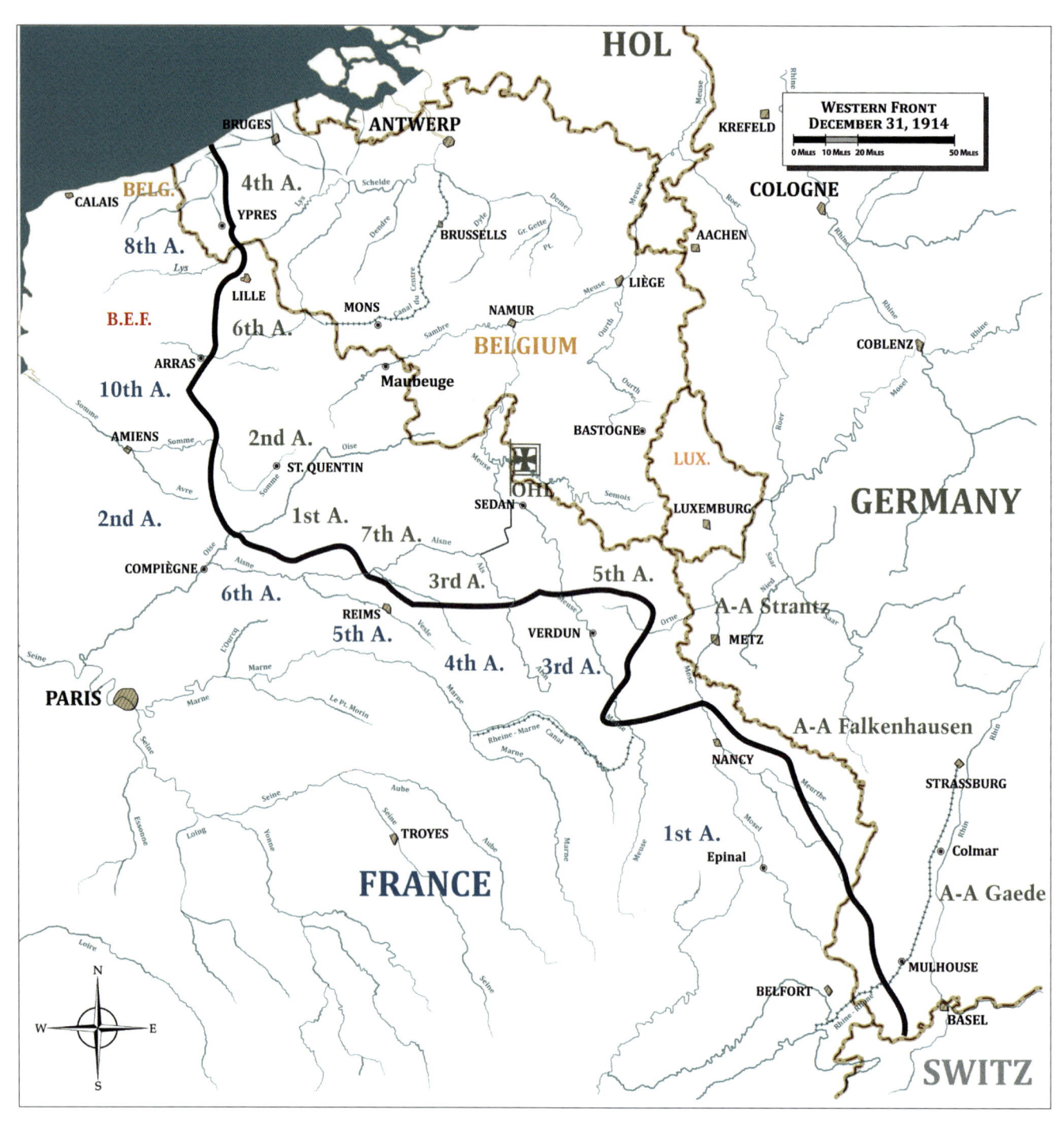

December battles. Poor weather and a passive attitude among higher commands kept the *Fliegertruppe* from discovering preparations for the allied attacks in the Ypres and Arras sectors. Fortunately, the situation was better in Champagne, where multiple reports of troop movements and the concentration of French field batteries alerted the German leadership in time to take action to repulse the initial attacks. These reports, delivered by FFA 22, prompted Third Army HQ to move reinforcements into the battle line opposite French Fourth Army's attacking left wing in the days prior to the attack. Additional flights performed later in the battle discovered de Langle's vulnerable defensive left wing in the Argonne. Using this information, the German leadership conceived and ordered XVI AK's attack, which brought the French offensive to a standstill.

Outside of the few aforementioned instances, the *Fliegertruppe* failed to obtain any meaningful intelligence during the month of December. Like most failures in German military aviation during the first year of the war, this was due mostly to poor

Above: XVI Armee Korps' counter-attack strikes French Fourth Army's right in the Argonne. (Author's Collection)

organization. The lack of a centralized command for the army's aviation forces caused each of the individual flying sections to adapt to the dramatically changing circumstances of trench warfare in their own way. Indeed, without any central leadership or direction, there were no means of prioritizing missions or assigning specific roles to the various flying sections as part of a sensible grand strategy. Instead, there was mass disorganization and confusion as many units repeatedly performed the same missions while other important tasks were never attempted.

The emergence of trench warfare necessitated a radical change in the employment of aircraft on the battlefield. With trench lines running from the Channel Coast to the Swiss Frontier, the German leadership already had knowledge of the allied armies' general strength and position. Thus, frequent operational reconnaissance flights were no longer needed to identify the closest allied army's latest position or to search for a gap or an open flank. The new situation required instead that the *Fliegertruppe* be frequently employed close to the front performing important tactical reconnaissance sorties with occasional long-range flights deep into the allied rear-area.

First among these new tactical missions were close reconnaissance flights focused directly over the allied front lines. The nature of trench warfare, however, made detailed observation of the front lines impossible with the naked eye. Operating around 6,000 feet, observers were simply unable to briefly note or sketch all of the necessary information regarding the enemy's elaborate trench systems and the troops that dwelled in them.[18] Fortunately, a simple solution to the problem presented itself through the use of aerial photography, as related by Ernest von Hoeppner in his history of the German Air Service:

When the operations stiffened into position warfare there was a decided change in the employment of our aviators. Instead of far reaching distant reconnaissance there arose the need for close reconnaissance. Ground observation did not extend beyond the hostile front line trenches and there was an increasing need of exactness and swiftness in transmitting information concerning the enemy.

Therefore, photographic intelligence was started. In the war of movement cameras were seldom used, for the large bodies of troops under observation stood out clearly enough to the naked eye. The camera now became the aviator's constant companion. The general extent of a hostile position was soon fixed and then the flyers would perform other missions to discover changes in the trace which might give the key to the enemy's intentions. Therefore, it was necessary to have in our hand all the details of the trench system and, while rapidly flying above it, to note the slightest changes which had been made. Visual reconnaissance was no longer effective when the enemy took pains to conceal his new works from the eyes of the aviators. The lens of the camera penetrated the veil which the enemy spread over his activities and it also retained in the negative everything that might have escaped the notice of the observer owing to the amount of detail which he had to examine.[19]

As Hoeppner describes, the camera quickly became the *Fliegertruppe's* most widely used asset. However, although each flying section had mobilized with at least one camera, it was widely recognized that more were needed at the front. Thus, orders were placed for large numbers of 25cm, 50cm, and 75cm cameras to be produced and dispersed amongst all the flying sections as quickly possible. Requests were also made to build photo labs near the airfields so that all photographs could be quickly produced and shared with local commands.[20]

The German Army leadership was slow to realize the camera's value. First, the poor weather throughout November and December, in combination with a small number of sorties flown, had resulted in the production of only a small number of photographs covering isolated, sometimes obscure, areas of the front for the army's consideration. For example, the most widely covered area during this period was in the vicinity of Belfort and Verdun—both areas were relatively quiet sectors that were of little immediate importance at the time.[21] Second, the current aircraft configuration that placed the observer in the front seat made it extremely difficult for photographs to be taken directly over a target.[22] This meant all photographs were taken from an oblique angle, which distorted the image and reduced its overall value. Finally, the general staff officers serving at the various army or corps headquarters often had no experience analyzing aerial photographs. As a result, the photographs often had to be verbally explained by the observer who took the photograph. Thus, with only a small number of poor quality photographs and very few officers who were trained to interpret them, the army originally viewed aerial photography with skepticism.[23] It would not be until the *Fliegertruppe's* reorganization in the spring of 1915, when the airmen were more experienced with their cameras and aviation staff officers were attached to each army headquarters, when aerial photography started to reach its true potential.

Arguably the most important tactical mission charged to the *Fliegertruppe* after the end of open warfare was in support of the artillery. Although a meaningful number of isolated artillery direction flights had been carried out during the Marne and Ypres Campaigns, the nature of static warfare demanded the mission now become a priority across the entire front. The army's leadership understood that the skillful and efficient employment of their artillery would be the key to maintaining Germany's defensive position on the Western Front against the numerically superior allied forces. Under the realities of trench warfare, any allied attack would be supported by massed artillery attempting to breakup and neutralize the German defenses before their own infantry were engaged. German batteries were therefore required to silence the allied guns and destroy the attacking allied infantry in order to keep the army's casualties comparatively low while keeping morale high. To that end, German aviators would be called upon to continuously reconnoiter the allied positions, identify the location of hostile batteries, and then direct friendly artillery fire against appropriate targets. During major allied attacks, the airmen were also expected to notify friendly batteries of important 'opportunity targets' that presented themselves during the course of the battle. For example, if an observer recognized a large mass of enemy infantry advancing in the open, he would promptly notify the artillery of its position and the target would be engaged accordingly.

Serving as a pilot in the Champagne Region with FFA 13 (VI AK), future fighter ace Oswald Boelcke described what it was like flying artillery direction sorties during the early stages of trench warfare: Although the passage is lengthy, the insight is extremely valuable.

October 12—We have three weeks here now, but the situation still remains unchanged. The two sets of opponents have dug themselves in up to the teeth; every now and then one or other makes a push forward, but, taken as a whole, the present situation is a sort of fortress warfare. And so—for good or for ill—we too have gradually accustomed ourselves to the idea that we must settle down here for some time.

It is no laughing matter for our troops; they are in the trenches day and night, and the nights are very cold already. One really feels ashamed of the good time we are having in contrast. So far I always have a

Above: The onset of trench warfare caused German field artillery to begin operating from concealed positions. This depiction shows a 105mm light howitzer operating in close proximity to a forward airfield. (Author's Collection)

bed—we have only once used sleeping bags—and we are better catered for than in peacetime.

All of which would be nice—if it was not for the horrible inactivity and boredom. I have not been over the enemy for the last ten days; the weather is bad, and there is hardly anything for us to reconnoiter since the war has frozen up…

October 25—At last we have something to do again in the last few days. The weather was so foggy for weeks that we might consider ourselves temporarily pensioned off. It was difficult work to kill the time. We had decent weather again for the first time on the 22nd; we made good use of it. We were in our machines at 9:00am and kept at work until 5:30pm. I took off five times. First Wilhelm [Oswald's brother—an observer whom Oswald regularly flew with] did a reconnaissance, and in the afternoon he directed our artillery fire. You see, there were several enemy batteries that our artillery could not locate. By arrangement with our own gunners we flew over the enemy batteries that our people had to shell, and Wilhelm noted whether their range was right, and then showed our artillerymen by means of various colored lights whether their shots were too short or too far or too much to the left or right, until at last they got on their targets.

At first the business did not go well, and I had to make two landings at Nauroy, because Wilhelm had to go into the matter more thoroughly with our artillerymen. But then they shot an enemy battery to pieces.

The next day the business worked splendidly straight off; we put our artillery onto three enemy batteries in three and a half hours. This kind of flying is strenuous work for both observer and pilot, because they must always keep such a close look out…

October 27—Wilhelm has already located nine enemy batteries to the south of Nauroy and southeast of Reims, and a heavy one in Reims itself, quite close to the cathedral!!! We would particularly like to direct our artillery's fire onto this battery, but the weather is too bad—always deep clouds. Yesterday morning we had to give up our flight as hopeless after ten minutes. We tried again early this morning, but got into clouds at one thousand six hundred meters

[5,250 feet] and had to turn back, because we cannot fly over the enemy at this height, unless we want to get shot down...

October 28—No luck again yesterday, because the clouds obstructed us; our flight was not quite useless as we were able to locate three new enemy batteries...

October 31—The weather is improving. We went off again at 2(pm) to put our guns onto the French by means of our colored lights. We could not pull off anything with Schroeter's battery, because the French had put up an anti-balloon gun on account of our frequent visits. We could not manage anything with our two other batteries either. We were furious at having tootled 'round for three and a half hours in vain. But our flight was not quite useless, because we located a battery near Reims and another in the town while flying up and down the front.

November 1—We succeeded splendidly today where we failed yesterday; five enemy batteries have to thank us for friendly greetings from our own artillery. We worked from 8:30 to 11:45am and from 3:00 to 5:45pm, i.e. six hours in all...

November 3—We really did not mean to fly today. But when we went out to [discuss matters with] our artillery at Fort Berru after lunch, we found that the French had fired at a neighboring village with shells of heavy caliber, and by chance they managed to hit a number of our men who were assembled to draw their rations. That called for revenge. We pelted back to our aerodrome in the car, took off and flew to Reims. Wilhelm found the guilty battery there, but —alas—our artillery could not reach it; they were just two hundred meters too short. When we got back at 4, the weather was so fine and calm that I got hold of that nice fellow, Jaenicke, who wants to get used to flying again, and took him to Sillery to drop four bombs. It was a nice little constitutional, in which I was boss of the show; all Jaenicke needed to do was to sit quiet and drop the bombs when I gave the signal.

November 5—As the weather often gets worse in the course of the morning, we fly immediately after sunrise whenever possible, somewhere around 7:30am. The enemy's artillery was active again in several spots and had to get a few smacks on the head. We were up soon after 7:30. Things went very well for us, so that we were finished in about an hour. Then we did another round for artillery. As a matter of fact we are now flying for four of our batteries, which only shoot when we direct their fire. Whenever we put them onto a target, we find it masked the next time we go up. So we had to go up twice more today, making three trips in all, and we put four enemy batteries (three near Nauroy and one near Reims) out of action. We are now doing this job on a large scale.

November 7—Thick fog yesterday and today. We utilized the time by going 'round to our artillery positions to ascertain what enemy batteries have been firing during the last few days. As the information was not sufficient for Wilhelm, we crept into the trenches today. Wilhelm tried to gain a clear picture of the situation from the rather confused statements of a reserve officer—apparently a senior master—and some volunteers. As it took him a long time, I amused myself on my own account. I scoured the country in front of me with field glasses, because I wanted to have a shot at a Frenchman for once in a way, but none were in sight. But the enemy kept a much sharper look out. As soon as we put our thick heads above the parapet, they started with a wuff, wuff, wuff, — I made an involuntary bow of acknowledgement to this attention, which amused our people mightily. Then French artillery also obliged us with a few shots. Some shells burst right over our heads... peng, peng—I made another bow of acknowledgement.[24]

As Boelcke's letters illustrate, artillery cooperation flights were of great value to the troops stuck in the trenches. Indeed, aerial artillery direction was a necessity for the Germans (more so than any other belligerent) due to a severe shortage of artillery shells that existed during the winter of 1914–15. Nevertheless, despite this fact, the army leadership refused to reorganize the *Fliegertruppe* or take immediate action in any other way to improve liaison between their air and ground forces. Thus, many attempts at artillery cooperation failed during November and December as the absence of a standardized doctrine within the *Fliegertruppe* created confusion and mistrust between the airmen and the artillerists in several sectors.

Yet another problem the airmen faced was that there was simply no means of communicating with the artillery fast enough in a positional warfare battle. With opposing trenches often times less than a thousand yards apart, infantry attacks materialized very quickly. Thus, by the time an observer saw the attack, wrote down a message, flew to the prearranged area, and dropped it over the side of the aircraft, the opportunity had already passed and it was too late for the artillery to intervene. Alternative methods such as using colored lights, signaling lamps, or flares produced inconsistent results and were considered too unreliable.[25]

A strong desire for a solution to this problem prompted several flying sections to experiment with the use of wireless radio transmitters on their aircraft. The use of these devices for aerial artillery direction purposes had been theorized and tested prior to the war with some promising results. However, the radios were extremely primitive and the airmen were unfamiliar with their use.[26] Test flights in the German rear-area were performed accordingly throughout December to give several crews the opportunity to become acquainted with the device and perfect its application. These experiments would ultimately lead to the transmitter's first use over allied territory in February, 1915.[27]

Meanwhile, the onset of trench warfare also gave new purpose to the men of the *Fliegertruppe's* various balloon sections. As discussed in the preceding chapters, the FLAs were rarely of assistance during the war of movement on account of their poor mobility. Circumstances changed however, once the front began to stabilize during the 'Race to the Sea.' Starting in late September, many balloon sections (particularly in the southern half of the front) were finally able to consistently begin regular tactical reconnaissance and artillery direction work. However, these first flights caused more harm to the local German ground forces than good. Despite numerous ascents over the span of six weeks, the FLAs were unable to produce any meaningful intelligence outside of several isolated reports concerning the movements of allied rail and vehicle traffic. This was because the Allied artillery would refuse to fire and give away their positions whenever a balloon was aloft, choosing instead to simply wait and then shell the area around the balloon immediately after it was lowered. Nearby German troops, as well as the men and animals of the balloon section itself, were therefore subjected to heavy allied fire whenever a FLA went into action.

The prime cause of the FLAs' poor performance throughout September and October was the "*Drachen*" balloon's inadequate observation distance. At 600cm, the *Drachen* was only able to reach an altitude of 1500–1800 feet under wartime conditions.[28] At that height, the radius of an observer's vision was limited to only 2½ to 3 miles, which was insufficient for the tasks demanded of it. Moreover, the FLAs were equipped with obsolete binoculars that further hindered their observers' vision. Attempts at aerial photography were similarly unsuccessful. Even in a mild breeze, the 600cm balloons were susceptible to severe rocking and swaying, making the observers sick while diminishing the clarity of the photographs.[29]

The FLA's collective lack of production had caused the army during the month of November to seriously consider the dissolution of all balloon sections. Fortunately, the initiative was quickly thwarted after an isolated report from the front displayed the balloon's potential value. The message stated that a FLA was compelled by circumstance to use a very rare 1000cm balloon that was able to reach 3600 feet. From that height, the balloon's observer spent the day directing heavy artillery fire with great success. In response to the incident, the Inspectorate of Airship Troops immediately placed orders for 800cm and 1000cm balloons to be constructed and dispersed amongst all existing FLAs. Shortly thereafter, the War Ministry approved the Inspectorate's request to establish five additional FLAs.[30]

Very few of the new balloons reached their units by the end of the year. Nonetheless, the balloon troops' rehabilitation was well under way. New cameras, designed specifically for use on the balloon were delivered to each FLA in November and December. These cameras (*Balloonkamer*) had a focal length of 70cm, which enabled the observers to photograph allied positions 11½ miles away. By contrast, the aircraft in the FFAs were equipped with only 25cm cameras and were experiencing serious problems with their use. Consequently, with the FLA's new larger balloons serving as a much steadier platform than their predecessor, the balloonists began 1915 as arguably the army's most effective reconnaissance gatherers.[31]

Yet another branch of the *Fliegertruppe* that experienced growth during the final months of the year was the anti-aircraft artillery. Back on September 25th, Falkenhayn wrote to members of the War Ministry in Berlin requesting additional guns stating: "Almost all armies report what greater assistance French artillery is receiving from air observation; they urgently request guns for defense."[32] Within one month, the army had doubled its number of specialized anti-aircraft guns from 18 at mobilization to 36, of which 22 were deployed on the Western Front. Experience had already shown however, that this number was grossly insufficient. Thus, the army turned to the artillery manufacturers of Krupp and Ehrhardt for help.

Above all, Falkenhayn and the rest of the army leadership did not want to affect the number of sorely needed light and heavy guns that had already been ordered from the manufacturers for regular use with the army by modifying them for anti-aircraft work. Therefore, with the factories already at full production, an alternative had to be found. Luckily the solution was quickly found by modifying captured allied artillery pieces for the task. The famous French 75mm gun for example was altered by increasing its elevation and deflection and reboring the barrel so

Above: German anti-aircraft gun converted from a captured French 75mm field gun. (Author's Collection)

that it could fire the German's 77mm ammunition. Meanwhile, Russian field guns (which the Germans had captured plenty of during the first months of the war) were not changed, but instead had special ammunition produced for it which caused delays in their employment.[33]

Falkenhayn's program to increase the army's anti-aircraft artillery was ordered in mid-October so that "each active infantry, reserve, *Landwehr*, and *Ersatz* Division would receive a horse-drawn anti-aircraft battery" consisting of two guns." By January, the army had accepted delivery of 216 guns and had established 108 active batteries at the front. 52 immobile guns were also deployed in the rear zone to defend key areas such as bridges, rail yards, airship hangers, and ammunition depots.[34]

The quantity of truck mounted anti-aircraft guns was also addressed. These highly mobile platforms carried a fully modified 77mm field gun, which Ernst von Hoeppner called "the best (A-A) gun in the army."[35] Recognizing their value, Falkenhayn openly called for each active and reserve corps HQ to have two of these vehicles while each army HQ was equipped with one. To that end, the War Ministry placed an order in December for 80 trucks to be produced at the rate of 10 per month during the following year.[36]

The Creation Of *BAO*: Germany's First Bomber Squadron

The most important organizational change adopted by the *Fliegertruppe* during the first year of the war occurred in late November with the establishment of their first bombing group. The concept of a dedicated bombing squadron was first proposed in a memo written by Major Wilhelm Siegert immediately after he was named OHL's aviation advisor on October 19th. Siegert argued that a large squadron dedicated to carrying out strategic bombing attacks against England was a necessity to the German war effort and should be formed as quickly as possible and placed under OHL's direct control. Falkenhayn, who believed Britain to be Germany's principal enemy, agreed with the report and ordered Siegert to proceed with the unit's formation.

With the assistance of his adjutant, *Ltn.* Fritz von Falkenhayn, as well as *Hptm.* Hermann Kastner, Siegert spent the first three weeks of November carrying out a multitude of tasks to prepare the squadron for operations. First was the issue of finding a sufficient number of suitable personnel. In accordance with Siegert's proposal, the new bomber group was to be comprised of 36 aircraft—the strength of six flying sections.[37] However, to maximize bomb payload, Siegert called for two-thirds of the aircraft to operate without an observer. Thus, it was necessary to extract 36 experienced pilots and 12 observers from the various flying sections at the front—a difficult task considering how under strength the *Fliegertruppe* already were at that time. Instead of simply making a selection off of a list, Siegert wanted to ensure the squadron was entirely composed of aggressive airmen who were agreeable with the unit's unique task. Therefore, the airmen of all the flying sections were informed of the new squadron's mission and were invited to apply as volunteers. In an encouraging response, Siegert and his aides received more applications than there were available positions. Thus, to avoid making a single flying section or area of the front too weak, the final selection of aircrews was made uniformly across the front using the list of volunteers.[38].

Next, an airfield capable of accommodating 36 aircraft had to be located. In order to successfully carry out attacks against England, it was initially thought that the squadron had to be at Calais. However, by the second week of November it had become clear that the harbor city would remain in allied possession for the foreseeable future. Siegert therefore chose to establish the group's airfield at Ghistelles (five miles south of Ostend) in order to carry out attacks against important British shipping and resupply targets around Dunkirk. To preserve secrecy and conceal the unit's mission, the new squadron was named: *Breiftauben Abteilung Ostend (BAO)* or "Carrier Pigeon Section Ostend."

Preparing the new field for 36 aircraft proved to be a monumental task. Boulders and large stones covered in sludge were found all over the field, which due to its location near the sea was susceptible to heavy flooding. Before flight operations could begin, these issues (among others) had to be solved. To this end, Siegert called upon two specialists, *Oblt* von Schröder and engineer Wegenast to immediately begin clearing and reconstructing the airfield and the surrounding area.[39] With a large number of engineers and workers at their disposal, the two men labored continuously throughout November and December to establish and improve the "Carrier Pigeon's" airfield. By mid-December, the field was ready for operations with additional improvements continuing well into the following spring.

While Wegenast's engineers were clearing and preparing the field, Schröder and his men constructed

Above: Wilhelm Siegert, OHL's aviation advisor and commander of the *BAO*. (Author's Collection)

permanent housing for the squadron's aircraft. Twenty hangars were originally completed, with each building capable of holding two or three machines. These hangars were to be painted inconspicuously and positioned in an irregular manner along the outer boundaries of the airfield in order to avoid detection by allied aircraft. Constructed with external wood paneling, the buildings were capable of withstanding strong winds during storms. Furthermore, for maximum efficiency, each hangar was also equipped with working electric lights, a telephone, and an access-road leading to the airfield in order to allow the aircraft to be quickly wheeled back and forth from the field without getting stuck in the soggy ground. Finally, a narrow-gauge rail line leading from the local rail station was laid around the entire airfield's perimeter in close proximity to each hanger.[40]

Finding housing for BAO's full complement of 50 officers and 200 enlisted men was also a challenge. Being directly under OHL's control, the squadron needed to remain highly mobile. Thus, with billeting in the small nearby village of Ghistelles out of the question, Siegert decided to house the entire unit's personnel in two modified passenger trains. The pilots and observers were given luxurious sleeping and dining cars while the enlisted men were comfortably placed in modified passenger cars. To avoid detection from allied aerial reconnaissance, the trains transported the crews at night to a major rail station several miles away at Bruges, followed by a return trip to the airfield the following morning.

Above all, Siegert's greatest difficulty in establishing "the pigeons" was obtaining a sufficient number modern aircraft capable of safely carrying

Above: Aerial photograph of Hangars 13 and 14 at the *BAO* airfield at Ghistelles. Note the four main rail sidings connected to the narrow gauge tracks which serviced each hangar at the airfield. (Ed Ferko Collection, Image 4EF-525-4-NB1, History of Aviation Collection, Special Collections and Archives Division, Eugene McDermott Library, The University of Texas at Dallas.)

out long distance bombing flights. Upon gaining permission to form the squadron, Siegert had placed an order with LVG for a group of the latest B-Types to be delivered to his squadron in northern Belgium. However, despite their completion in late November, the airplanes were not delivered to BAO until January. Consequently, Siegert was compelled to requisition existing aircraft from the frontline flying sections. In another attempt to maximize efficiency, each incoming pilot was given the power to select the type (and in many cases the exact model) of aircraft they wanted to fly. These aircraft were then withdrawn from existing flying sections and then shipped to Ghistelles where they were assembled and made ready for their crews throughout the month of December. A vast array of different aircraft was therefore attached to BAO at its founding. Among these were Aviatik, Albatros, LVG, DFW, and Otto B-Types as well as at least one Jeannin *Stahltaube* monoplane. To his credit, Siegert recognized that many of these aircraft were too outdated and underpowered to accomplish the squadron's assigned task.[41] Consequently, the pigeons' outspoken commander continued to lobby for newer model aircraft throughout the first months of 1915 until late spring when the unit was finally completely equipped with the latest aircraft.

BAO was officially formed at Ghistelles on November 27th as the engineers continued their work on the airfield and the surrounding hangars. Poor weather over the following two weeks caused serious delays to the work on the airfield and the arrival of a number of aircrews. Finally, by mid-December enough progress had been made to allow for the squadron's first raid—a sortie against Dunkirk by a flight of six aircraft. Two weeks later, a second raid on Dunkirk was staged using 12 machines. These early raids were performed using very primitive tactics, as *Ltn.* Andre Hug, a pilot with BAO at the time, later recalled:

Every second or third plane of the Staffel had an observer who was charged with the protection of one other plane that carried no observer. Instead of using automatic pistols and infantry rifles, the observer now carried a Mauser automatic rifle. Every pilot carried two bombs in his cockpit; the observer, sitting in the front cockpit, carried six more bombs. At that time the pilot was sitting in the rear cockpit and all planes without an observer had a device of tubes containing one bomb each besides the two bombs the pilot had behind his feet. When the visibility was poor, the only help for navigation, besides a map, was a compass.

As far as I can remember these bombs weighed about 10-15 kilograms each (about 20 pounds). Bombsights did not exist yet; the bombs were dropped when one thought that the right moment had come. The pear shaped bombs had a guiding mechanism at the tail which also ignited the bomb upon striking. Apparently the English pilots did not have bombsights either in those early days. One day an English plane dropped bombs on our airfield. They all fell between the hangars instead of hitting them.

Up to now formation flying was not yet known. We had to practice formation flying by attacking French ports and railway stations. When we started for the first attack to Dunkirk, the specific order for the first raid was: "Start in quick sequence at 1:00pm, rendezvous at 1:15 at 1,000 meters height over Ostend. Formation flight to Dunkirk over the sea, along the coast, for bombing military targets like the port, railway station, military barracks."

When I arrive on time at the point of departure, I saw no other aircraft. I circled the area for five minutes and since nobody showed up, I continued alone to drop our bombs over Dunkirk. The thirty-six [Sic][42] aircraft had taken off in quick succession, one after the other. Each dropped its bombs and landed separately within half an hour. Soon after that, a more orderly group flying was practiced.[43]

The BAO's first raids caused little damage and did nothing to change the strategic situation. Nonetheless, the squadron's formation was a critically important moment in the history of the German Air Service. For the first time, the army leadership was willing to admit the *Fliegertruppe's* offensive potential and concentrate valuable resources into a single unit to maximize their effect. As a result, the airmen of the BAO were able to go on to gain valuable experience throughout 1915, performing bombing and ground attack sorties as well as regularly engaging allied aircraft in aerial combat. The members of the BAO therefore formed an invaluable cadre that was later used to train and establish a doctrine for the *Fliegertruppe's* new 'battle squadrons' (*Kampfgeschwader*), which played a major role in German air operations during 1915 and 1916.

The Situation at the End of 1914

The first five months of war revealed a great deal about the untested *Fliegertruppe*—particularly regarding the quality of its personnel. With few exceptions, the airmen performed their assigned tasks brilliantly, giving the ground forces a wealth of valuable intelligence and support. Long range reconnaissance flights conducted throughout the Marne Campaign and the 'Race to the Sea' consistently located the allied armies in time to allow German troops to be maneuvered into advantageous positions prior to battle. Then, once an engagement had begun, the flying sections were successfully employed carrying out tactical reconnaissance and artillery direction missions which frequently impacted the battles' outcome. These actions repeatedly assisted the western armies win victories and minimize defeats during the critically important opening stages of the war, thus undoubtedly proving the airmen's worth above the battlefield.

However, despite their best efforts, the *Fliegertruppe* found themselves in a serious crisis by the end of the year. Severe organizational flaws, low aircraft production, and a trivial understanding of military aviation amongst many members of the army leadership had combined to limit the aviation forces' impact on the battlefield—resulting in a growing loss of confidence in the *Fliegertruppe* as 1915 began. Unfortunately, few people outside the aviation forces recognized the root causes for the *Fliegertruppe's* problems. The biggest issue, as has been discussed, was the army's substandard organizational model. Without trained aviation staff officers to advise the various army commanders, many flying sections had been misused or temporarily forgotten—creating numerous missed opportunities. Similarly, the lack of a centralized command for all aviation forces in the field resulted in the absence of a standardized liaison doctrine amongst the various flying sections. This created an atmosphere of mistrust and apathy between the army and the *Fliegertruppe* which served to further lower the quality of aerial support. Finally, the War Ministry's reluctance to sufficiently mobilize Germany's aircraft firms for wartime production upon the outbreak of hostilities had forced many flying sections to operate at half strength during the October and November battles when the ground forces sorely needed as much support as they could get.

By December, the *Fliegertruppe's* institutionalized problems had spread to infect the entire air service. The aviators' collective inability to quickly meet the unique demands of trench warfare had created the impression amongst the army that the *Fliegertruppe* were simply unable to support them any longer. Regrettably, the ground troops had once again failed to recognize the organizational causes of the problem and instead openly blamed the airmen themselves for the situation. This latest drop in confidence caused the *Fliegertruppe's* morale to plummet, thus further reducing the quantity and quality of the support flights provided by the various flying sections.

For Siegert and Eberhardt—German military aviation's two most influential voices in 1914—there could only be one solution: the complete reorganization of all aviation forces. This would both address the causes of the crisis and restore the situation.[44] As this narrative has demonstrated, the *Fliegertruppe's* greatest asset was their personnel. The aviation leadership recognized this and believed that under the correct circumstances, their airmen would flourish in the new tactical and operational roles required in trench warfare. All that was needed was the union of aviation forces at the front under a competent leader to standardize aerial doctrine, improve liaison with the army, and encourage aircraft production. However, with conditions rapidly deteriorating and an allied spring offensive fast approaching, the question was: when would this sorely needed reorganization finally take place?

Chapter 6 Endnotes

1. Hilmer *Frhr.* von Bülow, *Geschichte der Luftwaffe*, p.55.
2. Foley, *German Strategy and the Path to Verdun*, p.105.
3. The details of Falkenhayn's new strategy will be discussed in depth in Volume 2 of this series: *Great War's Finest: An Operational History of the German Air Service: Vol. 2 Western Front 1915*.
4. The German occupation of northeastern France had an incredible impact upon the French economy. The land under German control accounted for 64% of France's pre-war pig iron production, 58% of its steel, and 40% of its coal. These economic factors, in concert with the obvious desire to keep the front lines as far away from the German Frontier partly explain the German doctrine of stubbornly holding all territory won—no matter

how militarily insignificant. It was also later hoped that the high costs incurred in gaining such small pieces of land might be incentive to bring the allies to the peace table. Of course, as it happened, the high casualties simply drove the allies to double down on their losses and continue the war. Gerd Hardach, *The First World War 1914–1918* (London, 1977) pp.87–88; L.L. Farrar, *Divide and Conquer: German Efforts to Conclude a Separate Peace 1914–1918* (New York, 1978).
5 Greenhalgh, *Foch In Command*, p.74.
6 *Der Weltkrieg*, VI, p.380–385.
7 Foch, *The Memoirs of Marshal Foch*, pp.187–189.
8 Possession of the ridge was critically important for both sides. For the Germans, the heights allowed their artillery to support the infantry's defensive positions in the plain below. On the other hand, if the allies were able to place guns on the heights, the Germans would be forced to withdraw and concede a large amount of territory. Foch, *The Memoirs of Marshal Foch*, p.190.
9 Jack Sheldon, *German Army on Vimy Ridge 1914–1917* (South Yorkshire, 2008), pp.36–37.
10 Doughty, *Pyrrhic Victory*, pp.127–130.
11 In total, de Langle's army possessed 488 75mm, 144 90mm, 16 65mm, 14 80mm, 30 long range 120mm, and 50 short range 155mm pieces. Doughty, *Pyrrhic Victory*, p.131.
12 This strategy of slowly obtaining chunks of territory through successive attacks aimed at wearing down German resistance was to be the cornerstone of the unsuccessful allied strategy during the first three years of the war. Doughty, *Pyrrhic Victory*, pp.130–132.
13 *Der Weltkrieg*, VI, pp.389–390.
14 FOH, Tome 2, Annex No. 505.
15 FOH, Tome 2, pp.221–228; *Der Weltkrieg*, Vol. 6, p.390.
16 Joffre ultimately allowed Fourth Army's Champagne Offensive to continue for another two weeks until he finally suspended offensive activity there on January 13th. When it was finally over, the largest French gain was a mere 300 meters at the cost of an estimated 15,000 casualties. Gilles Bernard and Gérard Lachaux, *Batailles de Champagne 1914–1915*, (Paris, 2008) pp.12–17.
17 Recurring problems with aircraft and engine production ultimately kept many of the flying sections below full strength well into the following spring.
18 Advances in camouflage and other means of concealment only made an observer's task that much more difficult.
19 Hoeppner, *Germany's War in The Air*, p.18.
20 Hilmer *Frhr.* von Bülow, *Geschichte der Luftwaffe*, p.51; Hoeppner, *Germany's War in The Air*, pp.18–19.
21 Hilmer *Frhr.* von Bülow, *Geschichte der Luftwaffe*, p.55.
22 By sitting in the front seat, the aircraft's wing blocked the observer's downward view.
23 Cuneo, *The Air Weapon 1914–1916*, pp.154–155.
24 Entries taken from Boelcke's personal letters. Translated and published in: Werner, *Knight of Germany* (Havertown, 2009), pp.85–93.
25 Cuneo, *The Air Weapon 1914–1916*, p.156.
26 Helmut Förster, "Die Entwicklung der Fliegerei im Weltkrieg," Walter von Eberhardt (ed.) *Unsere Luftstreitkrafte 1914–18* (Berlin, 1930) pp.49–50.
27 Werner von Langsdorff, "Luftbildwesen, Funktelegraphie und Wetterkunde in ihrer Bedeutung für die Luftwaffe," Walter von Eberhardt (ed.) *Unsere Luftstreitkrafte 1914–18* (Berlin, 1930) pp.82–84; Cuneo, *The Air Weapon 1914–1916*, p.157.
28 In peacetime, the *Drachen* Balloons enjoyed a 2,600ft. ceiling. However, the lower quality of wartime mass-produced gas as well as routine "wear and tear" on the balloon's fabric combined to lower this ceiling dramatically. Hoeppner, *Germany's War in The Air*, p.26.
29 Cuneo, *The Air Weapon 1914–1916*, pp.157–158.
30 Hoeppner, *Germany's War in The Air*, pp.26–27.
31 The problem for the FLAs in the first months of 1915 was simply lack of numbers. In February 1915, the German Western Armies collectively possessed only 9 balloons. Richter, *Feldluftschiffer*, p.6; Cuneo, *The Air Weapon 1914–1916*, p.158.
32 Cuneo, *The Air Weapon 1914–1916*, p.198.
33 Hoeppner, *Germany's War in The Air*, pp.28–29.
34 Cuneo, *The Air Weapon 1914–1916*, pp.198–199.
35 Hoeppner, *Germany's War in The Air*, p.29.
36 Cuneo, *The Air Weapon 1914–1916*, p.199.
37 Although the squadron was originally envisioned to have 36 aircraft, it never reached this number. Due to losses and circumstance, the unit's strength peaked at 24 aircraft/crews in April 1915.
38 Ferko–UTD, Box 524, Folder 6, "History of the BAO."
39 Ferko–UTD, Box 524, Folder 6, "History of the BAO."
40 Ferko–UTD, Box 524, Folder 6, "History of the BAO."
41 Peter Kilduff, *Germany's First Air Force* (Osceola, 1991), p.21.
42 BAO never attempted any raid on Dunkirk using 36 aircraft. The squadron, in fact, never reached full strength. The biggest BAO raid against Dunkirk was in January 1915 when 14 airplanes were used. Hug wrote this after the war after he

had spent four years as a POW. It is most likely that he simply misremembered how many aircraft were used in this flight.

43 Hug, "Carrier Pigeon Flieger," p.304.
44 Apart from advocating an independent air service prior to the war, both men repeatedly suggested a reorganization during the conflict's first five months. The reader will recall that Eberhardt submitted the first proposal in late August after his troubling inspection of the Western Front flying sections. After his proposal was refused, Eberhardt carried on his work as *Inspekteur der Fliegertruppen* and continued to push for a reorganization throughout the winter of 1914/15. Meanwhile Siegert issued his first wartime recommendation for a reorganization in October upon being named "OHL's aviation advisor." Written within the same *Denkschrift* which served as the basis for the formation of BAO, Siegert recommended that all aviation units be reorganized under an independent command that was subordinated to OHL. Although the suggestion was ignored, Siegert continued to push for a reorganization during several meetings with Falkenhayn. These talks were ultimately successful as Falkenhayn wrote to the War Ministry in January proposing "all flying sections, airmen, balloon sections, and air defense formations be placed under the control of a unified command." Ferko–UTD, Box 524, Folder 6, "History of the BAO"; Cuneo, *The Air Weapon 1914–1916*, p.146; Hilmer *Frhr.* von Bülow, *Geschichte der Luftwaffe,* p.57.

Above & Below: Two views of Roland *Stahl-Pfeil-Eindeckers*. Only one Roland *Taube* was purchased by the German *Fliegertruppe*; it was given the designation A.157/13. Unlike many other *Taube* designs, the Roland had hinged elevators instead of warping elevators. (Aeronaut)

Appendix A:
Mobilization of Peacetime Aviation Formations

Flieger-Battalion #1
1. Kompanie (Döberitz): FFA 1, 11, 30
2. Kompanie (Döberitz): FFA 7, 12; Reserve Air Park #1
3. (Saxon) Kompanie (Grossenhain): FFA 23, 24, 29; Reserve Air Park #3

Flieger-Battalion #2
1. Kompanie (Posen): FFA 13, 19; Fest.FA 4; Reserve Air Park #8
2. Kompanie (Graudenz): FFA 16, 17; Fest. FA 6
3. Kompanie (Königsberg): FFA 14, 15; Fest. FA 5, 7

Flieger-Battalion #3
1. Kompanie (Cologne): FFAs 9, 10; Fest.FA 3; Reserve Air Park #4
2. Kompanie (Hannover): FFAs 21, 22, 28; Reserve Air Park #2
3. Kompanie (Darmstadt): FFAs 6, 18, 27; Reserve Air Park #5

Flieger-Battalion #4
1. Kompanie (Strasbourg): FFA 3, 4; Fest.FA 2
2. Kompanie (Metz): FFA 2, 5, 8; Fest. FA 1
3. Kompanie (Freiburg): FFA 20, 25, 26; Reserve Air Park #7

Bavarian Flieger-Battalion
1. Kompanie (Ober-Schleissheim): FFA(b) 1, 2, 3; FestFA(b) Germersheim; Reserve Air Park #6

Mobilization of Balloon Sections

Luftschiffer-Battalion # 1 (Berlin)	FLA 1
Luftschiffer-Battalion # 3 (Cologne)	FLA 2
Pioneer-Battalion # 21 (Mainz)	FLA 3
Infantry Regiment #46 (Posen)	FLA 4
Pioneer-Battalion #3 (Spandau)	FLA 5
Pioneer-Battalion #15 (Strasbourg)	FLA 6
Luftschiffer-Battalion #1 (Coblenz)	FLA 7
Luftschiffer-Battalion #5 (Königsberg)	FLA 8
Bavarian Luft.-Battalion (Munich)	FLA(b)

Below: The Rumpler 4C, normally powered by a 100 hp Mercedes D.I, was the last Rumpler *Taube* design. For improved maneuverability it had conventional ailerons and elevator. (Peter M. Bowers Collection/The Museum of Flight)

Appendix B:
Order of Battle: German Western Armies, August 1914

First Army (320,000 men, 164 Battalions, 796 Guns, 32 Aircraft)
Commander: Alexander von Kluck
Chief of Staff: Hermann von Kuhl

First Army HQ *Grevenboich*
FFA 12 (*Hauptmann* von Detten) — *Jülich*
FLA 1 (*Hauptmann* von Zychlinski) — *Jülich-Grevenbroich*
Etappen Flugpark #1 (*Major* Gundel) — *Düsseldorf*

   II AK (von Linsingen)
   FFA 30 (*Hauptmann* Wagenführ) — *Rheydt*

   IV AK (von Armin)
   FFA 9 (*Hauptmann* Musset) — *Aachen-Forest*

   III AK (von Lochow)
   FFA 7 (*Hauptmann* Grade) — *Elsdorf*

   IX AK (von Quast)
   FFA 11 (*Hauptmann* Wilberg) — *Aachen-Brand*

   III RK (von Beseler)

   IV RK (von Gronau)

   Attached: **2nd Cavalry Corps (HKK 2)** (von der Marwitz)
   3 Cavalry Divisions; 36 Guns; 72 Squadrons; 5 Jäger Battalions

Second Army (260,000 men, 147 Battalions, 693 Guns, 32 Aircraft)
Commander: Karl von Bülow
Chief of Staff: Otto von Lauenstein

Second Army HQ *Montjoie*
FFA 23 (*Oberleutnant* Vogel von Falckenstein) — *Höfen*
FLA 2 (*Hauptmann* Spangenberg) — *Aachen*
Etappen Flugpark #2 (*Major* Hohl) — *Hangelar*

   Guard Korps (von Plettenberg)
   FFA 1 (*Hauptmann* von Oertzen) — *Thirimont*

   X AK (von Emmich)
   FFA 21 (*Hauptmann* Geerdtz) — *Call*

   X RK (Count von Kirchbach)

   VII AK (von Einem)
   FFA 18 (*Hauptmann* von Gersdorff) — *Eupen*

   VII RK (von Zwehl)

Guard RK (von Gallwitz)

Attached: 1st Cavalry Corps (HKK 1) (von Richthofen)
2 Cavalry Divisions; 24 Guns; 48 Squadrons; 5 *Jäger* Battalions

Third Army (180,000 men, 97 Battalions, 549 Guns, 26 Aircraft)
Commander: Max von Hausen
Chief of Staff: Ernest von Hoeppner

Third Army HQ *Prüm*
FFA 22 (*Hauptmann* von Blomberg) — *St. Vith*
FLA 7 (*Hauptmann* Menzel) — *Niederprüm*
Etappen Flugpark #3 (*Major* Mardersteig) — *Niedermendig*

    XI AK (von Plüskow)
    FFA 28 (*Hauptmann* Freytag) — *Wallerode*

    XII RK (1st Saxon Reserve Korps) (von Kirchbach)

    XII AK (1st Saxon Korps) (D'Elsa)
    FFA 29 (*Hauptmann* von Jena) — *Ober-Beslingen*

    XIX AK (2nd Saxon Korps) (von Laffert)
    FFA 24 (*Hauptmann* von Minckwitz) — *Neuerburg*

Fourth Army (200,000 men, 122 Battalions, 621 Guns, 26 Aircraft)
Commander: Albrecht Duke of Württemberg
Chief of Staff: Walther von Lüttwitz

Fourth Army HQ *Treves*
FFA 6 (*Hauptmann* von Dewall) — *Trier-Euren*
FLA 3 (*Hauptmann* Schoof) — *Trier*
Etappen Flugpark #4 (*Major* Goebel) — *Trier*

    VIII AK (von Tschepe und Weidenbach)
    FFA 10 (*Hauptmann* Hantelmann) — *Trier-Euren*

    VIII RK (von Egloffstein)

    XVIII AK (von Schenck)
    FFA 27 (*Oberleutnant* Alfred Keller) — *Conz*

    XVIII RK (von Steuben)

    VI AK (von Pritzelwitz)
    FFA 13 (*Hauptmann* Alfred Streecius) — *Dillingen*

Fifth Army (200,000 men, 122 Battalions, 621 Guns, 26 Aircraft)
Commander: Crown Prince Wilhelm of Prussia
Chief of Staff: Konstantin Schmidt von Knobelsdorf

Fifth Army HQ *Sarrebruck*
FFA 25 (*Hauptmann* Blum) — *Dillingen*
FLA 4 (*Hauptmann* Stottmeister) — *Saarbrücken*

Etappen Flugpark #5 (Oberleutnant Pohl) Homburg/Pfalz

    V AK (Von Strantz)
    FFA 19 (*Hauptmann* von Poser und Großnädlitz) Beaumarais

    V RK (Von Gündell)

    VI RK (Von Gossler)

    XIII AK (Von Fabeck)
    FFA 4 (*Hauptmann* Haehnelt) Nieder-Jeutz

    XVI AK (Von Mudra)
    FFA 2 (*Hauptmann* Kirch) Metz

    Attached: 4th Cavalry Corps (HKK 4)—(Hollen)
    2 Cavalry Divisions; 24 Guns; 48 Squadrons; 5 Jäger Battalions

Sixth Army (220,000 men, 121 Battalions, 708 Guns, 32 Aircraft)
Commander: Crown Prince Rupprecht of Bavaria
Chief of Staff: Konrad Krafft von Dellmensingen

Sixth Army HQ *St. Avold*
FFA 5 (*Hauptmann* Kerksieck) *St. Avold*
Bavarian FLA 1 (*Hauptmann* Lochmüller) *St. Avold*
Bavarian *Etappen Flugpark #6* (*Hauptmann* Hiller) *Zweibrucken*

    XXI AK (Below)
    FFA 8 (*Hauptmann* Jerrmann) Bühl

    I Bavarian AK (Xylander)
    Bavarian FFA 1 (*Hauptmann* Erhard) Bühl

    II Bavarian AK (Martini)
    Bavarian FFA 2 (*Rittmeister* von Wolffskeel) Falkenberg

    III Bavarian AK (Gebsattel)
    Bavarian FFA 3 (*Hauptmann* Pohl) Urville

    I Bavarian RK (Fasbender)

    Attached: 3rd Cavalry Corps (HKK 3)—(Frommel)
    3 Cavalry Divisions; 24 Guns; 72 Squadrons; 5 Jäger Battalions

Seventh Army (125,000 men, 81 Battalions, 390 Guns, 20 Aircraft)
Commander: Josias von Heeringen
Chief of Staff: Karl Heinrich von Hänisch

Seventh Army HQ *Strasbourg*
FFA 26 (*Hauptmann* Walter) Strasbourg
FLA 6 (*Hauptmann* Kalsow) Strasbourg
Etappen Flugpark #7 (*Major* Siegert) Baden/Oos

    XIV AK (Hoiningen)
    FFA 20 (*Hauptmann* Barends) Freiburg

XIV RK (Schubert)

XV AK (Deimling)
FFA 3 (*Hauptmann* Genée) *Strasbourg*

19th & Bavarian Ersatz Divisions

OHL
Z VI (*Hauptmann* Kleinschmidt) *Cologne*
Z VII (*Hauptmann* Jacobi) *Baden-Oos*
Z VIII (*Hauptmann* Andrée) *Trier*
Z IX (*Hauptmann* Horn) *Düsseldorf*
"Victoria Lousie" (*Leutnant d. R.* Lempertz) *Frankfurt/Main*

Fortress Flying Sections
Fest.FA 1 (*Hauptmann* von Kleist) *Metz*
Fest.FA 2 (*Hauptmann* von Falkenhayn) *Strasbourg*
Fest.FA 3 (*Hauptmann* Volkmann) *Cologne*
Bavarian Fest.FA (*Hauptmann* Sorg) *Germersheim*

Fortress Balloon Troop
Fest.Luft.Trupp 10— Cologne
Fest.Luft.Trupp 13— Neu Breisach
Fest.Luft.Trupp 14, 15— Strasbourg
Fest.Luft.Trupp 18, 19, 20 , 21—Metz
Fest.Luft.Trupp 22—Diedenhofen
Fest.Luft.Trupp 29—Mainz
Bavarian Fest.Luft.Trupp— Germersheim

Note: Although 194 aircraft were needed in order to have the various Western Front units at full strength, the army was only able to mobilize 182 for the start of the war's opening campaign.

Left: A Pfalz-Otto pusher, normally used as a trainer, on operations. Pfalz and Otto were both Bavarian companies. Pfalz built this example of an Otto Pusher under license. In the Otto Pusher the pilot sat in front (note the elevator control horn at the feet of the crew member in front), which was unusual for a pusher. The engine was a 100 hp Rapp. Rapp was a modestly successful Bavarian engine company. It was purchased and reorganized in 1917 as the Bavarian Motor Works. BMW now makes well-known automobiles. (Aeronaut)

Appendix C:
Aircraft Types by FFA, August 1914*

First Army
FFA 7	Albatros B-Types
FFA 9	Gotha Taubes
FFA 11	LVG B-Types
FFA 12	Albatros B-Types
FFA 30	AEG B-Types

Second Army
FFA 1	LVG B-Types
FFA 18	LVG B-Types
FFA 21	5 B-Types, 1 Fokker A
FFA 23	DFW A/B-Types

Third Army
FFA 22	B-Types
FFA 24	B-Types
FFA 28	B-Types
FFA 29	B-Types

Fourth Army
FFA 6	Aviatik B-Types
FFA 10	Various Taubes
FFA 13	B-Types
FFA 27	B-Types

Fifth Army
FFA 2	Various Taubes
FFA 4	Gotha Taubes
FFA 19	B-Types
FFA 25	Aviatik B-Types

Sixth Army
FFA 5	B-Types
FFA 8	LVG B-Types
FFA 1(b)	Otto B-Types
FFA 2(b)	LVG B-Types
FFA 3(b)	LVG B-Types

Seventh Army
FFA 3	Various Taubes
FFA 20	Aviatik B-Types
FFA 26	Aviatik B-Types

Fortress Flying Sections
Fest.FA 4	Various A/B–types
Fest.FA 5	B-Types
Fest.FA 6	B-Types
Fest.FA 7	B-Types

* Ferko-UTD, Box 13, Folder 2, "Organization notes."

Above: Rumpler *Taube*. (Aeronaut)

Above: AEG B-types at a German flying school. (Aeronaut)

Appendix D:
Order of Battle: Western Allied Armies August 15, 1914

(From West to East)

4th Group of Reserve Divisions (51st, 53rd, & 69th)

British Expeditionary Force (Sir John French)
1st Corps (Haig)
2nd Corps (Smith-Dorrien)
Cavalry Division (Allenby)

Fifth Army (Lanrezac)
I Corps (d'Espèrey)
III Corps (Sauret); 3 divisions
X Corps (Defforges); 3 divisions
XVIII Corps (De Mas Latrie)
Cavalry Corps (Sordet)

Fourth Army (Langle de Cary)
52nd & 60th Reserve Divisions
IX Corps (Dubois)
XI Corps (Eydoux)
XII Corps (Roques)
XVII Corps (Poline)
Colonial Corps (Lefévre)
II Corps (Gérard)
4th & 9th Cavalry Divisions

Third Army (Ruffey)
IV Corps (Boëlle)
V Corps (Brochin)
VI Corps (Sarrail); 3 Divisions
3rd Group of Reserve Divisions (54th, 55th, 56th)
7th Cavalry Division

Second Army (De Castelnau)
XX Corps (Foch)
XV Corps (Espinasse)
XVI Corps (Taverna)
18th Infantry Division
2nd Group of Reserve Divisions (59th, 68th, & 70th)

First Army (Dubail)
VIII Corps (De Castelli)
XIII Corps (Alix)
XXI Corps (Legrand-Girarde)
XIV Corps (Pouradier-Duteil)
6th Cavalry Division

Army of Alsace (Pau)
VII Corps (Bonnier, then Vautier)
44th Division
1st Group of Reserve Divisions (58th, 63rd, 66th)
5 Groups of Battalions of Chasseurs
8th Cavalry Division
57th Reserve Division

Below: A Pfalz-Otto pusher with early national insignia but no other markings. The engine was a 100 hp Rapp. (Aeronaut)

Appendix E:
Order of Battle During the Battle of the Marne

German Armies
(From West to East)

First Army (Kluck)
4th Cavalry Division
IV RK (Gronau)
II AK (Linsingen)
IV AK (Arnim)
III AK (Lochow)
IX AK (Quast)
HKK 2 (Marwitz)

Second Army (Bülow)
HKK 1 (Richthofen)
VII AK (Einem)
X RK (Eben)
X AK (Emmich)
Guard (Plettenberg)

Third Army (Hausen)
32 ID, 24 RD, & 23 RD (Kirchbach's Group)

23 ID & XIX AK (D'Esla Group)

Fourth Army (Albrecht)
VIII AK (von Tschepe und Weidenbach)
VIII RK (Egloffstein)
XVIII AK (Schenck)
XVIII RK (Steuben)

Fifth Army (Crown Prince)
HKK 4 (Hollen)
VI AK (Pritzelwitz)
XIII AK (Fabeck)
XVI AK (Mudra)
VI RK (Gossler)
V RK (Gündell)
V AK (Strantz)

Sixth Army (Rupprecht)
Metz Garrison; 33 Reserve Division
Ersatz Divisions (Guard, 4th, 8th, 10th)
III Bavarian AK (Gebsattel)
I Bavarian RK (Fasbender)
II Bavarian AK (Martini)
XXI AK (Below)
I Bavarian AK (Xylander)
HKK 3 (Frommel)

Seventh Army (Heeringen)
XIV AK (Hoiningen)
XV AK (Schubert)
XIV RK (Deimling)
Ersatz Divisions (19th & Bavarian)
30th Reserve Division (Strasbourg Garrison)

Allied Armies
(From West to East)

Sixth Army (Maunoury); organized August 26th
Cavalry Corps (Sordet, then Bridoux)
VII Corps (Vautier); transferred from Alsace with 63rd Reserve Division replacing the 41st Division.
Group of Reserve Divisions (56th & 55th)
45th (Algerian) Division
Native Moroccan Brigade
Cavalry Brigade
61st Reserve Division (arriving September 7th)
IV Corps (Boëlle); arrived September 8th. Only 7th Division participated in the battle.

British Expeditionary Force (Sir John French)
3rd Corps (Pulteney)
Cavalry Detachment (3rd & 5th Cavalry Brigades)
2nd Corps (Smith-Dorrien)
1st Corps (Haig)
Cavalry Division (Allenby); 1st, 2nd, 4th Cavalry Brigades.

Fifth Army (d'Espèrey)
Cavalry Corps (Conneau)
XVIII Corps (De Maud'huy); 3 divisions
Group of Reserve Divisions (53rd & 69th)
III Corps (Hache); 3 divisions
I Corps (Deligny)
X Corps (Defforges); 3 divisions

Ninth Army (Foch); organized September 5th
42nd Division
IX Corps (Dubois)
52nd Reserve Division
XI Corps (Eydoux)
60th Reserve Division
9th Cavalry Division
18th Division (arrived from Second Army September 7th)

Fourth Army (Langle de Cary)
XXI Corps (Legrand-Girarde); arrived from First Army September 8th.
XVII (Dumas)
XII (Roques)
Colonial Corps (Lefévre)
II Corps (Gérard)

Third Army (Sarrail)
XV Corps (Espinasse) arrived from Second Army September 7th.
7th Cavalry Division
V Corps (Micheler)
VI Corps (Verraux); minus the 42nd Division
Group of Reserve Divisions (67th, 75th, 65th)
Verdun Garrison (72nd Reserve Division & brigade of 54th Reserve Division)

Second Army (De Castelnau)
2nd Cavalry Division
Toul Garrison (73rd Reserve Division)
Group of Reserve Divisions (59th, 68th, 70th)
XXV Corps (Balfourier)
74th Reserve Division
XVI Corps (Taverna)

First Army (Dubail)
6th Cavalry Division
VIII Corps (De Castelli)
XIII Corps (Alix)
Temporary Corps (44th Division, Colonial Brigade, & battalions of Chasseurs)
XIV Corps (Baret)
Group of the Vosges (41st Division & 58th Reserve Division)
Belfort Garrison (57th Reserve Division & active brigade)
66th Reserve Division
14th Dragoon Brigade

Above: Pfalz Parasol with Bavarian serial "P2". The national insignia were applied to all flying surfaces. In February 1914 the Pfalz company, located in Bavaria, signed a license agreement with the French Morane-Saulnier company to build the M-S Type H and Type L under license. The M-S Type L was built as the Pfalz Parasol; 60 were built with the first two being delivered to the *Fliegertruppe* in December 1914. Most were powered by the 80 hp, 7-cylinder Oberursel U-0 engines and some were built with the 100 hp, 9-cyliner Oberursel U-I engines. Later these types were retroactively designated the Pfalz A.I and A.II, respectively. (Aeronaut)

Appendix F:
Fliegertruppe Order of Battle: October 10, 1914*

Fourth Army (Reorganizing in Rear)
HQ— FFA 6
XXII RK
XXIII RK
XXVI RK
XXVII RK

Sixth Army
HQ— FFA 25
Guard Korps— FFA 1
IV AK— FFA 9
VII AK— FFA 18
XIII AK— FFA 4
XIV AK— FFA 20
XIX AK— FFA 24
I Bav. RK

Second Army
HQ– FFA 23
XVIII AK— FFA 27
XXI AK— FFA 8
I Bav. AK— FFA 1b
II Bav.AK— FFA 2b
XIV RK

First Army
HQ— FFA 12
II AK— FFA 30
III AK— FFA 7
IX AK— FFA 11
IV RK
IX RK

Seventh Army
HQ— FFA 26
X AK— FFA 21
XII AK— FFA 29
XV AK— FFA 3
VII RK— FFA 39
X RK

Third Army
HQ— FFA 22
VI AK— FFA 13
VIII AK— FFA 10
VIII RK
XII RK

Fifth Army
HQ— FFA 25
V AK—FFA 19
XVI AK—FFA 2
V RK— FFA 44
VI RK— FFA 34
XVIII RK
III Bav. RK

Army Abteilung Strantz

Army Abteilung Falkenhausen

Army Abteilung Gaede

* Ferko-UTD, Box 20, Folder 5, "Order of Battles."

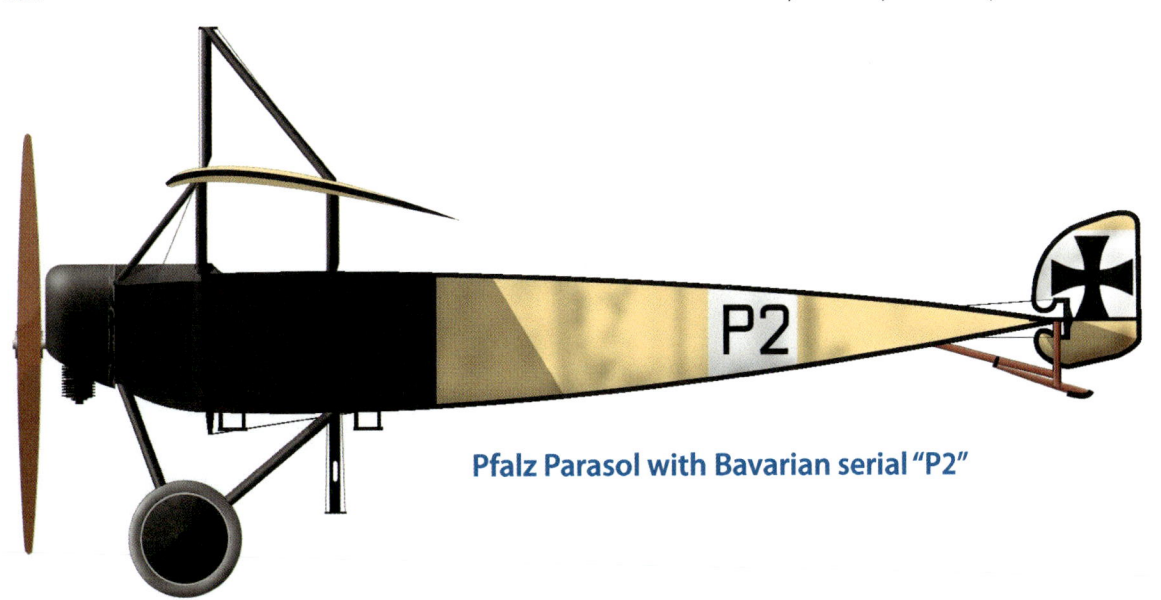

Pfalz Parasol with Bavarian serial "P2"

Appendix G:
Fliegertruppe Order of Battle: December 10, 1914*

OHL
Breiftauben Abteilung Ostend (BAO) [Carrier Pigeon Section Ostend]

Fourth Army
HQ–FFA 6 & 38
XV AK– FFA 3
XXII RK
XXIII RK– FFA 40
XXVI RK– FFA 41
XXVII RK
Marine Korps– Naval Air Station *Zeebrugge*

Sixth Army
HQ–FFA 5
Guard Korps– FFA 1
IV AK– FFA 9
VII AK– FFA 18
XIII AK– FFA 4
XIV AK– FFA 20
XIX AK– FFA 24
II Bav. AK– FFA 2b
6 Bav. RD– FFA 5b
I Bav. RK– FFA 4b

Second Army
HQ–FFA 23
XVIII AK– FFA 27
XXI AK– FFA 8
I Bav.– FFA 1b
XIV RK– FFA 32

First Army
HQ–FFA 12
III AK– FFA 7
IX AK– FFA 11
IV RK– FFA 33
IX RK

Seventh Army
HQ– FFA 26
X AK– FFA 21
XII AK– FFA 29
VII RK– FFA 39
X RK– FFA 26

Third Army
HQ–FFA 22
VI AK– FFA 13
VIII AK– FFA 10
VIII RK
XII RK

Fifth Army
HQ–FFA 25
XVI AK– FFA 2
V – FFA 44
VI RK– FFA 34
XVIII RK

Army *Abteilung* Strantz
V AK–FFA 19
III Bav. AK– FFA 33

Army *Abteilung* Falkenhausen
HQ–*Festung Flieger Abteilung* Metz
XV RK

Army *Abteilung* Gaede

* Ferko-UTD, Box 20, Folder 5, "Order of Battles."

Right: Otto-built LVG B.I in flight. Almost haflf the LVG B.I reconnaissance planes in Bavarian service in 1914 were built by Otto, a Bavarian company. (Aeronaut)

Appendix H:
Aircraft Data

Rumpler *Taube*

Engine:	86-hp Mercedes D. I	
Wing:	Span	14.3 m (46 ft. 11 in.)
	Area	32.5 m² (107 sq. ft.)
General:	Length	9.9m (32 ft. 6 in.)
	Empty Weight	650kg (1,433 lbs.)
	Loaded Weight	850 kg (1,874 lbs.)
Max Speed:		100 km/h (62mph)

Albatros B.I

Engine:	100-hp Mercedes D. I	
Wing:	Span Upper	14.48 m (47 ft. 6 in.)
	Chord Upper	1.80 m (5 ft. 11 in.)
	Chord Lower	1.80 m (5 ft. 11 in.)
	Gap	1.80 m (5 ft. 11 in.)
General:	Length	8.0 m (26 ft. 3 in.)
	(long fuselage)	8.56 m (28 ft. 1 in.)
	Height	3.15 m (10 ft. 4 in.)
	Empty Weight	752 kg (1,658 lbs.)
	Loaded Weight	1197 kg (2,639 lbs.)
Max Speed:	100 km/h (62mph)	
Climb:	800m (2,625 ft.)	10 min.
	2000m (6,562 ft.)	35 min.

Aviatik B.I

Engine:	100-hp Mercedes D. I	
Wing:	Span Upper	12 m (39 ft. 4 in.)
	Area	38 m² (125 sq. ft.)
General:	Length	7.5m (24 ft. 7 in.)
	Empty Weight	650kg (1,433 lbs.)
	Loaded Weight	980 kg (2,160 lbs.)
Max Speed:		115 km/h (71.5mph)
Climb:	1,200m (3,937 ft.)	15 min.

LVG B. I

Engine:	100-hp Mercedes D. I	
Wing:	Span Upper	14.5 m (47 ft. 6 in.)
	Span Lower	12.5 m (41 ft.)
	Dihedral Upper	2 deg
	Dihedral Lower	2 deg
	Sweepback	40 cm
	Gap	1.95 m (6 ft. 4 in.)
	Area	42.5 m² (139 sq. ft.)
General:	Length	9.00 m (29 ft. 6 in.)
	Empty Weight	765 kg (1,686 lbs.)
	Loaded Weight	1132 kg (2,496 lbs.)
Max Speed:		90 km/h (56mph)
Climb:	800m (2,625 ft.)	14 min.
	2,000m (6,562 ft.)	24.5 min

Appendix I:
Roster of *Breiftauben Abteilung* Ostend (BAO) as of December 1914

Rank	Name	Role	Joined	Left	Remarks
Major	Siegert	Commander/Observer	Nov. 27, 1914	Feb. 1915	
Leutnant	*Frhr.* von Falkenhayn	Adjutant/Pilot	Dec. 1914	Mid-April 1915	
Hauptmann	Kastner	Observer	Nov. 27, 1914	End of April 1915	Becomes commander of FFA
Hauptmann	von Ascheberg	Pilot	Nov. 27, 1914	June 1915	Becomes commander of FFA
Hauptmann	von Liebermann	Observer	Nov. 27, 1914	June 1915	Becomes commander of FFA
Hauptmann	Hempel	Pilot	Nov. 27, 1914	June 1915	Becomes commander of FFA
Oberleutnant	Adami	Pilot	Dec. 1914	Dec. 1914	Döberitz
Leutnant	von Blanc	Pilot	Nov. 27, 1914	May 1915	To *Abt.* Müller
Oberleutnant	Blume	Pilot	Nov. 27, 1914	June 1915	To *Abt.* Liebermann
Oberleutnant	Blumenbach	Pilot	Nov. 27, 1914	June 1915	To *Abt.* Liebermann
Oberleutnant	Bremer	Observer	Nov. 27, 1914	Dec. 22 1914	Captured at Dunkirk
Leutnant	Bonde	Pilot	Dec. 1914	April 1915	
Oberleutnant	Clemens	Pilot	Nov. 27, 1914	July 1915	To B.A.M.
Leutnant	Engwer	Pilot	Nov. 27, 1914	Unknown	
Oberleutnant	Emrich	Pilot	Nov. 27, 1914	May 16, 1915	Missing (dead)
Oberleutnant	*Frhr.* von Freyberg	Pilot	Nov. 27, 1914	April 1915	Commander FFA 5
Oberleutnant	Fink	Observer	Dec. 1914	June 1915	BAO HQ
Oberleutnant	Geyer	Pilot	Dec. 1914	December 1914	To Freiburg
Oberleutnant	Gravenstein	Observer	Nov. 27, 1914	June 1915	To *Abt.* Liebermann
Leutnant	Gröbedinkel	Pilot	Nov. 27, 1914	June 7 1915	Missing (captured)
Leutnant	*Frhr.* von Haller	Pilot	Nov. 27, 1914	Unknown	
Leutnant	Held	Pilot	Nov. 27, 1914	June 1915	To Schleissheim
Oberleutnant	Hell	Observer	Nov. 27, 1914	Unknown	
Leutnant	Hiddessen	Pilot	Nov. 27, 1914	Feb. 4 1915	Missing (captured)
Leutnant	Hug	Pilot	Dec. 1914	Dec. 22 1914	Captured at Dunkirk
Flieger	Ingold	Pilot	Nov. 27, 1914	Dec. 1915	To Freiburg
Oberleutnant	*Frhr.* von Könitz	Observer	Nov. 27, 1914	Unknown	
Leutnant	von Körber	Pilot	Nov. 27, 1914	May 1915	To *Abt.* Müller
Leutnant	von Ledebour	Pilot	Nov. 27, 1914	June 1915	FFA 13
Oberleutnant	Linke	Pilot	Nov. 27, 1914	Feb. 1915	*Flieger*-school

Rank	Name	Role	Joined	Left	Remarks
Oberleutnant	Müller, Kurt	Observer	Nov. 27, 1914	May 1915	Becomes FFA Commander
Leutnant	Müller, Fritz	Observer	Dec. 1914	Feb. 4, 1915	Missing (dead)
Leutnant	von Osterroht	Pilot	Nov. 27, 1914	Nov. 1915	FEA Döberitz
Leutnant	Parschau	Pilot	Nov. 27, 1914	May 1915	To *Abt.* Müller
Oberleutnant	Pfeifer	Pilot	Nov. 27, 1914	June 1915	To *Abt.* Liebermann
Oberleutnant	Prestien	Pilot	Nov. 27, 1914	June 1915	To *Abt.* Liebermann
Oberleutnant	Schregel	Observer	Nov. 27, 1914	January 1915	To War Ministry
Leutnant	Schwarzenberger	Pilot	Nov. 27, 1914	January 1915	
Leutnant	Seehagen	Pilot	Nov. 27, 1914	July 1915	To BAM
Oberleutnant	Stenzel	Pilot	Nov. 27, 1914	June 1915	To *Abt.* Liebermann
Oberleutnant	Wegener	Pilot	Dec. 1914	Feb. 1915	
Leutnant	Weyer	Observer	Nov. 27, 1914	April 1915	*Flieger*-school
Leutnant	Zeumer	Pilot	Nov. 27, 1914	May 1915	To *Abt.* Müller
Leutnant	Vierordt	Anti-Aircraft Gun Officer	Nov. 27, 1914	Sept. 1915	
Leutnant	Jürgens	Commander Vehicle Column	Nov. 27, 1914	June 1915	Home leave
Doctor	Gutmann	Surgeon	Dec. 1914		
Engineer	Wegenast		Nov. 27, 1914		Promoted to *Leutnant*—May 1915

Above: Rumpler B.I B.483/14 carries national insignia on both sides of all four wings. A 100 hp Mercedes D.I provided the power. Designed and placed into production in 1914, the Rumpler B.I was a good aircraft for its time. (Aeronaut)

Bibliography

Archival Materials

Bayerisches Hauptstaatsarchiv, IV, Kriegsarchiv [BHStA-KA].
Bundesarchiv, Coblenz [BA].
Bundesarchiv–Militärarchiv, Freiburg [BA–MA].
Ed Ferko Collection, History of Aviation Collection, Special Collections and Archives Division, Eugene McDermott Library, The University of Texas at Dallas [Ferko–UTD].
Landesarchiv Baden–Württemberg, Generallandesarchiv Karlshrue [GLA].
Landesarchiv Baden–Württemberg, Hauptstaatsarchiv Stuttgart [HSta].
Imperial War Museum, London [IWM].
Sächsisches Hauptstaatsarchiv—Dresden [SHStA].
The National Archives of the United Kingdom [TNA].
United States Army Heritage and Education Center. Carlisle, PA. [USAHEC].
United States National Archive [USNA] Record Group 242. Documents of the Royal Prussian Military Cabinet. M962.
United States Command & General Staff College, Combined Arms Research Library. Fort Leavenworth, Kansas [USCGSC–CARL].

Official Publications

Bose, Thilo von. *Das Marnedrama 1914*. 1 Abschnitt des III Teil. *Der Ausgang der Schlacht*. Bd. 24. *Schlachten des Weltkrieges, im Auftrage des Reichsarchivs*. Berlin, 1928.
———. *Das Marnedrama 1914*. 2 Abschnitt des III Teil. *Der Ausgang der Schlacht*. Bd. 25. *Schlachten des Weltkrieges, im Auftrage des Reichsarchivs*. Berlin, 1928.
———. *Das Marnedrama 1914*. I Teil. Bd. 22. *Schlachten des Weltkrieges, im Auftrage des Reichsarchivs*. Berlin, 1928.
———. *Das Marnedrama 1914*. II Teil. Bd. 23. *Schlachten des Weltkrieges, im Auftrage des Reichsarchivs*. Berlin, 1928.
Dahlmann, Reinhold. *Die Schlacht vor Paris: Das Marnedrama 4. Teil*. Bd. 26. *Schlachten des Weltkrieges, im Auftrage des Reichsarchivs*. Berlin, 1925.
Edmonds, J.E. *Military Operations France and Belgium*. 28 vols. [BOH] London, 1927–47.
Hendemann, Kurt. *Die Schlacht bei St. Quentin 1914: Der rechte Flügel der deutschen 2. Armee am 29. und 30. August*. Bd 7a. *Schlachten des Weltkrieges, im Auftrage des Reichsarchivs*. Berlin, 1926.
Hendemann, Kurt. *Die Schlacht bei St. Quentin 1914: Garde und hannoveraner vom 29. und 30. August*. Bd 7b. *Schlachten des Weltkrieges, im Auftrage des Reichsarchivs*. Berlin, 1926.
Reichsarchiv, *Der Weltkrieg 1914 bis 1918*, 14 vols. Berlin, 1925–1956.
Les Armées Françaises dans la Grande Guerre, 103 vols. [FOH]: Paris, 1922–1938.
Kriegswissenschaftliche Abteilung der Luftwaffe, ed. *Mobilmachung, Aufmarsch, und erster Einsatz der deutschen Luftstreitkräfte im August 1914*. (Dritte Einzelschrift der Kriegsgeschichtlichen Enzelschriften der Luftwaffe). Berlin, 1939.
———. *Die Militarluftfahrt bis zum Beginn des Weltkrieges 1914*. 3 vols. Militärgeschichtliches Forschungsamt. Freiburg im Breisgau, 1965–66.
Raleigh, Walter. *The War in The Air*. Vol.1. Nashville, 1998.
Schwink, Otto. *Ypres 1914: An Official Account Published by Order of The German General Staff*. Translated by G.C.W. Nashville, 1994.
Tschischwitz, Erich von. *Antwerpen 1914*. Bd. 3 *Schlachten des Weltkrieges, im Auftrage des Reichsarchivs*. Berlin, 1924.

Memoirs & Contemporary Accounts

Bauer, Max von. *Der grosse Krieg in Feld und Heimat: Erinnerungen und Betrachtungen*. Tübingen, 1921.
Boelcke, Oswald. *An Aviator's Field Book*. Nashville, 1991.
Bülow, Karl von. *Bülow's Report on the Battle of the Marne*. Translated by Captain F.G. Dumont. Fort Benning, 1936.
Coleman, Frederic. *From Mons to Ypres With General French*. New York, 1916.
Cossel, Maximilian von. "Artillerie-flieger 1914." In *der Luft unbesiegt: Erlebnisse im Weltkrieg erzählt von Luftkämpfern*. Munich, 1923.
Davout, Louis N. *Opérations du 3e Corps, 1806–1807. Rapport du maréchal Davout, duc d'Auerstadt*. Edited by Général Léopold Davout. Paris, 1896.
Eberhardt, Walter von. "Wie wir wurden: Tagebuchblätter aus dem Anfang des Weltkrieges." In *der Luft unbesiegt: Erlebnisse im Weltkrieg erzählt von Luftkämpfern*. Munich, 1923.
Eisenlohr, Roland. *Flugwesen und Flugzeugindustrie der Kriegführenden Staaten*. Berlin, 1915.
Falkenhayn, Erich von. *General Headquarters, 1914–1916, and its Critical Decisions*. Nashville, 2000.
Foch, Ferdinand. *The Memoirs of Marshal Foch*. Translated by Colonel T. Bentley Mott. New York, 1931.

French of Ypres, Field-Marshal Viscount. *1914*. London, 1919.
Galliéni, Joseph. *Mémoires de Général Galliéni*, Paris, 1920.
Groener, Wilhelm. *Commander Against His Will: Operative Studies of the World War*. Translated by Martin F. Schmitt. Washington DC, 1943.
Grossherzogliches Artilleriekorps, 1 Grossherzoglich Hessisches Feldartillerie-Regiment Nr. 25 im Weltkrieg 1914–1918 [FAR 25]. Berlin, 1935.
Hausen, Baron Max von. *Memoirs of the Marne Campaign*. Translated by John B. Murphy. Unpublished Text. USCGSC–CARL, 1922.
Heemskerck, Hans-Edouard von, "Das Gefecht der FFA 11 bei Mortefontaine 1914." *Ünsere Luftstreitkräfte 1914–18*. Berlin, 1930.
Hoeppner, Ernest von. *Germany's War in The Air*. Translated by J. Hawley Larned. Nashville, 1994.
Hug, Andre. "Carrier Pigeon Flieger." *Cross and Cockade USA Journal*. Vol.13, Nr. 4. 1972.
Kluck, Alexander von. *The March on Paris*. East Sussex, 2004.
Koeppen, Hans. *Im Auto Um Die Welt*. Berlin, 1909.
Koerber, Adolf-Victor von. "Heute vor 25 Jahren!" *Luftwelt*. 2 Jahrg. 1935, Nr. 4. 1935.
Kuhl, Hermann von. *The Marne 1914*. Fort Leavenworth, 1936.
———. "French Cavalry Raid in The Battle of The Marne." *The Cavalry Journal*. Vol. XXXI. Washington D.C., 1922.
Kuhl, Hermann von & Walther von Bergmann. *The Supply and Movement of German First Army During August & September 1914*. Fort Leavenworth, 1929.
Lanrezac, Charles. *French Plan of Campaign and First Month of the War*. Translated by Amico J. Barone. Unpublished Document. USAHEC. 1922.
Le Goffic, Charles. *Dixmude: French Marines in the Great War 1914–1918*. Leonaur, 2012.
Loßberg, Fritz von. *Mein Tätigkeit im Weltkriege*. Berlin, 1939.
Ludendorff, Erich von. *The General Staff and Its Problems (Vol. 1)*, New York, 1920.
McCudden, James. *Flying Fury*. London, 1933.
Müller-Loebnitz, Wilhelm. *The Mission of Lieutenant-Colonel Hentsch on Sept. 8–10, 1914*. Translated in the Office of the Military Attaché American Embassy, Berlin, Germany. USAHEC, 1933.
Private Letters of Josef Suwelack–FFA 24.
Poseck, Maximilian von. *The German Cavalry in Belgium and France 1914*. Essex, 2008.
Richthofen, Manfred von. *The Red Baron*. South Yorkshire, 2013.
Rupprecht, Crown Prince of Bavaria. *In Treue Fest: Kronprinz Rupprecht, Mein Kriegstagebuch*. 3 vols. Munich 1929.
Schell, Adolf von. *Battle Leadership*. Columbus, 1933.
Spears, Edward. *Liaison 1914*. London, 1999.
'T'. "Unsere Fliegertruppen im Kaisermanöver." *Deutsche Zeitschrift für Luftschiffahrt*. XVII. 1913.
Tappen, Gerhard. *Movements, transportation and supply of the German right wing in August and September, 1914*, 3 vols. Translated by W.C. Koenig. USAHEC.

Secondary Sources

Afflerbach, Holger. *Falkenhayn: Politisches Denken Und Handeln Im Kaiserreich*. Munich, 1996.
Arcq, Alain and Achille Van Yprezccle. *Leernes & Collarmont 22 Août 1914*. Fontaine-l'Évêque, 2013.
Armengaud, *La Renseignement Aérien Sauvegarde des Armées*. Paris, 1934.
Asprey, Robert B. *The German High Command at War*. New York, 1991.
Barry, Quintin. *The Franco-Prussian War 1870–1871. Volume 1*. West Midlands, 2009.
Baumgarten-Crusius, Artur. *German Generalship During the Marne Campaign in 1914: Contributions to a Determination of the Question of Responsibility*. Translation from German. Unpublished Text, 1922.
Beasley, Rex W. *A Critical Analysis of the French Colonial Corps in the Battle of the Ardennes with particular attention to the operations of the 3rd Colonial Division at Rossignol*. Command and General Staff School Student Papers Collection 1930–1936. Fort Leavenworth, 1933.
Beckett, Ian. *Ypres 1914: The First Battle*. New York, 2013.
Bergin, William E. *Principles and Methods of Conducting a Meeting Engagement as Illustrated by the French Colonial 5th Brigade at Neufchâteau, 22nd August 1914*. Command and General Staff School Student Papers Collection 1930–1936. Fort Leavenworth, 1934.
Bertrand, Charles. État *Actuel de l'Aéronautique Militaire & Navale en France et à l'Étranger*. Paris, 1913.
Bilton, David. *The Germans in Flanders 1914*. South Yorkshire, 2012.
Bowden, Scott. *Napoleon and Austerlitz*. Chicago, 1997.
British Army. *Handbook of the German Army 1914 (Home and Colonial) (Fourth Edition) 1912, Amended to August 1914*. Nashville, 2002.
Brose, Eric Dorn. *The Kaiser's Army: The Politics of Military Technology in Germany during the*

Machine Age 1870–1914. New York, 2001.
Buchholz, Frank & Joe and Janet Robinson. *The Great War Dawning*. Vienna, 2013.
Bülow, Hilmer Frhr. Von. *Geschichte Der Luftwaffe*. Frankfurt am Main, 1934.
Carlswärd, Tage. *Strategic Signal Communications with the German Right Wing*. Translated by Sigurd N. Ronning. Fort Monmouth, 1933.
Carruth, John H. *Study of the Operations of the French XI Corps at the Battle of Maissin, World War, 22 August, 1914*. Command and General Staff School Student Papers Collection 1930–1936. Fort Leavenworth, 1932.
Cave, Nigel and Jack Sheldon. *Ypres 1914: Messines*. South Yorkshire, 2015.
Citino, Robert M. *The German Way of War: From the Thirty Years' War to the Third Reich*. Lawrence, 2005.
Cordonnier. *L'Obéissance aux Armées*. Paris, 1924.
Corum, James S. & Richard R. Muller. *The Luftwaffe's Way of War: German Air Force Doctrine 1911–1945*. Baltimore, 1998.
Crevald, Martin van. *Supplying War: Logistics from Wallenstein to Patton*. New York, 2004.
Cron, Hermann. *Imperial Germany Army 1914–18: Organization Structure, Orders of Battle*. Translated by Duncan Rogers. Eastbourne, 2001.
Cuneo, John R. *The Air Weapon 1914–1916*. Harrisburg, 1947.
———. *Winged Mars: The German Air Weapon 1870–1914*. Harrisburg, 1942.
d'Ydewalle, Charles. *Albert: King of the Belgians*. Translated by Phyllis Mégroz. London, 1935.
Dahlquist, John E. *Study of The Mission of Lieutenant-Colonel Hentsch, German General Staff, on 8th–10th September 1914 During the First Battle of The Marne*. Command and General Staff School Student Papers Collection 1930–1936. Fort Leavenworth, 1931.
Demerre, Guido and Johan Termote. "The Flooding of the Yser Plain." *De Grote Rede 36: The Great War and the Sea. De Grote Rede: Nieuws over onze Kust en Zee*, 2013.
Deneckere, Bernard. *Above Ypres: The German Air Force in Flanders 1914–1918*. Brighton, 2013.
de Syon, Guillaume. *Zeppelin! Germany and the Airship, 1900–1939*. Baltimore, 2002.
Deuringer, Karl. *The First Battle of the First World War: Alsace-Lorraine*. Translated and Edited by Terence Zuber. Gloucestershire, 2014.
Donnell, Clayton. *Breaking the Fortress Line*. South Yorkshire, 2013.
Doughty, Robert A. *Pyrrhic Victory*. Cambridge, 2008.
During, Fred. *A Critical Analysis of the Battle of Virton, August 22 1914*. Command and General Staff School Student Papers Collection 1930–1936. Fort Leavenworth, 1932.
Edmonds, J.E. "The Scapegoat of the Battle of the Marne, 1914. Lieutenant-Colonel Hentsch, and the Order for the German Retreat." *Field Artillery Journal*. Vol. XII. 1922.
Farrar, L.L. *Divide and Conquer: German Efforts to Conclude a Separate Peace, 1914–1918*. New York, 1978.
Fenster, Julie M. *Race of The Century: The Heroic True Story of the 1908 New York To Paris Auto Race*. New York, 2005.
Foley, Robert T. *German Strategy and the Path to Verdun*. New York, 2007.
———. "Hermann von Kuhl." *Chief of Staff: The Principal Officers Behind History's Great Commanders*. Vol.1. Annapolis, 2008.
Foley, Robert T., Translator & editor. *Alfred von Schlieffen's Military Writings*. London, 2003.
Förster, Helmut. "Die Entwicklung der Fliegerei im Weltkrieg." *Unsere Luftstreitkräfte*. Edited by Walter von Eberhardt. Berlin, 1930.
Fortier, Louis J. *The Defense of the Kluck–Bülow Gap by the Marwitz and Richthofen Cavalry Corps (September 6–noon September 8 1914)*. Unpublished paper. USAHEC, 1934.
Franks, Norman, Frank Bailey and Rick Duvien. *Casualties of the German Air Service 1914–1920*. London, 1999.
"The French Official Account of the Marne, 1914. Franchet d'Espèrey's Army." *The Army Quarterly*. Vol. XXVII, No. 1. 1933.
Germany. *The German Army in Belgium: The White Book of May 1915*. Translated by: E.N. Bennett. London, 1921.
Gilbert, Adrian. *Challenge of Battle: The Real Story of the British Army in 1914*. Oxford, 2014.
Glasgow, Lawrence B. *Why Was German GHQ Unable to Locate the BEF Prior To Mons, And Why Was It Surprised to Meet the BEF At Mons?* Command and General Staff School Student Papers Collection 1930–1936. Fort Leavenworth, 1931.
Greenhalgh, Elizabeth. *Foch in Command: The Forging of a First World War General*. New York, 2013.
Gross, Gerhard P. *The Myth and Reality of German Warfare: Operational Thinking from Moltke the Elder to Heusinger*. Lexington, 2016.
Grosz, P M. *Albatros B.I: Windsock Datafile 87*. Hertfordshire, 2001.
———. *Aviatik B-Types: Windsock Datafile 102*. Hertfordshire, 2003.
———. *The LVG B.I: Windsock Datafile 98*. Hertfordshire, 2003.
———. *The Taube at War: Windsock Datafile 104*. Hertfordshire, 2004.
Hardach, Gerd. *The First World War 1914–1918*.

London, 1977.
Hart, Peter. *Fire and Movement*. New York, 2015.
Herrick, Hugh H. *A Critical Analysis of the Operations of the XVII French Corps in the Battle of the Ardennes, 21 and 22 August 1914*. Command and General Staff School Student Papers Collection 1930–1936. Fort Leavenworth, 1933.
Herris, Jack. Albatros Aircraft of WWI, Volume 1 Early Two-Seaters. Reno, 2016.
———. *Pfalz Aircraft of WWI*. Reno, 2012.
———. *Development of German Warplanes in WWI*. Reno, 2012.
———. *Rumpler Aircraft of WWI*. Reno, 2014.
Herwig, Holger. *The Battle of the Marne: The opening of World War I and the Battle That Changed the World*. New York, 2011.
Hh. "Deutsche Flieger in der Schlacht von St. Quentin 28. bis 30. August 1914," *Wissen und Wehr*. Jahrgang 1923, Heft 2. *1923*.
Head, R.G. *Oswald Boelcke: Germany's First Fighter Ace and Father of Air Combat*. London, 2016.
Hierl, Constantin. *Strategic and Tactical Problems for the Study of the Marne Campaign Vol. 1—Studies of the Command of German 3rd Army August 27–29 1914*. Translation. Berlin, 1927.
Holmes, Terence M. "Asking Schlieffen: A Further Reply to Terence Zuber." Volume 10, Issue 4. 2003.
———. "The Reluctant March on Paris: A Reply to Terence Zuber's 'The Schlieffen Pan Reconsidered'." *War in History*. Volume 8, Issue 2. Essex, 2001.
Hooper, George. *The Campaign of Sedan*. London, 1914.
Hösslin, Hubert von. *Die Organisation der K.B. Fliegertruppe, 1912–1919*. Munich, 1924.
House, Simon. *The Battle of the Ardennes August 22*. Thesis Paper. King's College London, 2012.
Hughes, Daniel J., editor. *Moltke on the Art of War*. New York, 1993.
Humphries, Mark Osborne. & John Maker (ed.), *Germany's Western Front 1914, Part 1*. Waterloo, 2013.
Imrie, Alex. *Pictorial History of The German Army Air Service*. Chicago, 1971.
Johnston, Edward S. *Study of The German Fifth Army on 21 August 1914, Prior to the Battle of the Ardennes*. Command and General Staff School Student Papers Collection 1930–1936. Fort Leavenworth, 1931.
"Kaisermanöver 1911: Erster Einsatz der deutschen Fliegertruppe." *Luftwelt*. Bd.3 Nr.11. 1936.
Kehler, Richard von. "Frei- und Fesselballon im Dienste der Kriegführung vor dem Weltkrieg." *Unsere Luftstreitkräfte*. Edited by Walter von Eberhardt. Berlin, 1930.
Kilduff, Peter. *Germany's First Air Force: 1914–1918*. Osceola, 1991.
———. *Iron Man, Rudolf Berthold: Germany's Indomitable Fighter Ace of World War I*. London, 2012.
Koeltz, Louis Marie. *Le G.Q.G. allemand et la bataille de la Marne*. Paris, 1931.
Koenig, Maj. Egmont F. *The Battle of Morhange-Sarrebourg 20 August 1914*. Command and General Staff School Student Papers Collection 1930–1936, Fort Leavenworth, 1933.
Langevin, "Une etude allemande sur les reconnaissances aériennes dans l'armée von Kluck à la bataille de la Marne de 1914" *Revue des Forces Aériennes*, Vol. II. December, 1931.
Langsdorff, Werner von. "Luftbildwesen, Funkentelegraphie und Wetterkunde in ihrer Bedeutung für die Luftwaffe."
Unsere Luftstreitkräfte. Edited by Walter von Eberhardt. Berlin, 1930.
Lawson, Eric and Jane. *The First Air Campaign: August 1914–November 1918*. Cambridge, 1996.
Lehmann, Ernst A. *Zeppelin: The Story of Lighter Than Air Craft*. Charleston, 2015.
Liebach, Johannes. *Bataillons-, Regiments- und Brigade-Uebungen und Besichtigungten der Infanterie in praktischen Beispielen*. Berlin, 1914.
Loewenstern, Elard von. *Die Fliegersichterkundung im Weltkriege*. Berlin, 1937.
———. *Mobilmachung, Aufmarsch und erster Einsatz der deutschen Luftstreitkräfte im August 1914*. Stuttgart, 1939.
Lomas, David. *First Ypres 1914: The Birth of Trench Warfare*. Westport, 2004.
Lucas, Andrew and Jurgen Schmieschek. *Fighting the Kaiser's War: The Saxons In Flanders 1914–18*. South Yorkshire, 2015.
Mahnke, Alfred. "25 Jahre Deutsche Luftwaffe: Zum Ersten Einsatz deutscher Heeresflugzuege im Kaisermanöver am 11. September 1911," *Der Deutsche Sport Flieger*. 3 Jahrgang 1936, Heft 10. 1936.
Mombauer, Annika. *Helmuth von Moltke and the Origins of the First World War*. New York, 2005.
Morton-Jack, George. *The Indian Army on The Western Front*. New York, 2014.
Mosier, John. *The Myth of the Great War: How the Germans Won the Battles and How the Americans Saved the Allies*. New York, 2002.
"More Marne Through German Spectacles." *The Army Quarterly*. Edited by Major-General G. P. Dawnay. Vol. XVII, No.2. 1929.
"More Marne Through German Spectacles: The Actions of the Guard Korps and of the Right Wing of the Third Army from the 5th to the 8th of September, 1914," *The Army Quarterly*. Edited by Major-General G. P. Dawnay. Vol. XVIII, No.1.

1929.

Morrow, John H., Jr. *Building German Air Power 1909–1914.* Knoxville, 1976.

———. *The Great War in The Air.* Washington D.C., 1993.

Murland, Jerry. *Battle on the Aisne 1914–The BEF and the Birth of the Western Front.* South Yorkshire, 2012.

———. *Retreat and Rearguard 1914.* South Yorkshire, 2014.

Neumann, Georg Paul. *Die Deutschen Luftstreitkrafte im Weltkriege.* Berlin, 1920.

———. (ed.) *In der Luft unbesiegt: Erlebnisse im Weltkrieg erzählt von Luftkämpfern.* Munich, 1923.

Notestein, Wallace & Elmer E. Stoll. *Conquest & Kultur: Aims of The Germans In Their Own Words.* Washington D.C., 1918.

Nowarra, H.J. *50 Jahre Deutsche Luftwaffe.* 3 vols. Berlin, 1961.

Paret, Peter. *Yorck and the Era of Prussian Reform 1807–1815.* Princeton, 1966.

Porch, Douglas. *The March to the Marne: The French Army 1871–1914.* Cambridge, 2003.

Potempa, Harald. *Die Königlich–Bayerische Fliegertruppe 1914–18.* Frankfurt am Main, 1997.

———. "Die Königlich-Bayerische Fliegertruppe als Teil der Deutschen Luftstreitkräfte bis 1918." *Blätter zur Geschichte der Deutschen Luft- und Raumfahrt.* XIX. 2013.

Richter, Oliver. *Feldluftschiffer: The German Balloon Corps and Aerial Reconnaissance.* Erlangen, 2013.

Ritter, Hans. *Der Luftkrieg.* Berlin, 1926.

Robinson, Joe & Janet. *Handbook of Imperial Germany.* Bloomington, 2009.

Rottgardt, Dirk. *German Armies' Establishments 1914/18, Vol.1.* West Chester, 2009.

Samuels, Martin. *Command or Control? Command, Training and Tactics in the British and German Armies, 1888–1918.* London, 1995.

Scott, John L. *Operations on the Right of French Fourth Army, August 22 1914.* Command and General Staff School Student Papers Collection 1930–1936. Fort Leavenworth, 1932.

Schuette, Robert C. *Effects of Decentralized Execution on the German Army During the Marne Campaign of 1914.* Master of Military Art and Science Thesis Paper. US Army Command and General Staff College, 2014.

Senior, Ian. *Home Before the Leaves Fall.* Oxford 2012.

Skerry, Harry A. *German XVIII Reserve Corps at the Battle of Neufchâteau on August 22, 1914.* Command and General Staff School Student Papers Collection 1930–1936. Fort Leavenworth, 1932.

Sheldon, Jack. *The German Army at Ypres 1914.* South Yorkshire, 2010.

———. *German Army on Vimy Ridge 1914–1917.* South Yorkshire, 2008.

Sheldon, Jack and Nigel Cave. *Ypres 1914: Langemarck.* South Yorkshire, 2014.

Snyder, Jack. *The Ideology of the Offensive–Military Decision Making and the Disasters of 1914.* Ithaca, 1984.

Stone, Norman. *The Eastern Front 1914–1917.* New York, 1998.

Storz, Dieter. "This Trench and Fortress Warfare is Horrible!" *The Schlieffen Plan: International Perspectives on the German Strategy for World War 1,* Edited by Hans Ehlert, Michael Epkenhans & Gerhard P. Gross. Lexington, 2014.

Strachan, Hew. *The First World War: Volume I To Arms.* New York, 2003.

Sumner, Ian. *German Air Forces 1914–1918.* Long Island, 2005.

Supf, Peter. *Das Buch der deutschen Fluggeschichte.* zweiter Band. Berlin, 1935.

Tyng, Sewell. *The Campaign of the Marne.* Yardley, 2007.

Unruh, Karl. *Langemarck: Legende und Wirklichkeit.* Koblenz, 1986.

Wallach, Jehuda. *The Dogma of The Battle of Annihilation.* Westport, 1986.

Werner, Johannes. *Knight of Germany: Oswald Boelcke German Ace.* Havertown, 2009.

Whitehead, Ralph J. *The Other Side of the Wire: With the German XIV Reserve Corps on the Somme, September 1914–June 1916.* Volume 1. West Midlands, 2013.

Zuber, Terence. *Ardennes 1914.* Gloucestershire, 2009.

———. *The Real German War Plan 1904–1914.* Gloucestershire, 2012.

———. *The Mons Myth.* Gloucestershire, 2010.

———. *Ten Days in August.* Gloucestershire, 2014.

Websites

Stimson, Dr. Richard. *Orville Flies in Germany.* http://wrightstories.com/orville-flies-in-germany/

Index

Aerial Bombing   26, 33–35, 37–38, 49, 53, 56–57, 67, 75, 82, 155–156, 304, 310–312, 325, 346, 348–351
Aerial Photography/cameras   17, 26, 37, 49, 68, 343–344, 347
Air Fleet League   10–11
Airship   8–10, 12, 14–21, 26–28, 33–34, 39, 41, 42, 52, 82, 348
Aisne (battle)   261–264, 278
Aisne (river)   133, 135–136, 156, 158, 169, 223, 236, 238–241, 250–251, 259, 261–263, 267–268, 281, 290–291, 294, 296
Albatros Works   12–13, 17, 24, 29, 30–33, 37, 39, 49, 50, 59, 62, 114, 117, 118, 130, 155, 178, 184, 199, 205, 218, 229, 237, 271, 273, 277, 279, 309, 311, 328, 335, 339, 350
Albert I, King of the Belgians   102, 282–283, 331
Albrecht, Duke of Württemberg   84, 88–90, 95–98, 244–245, 285, 290, 293, 295–296, 298–299, 301–304, 306–307, 314, 320–322, 324, 333
Allenby, Edmund 124
Alsace   70–71, 73, 81–82, 135, 146, 147, 154, 234
Amiens   133, 139, 154, 160, 169, 174, 268–269
Anti-Aircraft artillery   33, 53, 57, 128, 158, 249, 346–348
Antwerp   68, 70, 101–104, 134, 135, 150, 264, 281–287, 289
Ardennes   73, 83–85, 87, 89–90, 92, 97–98, 105, 115, 124, 148–149
Argonne   135, 244, 332, 341–343
Armin, Friedrich Sixt von   130, 188, 201, 214, 229–230, 236, 238
Arras   133, 265, 268–269, 279, 286, 290, 298, 302, 305, 307, 312, 334, 336, 340
Artillery direction/cooperation flights   24, 35, 37–38, 40, 54–56, 62, 75, 81, 117–118, 121, 148, 158, 215, 217, 219–220, 225, 230, 244, 248, 257, 302–303, 307–308, 317–318, 320, 322–326, 330, 334, 337, 344, 346–347
Auftragstaktik   59–60
Aviatik   12, 29–31, 39, 49, 50, 63, 259, 271, 273, 279–280, 350
Bahrends, *Oblt.*   18
Balloons   8–9, 23, 26, 28, 33–34, 51–52, 62, 64, 77, 80–81, 88, 245, 249, 318, 325, 337, 346–347, 352
Bavarian Flying Service   25, 28, 31–33, 35, 42, 43, 75, 77, 79–81, 83, 147–148, 235, 248–249, 268–269, 273–274, 279, 287, 318–319, 323, 325, 335
Bavarian War Ministry   32–33
Behn, *Oblt.*   136, 153
Belgian Army   65–68, 101–103, 134, 145–146, 150, 151, 264, 282, 285–286, 292–296, 298–299, 301–302, 306, 308, 314, 320–321, 331, 333, 338

Bergmann, Walter von   50, 144, 221, 234–235, 236, 255
Berthold, Rudolf 63, 115, 169, 226–227, 257
Beseler, Hans von 102, 282–283, 285–286, 289–290, 292–293, 295–296, 298, 300–301, 306–307, 320
Bewegungskrieg   34, 37, 39, 53, 303, 331
Blériot, Louis   11–12
Blumenbach, *Oblt.* 167, 221
Boehn, Max von   268, 282
Boelcke, Oswald   344–346, 352
Böhmer, *Ltn.*   155
Bornstedt, Ernst von   115
Brachtenbrock, *Ltn.*   226
Braun, *Ltn.*   17–18
Breiftauben Abteilung Ostend (BAO)   348–351, 352–353
British Expeditionary Force (BEF)   64–65, 83, 101, 103, 105, 107, 109, 121–131, 133, 138–139, 150, 151, 152, 154, 160–161, 171–172, 176–177, 181, 184–189, 191–193, 200–201, 209, 212, 213, 215, 220–223, 226, 230–234, 236–237, 239, 254, 259, 268, 275, 290, 292, 295, 296, 299, 321, 322, 327, 332, 333, 340

Corps
 I Corps   124–125, 130, 294, 296, 299, 301–302, 320
 II Corps   124–126, 129–131, 275, 290–292, 296, 299, 303, 306–308, 322, 332,
 III Corps   290–292, 296, 299, 332
 IV Corps   286, 292, 296, 299
 Indian Corps   302–303, 307–308, 321–322
Brussels   68, 70, 101, 103, 214, 222, 229, 284–286, 295
Bülow, Karl von   99–103, 106–113, 115, 119–122, 131, 134, 136, 138, 140–143, 146, 151, 153, 154–156, 158, 163–164, 169, 171, 180–183, 187–189, 191–194, 201, 206, 208–214, 220–223, 225–228, 234, 236, 240–241, 251, 254, 257, 258, 263–264, 267–268, 275, 278, 281
Bürhmann, *Hptm.* 233
Camp des Romains   242, 265
Caspar, Karl von 310–311
Charleroi   101, 103–105, 107–108, 110, 113, 115, 289
Charleville   87–88, 95, 206, 264
Château Montmort   208–209, 211, 223, 226, 234
Cologne Commission   14–15, 18
Coulommiers   165, 167, 171–172, 175, 181, 184, 187–189, 200–201, 209, 212–215, 220, 223, 226, 230, 236–237, 253
Crown Prince Wilhelm   84, 88, 211
Cuneo, John   5, 38–39, 151, 152, 260, 274
D'Espèrey, Louis Franchet   140, 176, 181, 184, 191–192, 194, 206, 241, 254
Dalwig, Friedrich 115
de Castelnau, Edouard   73, 77–79, 81, 132, 247–248, 265, 267–269, 281

de Langle de Cary, Fernand   83–85, 98, 118, 133, 242, 341–342, 352
Deimling, Berthold von   72, 314, 319, 322, 324, 331, 337
Dellmensingen, Krafft von   76, 147, 246, 336
Detten, Max von 27, 213, 220–222, 237
Deutsche Flugzeug-Werke (DFW)   29–30, 38, 40, 43, 49, 82, 113, 205, 226, 271, 273, 279, 350
Dietze, *Oblt.*   309, 311–312, 335
Dixmude   283, 285–286, 289–290, 292, 295–296, 298–299, 301–302, 306–308, 320, 326, 329, 338
Dommes, Wilhelm von   210, 251
Dubail, Auguste 73–74, 77, 81, 132
Dunkirk   284, 286–288, 290, 293–294, 301, 304, 308, 313, 320, 326, 336, 349–350, 352
Eben, Johannes von   183
Eberhardt, Walter 11, 28, 260, 276–278, 280, 351, 353
Einem, Karl von   66, 68, 70, 73, 260–261, 341
Emmich, Otto   65–67, 107–108, 112, 194, 204–205
Épinal   75, 82–83, 133, 135–136, 173–175, 242–243, 245, 247–249
Euler Works   12, 16, 19–20, 24, 29, 32–33, 49, 184, 232, 256, 273, 279, 328
Fabeck, Max von 304, 311–314, 316–331, 336, 337, 338
Falckenstein, Vogel von   63, 111, 115, 226
Falkenhayn, Erich von   26, 28, 30, 33–35, 260–261, 263–265, 267–270, 277–278, 279, 280, 281–295, 298, 300, 305, 307–308, 312–314, 326, 331, 332, 333, 336, 338, 339, 340, 347–348, 351, 353
Falkenhayn, Eugen von   296, 333
Falkenhayn, Fritz von   348, 358, 366
Fink, *Ltn.*   17, 27, 184, 188–189, 254
Flanders   270, 281–283, 285–290, 292–296, 298–299, 302, 304–305, 308, 310, 313–314, 321, 323, 326–327, 331, 332, 336, 338, 339, 340
Flight Command Döberitz   13–14, 16–18, 27, 28, 38–40, 41, 42, 44
Fliegertruppe (organization)   5, 12–14, 18, 21, 23–26, 28–29, 33–37, 40–41, 49, 55, 123, 129, 153, 168, 252, 260, 274–278, 309, 343–344, 348, 351, 353
Foch, Ferdinand   75, 78, 81, 148, 176, 181, 192, 202–203, 205, 209, 222, 223, 225, 227, 240–241, 267, 279, 286, 291–292, 306, 314, 324–326, 339, 340–341
Franco–Prussian War   8, 31, 65
French Army
  Army
    First Army   64, 73–75, 77–78, 81, 83
    Second Army 64, 73, 77–79, 81, 147, 247, 265, 268–269, 332
    Third Army   64, 83–85, 87, 88–91, 96–97, 154, 176, 242–245
    Fourth Army   64, 83–85, 87, 90, 92–94, 98, 107, 118, 124, 132–133, 136, 149, 154, 156, 158, 176, 240, 242, 244–245, 251, 341–343, 352
    Fifth Army   64, 83, 87, 101, 103–105, 107–110, 112–113, 115, 117–119, 121, 124, 126, 131–136, 138–139, 141–142
    Sixth Army   139, 141, 175–176, 178–181, 186–187, 189, 197–201, 212, 214, 216–218, 221–222, 227–228, 230–231, 233–234, 237–240, 244, 258, 267
    Ninth Army   176, 181, 183, 192–194, 201–205, 209, 222–224, 227, 240–241
    Tenth Army   267, 269, 279, 286, 290, 332, 340–341
    Army of Alsace   73
    Army of Lorraine   148
  Corps
    I CA   105, 108–109, 114, 139–141, 184, 341
    II CA   92, 341
    III CA   108, 113, 141
    IV CA   91–92, 97–98
    V CA   91
    VI CA   96
    VII CA   70, 73, 133, 154–155
    VIII CA   78, 80, 148
    IX CA   158, 302–303, 306, 314
    X CA   108, 139–141, 183, 192, 205, 225
    XI CA   94, 202–203, 269
    XII CA   94
    XV CA   78, 245
    XVI CA   78, 314, 337
    XVII CA   94–95
    XVIII CA   109–110, 140, 184, 208
    XX CA   75, 78, 81, 147, 148
    XXI CA   291
    XXXII CA   314, 320
    Cavalry Corps (Conneau)   175
    Cavalry Corps (Sordet) 108
    Colonial Corps   87, 92–94, 98, 148, 158, 245, 341
French, Sir John   124–126, 129–131, 133, 176, 181, 185, 222, 233, 290–292, 296, 299, 306
Fulda, *Oblt.*   189
Galliéni, Joseph 267
Geerdtz, Franz   17, 18, 24, 30, 41, 137–138, 205
German Airship Association   10
German Army
  Army
    First Army   46, 50, 52, 99–104, 110, 113–114, 121–131, 134, 135, 138, 144, 149, 150, 151, 152, 154–156, 158–179, 181, 184–191, 194–201, 210, 212–241, 250–251, 252, 253, 254, 255, 257, 258, 262, 264, 267–268, 275–276, 280, 282, 305, 313, 323
    Second Army   46, 68, 70, 99–104, 106, 108–116, 118–122, 131, 133–134, 135, 136–143, 144, 150, 151, 153, 154–156, 163–164, 166, 169–171, 173, 174, 176, 180, 181–184, 186–196, 200–201, 204, 206–214, 220–228, 232–241, 250–251, 252, 258, 263, 268, 275–276, 288
    Third Army   46, 87, 95, 97–99, 101, 103–104, 106, 108–109, 111, 113–121, 131–134, 135, 136, 143,

	151, 153, 156, 158, 170–171, 173, 174, 176, 183, 187–188, 192–196, 201–205, 208–209, 211, 222–225, 240–241, 250–251, 258, 260, 341–342
Fourth Army	46, 84–85, 87–90, 92, 94–98, 104, 115, 120, 124, 134–135, 136, 147, 148, 149, 154, 156, 158, 193–194, 196, 211, 240–242, 244–245, 250, 283–296, 298–303, 306–308, 312–314, 316–317, 319–322, 324, 326, 328–329, 331, 333, 338, 339
Fifth Army	46, 76, 84–85, 87–91, 95–97, 134–135, 144, 147, 150, 151, 154, 174, 241–245, 250, 264
Sixth Army	46–47, 73–79, 81–83, 134, 135–136, 144, 145, 146, 147, 173–175, 243–246, 248–249, 264–265, 268–270, 276, 279, 281, 285–295, 298–300, 302, 304–314, 316–318, 321–323, 325, 328, 333, 334, 335, 336, 337
Seventh Army	46–47, 63, 70–77, 81–82, 134, 135–136, 144, 145, 146, 173, 174, 234, 245–251, 262–264, 268, 276, 281, 312

Army Corps

Guard Korps	16, 68, 104, 106–110, 112–114, 120, 134, 137–142, 150, 181, 183, 187, 192, 194, 196, 201–204, 209, 212, 223–225, 232, 241, 269, 287, 293, 298, 328
I Bavarian AK	74–75, 77–78, 80, 148, 248, 265, 269, 279
II Bavarian AK	74, 77–80, 83, 248, 269, 312, 314, 317–319, 323–324, 326
III Bavarian AK	74, 83, 248
II AK	102–103, 126, 129, 154, 159, 161, 163, 166–168, 171–172, 178–179, 185, 187, 197–198, 216, 232, 252–253, 268, 275–276, 279, 280, 328
III AK	124–126, 128–130, 159, 161, 163, 171, 184, 188, 191, 199, 213–214, 217, 305
IV AK	126, 129–130, 159, 163, 171–172, 185–189, 196–197, 229, 269, 287, 293, 302
V AK	89, 91, 96, 242, 250
VI AK	85, 89, 93–95, 98, 344
VII AK	66, 68, 108–109, 113, 138, 140, 142, 192, 287, 293, 322
VIII AK	95, 194, 196
IX AK	16, 102–103, 114, 123–127, 129, 155, 159, 161–167, 171, 184, 186–187, 191, 214, 220–222, 231, 252, 254, 257
X AK	66, 104, 106, 109–110, 112–113, 120, 138–142, 151, 170, 183, 187, 194, 201, 204–205, 209, 212, 225
XI AK	120, 134
XII AK	116, 118, 298
XIII AK	89, 289–290, 293
XIV AK	71–72, 76, 145, 146, 287, 293, 322
XV AK	53, 71–72, 145, 146, 247, 250, 305, 312, 314, 317, 319–320, 322, 324, 328, 331, 337
XVI AK	89, 96–97, 243, 245, 341
XVIII AK	85, 94–96, 268, 279
XIX AK	116–119, 156, 194, 240, 244, 287, 289–290, 293, 298, 300, 321, 328, 332, 337
XXI AK	74–75, 78–79, 265, 279

Reserve Corps

Guard RK	104, 134
I Bavarian RK	74, 80, 148, 269, 287, 298, 305
III RK	102, 134, 282–286, 288–290, 292–296, 298–303, 306–308, 320, 329, 333, 334
IV RK	152, 163, 166–167, 171, 177–179, 185–187, 197, 214–216, 231
V RK	89
VI RK	89
VII RK	134, 250
VIII RK	97–98, 149
X RK	108–109, 113–115, 120, 137–138, 140, 142, 183, 188, 191–193, 206–207, 209, 227
XII RK	114, 119, 134, 170, 202
XIV RK	74, 269
XVIII RK	93–95, 98
XXII RK	295–296, 298, 303, 307, 329, 333
XXIII RK	295–296, 298, 301, 303, 329, 333
XXIV RK	302, 317, 336
XXVI RK	295–296, 298–299, 301, 304, 333
XXVII RK	283, 295–296, 298, 301, 306, 317, 319, 324, 333

Cavalry Corps

HKK 1	100, 104, 115, 151, 187, 191–193, 210, 212, 232–233, 298, 300, 321, 337
HKK 2	66, 100, 122, 124–125, 129–130, 149, 151, 152, 185, 187, 191, 213–214, 220, 222, 226, 229–233, 238, 241, 312
HKK 3	74, 248
HKK 4	85, 88, 97, 270, 279, 287, 290, 294, 298
HKK 5 (Group Stetten)	307

Fliegertruppe

AFP 1	199
AFP 2	70, 204, 276
AFP 6 (Bavarian)	268, 311
AFP 7	146
Fest. FA Cologne	67, 146, 285, 334
Fest. FA Germersheim	287
FFA 1	24, 30, 68, 70, 110, 112–115, 138, 140, 151, 183, 204–205, 225, 257, 287, 293, 298, 328
FFA 1b	75, 79–80, 83, 147, 148, 269
FFA 2	24, 30, 96–97, 245, 248
FFA 2b	75, 79, 83, 269, 317, 319, 323–325
FFA 3	24, 30, 71, 73, 146, 248, 317, 320, 323–325, 337
FFA 3b	75, 83, 248
FFA 4	24, 30, 85, 87–88, 269, 289, 293–294, 300, 304, 308, 317, 323, 325–326, 332
FFA 4b	287, 294
FFA 5	30, 74–75, 79, 81–82, 145, 269, 287, 293–294, 301, 304, 308, 313, 317, 322–323
FFA 6	30, 85, 87–88, 95, 98, 244–245, 286, 295, 303, 307, 313, 321, 332, 333, 334

FFA 7	128–131, 159, 172, 184, 188, 200, 221, 229, 230, 237, 239, 275
FFA 8	75, 78–79
FFA 9	67–68, 127–128, 130, 146, 159, 167, 172, 177, 189, 197, 218, 221, 229–230, 237, 239, 287, 293, 302, 304, 311
FFA 10	87–88, 95, 97, 244
FFA 11	114, 122–124, 126–127, 155, 159–162, 165, 171, 184, 188, 217, 229–230, 237, 239, 275
FFA 12	103, 123, 126, 129, 151, 159–160, 165, 171, 177, 185–186, 188, 199–201, 212–213, 215, 218, 220–221, 229–230, 232, 236–237, 239, 267–268
FFA 13	93–94, 344
FFA 18	110, 113, 138, 140, 143, 146, 169, 192, 206–207, 225, 287, 293, 308–309, 317
FFA 19	96–97
FFA 20	71–72, 74, 76–77, 81, 145, 146, 287, 293–294
FFA 21	103, 110, 112–113, 115, 136–140, 153, 205, 225, 275
FFA 22	115, 117, 120–121, 131, 170, 193, 202, 240, 342
FFA 23	63, 110–111, 113, 115, 136, 138, 141–143, 156, 169, 194, 209, 226–227
FFA 24	116, 118–119, 156, 158, 240, 251, 281, 287, 293, 300, 308, 311, 317, 322, 337
FFA 25	87–88, 96, 245
FFA 26	63, 71, 75, 82, 145, 146, 248
FFA 27	87–88, 95, 98, 245
FFA 28	121
FFA 29	116, 118, 120, 202
FFA 30	102–103, 128, 154, 159–160, 167–169, 171–172, 177–179, 216–217, 219–220, 229–230, 232–233, 237, 239, 257, 275–276, 280, 328
FFA 38	285–286, 289, 293, 295, 299, 302, 303, 307–308, 320, 333, 334
FFA 40	307–308, 329
FLA 1	52
FLA 1b	77, 80–81, 148, 249, 318, 325, 347
FLA 3	245
FLA 4	88
'Flying Section St. Amand'	287, 294
German War Plan	18, 46–47, 49, 58, 65, 70, 73–74, 85, 100, 104, 121, 126, 131, 134, 143, 147, 154, 174, 274–275
Gersdorff, Ernst von	206–207
Gette (river)	66, 68, 101
Ghistelles	349–350
Giebenhain, Hans Jürgen von	17, 42, 44
Goltz, Colmar von der	16, 18
Gotha	24, 29–30, 45, 49, 221, 229, 230, 273, 280, 305
Givet	85, 87, 95, 103–105, 110, 117, 119, 121, 134, 204
Grade, Wilhelm	27, 28, 30, 130
Grand Couronné de Nancy	80, 82–83, 245, 248
Grand Morin (river)	171, 185, 187, 189, 191–194, 201, 210, 214
Gronau, Hans von	176–177, 179, 186, 197, 199, 215–216, 222
Guise-St. Quentin (battle)	138–143, 154, 156, 164, 169, 175, 275, 289
Haig, Sir Douglas	124–125, 130, 296, 299, 319
Hagen, Oskar von dem	186, 199–200, 218, 236–237
Hanisch, Karl	71
Haupt, *Ltn.*	185, 199, 218, 236
Hausen, Baron Max von	98–99, 109, 115–121, 136, 156, 158, 170, 188, 193–196, 201, 203, 240–241, 244, 260
Hazebrouck	269–270, 287–288, 293–294, 300, 304, 308, 311, 313, 335
Heeringen, Josias von	15, 28, 63, 70–72, 146, 276
Hentsch, Richard	174–175, 209–214, 220–223, 233–236, 239, 241, 249, 256, 258
Hoeppner, Ernst von	202–203, 211, 248, 343, 344, 348
Hofer, *Ltn.*	17
Hollweg, Bethmann	20
Hug, Andre	37, 63, 71, 147, 350, 352–252
Instruction and Research Institute for Military Aviation	18
Inspectorate of Airship Troops	28, 34, 347
Inspectorate of Flying Troops (*Idflieg*)	28–29, 33–37, 41, 44, 271, 273–274, 276, 280
Inspectorate of Military Aviation and Motor Vehicles	14, 18, 29
Inspectorate of Transportation Troops	8, 10, 15, 18, 20, 23, 29, 31, 35–37, 41, 273
Janson, Erich	115
Jeannin	24, 29, 49, 131, 271, 273, 279, 310–311, 350
Joffre, Joseph	64–65, 70, 73–74, 83–85, 87, 91, 98, 104–105, 108, 124, 131, 133, 138–139, 141–142, 148, 152, 164, 173, 175–176, 179–182, 185–187, 189, 244, 247–248, 261, 264–265, 267–269, 276, 281, 290–291, 326, 340–341, 352
Kaiser Manöver	8, 16–18, 21, 24, 30, 33, 38–39
Kaiser Wilhelm I	31
Kaiser Wilhelm II	16, 18, 20, 46–47, 59, 99, 162, 140, 240, 260–261
Kastner, Hermann	27, 348
Kirchbach, Günther Graf von	108, 202–205, 209, 211, 240, 258–259
Klein, *Ltn.*	123–124, 126
Kleist, *Ltn.*	185, 213, 221, 237
Kluck, Alexander von	99–101, 104, 121–122, 124–131, 146, 154–155, 158–164, 166, 168, 171–179, 184–192, 196, 199–201, 212–214, 216–217, 220–223, 227–232
Koerber, Adolf-Victor	11
Koppen, *Hptm.*	17–18
Knobelsdorf, Schmidt von	89
Knobelsdorff, Viktor von	178, 197–199
Kuhl, Hermann von	101–102, 122, 123, 124–127, 129–131, 149, 150, 152, 154, 155, 158–162, 164, 166, 168, 170–173, 175, 178–179, 185–191, 196, 199–201, 212–214, 216–220, 222–223, 228–230, 232–239, 241, 251, 252, 253,

255,	258, 262, 267, 275–276
Langemarck	294, 298, 301–302, 329, 334
Lanrezac, Charles	104–110, 112–113, 115, 118–121, 124–125, 131–133, 136, 138–143, 150, 151, 152, 153, 156, 158–159, 161–164, 169, 176, 181, 252, 275
Lauenstein, Otto von	133, 170, 212, 223, 225, 226–227, 234
Le Cateau	124, 127, 129–130, 133, 139, 152, 154, 275, 289
Leman, Gérard	65–66
Leonhardi, Ernst	142–143
LFG	24, 29
Longwy	61, 88–89, 91
Liège	65–68, 70, 73, 82–83, 98, 101, 105, 142, 145–146, 150, 264, 276
Lille	101, 104, 122, 125–126, 152, 174, 253, 268–269, 275, 284, 290–293, 304, 311–313, 317, 324, 326, 332, 336, 337
Linsingen, Alexander	179, 198, 214–216, 229, 231, 233, 236, 238, 328–330
Lochow, Ewald von	183–184, 236, 238
Lorraine	46, 73–76, 82–83, 85, 90, 118, 133–135, 143, 146, 147, 148, 247–248, 265, 268
Loßberg, Fritz von	314, 316, 326
Ludendorff, Erich	9–10, 15, 24–25, 40, 66, 134, 147, 150, 210, 339
Luxembourg	46, 84–85, 87, 89, 164, 174, 210–211, 240–241, 246, 250–251, 260–261, 264, 277
LVG	29–33, 49, 50, 79, 116, 123, 127, 136, 183–185, 204, 207, 213, 225, 229, 235, 271, 273, 279, 328, 350
Lyncker, Alfred von	9
Lyncker, Moriz	261
Lys (river)	290–292, 298–300, 306, 322, 325
Mackenthun, Walter	18, 41
Marne (battle)	123, 134, 176–241, 260–262, 276, 336, 344
Marne (river)	134–136, 154, 160, 162–164, 166–176, 179, 184–185, 187–191, 194, 197–199, 201–202, 208, 211–214, 220, 222–223, 225–227, 230–238, 240–242, 249–251, 260–264, 267, 270, 273, 275
Marwitz, Georg von der	66, 222, 231
Maud'huy, Louis de	208, 213, 267, 269, 281, 286, 290–291
Matthes, Arthur	212, 223, 225–226, 234, 256
Maunoury, Michel–Joseph	176, 179, 189, 198–199, 214–219, 221–222, 228, 230, 233–235, 238
McCudden, James	50
Meaux	165, 171, 175, 178–179, 189, 197–198, 200–201, 213–214, 221, 233
Mengelbier, Rudolf	168, 252
Menin (city)	288, 290, 292–296, 298, 307, 314, 326–327, 330
Menzel, *Ltn.*	185–187
Mercedes	17, 50, 59, 62–63, 141, 271, 273, 305, 339
Messines	294, 300, 313–314, 317–318, 323–326, 329, 336, 337
Metz	15, 22, 25, 27, 46, 63–64, 73, 76, 78, 83–85, 134–135, 147, 148
Meurthe (river)	81–83, 248
Meuse (river)	46, 66, 85, 87–88, 95, 97, 99, 101, 103–105, 108–110, 113–119, 121, 124, 132–133, 135, 136, 144, 146, 147, 154, 156, 158, 178, 242, 244
Mézières	95, 105, 134, 158, 264, 285–286, 288, 312
Moltke, Helmuth von	14–15, 18, 20–21, 23–25, 27, 35, 39–40, 42, 47, 48, 49, 65, 74, 76, 87, 97–98, 100–101, 104, 122, 131, 133, 134, 136, 143, 144, 149, 154–155, 162, 164, 166–167, 171, 173–175, 182, 187, 196, 201, 210–212, 214, 222, 234, 236, 240–251, 252, 258, 259, 260–264, 274, 276, 278, 332
Moltke, Helmuth von (elder)	149
Mons	109, 122–127, 130–131, 150, 152, 172, 190, 275, 289
Montmirail	166–167, 171–172, 175, 183, 186, 188, 191, 206–208, 212–214, 225–227
Montmort	182, 187, 208–209, 211, 213–214, 223, 226, 234
Morhange-Sarrebourg	73, 77–81, 132, 134, 146, 147, 245, 248
Moselle (river)	135–136, 173–174, 242, 245–249
Mt. Kemmel	314, 323–326, 328, 331
Mudra, Bruno von	96
Mulhouse	70–73, 76, 146,
Musset, *Hptm.*	67, 311
Namur	64, 68, 70, 99–101, 103–107, 109–110, 113, 114, 119, 121, 133, 134, 136, 150, 151
Nancy	75, 81–82, 132, 175, 245–249
Nanteuil-le-Haudouin	160, 163, 167–168, 171, 185, 189, 197–199, 201, 214, 217–218, 221, 230–231, 233–234, 238, 241
National Aviation Fund	19–20, 23–24, 29, 31, 39
Neufchâteau	75, 84–85, 87, 91–95, 98, 135
Neuve-Chapelle	307–308
Niemöller, Oblt.	216–217, 232, 239
Nieuport (city)	103, 253, 286, 289–290, 294–296, 298–299, 301–303, 306–308, 320, 334
Nivelle, Robert	199
Oertzen, Jasper von	27, 28, 30, 68
OHL	47, 52, 71, 82–83, 85, 87, 98, 101, 103–104, 116–119, 122, 123, 134, 135, 143, 152, 156, 160–161, 164, 167–168, 173, 187, 194, 196, 201, 209–214, 235, 238, 240–241, 243–244, 246, 248, 250, 252, 260, 263, 267–268, 270, 277–278, 279, 280, 281, 285, 287–288, 292–294, 300, 302 305, 307–308, 312–314, 322, 326, 333, 335, 336, 353
Oise (river)	135–140, 142, 143, 153, 155, 159, 169, 173–174, 265, 267–268, 275
Oriola, Count von	137, 153, 205
Otto Works	32–33, 79, 248, 273, 350, 360
Ourcq (river)	175–176, 181, 185–191, 196–198, 200–201, 210, 214–215, 220–223, 233–234, 238–239, 242, 255, 256, 257, 258
Paris	25, 82, 134, 135, 136, 139, 143, 147, 154–156,

159–160, 164, 166–169, 171–180, 182, 185, 187, 189, 193, 195, 199–200, 212–213, 215, 221–223, 228, 230, 238, 244–246, 248–249, 252, 253, 254, 256, 258, 275–276, 282
Pelzer, *Oblt.* 230
Petit Morin (river) 183, 187–188, 191–195, 199, 201, 206–207, 209, 213, 254
Petri, *Hptm.* 28, 185
Pfähler, *Ltn.* 200
Pfalz 32, 273
Plan XVII 64, 83, 98
Plettenberg, Karl von 112, 183, 328
Poser, Florian von 28, 30, 96
Prince Heinrich of Prussia 19–20
Prussian General Staff 8, 10, 14–15, 17–18, 20, 23–25, 27–30, 32–25, 37, 39–40, 41, 46–47, 55, 56–57, 58, 99, 101, 120, 123, 134, 144, 147, 149, 151, 152, 153, 164, 168, 210, 221, 241, 247, 256, 260–261, 263–264, 278, 301, 335, 339, 344
Prussian War Ministry 8–10, 12, 14, 16, 18–21, 23–25, 27–31, 35–37, 40, 41, 42, 63, 270–274, 276–277, 347–348, 353
Quast, Ferdinand von 161–162, 171, 184, 220, 228, 230–231, 233, 236–238
'Race to the Sea' 260, 264–268, 279, 281, 347, 351
Reims 133, 135, 141, 154, 158, 170–171, 251, 264, 345–346
Richthofen, Manfred von 91
Ritter, *Oblt.* 200–201, 246
Rohdewadd, *Ltn.* 189
Roi, Wolfram de le 8–9, 12–13, 17, 41
Roos, *Oblt.* 172, 197, 255, 311
Rosenmüller, *Oblt.* 309, 311
Royal Prussian Air Service, creation of 23–31
Ruffey, Pierre 85
Rumpler 12, 24, 29–30, 33, 49, 67, 103, 204, 271, 273, 304
Rundstedt, Gerd von 17

Ruville, Hans-Carl von 178
Rupprecht, Crown Prince 74–77, 81–83, 147, 245–247, 249, 268–269, 279, 288, 294–295, 298, 300, 302, 305, 307–308, 312, 325, 334
Sambre (river) 100, 102–108, 110, 112–113, 118–119, 122, 124, 132–134, 151
Sarrail, Maurice 242
Schauenburg, *Gefr.* 216–217, 232,
Schinzing, *Oblt.* 304–305
Schlichting, Wilhelm 309
Schlieffen, Alfred von 58, 99–101, 143–144, 147, 149, 233, 253, 274, 281
Schwab, *Ltn.* 184
Schwarzenberger, *Ltn.* 200
Sedlmayr, Gerhard 116–117
siege artillery 66, 68, 73, 88, 282
Siegert, Wilhelm 27, 28, 348–351, 353
Smith-Dorrien, Sir Horace 124–125, 129–130, 292
Soissons 159–160, 163, 210, 236–238, 251–253
Sperrle, *Oblt.* 317
Stellungskrieg 283, 340, 346
Stietenkron, *Ltn.* 115–116
Streccius, Alfred 94
Suwelack, Josef 158, 280, 311
Tappen, Gerhard 133, 210, 240, 250, 264, 278
Taube (aircraft) 17, 24, 30–31, 44, 45, 50, 54, 59, 67, 82, 131, 141, 155–156, 172, 185, 197, 199, 204, 218, 221, 280, 285, 304, 319, 333, 353
Thomsen, Hermann von der Lieth 9, 15, 41, 123
Verdun 64, 87–88, 96–97, 132–135, 148, 173, 175, 242–243, 245, 251, 254, 264–265, 268, 344
Virton 88–89, 91, 96
Völkers, Kurt 116–117
Volkmann, Bruno 285, 302
Wagenführ, *Hptm.* 24, 27, 28, 30, 178
Wandel, Franz von 272
Wentscher, *Ltn.* 155
Wilberg, Helmuth 123
Wilhelm, Crown Prince 84, 88, 211, 245
wireless radio 347
Wright, Orville 10–12, 16, 29, 38, 41
Xylander, Oskar von 80, 246
Ypres 270, 278, 281, 285, 287–288, 290, 292–294, 296, 298, 301–302, 304–308, 311–314, 316–317, 320–327, 329–330, 333, 335, 336, 337, 338, 339, 340, 342, 344
Yser (river) 293, 295–296, 298, 300–302, 306–307, 313, 316, 320–322, 324, 329, 334
Zeppelin
  Z VI 52, 67, 82
  Z VII 52, 82
  Z VIII 52, 82
  Z IX 52
Zagen, Fritz von 309
Zeumer, *Ltn.* 304

Above: Otto Pusher to-seater coming in to land. Primarily used as trainers, these primitive aircraft were also used operationally early in the war. (Aeronaut)

Printed in Great Britain
by Amazon